GUIDE TO PARTS

PCI design handbook

precast prestressed concrete

SECOND EDITION

ARCHITECTURAL

STRUCTURAL

PRESTRESSED CONCRETE INSTITUTE®

201 NORTH WELLS STREET / CHICAGO, ILLINOIS 60606

Library of Congress Catalog Card Number 78-69931

ISBN 0-937040-12-6

Printed in U.S.A.

FOREWORD

The Prestressed Concrete Institute, a non-profit corporation, was founded in 1954 for the purpose of advancing the design, manufacture and use of prestressed and precast concrete. The Institute represents the prestressed concrete industry, and a large segment of the architectural precast concrete industry, in the United States and Canada.

The many technical, research and development programs of the Institute are financially supported, primarily, by about 200 Producer Member companies, representing some 300 plants throughout the United States and Canada and in many foreign countries. These firms are engaged in the manufacture of precast and prestressed concrete products for the construction industry.

Important financial and technical support is also received from Associate Member companies, supplying materials and services related to the industry, and from individual Professional Members engaged in the practice of engineering, architecture or related professions. Affiliate and Student Members are other important categories of individual PCI membership.

As the spokesman for the prestressed and precast concrete industry, PCI continually disseminates information on the latest concepts, techniques and design data pertinent to the industry and to the architectural and engineering professions through regional and national programs and technical publications. These programs and publications are aimed at advancing the state of the art for the entire industry.

Engineers, architects, contractors and owners interested in prestressed and precast concrete design and construction, should contact PCI headquarters in Chicago for information on all aspects of such construction as well as for information on PCI membership.

PREFACE TO FIRST EDITION

This, the First Edition of the PCI Design Handbook, is the culmination of a substantial effort on the part of many individuals and committees within PCI dedicated to the promulgation of knowledge of precast and prestressed concrete.

The primary objective of this Handbook is to make it easier for architects and engineers to use prestressed and precast concrete. It is intended to be a working tool, assisting the designer in achieving optimum solutions in minimum time. The Handbook is primarily for use in the design and construction of buildings and is based on the ACI Building Code (ACI 318-71).

Although this Handbook will become a valuable aid to all designers, it is not intended to replace, but rather to supplement the many manuals and other technical data currently distributed by precast and prestressed concrete manufacturers.

It is strongly recommended that all users of the Handbook contact local manufacturers in the vicinity of their project in the initial design stages to determine availability of sections shown in the Handbook, optimum concrete strengths, strand type and placement, and other pertinent information. This local information, together with the valuable new material in this Handbook, should enable the designer to greatly improve and shorten his procedures for designing prestressed and precast concrete.

Extreme care has been taken to have all data and information in the PCI Design Handbook as accurate as possible. However, as PCI does not actually prepare engineering plans, it cannot accept responsibility for any errors or oversights in the use of the Handbook material or in the preparation of engineering plans.

While every effort has been made to prepare this publication as the national standard for the industry, it is possible that, due to differences of opinion, there may be some conflicts between the material herein and local practices. It is expected that future editions of this Handbook will resolve and minimize any such differences.

Users of this Handbook are encouraged to offer comments to PCI on the contents of this publication and to offer suggestions for changes to be incorporated in the Second Edition. Questions concerning the source and derivation of any material in the Handbook should be directed to PCI.

PREFACE TO SECOND EDITION

The Second Edition of the PCI Design Handbook is a significant revision of the First Edition, not only in content and arrangement, but also reflecting the changes and advancements in design of prestressed concrete. Furthermore, it is based on the latest ACI Building Code (ACI 318-77). Thus, the PCI Design Handbook Committee recommends that users of the Handbook replace the previous edition with this Second Edition.

Material relating to architectural precast concrete and to post-tensioned prestressed concrete has been deleted because of the availability of separate publications — *PCI Manual for Structural Design of Architectural Precast Concrete,* Prestressed Concrete Institute, 1977, and *Post-Tensioning Manual,* Post-Tensioning Institute, 1976. New and up-dated information on design of connections, assembly of all information on analysis and design of precast, prestressed concrete buildings into one chapter (Part 4), and a revised listing of standard products, based on an industry survey, along with new types of load tables (Part 2) represent a few of the changes.

In the seven years since the First Edition was produced, PCI technical committees developed reports and recommendations on which many Handbook presentations are based. Further, as the Second Edition was being developed, various committees were involved in reviewing new material. As with the First Edition, extreme care has been taken to be as accurate as possible with the information presented. However, as PCI does not actually prepare engineering plans, it cannot accept responsibility for any errors or oversights in the use of the Handbook material or in the preparation of engineering plans.

The PCI Design Handbook Committee wishes to acknowledge the help received from its consultants — The Consulting Engineers Group, Glenview, Illinois; Wiss, Janney, Elstner and Associates, Inc., Northbrook, Illinois; Raths, Raths & Johnson, Inc., Hinsdale, Illinois; Portland Cement Association, Skokie, Illinois; Shiner & Associates, Skokie, Illinois; Paul Zia, Consultant, Raleigh, North Carolina — and the many individuals on PCI technical committees, too numerous to identify, who provided extra effort in their reviews. And the prestressing industry is indebted to the experienced and knowledgeable engineers on the PCI Design Handbook Committee which administered the project:

Dan E. Branson	Samuel N. Payne
Vernon Coenen	Charles H. Raths
Wesley Dolhun	H. W. Reinking, Jr.
Edward S. Hoffman	Fromy Rosenberg
Daniel P. Jenny[1]	Virendra Sharma
Leslie D. Martin	Robert E. Smith
McLeod C. Nigels[2]	Charles F. Terry
Alfred L. Parme	Wallace Weihing[3]

(1) Chairman
(2) Past Chairman
(3) Deceased

PART 1
ASSEMBLY CONCEPTS

SINGLE STORY CONSTRUCTION
Wall Bearing

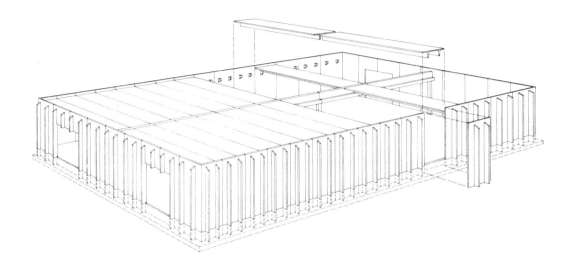

The use of precast prestressed concrete for bearing walls can provide economies by eliminating the need for a structural frame at the perimeter of a building. This economy is most apparent in buildings with a larger ratio of wall to floor area.

The connection between the roof framing member and the wall member can be handled by either bearing the roof member on a haunch cast into the back face of the panel, or by bearing directly on the top of the wall panel. The former method allows the wall panel to extend above the roof forming a parapet, and the latter method allows the roof members to project beyond the walls forming an overhang.

The wall panels themselves can be selected from a variety of standard sections or flat panels, and specially formed architectural precast shapes depending on the intended use, location, and budget for the building. Any of the standard precast deck units can be used for roofs of single-story buildings.

Sketches of connections in Part 1 are conceptual and not to be used unless properly designed and detailed.

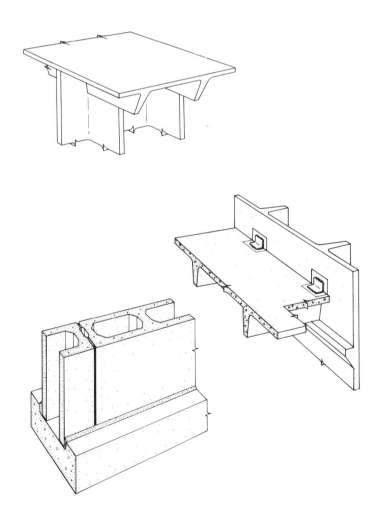

8 ft. wide double-tee wall panels support double-tee roof units of this industrial building.

Precast and prestressed concrete flat panels used for warehouse walls. Hollow-core slabs are often used in this application.

Smooth precast concrete fascia contrast with exposed aggregate wall panels that support double tee roof.

Slanted wall tee stems give battered effect, support prestressed double tees in roof.

SINGLE STORY CONSTRUCTION
Beam-Column Framing

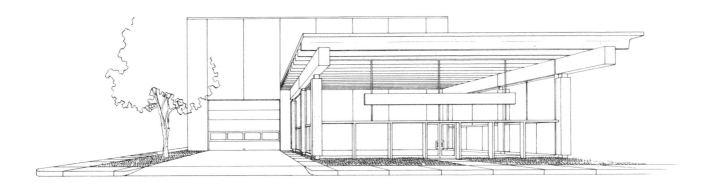

All of the standard precast beam and column shapes shown in Part 2 can be used for single story structures. Selection of the type of beam to be used depends on engineering considerations such as span length and superimposed loads, and architectural considerations such as depth and ceiling construction, if any. Details and design methods for connections are shown in Part 5.

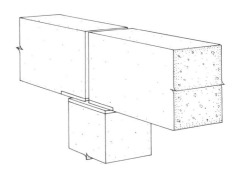

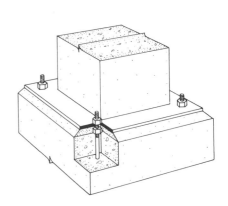

Structure built entirely with precast prestressed concrete. Hollow-core slabs provide office floors. Unique precast trusses support roof.

This huge airport complex uses precast prestressed concrete framing and post-tensioned connections.

Precast prestressed concrete members frame loading docks.

Double tees spanning 60 ft. supported by precast beams and columns.

Roof, floor and wall double tees, as well as all beams and columns make up huge production/administrative complex.

SINGLE STORY CONSTRUCTION
Long Span

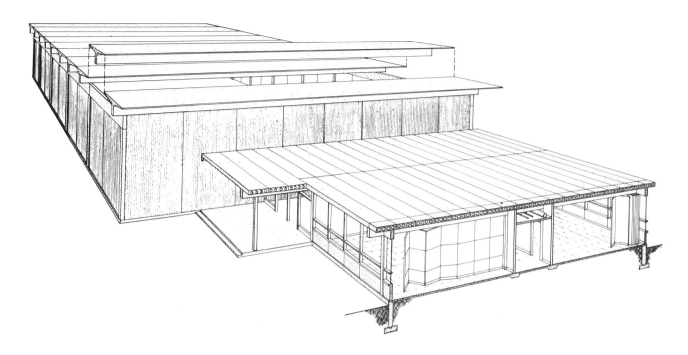

Standard precast prestressed sections that will span 100 feet or more are available in most parts of the United States and Canada. Even longer spans can be achieved by combining post-tensioning with precast or cast-in-place concrete. Special consideration must be given to camber and deflection when designing structures of this type.

Precast prestressed bridge beams span up to 114 ft. across the pool. Both roof and walls are composed of 4 ft. wide hollow-core units from 20 to 34 ft. long.

Exposed prestressed concrete channel joists permit integration of electrical and mechanical systems. Hollow-core slabs are used for exterior wall panels.

Long clear spans are provided by prestressed single and double tees.

Precast prestressed concrete single tees span 90 ft. in spite of loads of up to 150 psf.

High quality finish of the precast prestressed units made suspended ceiling unnecessary. Exposed tees provide ready support for mechanical and electrical systems.

A clear span for open class room area was achieved with double tees.

MULTI-STORY CONSTRUCTION
Wall Bearing — Low Rise

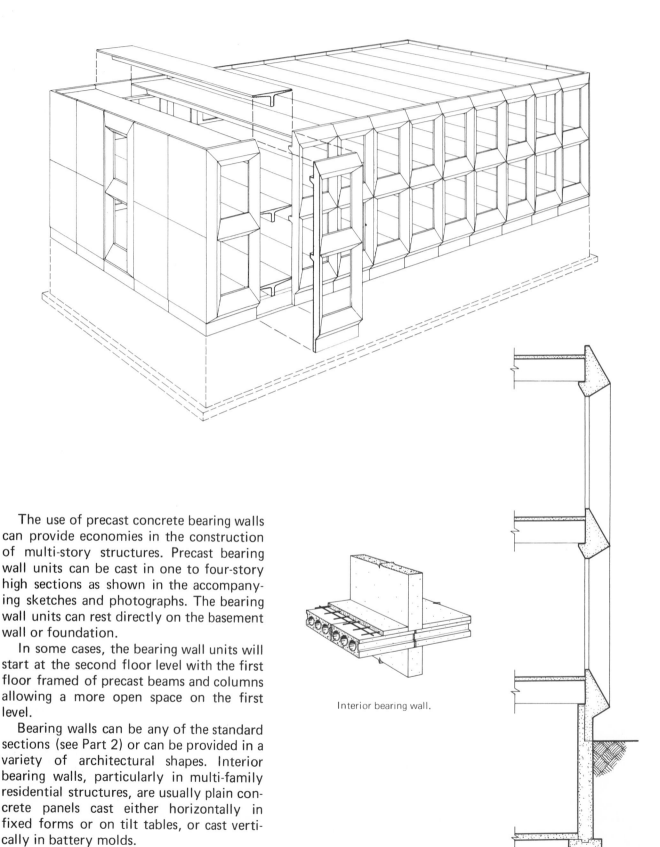

Interior bearing wall.

The use of precast concrete bearing walls can provide economies in the construction of multi-story structures. Precast bearing wall units can be cast in one to four-story high sections as shown in the accompanying sketches and photographs. The bearing wall units can rest directly on the basement wall or foundation.

In some cases, the bearing wall units will start at the second floor level with the first floor framed of precast beams and columns allowing a more open space on the first level.

Bearing walls can be any of the standard sections (see Part 2) or can be provided in a variety of architectural shapes. Interior bearing walls, particularly in multi-family residential structures, are usually plain concrete panels cast either horizontally in fixed forms or on tilt tables, or cast vertically in battery molds.

Precast wall panels are combined with other prestressed concrete elements to provide an economical structure.

High density concrete in precast concrete walls and prestressed double tee floors provide sound resistance and reduced maintenance cost in low rise housing.

Precast prestressed concrete hollow-core slabs are attractively combined with other materials into low rise buildings. Balconies are of precast hollow-core slabs.

Hollow-core slabs supported on precast bearing walls.

MULTI-STORY CONSTRUCTION
Wall Bearing — High Rise

Vertical post-tensioning is used to tie in shear walls in an entirely precast and prestressed concrete system building.

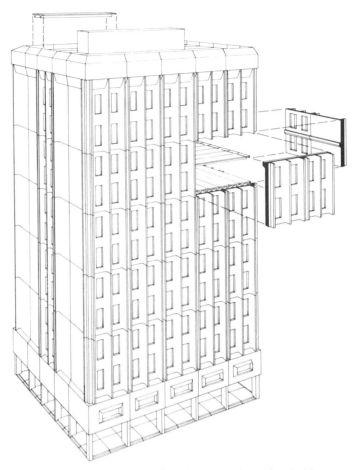

A variety of high-rise facades and floor layouts can be achieved with wall bearing construction. Here, perimeter walls and a central core support the floors and roof for column-free interior space.

Hollow-core floors and roofs bear on precast exterior and interior walls; balconies are also precast.

Mass production of repetitive units lowered the cost for this thirteen-story tower. The building is designed to accommodate five more floors.

118,000 sq. ft. of prestressed concrete hollow-core floor and roof planks and precast wall panels were erected in 41 working days.

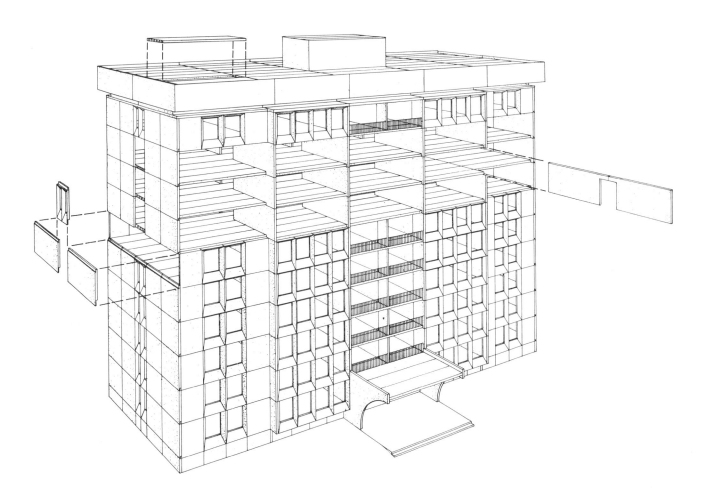

Transverse load bearing wall panels can be used effectively to resist lateral forces and to support the floors and roof.

MULTI-STORY CONSTRUCTION
Beam-Column Framing

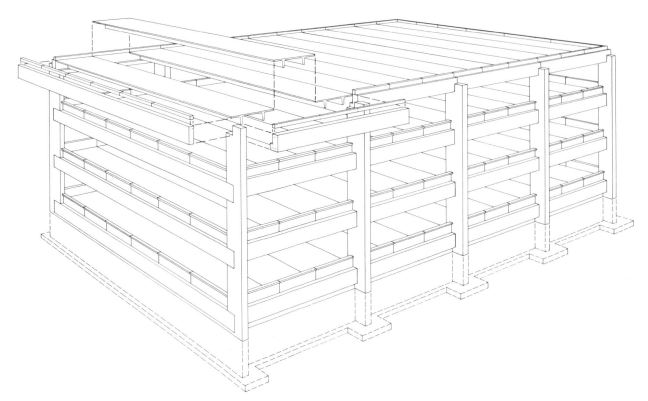

Precast prestressed beams and columns can be used for any type of low or high rise structure. In many buildings, the beams and columns become an architectural feature, and aesthetic consideration is given to the connection between the beams and columns. Architectural and engineering considerations dictate whether the beams are continuous with single story columns, or whether multi-story columns are used with single span beams. In some cases, so-called "tree columns" have been used so that the beam connection is made away from the column, usually about the quarter point or at mid-span. These "tree columns" have been cast in single-story or multi-story units.

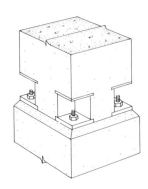

Multi-story column connection.

Precast concrete beams 37½ in. deep are pierced for ductwork, large utility piping and lines for water and gas systems in hospital.

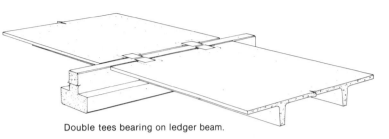

Double tees bearing on ledger beam.

Precast columns, up to 77 ft. long, support L-girders, rectangular beams and double-tees.

This four-story medical facility includes more than 1000 precast prestressed tees, varying in length from 30 to 85 ft., and other precast concrete structural elements.

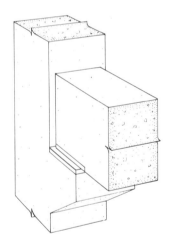

Beam-column connection.

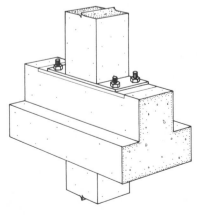

Continuous ledger beam with one-story columns.

Hollow-core slabs used in all precast systems building.

MULTI-STORY CONSTRUCTION
Beam-Column Framing

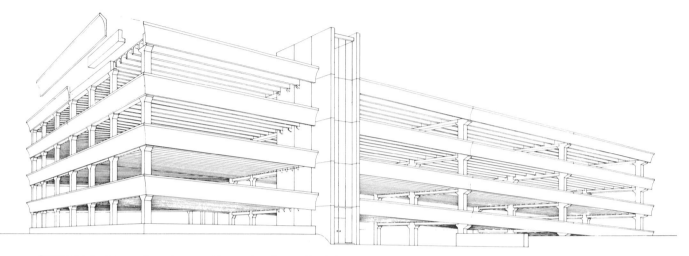

Parking structures generally employ precast beam-column frames. The deck may consist of prestressed concrete single tee, double tee or hollow-core slabs. Generally they have topping slabs. If precast units are spaced, the cast-in-place deck slab is often post-tensioned transversely.

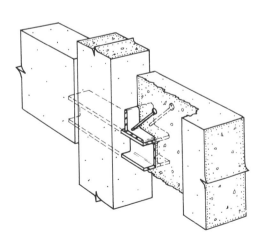

This cut-away drawing shows a method of concealing the beam-to-column connection.

Precast column trees and haunches on load-bearing panels support double tee floor and roof units.

Large 4-level parking facility has direct access at each level, uses hollow-core floor slabs throughout.

Ramped floors allow more efficient use of the structure, resulting in lower cost per parking space.

This five level parking structure consists of literally miles of precast and prestressed concrete elements.

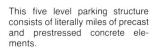

SPECIAL STRUCTURES

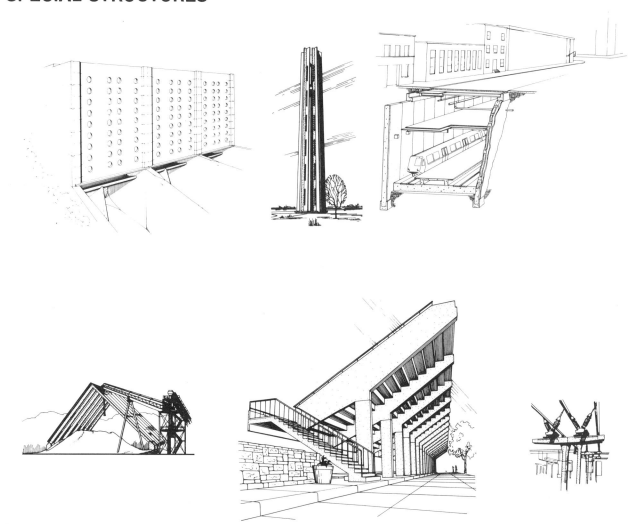

In addition to the wide variety of buildings and bridges that are built using standard components, precast prestressed concrete is often selected for unusual and special applications. Construction speed, economy, long-span capability, all-weather construction, fire safety and reserve strength are the qualities most often cited for selection of precast and prestressed concrete. Transmission towers, stadia, marine structures, retaining walls and others shown on these pages are but a few of the many unusual applications that have been found for precast, prestressed concrete. The use of post-tensioning in precast segmental construction has been adapted to special framing problems in buildings as well as bridges.

Famous amusement park building in background has precast, prestressed roof and walls. Monorail guide is also precast, prestressed concrete.

Three-tread seating elements measure up to 54 ft long and weigh 38,000 lbs.

Precast concrete retaining wall designed to blend with mountain terrain.

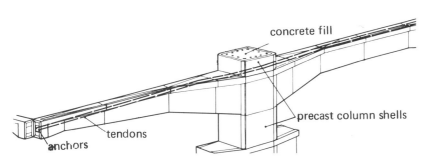

Segmental precast frame assembled by post-tensioning. Frame and central core support column-free double-tee floors on building shown at right.

This unique structure stores 120,000 tons of cement clinker.

PART 2
PRODUCT INFORMATION AND CAPABILITY

PRODUCT INFORMATION AND CAPABILITY

2.1 General

2.1.1 Notation

A = cross sectional area

A_g = gross cross-sectional area

A_{ps} = area of prestressed reinforcement

b_w = web width

D = unfactored dead loads

E_c = modulus of elasticity of concrete

e_c = eccentricity of prestress force from the centroid of the section at the center of the span

e_e = eccentricity of prestress force from the centroid of the section at the end of the span

f'_c = specified compressive strength of concrete

f'_{ci} = compressive strength of concrete at time of initial prestress

f_{pc} = average compressive stress in concrete due to effective prestress force only

f_{pu} = ultimate strength of prestressing steel

h = overall depth of member

I = moment of inertia

L = unfactored live loads

ℓ = span

M_n = nominal moment strength of a member

M_{nb} = nominal moment strength under balanced conditions

P_n = nominal axial load strength of a compression member at given eccentricity

P_{nb} = nominal axial load strength under balanced conditions

P_o = axial load nominal strength of a compression member with zero eccentricity

P_u = factored axial load

t = thickness

V_{ci} = nominal shear strength provided by concrete when diagonal cracking results from combined shear and moment

V_{cw} = nominal shear strength provided by concrete when diagonal cracking results from excessive principal tensile stress in the web

V_u = factored shear force

V/S = volume-surface ratio

y_b = distance from bottom fiber to center of gravity of section

y_t = distance from top fiber to center of gravity of section

Z = section modulus

Z_b = section modulus with respect to the bottom fiber of a cross section

Z_t = section modulus with respect to the top fiber of a cross section

δ = moment magnification factor

ϕ = strength reduction factor

2.1.2 Introduction

This part of the Handbook is devoted to technical data on the shapes that are standard in the precast prestressed concrete industry today. The use of this part of the Handbook as a guide for selection of a particular shape together with information on design aids and techniques provided in other parts of this Handbook should enable the designer to quickly and expeditiously complete his design.

The load tables on the following pages show dimensions, section properties and engineering capabilities of the shapes most commonly used throughout the industry. These shapes include double and single tees, hollow-core and solid flat slabs, beams, girders, columns, piles and wall panels. The dimensions of the shapes shown in the tables may vary among manufacturers. Adjustment of these minor variations can be made by the designer. Hollow-core slabs of different thicknesses, core sizes and shapes are available in the market under various trade names. Cross-sections and section properties of proprietary hollow-core slabs are shown on pages 2—33 through 2—43. Load tables, on pages 2—25 through 2—32, are developed for non-proprietary hollow-core sections of thicknesses most commonly used in the industry.

Designers making use of these load tables should contact the manufacturers in the geographic area of the proposed structure to determine availability and exact dimensions of products shown here. Manufacturers will usually have their own load tables for sections which are not included on the following pages.

2.2 Explanation of load tables

Load tables for stemmed deck members, flat deck members and beams show the allowable superimposed service load, estimated camber at the time of erection and the estimated long-time camber after the member has essentially stabilized. For the deck members, the table at the top gives the information for the member with no topping, and the table at the bottom of the page is for the same member with two inches of normal weight concrete topping acting compositely with the precast section. Values in the tables assume a uniform 2 in. topping the full span length, and assume the member to be unshored at the time the topping is placed. Safe loads and cambers shown in the tables are based on the dimensions and section properties shown on the page, and will vary for members with different dimensions.

For beams, a single table is used for several sizes of members. The values shown are based on sections containing the maximum practical number of prestressing strands, but in some cases, more strands could be used.

2.2.1 Safe Superimposed Load

The values for safe superimposed service load are based on the capacity of the member as governed by the ACI Building Code limitations on flexural strength, service load flexural stresses, or, in the case of flat deck members, shear strength. A portion of the safe load shown is assumed to be dead load for the purpose of applying load factors and determining time-dependent cambers and deflections. For untopped deck members, 10 psf of the capacity shown is assumed as superimposed dead load, typical for roof members. For topped deck members, 15 psf of the capacity shown is assumed as superimposed dead load, typical for floor members. The capacity shown is in addition to the weight of the topping. For beams, 50 percent of the capacity shown is assumed as dead load, normally conservative for beams which support concrete decks.

Example: For an 8DT24/88-D1 (p. 2—16) with a 52 ft span, the capacity shown is 68 psf. The member can safely carry service loads of 10 psf dead and 58 psf live.

2.2.2 Limiting Criteria

The criteria used to determine the safe superimposed load and strand placement are based on "Building Code Requirements for Reinforced Concrete (ACI 318-77)." For design procedures, see Part 3 of this Handbook. A summary of the Code provisions used in the development of these load tables is as follows:

1. Capacity governed by design flexural strength:

   Load factors: $1.4D + 1.7L$

   Strength reduction factor, $\phi = 0.90$

   Calculation of design moments assumes simple spans with roller supports. If the strands are fully developed (see Sect. 3.2.3), the critical moment is assumed to be at midspan in members with straight strands, and at 0.4ℓ (ℓ = span) in products with strands depressed at midspan. (Note: The actual critical point can be determined by analysis, but will seldom vary significantly from 0.4ℓ.) Flexural strength is calculated using strain compatibility as discussed in Part 3.

2. Capacity governed by service load stresses:

   Flexural stresses immediately after transfer of prestress, before long time losses (Note: It is assumed that strands are always initially tensioned to $0.7 f_{pu}$):

a) Compression — $0.6 f'_{ci}$

b) End tension — $6\sqrt{f'_{ci}}$

Midspan tension — $3\sqrt{f'_{ci}}$

Note 1: These stresses are calculated 50 strand diameters from the end of the member, the theoretical point of full transfer, and at midspan.

Note 2: Release tension is not used as a limiting criterion for beams. Supplemental top reinforcement must be provided, designed as described in Sect. 3.2.2 of this Handbook.

Stresses at service loads, after all losses:

a) Compression — $0.45 f'_c$

b) Tension:

Stemmed deck members and beams — $12\sqrt{f'_c}$

Note that in final design, deflections must be determined based on bilinear moment-deflection relationships. See Sect. 3.4.2.

Flat deck members — $6\sqrt{f'_c}$

The critical point for service load moment is assumed at midspan for members with straight strands and at 0.4ℓ for members with strands depressed at midspan, as described above.

3. For *flat deck members,* the capacity may be limited by the design shear strength. In this case, the safe superimposed load is that which will yield a factored shear force V_u of no more than ϕV_{ci} or ϕV_{cw}, as permitted by ACI 318-77 for slabs without shear reinforcement. See Sect. 3.3 for the design procedures.

Stemmed deck members and beams do not have this limitation. It may be necessary to provide shear reinforcement, designed as described in Sect. 3.3. For the majority of such deck members, minimum or no reinforcement may be required as provided by Sect. 11.5.5 of ACI 318-77. For loads which are heavier than normal, special transverse reinforcement may be required to resist the moment in the overhanging flange.

4. *Flat deck members* show no values beyond a span/depth ratio of 50 for untopped members and 40 for topped members. These are the suggested maximums for roof and floor members respectively, unless a detailed analysis is made.

2.2.3 Estimated Cambers

The estimated cambers shown are calculated using the multipliers shown in Sect. 3.4.3 of this Handbook. *These values are estimates, and should not be used as absolute values.* Non-structural components attached to members which could be affected by camber variations, such as partitions or folding doors, should be placed with adequate allowance for error. Calculation of topping quantities should also recognize that the values can vary.

2.2.4 Concrete Strengths and Unit Weights

Twenty-eight day cylinder strength for concrete in the prestressed units is assumed to be 5000 psi. Tables for units with composite topping are based on the topping concrete being normal weight concrete with a cylinder strength of 3000 psi.

Concrete strength at time of strand tension release is 3500 psi unless the value falls below the heavy line shown in the load table, indicating that a cylinder strength greater than 3500 psi is required.

No values are shown when the required release strength exceeds 4500 psi. The designer should recognize that it is sometimes difficult to obtain a release strength higher than 3500 psi on a one-day casting cyle. In such cases, the cost of production will be increased and the designer should consult with prospective producers when required release strengths are above 3500 psi.

Many prestressing plants prefer to use higher strength concretes, resulting in somewhat higher allowable loads or greater spans than indicated in the load tables contained herein.

Unit weights of concrete are assumed to be 150 lb per cu ft for normal weight and 115 lb per cu ft for lightweight.

2.2.5 Prestressing Strands

For the *stemmed deck members* (p. 2—6 to 2—24 and *beams* (p. 2—50 to 2—53) prestressing strands used in the load tables are 1/2 inch diameter strands with an ultimate strength of 270,000 psi. Quantity, size, and profile of strands are shown in the load tables under the column headed "Strand Pattern." In the double tee load tables, for example, "88-S" indicates 8 — 1/2 inch diameter 270 K strands (4 per double tee stem) and the "S" indicates that the strands are straight. "88-D1" indicates 8 — 1/2 inch diameter 270 K strands depressed at one point at the midspan of the double tee.

For the *flat deck members* (p. 2—25 to 2—49), the manufacturer is allowed some flexibility in choice of strand size and ultimate strength. The "Strand Designation Code" number shown in the first column of the load table is the product of the strand area (A_{ps}) and the ultimate strand strength (f_{pu}), expressed in kips per foot of width.

Example: For a 4HC8/50-S (p. 2–27) the total value of $A_{ps} \times f_{pu}$ required would be 4 x 50 = 200 kips. If the producer uses 250K strand, the total strand area required would be 200/250 = 0.80 sq in. The producer might then choose, for example, to use 6 – 1/2 in. diameter strands (A_{ps} = 6 x 0.144 = 0.86 sq in.) or 4 – 1/2 in. diameter and 3 – 3/8 in. diameter strands (A_{ps} = 4 x 0.144 + 3 x 0.080 = 0.82 sq in.), or another combination of size and strength to produce a total value of $A_{ps} \times f_{pu}$ of at least 200 kips.

Note: For development length (see Sect. 3.2.3); Strand Designation code number of 30S, assume 3/8 in. diameter strands, others assume 1/2 in. diameter strands.

2.2.6 Losses

Losses assumed in computing the required concrete strength at time of strand release are 10%. Total losses are assumed to be 22% for normal weight concrete and 25% for lightweight concrete. For long span, heavily prestressed products, losses may be somewhat higher than these assumed values, and for shorter spans with less prestressing they may be lower. However, these values will usually be adequate for member selection. Additional information on losses is given in Part 3 of this Handbook.

2.2.7 Strand Placement

For *stemmed deck members* and *beams,* the eccentricities of strands at the ends and midspan are shown in the load tables. Strands have been placed so that the stress at 50 strand diameters from the end (theoretical transfer point) will not exceed those specified above, with a concrete strength at release of 3500 psi, except as noted.

For *flat deck members,* the load table values are based on strand centered 1-1/2 in. from the bottom of the slab. Strand placement can vary from as low as 7/8 in. to as high as 2-1/8 in. from the bottom, which will change the capacity and camber values shown. The higher strand placements give improved fire resistance ratings (see Part 6 of this Handbook for more information on fire resistance). The lower strand placement may require higher release strengths, or top tension reinforcement at the ends. **The designer should contact the local supplier of flat deck members for available and recommended strand placement locations.**

2.2.8 Columns and Load-Bearing Wall Panels

Interaction curves for selected precast, prestressed columns, precast reinforced columns and various types of commonly used wall panels are provided on p. 2–54 to p. 2–61.

These interaction curves are for strength design loads and moments and the appropriate load factors must be applied to the service loads and moments before entering the charts. Also, the curves are for *short* members. Moment magnifiers caused by slenderness effects must be calculated and applied to the design moments before using the curves for final member selection (see Part 4).

The column curves are terminated at a value of $P_u = 0.80 P_o$, the maximum allowable load for tied columns under ACI 318-77. Most of the wall panel curves show the lower portion of the curve only (flexure controlling). Actual design loads will rarely exceed the values shown.

The curves for double tee wall panels are for bending in the direction that causes tension in the stem. They are conservative, in the range shown, for bending in the opposite direction.

The curves for hollow-core wall panels are based on a generic section as shown. They can be used with small error for all sections commonly marketed for wall panel use.

2.2.9 Piles

Allowable concentric service loads on prestressed concrete piles, based on the structural capacity of the pile alone are shown in Table 2.7.1. The ability of the soil to carry these loads must be evaluated by soils engineers. Values for concrete strengths up to 8000 psi are shown. Available strengths should be checked with local manufacturers. The design of prestressed concrete piles is discussed in Sect. 3.6.7 of this Handbook. book.

Section properties and allowable service load bending moments for prestressed concrete sheet pile units are shown in Table 2.7.2. These units are available in some areas for use in earth retaining structures.

DOUBLE TEE

8'-0" x 12"
Normal Weight Concrete

Strand Pattern Designation

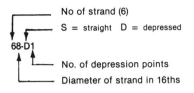

- No of strand (6)
- S = straight   D = depressed
- 68-D1
- No. of depression points
- Diameter of strand in 16ths

Safe loads shown include dead load of 10 psf for untopped members and 15 psf for topped members. Remainder is live load. Long-time cambers include superimposed dead load but do not include live load.

Key

- 178 — Safe superimposed service load, psf
- 0.1 — Estimated camber at erection, in.
- 0.2 — Estimated long-time camber, in.

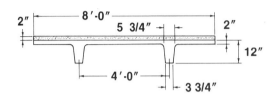

$f'_c = 5000$ psi
$f_{pu} = 270,000$ psi

Section Properties

	Untopped	Topped
A =	287 in.²	—
I =	2872 in.⁴	4389 in.⁴
Y_b =	9.13 in.	10.45 in.
Y_t =	2.87 in.	3.55 in.
Z_b =	315 in.³	420 in.³
Z_t =	1001 in.³	1236 in.³
wt =	299 plf	499 plf
	37 psf	62 psf
V/S =	1.22 in.	

8DT12 — No Topping

Table of safe superimposed service load (psf) and cambers

Strand Pattern	e_e / e_c	12	14	16	18	20	22	24	26	28	30	32	34	36	38	40
28-S	7.13	178	136	108	81	60	45	33								
	7.13	0.1	0.2	0.2	0.3	0.3	0.3	0.3								
		0.2	0.2	0.3	0.3	0.3	0.3	0.2								
48-S	5.13			187	143	110	86	68	53	42	33					
	5.13			0.4	0.4	0.5	0.5	0.6	0.6	0.6	0.5					
				0.5	0.5	0.6	0.6	0.6	0.5	0.3						
68-S	3.13			159	123	97	77	61	49	39	31					
	3.13			0.4	0.4	0.5	0.5	0.5	0.5	0.4	0.3					
				0.5	0.5	0.5	0.5	0.4	0.3	0.1	0.0					
68-D1	3.13										76	64	53	**44**	**36**	
	6.63										1.3	1.4	1.5	1.4	1.4	
											1.4	1.4	1.3	1.1	0.8	
88-D1	1.13															**40**
	6.38															1.6
																0.9

8DT12 + 2 — 2" Normal Weight Topping

Table of safe superimposed service load (psf) and cambers

Strand Pattern	e_e / e_c	12	14	16	18	20	22	24	26	28	30	32	34
28-S	7.13	199	149	115	83	58	39						
	7.13	0.1	0.2	0.2	0.3	0.3	0.3						
		0.1	0.2	0.2	0.2	0.1	0.0						
48-S	5.13				164	124	94	71	53				
	5.13				0.4	0.5	0.5	0.6	0.6				
					0.4	0.4	0.3	0.2	0.1				
68-S	3.13				**198**	**150**	110	**79**	**55**				
	3.13				0.4	0.4	0.5	0.5	0.5				
					0.3	0.3	0.2	0.1	0.0				
68-D1	3.13								74	**55**	39		
	6.63								1.3	1.4	1.5		
									0.5	0.3	0.0		

Bold type — *Capacity governed by stresses, others governed by flexural strength*

Values below heavy line require release strengths higher than 3500 psi.

DOUBLE TEE

8'-0" x 12"
Lightweight Concrete

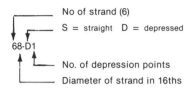

Strand Pattern Designation

No of strand (6)

S = straight D = depressed

68-D1

No. of depression points

Diameter of strand in 16ths

Safe loads shown include dead load of 10 psf for untopped members and 15 psf for topped members. Remainder is live load. Long-time cambers include superimposed dead load but do not include live load.

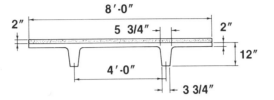

f'_c = 5000 psi
f_{pu} = 270,000 psi

Section Properties

	Untopped	Topped
A =	287 in.²	—
I =	2872 in.⁴	4819 in.⁴
Y_b =	9.13 in.	10.82 in.
Y_t =	2.87 in.	3.18 in.
Z_b =	315 in.³	445 in.³
Z_t =	1001 in.³	1515 in.³
wt =	229 plf	429 plf
	29 psf	54 psf
V/S =	1.22 in.	

Key

181 — Safe superimposed service load, psf
0.2 — Estimated camber at erection, in.
0.3 — Estimated long-time camber, in.

Table of safe superimposed service load (psf) and cambers

8LDT12
No Topping

Strand Pattern	e_e e_c	12	14	16	18	20	22	24	26	28	30	32	34	36	38	40
28-S	7.13 7.13	181 0.2 0.3	141 0.3 0.4	113 0.4 0.4	89 0.4 0.5	68 0.5 0.5	52 0.5 0.5	40 0.5 0.5	31 0.5 0.4							
48-S	5.13 5.13			191 0.6 0.7	150 0.7 0.8	118 0.8 1.0	93 0.9 1.1	75 1.0 1.1	61 1.0 1.1	49 1.1 1.1	40 1.1 1.0	33 1.1 0.8				
68-S	3.13 3.13			166 0.6 0.8	131 0.7 0.8	104 0.8 0.9	84 0.9 0.9	68 0.9 0.9	56 0.9 0.9	46 0.9 0.7	38 0.9 0.5	30 0.8 0.1				
68-D1	3.13 6.63									60 2.5 2.5	50 2.7 2.5	42 2.8 2.3	35 2.9 2.1			

Table of safe superimposed service load (psf) and cambers

8LDT12 + 2
2" Normal Weight Topping

Strand Pattern	e_e e_c	14	16	18	20	22	24	26	28	30	32	34	36
28-S	7.13 7.13	153 0.3 0.3	120 0.4 0.3	90 0.4 0.3	65 0.5 0.3	47 0.5 0.2	32 0.5 0.0						
48-S	5.13 5.13			172 0.7 0.6	131 0.8 0.6	101 0.9 0.6	78 1.0 0.6	61 1.1 0.4	45 1.1 0.2				
68-S	3.13 3.13			158 0.7 0.5	123 0.8 0.5	91 0.9 0.4	67 0.9 0.2	47 0.9 0.0					
68-D1	3.13 6.63								50 2.5 0.5	36 2.7 0.0			

Bold type — *Capacity governed by stresses, others governed by flexural strength*

Values below heavy line require release strengths higher than 3500 psi.

DOUBLE TEE

8'-0" x 14"
Normal Weight Concrete

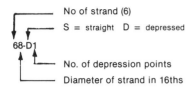

Strand Pattern Designation

68-D1

- No of strand (6)
- S = straight D = depressed
- No. of depression points
- Diameter of strand in 16ths

Safe loads shown include dead load of 10 psf for untopped members and 15 psf for topped members. Remainder is live load. Long-time cambers include superimposed dead load but do not include live load.

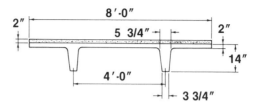

f'_c = 5000 psi
f_{pu} = 270,000 psi

Section Properties

		Untopped	Topped
A	=	306 in.²	—
I	=	4508 in.⁴	6539 in.⁴
Y_b	=	10.51 in.	11.97 in.
Y_t	=	3.49 in.	4.03 in.
Z_b	=	429 in.³	546 in.³
Z_t	=	1292 in.³	1623 in.³
wt	=	319 plf	519 plf
		40 psf	65 psf
V/S	=	1.25 in.	

Key

168 — Safe superimposed service load, psf
0.1 — Estimated camber at erection, in.
0.2 — Estimated long-time camber, in.

8DT14

Table of safe superimposed service load (psf) and cambers
No Topping

Strand Pattern	e_e e_c	Span, ft.																	
		14	16	18	20	22	24	26	28	30	32	34	36	38	40	42	44	46	
28-S	8.51 8.51	168 0.1 0.2	134 0.2 0.2	102 0.2 0.2	77 0.2 0.3	58 0.2 0.3	44 0.2 0.2	33 0.2 0.2											
48-S	7.51 7.51				163 0.5 0.6	129 0.6 0.7	103 0.6 0.8	83 0.7 0.8	68 0.7 0.8	55 0.7 0.8	45 0.7 0.7	36 0.7 0.5							
68-S	4.51 4.51				175 0.4 0.5	139 0.5 0.6	112 0.5 0.6	91 0.6 0.7	74 0.6 0.6	61 0.6 0.6	50 0.6 0.5	40 0.5 0.3	33 0.5 0.1						
68-D1	4.51 8.01							142 1.0 1.2	118 1.1 1.3	99 1.2 1.4	83 1.3 1.4	70 1.4 1.4	59 1.4 1.4	50 1.4 1.2	42 1.3 1.0	35 1.2 0.7			
88-D1	2.51 7.76														59 1.7 1.5	50 1.7 1.4	42 1.7 1.1	35 1.6 0.8	

8DT14+2

Table of safe superimposed service load (psf) and cambers
2" Normal Weight Topping

Strand Pattern	e_e e_c	Span, ft.											
		14	16	18	20	22	24	26	28	30	32	34	36
28-S	8.51 8.51	181 0.1 0.1	141 0.2 0.2	103 0.2 0.2	74 0.2 0.1	52 0.2 0.1	36 0.2 0.0						
48-S	7.51 7.51				176 0.5 0.5	136 0.6 0.5	106 0.6 0.5	83 0.7 0.4	65 0.7 0.3	50 0.7 0.2	38 0.7 0.0		
68-S	4.51 4.51					159 0.5 0.4	126 0.5 0.4	100 0.6 0.3	75 0.6 0.2	55 0.6 0.0			
68-D1	4.51 8.01							152 1.0 0.8	124 1.1 0.8	102 1.2 0.7	83 1.3 0.6	65 1.4 0.5	49 1.4 0.2

Bold type — *Capacity governed by stresses, others governed by flexural strength*

Values below heavy line require release strengths higher than 3500 psi.

DOUBLE TEE

Strand Pattern Designation

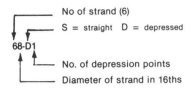

No of strand (6)

S = straight   D = depressed

68-D1

No. of depression points

Diameter of strand in 16ths

Safe loads shown include dead load of 10 psf for untopped members and 15 psf for topped members. Remainder is live load. Long-time cambers include superimposed dead load but do not include live load.

8'-0" x 14"
Lightweight Concrete

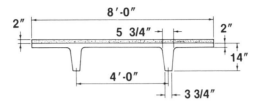

f'_c = 5000 psi
f_{pu} = 270,000 psi

Section Properties

	Untopped	Topped
A =	306 in.²	—
I =	4508 in.⁴	7173 in.⁴
Y_b =	10.51 in.	12.40 in.
Y_t =	3.49 in.	3.60 in.
Z_b =	429 in.³	578 in.³
Z_t =	1292 in.³	1992 in.³
wt =	244 plf	444 plf
	31 psf	56 psf
V/S =	1.25 in.	

Key

172 — Safe superimposed service load, psf
0.2 — Estimated camber at erection, in.
0.3 — Estimated long-time camber, in.

8LDT14

No Topping

Table of safe superimposed service load (psf) and cambers

Strand Pattern	e_e e_c	Span, ft.																
		14	16	18	20	22	24	26	28	30	32	34	36	38	40	42	44	46
28-S	8.51 8.51	172 0.2 0.3	139 0.3 0.3	109 0.3 0.4	84 0.4 0.4	66 0.4 0.5	51 0.5 0.4	40 0.5 0.4	32 0.5 0.3									
48-S	7.51 7.51				170 0.8 1.0	137 0.9 1.1	111 1.0 1.2	91 1.1 1.3	75 1.3 1.4	63 1.3 1.4	52 1.4 1.4	44 1.4 1.3	36 1.5 1.1	30 1.4 0.9				
68-S	4.51 4.51				183 0.7 0.9	147 0.8 1.0	120 0.9 1.1	99 1.0 1.1	82 1.1 1.1	68 1.1 1.1	57 1.2 1.1	48 1.2 0.9	40 1.2 0.7	34 1.1 0.4				
68-D1	4.51 8.01									107 1.9 2.2	91 2.1 2.3	78 2.3 2.4	67 2.4 2.5	58 2.6 2.5	50 2.7 2.4	43 2.7 2.1	37 2.7 1.7	31 2.6 1.2

8LDT14+2

2" Normal Weight Topping

Table of safe superimposed service load (psf) and cambers

Strand Pattern	e_e e_c	Span, ft.													
		14	16	18	20	22	24	26	28	30	32	34	36	38	40
28-S	8.51 8.51	184 0.2 0.2	146 0.3 0.3	111 0.3 0.3	82 0.4 0.3	60 0.4 0.2	44 0.5 0.1	31 0.5 0.0							
48-S	7.51 7.51				183 0.8 0.8	144 0.9 0.8	114 1.0 0.8	91 1.1 0.8	72 1.3 0.8	57 1.3 0.6	45 1.4 0.4	35 1.4 0.1			
68-S	4.51 4.51				167 0.8 0.7	134 0.9 0.7	107 1.0 0.6	87 1.1 0.5	**67** 1.1 0.4	49 1.2 0.1					
68-D1	4.51 8.01								110 1.9 1.3	91 2.1 1.2	76 2.3 1.1	**60** 2.4 0.9	47 2.6 0.5	35 2.7 0.1	

Bold type — *Capacity governed by stresses, others governed by flexural strength*

Values below heavy line require release strengths higher than 3500 psi.

DOUBLE TEE

8'-0" x 16"
Normal Weight Concrete

Strand Pattern Designation

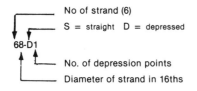

No of strand (6)
S = straight D = depressed

68-D1

No. of depression points
Diameter of strand in 16ths

Safe loads shown include dead load of 10 psf for untopped members and 15 psf for topped members. Remainder is live load. Long-time cambers include superimposed dead load but do not include live load.

Key

199 — Safe superimposed service load, psf
0.1 — Estimated camber at erection, in.
0.1 — Estimated long-time camber, in.

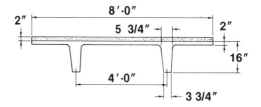

f'_c = 5000 psi
f_{pu} = 270,000 psi

Section Properties

		Untopped	Topped
A	=	325 in.²	—
I	=	6634 in.⁴	9306 in.⁴
Y_b	=	11.93 in.	13.52 in.
Y_t	=	4.07 in.	4.48 in.
Z_b	=	556 in.³	688 in.³
Z_t	=	1630 in.³	2077 in.³
wt	=	339 plf	539 plf
		42 psf	67 psf
V/S	=	1.29 in.	

8DT16
No Topping

Table of safe superimposed service load (psf) and cambers

Strand Pattern	e_e / e_c	14	16	18	20	22	24	26	28	30	32	34	36	38	40	42	44	46	48	50	52
28-S	9.93 / 9.93	199	159	122	93	71	54	41	31												
		0.1	0.1	0.2	0.2	0.2	0.2	0.2	0.2												
		0.1	0.2	0.2	0.2	0.2	0.2	0.2	0.1												
48-S	8.93 / 8.93				197	157	127	103	84	69	57	47	38	31							
					0.4	0.5	0.5	0.6	0.6	0.7	0.7	0.7	0.6	0.6							
					0.5	0.6	0.6	0.7	0.7	0.7	0.7	0.6	0.5	0.3							
68-S	5.93 / 5.93						182	147	121	100	82	68	57	47	39	32					
							0.5	0.5	0.6	0.6	0.7	0.7	0.7	0.7	0.6	0.6	0.5				
							0.6	0.6	0.7	0.7	0.7	0.7	0.6	0.5	0.3	0.0					
68-D1	5.93 / 9.43							173	145	122	103	88	75	64	54	46	39	33			
								0.8	0.9	1.1	1.2	1.2	1.3	1.3	1.3	1.3	1.2	1.1			
								1.0	1.1	1.2	1.3	1.4	1.3	1.3	1.1	0.9	0.7	0.3			
88-D1	3.93 / 9.18														77	67	58	50	**42**	**36**	
															1.7	1.7	1.8	1.8	1.7	1.6	
															1.6	1.6	1.5	1.3	1.0	0.6	
108-D1	1.93 / 8.93																		46	40	
																			1.9	1.8	
																			1.1	0.7	

8DT16 + 2
2" Normal Weight Topping

Table of safe superimposed service load (psf) and cambers

Strand Pattern	e_e / e_c	16	18	20	22	24	26	28	30	32	34	36	38	40	42
28-S	9.93 / 9.93	166	124	90	65	47	32								
		0.1	0.2	0.2	0.2	0.2	0.2								
		0.1	0.1	0.1	0.1	0.1	0.0								
48-S	8.93 / 8.93			164	129	102	81	64	50	**38**					
				0.5	0.5	0.6	0.6	0.7	0.7	0.7					
				0.4	0.4	0.4	0.4	0.3	0.2	0.0					
68-S	5.93 / 5.93					161	130	104	84	68	**51**				
						0.5	0.6	0.6	0.7	0.7	0.7				
						0.4	0.4	0.4	0.3	0.2	0.0				
68-D1	5.93 / 9.43						183	151	125	103	**85**	70	57	43	
							0.8	0.9	1.1	1.2	1.2	1.3	1.3	1.3	
							0.7	0.8	0.8	0.8	0.7	0.5	0.3	0.0	
88-D1	3.93 / 9.18										68	54			
											1.7	1.7			
											0.5	0.2			

Bold type — *Capacity governed by stresses, others governed by flexural strength*

Values below heavy line require release strengths higher than 3500 psi.

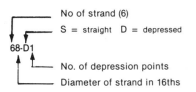
DOUBLE TEE

8'-0" x 16"
Lightweight Concrete

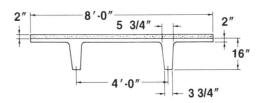

Safe loads shown include dead load of 10 psf for untopped members and 15 psf for topped members. Remainder is live load. Long-time cambers include superimposed dead load but do not include live load.

Key

- 165 — Safe superimposed service load, psf
- 0.2 — Estimated camber at erection, in.
- 0.3 — Estimated long-time camber, in.

$f'_c = 5000$ psi
$f_{pu} = 270{,}000$ psi

Section Properties

	Untopped	Topped
A =	325 in.²	—
I =	6634 in.⁴	10,094 in.⁴
Y_b =	11.93 in.	13.99 in.
Y_t =	4.07 in.	4.01 in.
Z_b =	556 in.³	721 in.³
Z_t =	1630 in.³	2517 in.³
wt =	260 plf	460 plf
	33 psf	58 psf
V/S =	1.29 in.	

8LDT16
No Topping

Table of safe superimposed service load (psf) and cambers

Strand Pattern	e_e / e_c	16	18	20	22	24	26	28	30	32	34	36	38	40	42	44	46	48	50	52
28-S	9.93	165	130	101	79	62	50	39	31											
	9.93	0.2	0.3	0.3	0.3	0.4	0.4	0.4	0.4											
		0.3	0.3	0.4	0.4	0.4	0.4	0.3	0.3											
48-S	8.93				165	135	111	92	77	65	55	46	39	33						
	8.93				0.8	0.9	1.0	1.1	1.1	1.2	1.3	1.3	1.3	1.3						
					0.9	1.1	1.2	1.2	1.3	1.3	1.3	1.2	1.1	0.9						
68-S	5.93				190	156	129	108	91	77	65	55	47	40	34					
	5.93				0.8	0.9	1.0	1.1	1.1	1.2	1.3	1.3	1.3	1.3	1.2					
					0.9	1.1	1.1	1.2	1.3	1.3	1.3	1.2	1.0	0.8	0.6					
68-D1	5.93						181	153	130	111	96	83	72	62	54	47	41	36	31	
	9.43						1.3	1.5	1.7	1.8	2.0	2.2	2.4	2.5	2.5	2.5	2.5	2.4	2.3	
							1.6	1.8	2.0	2.1	2.3	2.4	2.5	2.4	2.3	2.1	1.9	1.5	1.0	
88-D1	3.93																57	50	43	38
	9.18																3.2	3.4	3.4	3.5
																	2.9	2.7	2.5	2.2

8LDT16 + 2
2" Normal Weight Topping

Table of safe superimposed service load (psf) and cambers

Strand Pattern	e_e / e_c	16	18	20	22	24	26	28	30	32	34	36	38	40	42	44	46
28-S	9.93	171	132	98	74	55	40										
	9.93	0.2	0.3	0.3	0.3	0.4	0.4										
		0.2	0.2	0.2	0.2	0.2	0.1										
48-S	8.93				172	137	111	89	72	58	46	36					
	8.93				0.8	0.9	1.0	1.1	1.1	1.2	1.3	1.3					
					0.7	0.8	0.8	0.8	0.7	0.6	0.4	0.1					
68-S	5.93					169	138	113	92	76	62	**48**					
	5.93					0.9	1.0	1.1	1.1	1.2	1.3	1.3					
						0.8	0.8	0.8	0.7	0.6	0.4	0.1					
68-D1	5.93						192	159	133	111	93	79	66	54	42		
	9.43						1.3	1.5	1.7	1.8	2.0	2.2	2.4	2.5	2.5		
							1.2	1.3	1.3	1.3	1.3	1.2	1.1	0.8	0.4		
88-D1	3.93															41	
	9.18															3.2	
																0.1	

Bold type — *Capacity governed by stresses, others governed by flexural strength*

Values below heavy line require release strengths higher than 3500 psi.

DOUBLE TEE

8'-0" x 18"
Normal Weight Concrete

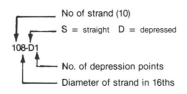

Strand Pattern Designation

108-D1
- No of strand (10)
- S = straight   D = depressed
- No. of depression points
- Diameter of strand in 16ths

Safe loads shown include dead load of 10 psf for untopped members and 15 psf for topped members. Remainder is live load. Long-time cambers include superimposed dead load but do not include live load.

Key
- 185 — Safe superimposed service load, psf
- 0.1 — Estimated camber at erection, in.
- 0.1 — Estimated long-time camber, in.

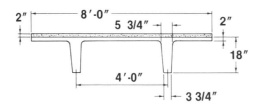

f'_c = 5000 psi
f_{pu} = 270,000 psi

Section Properties

	Untopped	Topped
A =	344 in.²	—
I =	9300 in.⁴	12,749 in.⁴
Y_b =	13.27 in.	15.00 in.
Y_t =	4.73 in.	5.00 in.
Z_b =	701 in.³	850 in.³
Z_t =	1966 in.³	2550 in.³
wt =	358 plf	558 plf
	45 psf	70 psf
V/S =	1.32 in.	

8DT18

Table of safe superimposed service load (psf) and cambers — No Topping

Strand Pattern	e_e / e_c	16	18	20	22	24	26	28	30	32	34	36	38	40	42	44	46	48	50	52	54	56	58
28-S	11.27 / 11.27	185	143	109	84	65	50	38															
		0.1	0.1	0.2	0.2	0.2	0.2	0.2															
		0.1	0.2	0.2	0.2	0.2	0.2	0.1															
48-S	10.27 / 10.27				185	150	123	101	83	69	57	47	39	31									
					0.4	0.4	0.5	0.5	0.6	0.6	0.6	0.6	0.6	0.5									
					0.5	0.6	0.6	0.6	0.6	0.6	0.6	0.5	0.4	0.2									
68-S	7.27 / 7.27					183	150	125	104	87	73	62	52	43	36								
						0.5	0.5	0.6	0.6	0.7	0.7	0.7	0.7	0.6	0.6								
						0.6	0.7	0.7	0.7	0.7	0.7	0.6	0.5	0.4	0.2								
68-D1	7.27 / 10.77							172	145	**123**	**105**	**90**	**77**	**66**	**57**	**49**	**41**	**35**					
								0.8	0.9	1.0	1.1	1.1	1.2	1.2	1.2	1.1	1.1	1.0					
								1.0	1.1	1.2	1.2	1.2	1.2	1.2	1.0	0.9	0.7	0.4					
88-D1	5.27 / 10.52												107	93	81	71	62	**54**	**47**	**41**	**35**		
													1.5	1.6	1.6	1.7	1.7	1.7	1.6	1.5	1.4		
													1.6	1.7	1.7	1.6	1.6	1.4	1.1	0.8	0.3		
108-D1	3.27 / 10.27																		61	53	46	40	**34**
																			2.0	2.0	1.9	1.9	1.7
																			1.6	1.4	1.1	0.8	0.3

8DT18 + 2

Table of safe superimposed service load (psf) and cambers — 2" Normal Weight Topping

Strand Pattern	e_e / e_c	16	18	20	22	24	26	28	30	32	34	36	38	40	42	44	46
28-S	11.27 / 11.27	192	144	106	78	57	41										
		0.1	0.1	0.2	0.2	0.2	0.2										
		0.1	0.1	0.1	0.1	0.1	0.0										
48-S	10.27 / 10.27				192	153	122	98	78	62	49	37					
					0.4	0.4	0.5	0.5	0.6	0.6	0.6	0.6					
					0.4	0.4	0.4	0.4	0.3	0.3	0.1	0.0					
68-S	7.27 / 7.27					196	159	130	106	86	70	57					
						0.5	0.5	0.6	0.6	0.7	0.7	0.7					
						0.4	0.5	0.6	0.6	0.4	0.2	0.1					
68-D1	7.27 / 10.77							177	147	**123**	**103**	**86**	**71**	**59**	**48**		
								0.8	0.9	1.0	1.1	1.1	1.2	1.2	1.2		
								0.7	0.8	0.8	0.7	0.6	0.5	0.3	0.1		
88-D1	5.27 / 10.52												105	90	75	61	48
													1.5	1.6	1.6	1.7	1.7
													0.9	0.8	0.6	0.4	0.2

Bold type — *Capacity governed by stresses, others governed by flexural strength*

Values below heavy line require release strengths higher than 3500 psi.

DOUBLE TEE

8'-0" x 18"
Lightweight Concrete

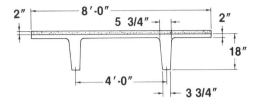

f'_c = 5000 psi
f_{pu} = 270,000 psi

Strand Pattern Designation

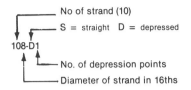

No of strand (10)
S = straight   D = depressed
108-D1
No. of depression points
Diameter of strand in 16ths

Safe loads shown include dead load of 10 psf for untopped members and 15 psf for topped members. Remainder is live load. Long-time cambers include superimposed dead load but do not include live load.

Key

190 — Safe superimposed service load, psf
0.2 — Estimated camber at erection, in.
0.2 — Estimated long-time camber, in.

Section Properties

	Untopped	Topped
A =	344 in.²	—
I =	9300 in.⁴	13,799 in.⁴
Y_b =	13.27 in.	15.51 in.
Y_t =	4.73 in.	4.49 in.
Z_b =	701 in.³	890 in.³
Z_t =	1966 in.³	3073 in.³
wt =	275 plf	475 plf
	34 psf	59 psf
V/S =	1.32 in.	

8LDT18

No Topping

Table of safe superimposed service load (psf) and cambers

Strand Pattern	e_e e_c	16	18	20	22	24	26	28	30	32	34	36	38	40	42	44	46	48	50	52	54	56	58	60
28-S	11.27	190	151	118	93	74	59	47	38															
	11.27	0.2	0.2	0.3	0.3	0.3	0.3	0.4	0.4															
		0.2	0.3	0.3	0.3	0.4	0.3	0.3	0.3															
48-S	10.27				194	159	131	109	92	78	66	56	47	40	34									
	10.27				0.6	0.7	0.9	1.0	1.0	1.1	1.1	1.2	1.2	1.2	1.1									
					0.8	0.9	1.0	1.1	1.1	1.2	1.2	1.1	1.1	0.9	0.8									
68-S	7.27				191	159	133	113	96	82	70	60	52	45	38	33								
	7.27				0.8	0.9	1.0	1.1	1.1	1.2	1.3	1.3	1.3	1.3	1.3	1.2								
					1.0	1.1	1.2	1.2	1.3	1.3	1.3	1.3	1.2	1.0	0.8	0.5								
68-D1	7.27							180	154	132	114	99	86	75	65	57	50	44	38	33				
	10.77							1.3	1.5	1.6	1.8	2.0	2.1	2.2	2.2	2.3	2.3	2.3	2.2	2.1				
								1.6	1.8	1.9	2.1	2.2	2.3	2.3	2.3	2.2	2.0	1.8	1.5	1.1				
88-D1	5.27														90	80	71	63	56	49	43	37	33	
	10.52														2.7	2.9	3.0	3.2	3.3	3.4	3.4	3.3	3.1	
															2.9	3.0	3.0	3.0	2.9	2.8	2.4	2.0	1.4	
108-D1	3.27																						42	37
	10.27																						3.8	3.8
																							2.4	2.0

8LDT18 + 2

2" Normal Weight Topping

Table of safe superimposed service load (psf) and cambers

Strand Pattern	e_e e_c	16	18	20	22	24	26	28	30	32	34	36	38	40	42	44	46	48	50
28-S	11.27	197	153	115	87	66	49	36											
	11.27	0.2	0.2	0.3	0.3	0.3	0.3	0.4											
		0.2	0.2	0.2	0.2	0.2	0.1	0.0											
48-S	10.27				161	130	106	87	70	57	46	37							
	10.27				0.7	0.8	0.9	1.0	1.0	1.1	1.1	1.2							
					0.7	0.7	0.7	0.7	0.6	0.5	0.3	0.1							
68-S	7.27				168	138	115	95	79	65	54	44							
	7.27				0.9	1.0	1.1	1.1	1.2	1.3	1.3	1.3							
					0.8	0.8	0.8	0.7	0.6	0.5	0.3	0.0							
68-D1	7.27						186	156	132	111	94	80	68	57	48	**38**			
	10.77						1.3	1.5	1.6	1.8	2.0	2.1	2.2	2.2	2.3	2.3			
							1.2	1.3	1.3	1.4	1.3	1.2	1.0	0.8	0.4	0.0			
88-D1	5.27														85	**72**	**60**	**49**	**39**
	10.52														2.7	2.9	3.0	3.2	3.3
															1.4	1.2	0.9	0.6	0.1

Bold type — *Capacity governed by stresses, others governed by flexural strength*

Values below heavy line require release strengths higher than 3500 psi.

DOUBLE TEE

8'-0" x 20"
Normal Weight Concrete

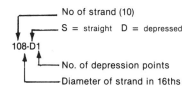

Strand Pattern Designation

No of strand (10)
S = straight   D = depressed

108-D1

No. of depression points
Diameter of strand in 16ths

Safe loads shown include dead load of 10 psf for untopped members and 15 psf for topped members. Remainder is live load. Long-time cambers include superimposed dead load but do not include live load.

Key

173 — Safe superimposed service load, psf
0.4 — Estimated camber at erection, in.
0.5 — Estimated long-time camber, in.

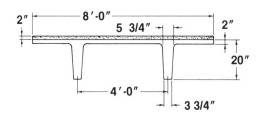

f'_c = 5000 psi
f_{pu} = 270,000 psi

Section Properties

	Untopped	Topped
A =	363 in.²	—
I =	12,551 in.⁴	16,935 in.⁴
Y_b =	14.59 in.	16.45 in.
Y_t =	5.41 in.	5.55 in.
Z_b =	860 in.³	1029 in.³
Z_t =	2320 in.³	3051 in.³
wt =	378 plf	578 plf
	47 psf	72 psf
V/S =	1.35 in.	

8DT20 — No Topping

Table of safe superimposed service load (psf) and cambers

Strand Pattern	e_e / e_c	24	26	28	30	32	34	36	38	40	42	44	46	48	50	52	54	56	58	60	62	64	66	68
48-S	11.59 11.59	173 0.4 0.5	142 0.4 0.5	117 0.5 0.6	97 0.5 0.6	81 0.5 0.6	68 0.5 0.6	56 0.5 0.5	47 0.5 0.4	39 0.5 0.3	32 0.4 0.2													
68-S	8.59 8.59		180 0.5 0.6	150 0.5 0.7	126 0.6 0.7	106 0.6 0.7	90 0.6 0.7	76 0.7 0.7	65 0.7 0.6	55 0.6 0.4	46 0.6 0.3	39 0.6 0.0	32 0.5											
68-D1	8.59 12.09			199 0.7 0.9	168 0.8 1.0	143 0.9 1.1	123 1.0 1.1	106 1.0 1.1	91 1.0 1.1	78 1.1 1.1	68 1.1 1.0	58 1.1 0.9	50 1.0 0.8	43 1.0 0.6	37 0.9 0.4	31 0.8 0.1								
88-D1	6.59 11.84						166 1.1 1.4	144 1.2 1.5	126 1.3 1.5	110 1.4 1.6	96 1.5 1.6	84 1.6 2.7	74 1.6 1.6	65 1.6 1.5	57 1.6 1.3	50 1.5 1.1	44 1.4 0.8	38 1.3 0.4	33 1.1 0.0					
108-D1	4.59 11.59														75 1.9 1.8	67 2.0 1.7	59 2.0 1.6	51 2.0 1.4	45 1.9 1.0	39 1.7 0.6	33 1.5 0.0			
128-D1	2.92 11.34																			49 2.1 1.2	43 2.0 0.8	37 1.9 0.4		

8DT20 + 2 — 2" Normal Weight Topping

Table of safe superimposed service load (psf) and cambers

Strand Pattern	e_e / e_c	24	26	28	30	32	34	36	38	40	42	44	46	48	50	52
48-S	11.59 11.59	176 0.4 0.4	141 0.4 0.4	114 0.5 0.4	92 0.5 0.3	74 0.5 0.3	59 0.5 0.2	47 0.5 0.1	36 0.5 0.0							
68-S	8.59 8.59		189 0.5 0.5	155 0.5 0.5	128 0.6 0.4	105 0.6 0.4	87 0.6 0.4	71 0.7 0.3	58 0.7 0.1							
68-D1	8.59 12.09			169 0.8 0.7	142 0.9 0.7	119 1.0 0.7	100 1.0 0.7	84 1.0 0.6	70 1.1 0.5	59 1.1 0.3	48 1.1 0.1					
88-D1	6.59 11.84						169 1.1 0.9	145 1.2 1.0	124 1.3 0.9	106 1.4 0.9	91 1.5 0.8	78 1.6 0.7	66 1.6 0.5	53 1.6 0.2		
108-D1	4.59 11.59													62 1.9 0.3	50 2.0 0.0	

Bold type — *Capacity governed by stresses, others governed by flexural strength*

Values below heavy line require release strengths higher than 3500 psi.

PCI Design Handbook

DOUBLE TEE

8'-0" x 20"
Lightweight Concrete

Strand Pattern Designation

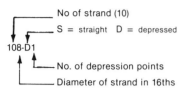

No of strand (10)
S = straight D = depressed

108-D1

No. of depression points
Diameter of strand in 16ths

Safe loads shown include dead load of 10 psf for untopped members and 15 psf for topped members. Remainder is live load. Long-time cambers include superimposed dead load but do not include live load.

Key
182 — Safe superimposed service load, psf
0.6 — Estimated camber at erection, in.
0.8 — Estimated long-time camber, in.

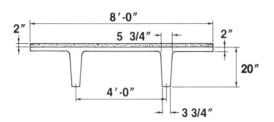

f'_c = 5000 psi
f_{pu} = 270,000 psi

Section Properties

		Untopped	Topped
A	=	363 in.²	—
I	=	12,551 in.⁴	18,278 in.⁴
Y_b	=	14.59 in.	17.02 in.
Y_t	=	5.41 in.	4.98 in.
Z_b	=	860 in.³	1074 in.³
Z_t	=	2320 in.³	3670 in.³
wt	=	290 plf	490 plf
		36 psf	61 psf
V/S	=	1.35 in.	

8LDT20

Table of safe superimposed service load (psf) and cambers — No Topping

Strand Pattern	e_e / e_c	24	26	28	30	32	34	36	38	40	42	44	46	48	50	52	54	56	58	60	62	64	66
48-S	11.59 / 11.59	182	151	126	107	90	77	65	56	48	41	35											
		0.6	0.7	0.8	0.8	0.9	1.0	1.0	1.0	1.0	1.0	1.0											
		0.8	0.9	0.9	1.0	1.0	1.1	1.1	1.0	0.9	0.8	0.7											
68-S	8.59 / 8.59		189	159	135	115	99	85	74	64	55	48	41	36	31								
			0.8	0.9	1.0	1.1	1.1	1.2	1.2	1.3	1.3	1.3	1.3	1.2	1.1								
			1.0	1.1	1.2	1.2	1.3	1.3	1.3	1.3	1.2	1.0	0.9	0.6	0.3								
68-D1	8.59 / 12.09				177	152	132	115	100	87	77	67	59	52	46	40	35	31					
					1.3	1.3	1.6	1.7	1.8	1.9	2.0	2.0	2.1	2.1	2.1	2.0	1.9	1.8					
					1.6	1.6	1.9	2.0	2.1	2.1	2.1	2.1	2.0	1.8	1.6	1.4	1.0	0.6					
88-D1	6.59 / 11.84									119	105	93	83	74	66	59	53	47	42	36	32		
										2.3	2.5	2.6	2.8	2.9	3.1	3.2	3.2	3.1	3.1	2.9	2.8		
										2.7	2.8	2.9	3.0	3.0	3.0	2.9	2.7	2.4	2.0	1.5	1.0		
108-D1	4.59 / 11.59																	59	53	47	42	37	33
																		3.7	3.8	3.9	3.9	3.8	3.7
																		3.3	3.1	2.8	2.5	2.1	1.4

8LDT20 + 2

Table of safe superimposed service load (psf) and cambers — 2" Normal Weight Topping

Strand Pattern	e_e / e_c	24	26	28	30	32	34	36	38	40	42	44	46	48	50	52	54	56
48-S	11.59 / 11.59	185	150	123	101	83	68	56	45	36								
		0.6	0.7	0.8	0.8	0.9	1.0	1.0	1.0	1.0								
		0.6	0.6	0.7	0.6	0.6	0.5	0.4	0.3	0.1								
68-S	8.59 / 8.59		198	164	137	114	96	80	67	56	46							
			0.8	0.9	1.0	1.1	1.1	1.2	1.2	1.3	1.3							
			0.8	0.8	0.8	0.8	0.8	0.7	0.5	0.4	0.1							
68-D1	8.59 / 12.09				178	151	128	109	93	79	68	57	48	41				
					1.3	1.3	1.6	1.7	1.8	1.9	2.0	2.0	2.1	2.1				
					1.2	1.2	1.3	1.3	1.2	1.1	1.0	0.7	0.4	0.1				
88-D1	6.59 / 11.84									115	100	87	76	65	54	44		
										2.3	2.5	2.6	2.8	2.9	3.1	3.2		
										1.6	1.6	1.5	1.3	1.1	0.8	0.4		
108-D1	4.59 / 11.59																	43
																		3.7
																		0.0

Bold type — *Capacity governed by stresses, others governed by flexural strength*

Values below heavy line require release strengths higher than 3500 psi.

DOUBLE TEE

8'-0" x 24"
Normal Weight Concrete

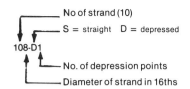

- No of strand (10)
- S = straight   D = depressed
- 108-D1
- No. of depression points
- Diameter of strand in 16ths

Safe loads shown include dead load of 10 psf for untopped members and 15 psf for topped members. Remainder is live load. Long-time cambers include superimposed dead load but do not include live load.

Key
- 171 — Safe superimposed service load, psf
- 0.5 — Estimated camber at erection, in.
- 0.6 — Estimated long-time camber, in.

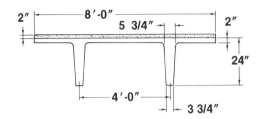

f'_c = 5000 psi
f_{pu} = 270,000 psi

Section Properties

	Untopped	Topped
A =	401 in.²	—
I =	20,985 in.⁴	27,720 in.⁴
Y_b =	17.15 in.	19.27 in.
Y_t =	6.85 in.	6.73 in.
Z_b =	1224 in.³	1438 in.³
Z_t =	3063 in.³	4119 in.³
wt =	418 plf	618 plf
	52 psf	77 psf
V/S =	1.41 in.	

8DT24 — No Topping

Table of safe superimposed service load (psf) and cambers

Strand Pattern	e_e / e_c	30	32	34	36	38	40	42	44	46	48	50	52	54	56	58	60	62	64	66	68	70	72	74
68-S	11.15	171	145	124	106	91	78	67	57	49	41	35												
	11.15	0.5	0.5	0.6	0.6	0.6	0.6	0.6	0.6	0.6	0.5	0.4												
		0.6	0.6	0.6	0.7	0.6	0.6	0.6	0.5	0.4	0.2	0.1												
88-S	9.15		176	152	131	113	98	85	74	64	56	48	41											
	9.15		0.6	0.6	0.7	0.7	0.7	0.7	0.7	0.7	0.7	0.6	0.5											
			0.7	0.8	0.8	0.8	0.8	0.7	0.7	0.6	0.5	0.3	0.1											
88-D1	9.15			187	163	143	**126**	111	98	87	77	68	60	53	47	41	36							
	14.40			1.0	1.1	1.2	1.3	1.3	1.4	1.4	1.4	1.4	1.3	1.3	1.2	1.1	0.9							
				1.2	1.4	1.4	1.5	1.5	1.5	1.4	1.4	1.3	1.1	0.9	0.7	0.4	0.1							
108-D1	7.15							142	127	113	101	90	**81**	72	64	57	51	45	40					
	14.15							1.5	1.6	1.7	1.8	1.8	1.9	1.9	1.8	1.8	1.7	1.5	1.4					
								1.7	1.8	1.9	1.9	1.8	1.7	1.7	1.6	1.3	1.1	0.7	0.3					
128-D1	5.48														81	72	65	57	**51**	45	39			
	13.90														2.2	2.2	2.2	2.2	2.1	2.0	1.8			
															2.0	1.9	1.8	1.6	1.3	0.9	0.4			
148-D1	4.29																				55	49	44	38
	13.65																				2.4	2.4	2.2	2.1
																					1.6	1.3	0.9	0.4

8DT24 + 2 — 2" Normal Weight Topping

Table of safe superimposed service load (psf) and cambers

Strand Pattern	e_e / e_c	26	28	30	32	34	36	38	40	42	44	46	48	50	52	54	56	58	60
48-S	14.15	180	147	120	98	80	65	52	41										
	14.15	0.3	0.4	0.4	0.3	0.4	0.2	0.1	0.0										
		0.3	0.3	0.3	0.3	0.3	0.2	0.1	0.0										
68-S	11.15			171	143	120	100	84	70	58	47								
	11.15			0.5	0.5	0.6	0.6	0.6	0.6	0.6	0.6								
				0.4	0.4	0.4	0.4	0.3	0.2	0.1	0.0								
68-D1	11.15				180	153	130	110	93	79	67	56	46						
	14.65				0.7	0.7	0.8	0.8	0.9	0.9	0.9	0.9	0.9						
					0.6	0.6	0.6	0.6	0.5	0.5	0.3	0.2	0.0						
88-D1	9.15						186	161	139	**121**	105	91	78	67	58				
	14.40						1.0	1.1	1.2	1.3	1.3	1.4	1.4	1.4	1.4				
							0.9	1.0	1.0	0.9	0.9	0.8	0.6	0.4	0.2				
108-D1	7.15									139	122	107	94	82	**70**	59	49		
	14.15									1.5	1.6	1.7	1.8	1.8	1.9	1.9	1.8		
										1.1	1.1	1.0	0.9	0.8	0.6	0.3	0.0		
128-D1	5.48																66	56	
	13.90																2.2	2.2	
																	0.4	0.1	

Bold type — *Capacity governed by stresses, others governed by flexural strength*

Values below heavy line require release strengths higher than 3500 psi.

DOUBLE TEE

8'-0" x 24"
Lightweight Concrete

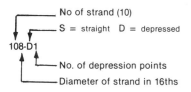

Strand Pattern Designation

No of strand (10)

S = straight D = depressed

108-D1

No. of depression points

Diameter of strand in 16ths

Safe loads shown include dead load of 10 psf for untopped members and 15 psf for topped members. Remainder is live load. Long-time cambers include superimposed dead load but do not include live load.

Key

116 — Safe superimposed service load, psf
1.0 — Estimated camber at erection, in.
1.2 — Estimated long-time camber, in.

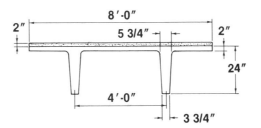

f'_c = 5000 psi
f_{pu} = 270,000 psi

Section Properties

	Untopped	Topped
A =	401 in.²	—
I =	20,985 in.⁴	29,853 in.⁴
Y_b =	17.15 in.	19.94 in.
Y_t =	6.85 in.	6.06 in.
Z_b =	1224 in.³	1497 in.³
Z_t =	3063 in.³	4926 in.³
wt =	320 plf	520 plf
	40 psf	65 psf
V/S =	1.41 in.	

8LDT24

No Topping

Table of safe superimposed service load (psf) and cambers

Strand Pattern	e_e / e_c	36	38	40	42	44	46	48	50	52	54	56	58	60	62	64	66	68	70	72	74	76	78	80
68-S	11.15 / 11.15	116 1.0 1.2	101 1.1 1.2	88 1.1 1.2	77 1.1 1.2	67 1.2 1.2	59 1.2 1.1	52 1.2 1.0	45 1.1 0.8	39 1.1 0.6	34 1.0 0.4													
88-S	9.15 / 9.15	141 1.1 1.4	123 1.2 1.4	108 1.3 1.4	95 1.3 1.5	84 1.4 1.4	74 1.4 1.4	66 1.4 1.3	58 1.4 1.2	51 1.4 1.0	45 1.3 0.8	40 1.2 0.5	35 1.1 0.2											
88-D1	9.15 / 14.40	197 1.6 2.0	173 1.8 2.2	153 1.9 2.3	136 2.1 2.5	121 2.2 2.6	101 2.4 2.7	97 2.5 2.8	87 2.6 2.8	78 2.6 2.7	70 2.7 2.7	63 2.7 2.5	57 2.7 2.4	51 2.7 2.1	46 2.6 1.8	41 2.5 1.5	37 2.4 1.0	33 2.2 0.5						
108-D1	7.15 / 14.15								111 2.9 3.3	100 3.1 3.3	91 3.2 3.4	82 3.3 3.4	74 3.5 3.4	68 3.6 3.4	61 3.6 3.3	55 3.7 3.0	49 3.6 2.7	44 3.5 2.3	39 3.3 1.8	35 3.1 1.2	31 2.9 0.5			
128-D1	5.48 / 13.90															60 4.2 3.6	54 4.2 3.3	49 4.3 3.1	44 4.2 2.7	39 4.2 2.2	35 4.0 1.6	31 3.8 0.8		
148-D1	4.29 / 13.65																						39 4.7 4.6	35 2.2 1.6

8LDT24 + 2

2" Normal Weight Topping

Table of safe superimposed service load (psf) and cambers

Strand Pattern	e_e / e_c	26	28	30	32	34	36	38	40	42	44	46	48	50	52	54	56	58	60	62
48-S	14.15 / 14.15	190 0.5 0.5	157 0.6 0.5	130 0.6 0.5	108 0.7 0.5	90 0.7 0.5	75 0.8 0.5	62 0.8 0.4	51 0.8 0.3	42 0.8 0.1	33 0.8 0.0									
68-S	11.15 / 11.15			181 0.8 0.7	153 0.9 0.8	130 0.9 0.8	110 1.0 0.7	94 1.1 0.7	80 1.1 0.6	68 1.1 0.5	57 1.2 0.4	48 1.2 0.2								
88-S	9.15 / 9.15				191 1.0 0.9	163 1.1 0.9	140 1.1 0.9	121 1.2 0.9	104 1.3 0.8	90 1.3 0.7	77 1.4 0.6	66 1.4 0.4	57 1.4 0.2							
68-D1	11.15 / 14.65				190 1.0 0.9	163 1.2 1.1	140 1.3 1.1	120 1.4 1.1	103 1.5 1.1	89 1.6 1.1	77 1.6 1.0	66 1.7 0.9	56 1.7 0.8	48 1.7 0.3	40 1.7 0.0					
88-D1	9.15 / 14.40						196 1.6 1.5	171 1.8 1.6	149 1.9 1.6	131 2.1 1.6	115 2.2 1.6	101 2.4 1.6	88 2.5 1.5	77 2.6 1.4	68 2.6 1.1	59 2.7 0.9	51 2.7 0.5	43 2.7 0.1		
108-D1	7.15 / 14.15													104 2.9 1.8	92 3.1 1.7	82 3.2 1.5	72 3.3 1.3	61 3.5 1.0	52 3.6 0.7	44 3.6 0.2

Bold type — *Capacity governed by stresses, others governed by flexural strength*

Values below heavy line require release strengths higher than 3500 psi.

DOUBLE TEE

8'-0" x 32"
Normal Weight Concrete

Strand Pattern Designation

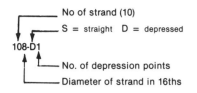

- No of strand (10)
- S = straight   D = depressed
- 108-D1
- No. of depression points
- Diameter of strand in 16ths

Safe loads shown include dead load of 10 psf for untopped members and 15 psf for topped members. Remainder is live load. Long-time cambers include superimposed dead load but do not include live load.

Key
- 182 — Safe superimposed service load, psf
- 1.3 — Estimated camber at erection, in.
- 1.5 — Estimated long-time camber, in.

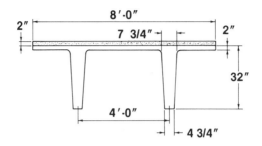

f'_c = 5000 psi
f_{pu} = 270,000 psi

Section Properties

	Untopped	Topped
A =	567 in.²	—
I =	55,464 in.⁴	71,886 in.⁴
Y_b =	21.21 in.	23.66 in.
Y_t =	10.79 in.	10.34 in.
Z_b =	2615 in.³	3038 in.³
Z_t =	5140 in.³	6952 in.³
wt =	591 plf	791 plf
	74 psf	99 psf
V/S =	1.79 in.	

8DT32 — No Topping

Table of safe superimposed service load (psf) and cambers

Strand Pattern	e_e / e_c	50	52	54	56	58	60	62	64	66	68	70	72	74	76	78	80	82	84	86	88
128-D1	12.04 / 17.96	182	164	147	133	120	108	98	88	79	71	64	57	51							
		1.3	1.3	1.3	1.3	1.3	1.3	1.3	1.3	1.2	1.1	1.0	0.9	0.7							
		1.5	1.5	1.5	1.5	1.4	1.4	1.3	1.3	1.1	1.0	0.5	0.3	0.0							
148-D1	9.71 / 17.71		193	175	159	144	131	119	108	98	89	80	73	65	58						
			1.5	1.5	1.6	1.6	1.6	1.6	1.6	1.5	1.5	1.4	1.3	1.2	1.0						
			1.8	1.8	1.8	1.8	1.7	1.7	1.6	1.4	1.3	1.1	0.8	0.6	0.2						
168-D1	8.21 / 17.46				184	168	153	139	127	116	105	95	86	77	69	62	55				
					1.7	1.8	1.8	1.9	1.9	1.9	1.9	1.8	1.7	1.6	1.5	1.4	1.2				
					2.0	2.0	2.0	2.0	2.0	1.9	1.8	1.6	1.4	1.2	0.9	0.6	0.2				
188-D1	6.82 / 17.21						173	157	143	130	118	107	98	88	80	72	65	58			
							1.9	2.0	2.0	2.1	2.1	2.1	2.1	2.0	1.9	1.8	1.7	1.5			
							2.2	2.2	2.2	2.2	2.1	2.0	1.9	1.7	1.5			0.4			
208-D1	5.71 / 16.96											120	109	100	91	82	74	67	61	55	
												2.3	2.3	2.3	2.2	2.2	2.1	1.9	1.7	1.5	
												2.3	2.2	2.0	1.9	1.7	1.4	1.1	0.6	0.2	
228-D1	4.80 / 16.71														92	84	76	69	63	56	
															2.4	2.3	2.3	2.1	2.0	1.8	
															2.0	1.8	1.5	1.2	0.8	0.3	

8DT32+2 — 2" Normal Weight Topping

Table of safe superimposed service load (psf) and cambers

Strand Pattern	e_e / e_c	44	46	48	50	52	54	56	58	60	62	64	66	68	70	72	74	76
108-D1	15.31 / 18.21	200	176	155	137	121	106	93	81	71	61							
		1.0	1.0	1.0	1.1	1.1	1.1	1.1	1.1	1.0	1.0							
		0.9	0.9	0.8	0.8	0.7	0.6	0.5	0.4	0.2	0.0							
128-D1	12.04 / 17.96			198	176	157	140	125	111	98	87	77	68					
				1.2	1.3	1.3	1.3	1.3	1.3	1.3	1.3	1.3	1.2					
				1.1	1.0	1.0	0.9	0.9	0.7	0.6	0.4	0.2	0.0					
148-D1	9.71 / 17.71					188	169	152	136	122	109	98	87	76				
						1.5	1.5	1.6	1.6	1.6	1.6	1.6	1.5	1.5				
						1.2	1.2	1.2	1.1	0.9	0.8	0.6	0.4	0.1				
168-D1	8.21 / 17.46						198	178	161	145	131	118	106	93	81	70		
							1.6	1.7	1.8	1.8	1.9	1.9	1.9	1.9	1.8	1.7		
							1.3	1.3	1.3	1.2	1.1	1.0	0.8	0.6	0.3	0.0		
188-D1	6.82 / 17.21								167	151	137	122	108	96	84	74		
									1.9	2.0	2.0	2.1	2.1	2.1	2.1	2.0		
									1.3	1.3	1.2	1.0	0.9	0.7	0.4	0.1		
208-D1	5.71 / 16.96														110	98	87	76
															2.3	2.3	2.3	2.2
															0.9	0.7	0.4	0.1

Bold type — Capacity governed by stresses, others governed by flexural strength

Values below heavy line require release strengths higher than 3500 psi.

DOUBLE TEE

8'-0" x 32"
Lightweight Concrete

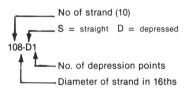

Strand Pattern Designation

108-D1
- No of strand (10)
- S = straight   D = depressed
- No. of depression points
- Diameter of strand in 16ths

Safe loads shown include dead load of 10 psf for untopped members and 15 psf for topped members. Remainder is live load. Long-time cambers include superimposed dead load but do not include live load.

Key
- 147 — Safe superimposed service load, psf
- 2.4 — Estimated camber at erection, in.
- 2.8 — Estimated long-time camber, in.

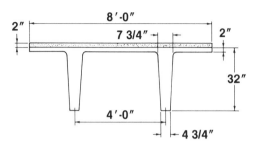

f'_c = 5000 psi
f_{pu} = 270,000 psi

Section Properties

	Untopped	Topped
A =	567 in.²	—
I =	55,464 in.⁴	77,617 in.⁴
Y_b =	21.21 in.	24.55 in.
Y_t =	10.79 in.	9.45 in.
Z_b =	2615 in.³	3167 in.³
Z_t =	5140 in.³	8213 in.³
wt =	453 plf	653 plf
	57 psf	82 psf
V/S =	1.79 in.	

8LDT32

Table of safe superimposed service load (psf) and cambers — No Topping

Strand Pattern	e_e / e_c	56	58	60	62	64	66	68	70	72	74	76	78	80	82	84	86	88	90	92	94	96	98	100
128-D1	12.04	147	134	122	112	102	93	85	78	71	65	59	54	49	44									
	17.96	2.4	2.5	2.5	2.6	2.6	2.6	2.6	2.5	2.5	2.4	2.3	2.1	2.0	1.8									
		2.8	2.8	2.8	2.8	2.7	2.6	2.5	2.3	2.1	1.8	1.5	1.2	0.7	0.2									
148-D1	9.71	173	158	145	133	122	112	103	94	86	78	71	65	59	53	48	43							
	17.71	2.5	2.8	2.9	3.0	3.1	3.1	3.1	3.1	3.1	3.0	3.0	2.9	2.7	2.6	2.4	2.1							
		2.9	3.3	3.3	3.4	3.4	3.3	3.2	3.1	2.9	2.7	2.5	2.1	1.8	1.3	0.8	0.3							
168-D1	8.21			182	166	152	139	127	117	107	98	90	82	75	69	63	57	52	47	42				
	17.46			2.9	3.1	3.2	3.3	3.4	3.5	3.6	3.7	3.7	3.7	3.6	3.5	3.4	3.2	3.0	2.8	2.5				
				3.5	3.6	3.7	3.7	3.8	3.8	3.8	3.7	3.6	3.4	3.2	2.8	2.5	2.0	1.5	0.9	0.3				
188-D1	6.82							119	110	101	93	85	78	72	66	60	55	50	45	41				
	17.21							3.8	3.9	4.0	4.0	4.1	4.1	4.1	4.0	3.8	3.6	3.4	3.1	2.8				
								4.1	4.1	4.0	3.9	3.8	3.6	3.4	3.1	2.6	2.1	1.5	0.8	0.1				
208-D1	5.71													87	80	74	68	62	57	52	48	43		
	16.96													4.4	4.4	4.4	4.3	4.3	4.2	4.0	3.7	3.4		
														4.1	3.9	3.6	3.3	3.0	2.6	2.0	1.3	0.6		
228-D1	4.80																	70	64	59	54	49	45	41
	16.71																	4.7	4.6	4.5	4.3	4.2	3.9	3.6
																		3.6	3.2	2.8	2.3	1.6	0.9	0.1

8LDT32 + 2

Table of safe superimposed service load (psf) and cambers — 2" Normal Weight Topping

Strand Pattern	e_e / e_c	46	48	50	52	54	56	58	60	62	64	66	68	70	72	74	76	78	80	82
108-D1	15.31	190	169	151	135	120	107	96	85	76	67	59	52							
	18.21	1.6	1.8	1.8	1.9	2.0	2.0	2.1	2.1	2.1	2.1	2.1	2.0							
		1.5	1.6	1.6	1.5	1.5	1.4	1.3	1.1	0.9	0.7	0.4	0.1							
128-D1	12.04			191	171	154	139	125	113	101	91	82	73	66	58					
	17.96			2.1	2.2	2.3	2.4	2.5	2.5	2.6	2.6	2.6	2.6	2.5	2.5					
				1.9	1.8	1.9	1.8	1.8	1.6	1.5	1.3	1.1	0.8	0.4	0.0					
148-D1	9.71					183	166	150	136	123	112	101	92	83	74	64				
	17.71					2.5	2.5	2.8	2.9	3.0	3.1	3.1	3.1	3.1	3.1	3.0				
						2.1	2.0	2.1	2.1	2.0	1.9	1.7	1.4	1.1	0.8	0.4				
168-D1	8.21						175	159	145	132	120	110	99	88	78	69	61			
	17.46						2.9	3.1	3.2	3.3	3.4	3.5	3.6	3.7	3.7	3.7	3.6			
							2.3	2.3	2.3	2.2	2.1	1.9	1.7	1.5	1.2	0.8	0.3			
188-D1	6.82											114	102	92	82	73	64			
	17.21											3.8	3.9	4.0	4.0	4.1	4.1			
												2.0	1.8	1.5	1.2	0.9	0.4			
208-D1	5.71															75	67			
	16.96															4.4	4.4			
																0.8	0.4			

Bold type — *Capacity governed by stresses, others governed by flexural strength*

Values below heavy line require release strengths higher than 3500 psi.

DOUBLE TEE

10'-0" x 32"
Normal Weight Concrete

Strand Pattern Designation

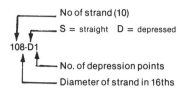

No of strand (10)
S = straight D = depressed
No. of depression points
Diameter of strand in 16ths

108-D1

Safe loads shown include dead load of 10 psf for untopped members and 15 psf for topped members. Remainder is live load. Long-time cambers include superimposed dead load but do not include live load.

Key

200 — Safe superimposed service load, psf
1.1 — Estimated camber at erection, in.
1.3 — Estimated long-time camber, in.

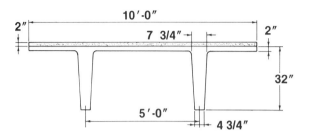

$f'_c = 5000$ psi
$f_{pu} = 270{,}000$ psi

Section Properties

	Untopped	Topped
A =	615 in.²	—
I =	59,720 in.⁴	77,118 in.⁴
Y_b =	21.98 in.	24.54 in.
Y_t =	10.02 in.	9.46 in.
Z_b =	2717 in.³	3142 in.³
Z_t =	5960 in.³	8152 in.³
wt =	641 plf	891 plf
	64 psf	89 psf
V/S =	1.69 in.	

10DT32 — No Topping

Table of safe superimposed service load (psf) and cambers

Strand Pattern	e_e / e_c	44	46	48	50	52	54	56	58	60	62	64	66	68	70	72	74	76	78	80	82	84	86
128-D1	12.81	200	179	160	143	129	116	104	93	84	75	68	61	54	48	43							
	18.73	1.1	1.1	1.2	1.2	1.2	1.2	1.3	1.3	1.2	1.2	1.1	1.1	1.0	0.9	0.8							
		1.3	1.3	1.4	1.4	1.4	1.4	1.3	1.3	1.2	1.1	0.9	0.7	0.5	0.3	0.0							
148-D1	10.48			187	169	152	137	124	112	102	92	83	75	68	61	55	49						
	18.48			1.3	1.4	1.4	1.5	1.5	1.5	1.5	1.5	1.5	1.4	1.4	1.3	1.2	1.0						
				1.6	1.6	1.7	1.7	1.7	1.6	1.6	1.5	1.4	1.2	1.0	0.8	0.6	0.3						
168-D1	8.98				194	176	159	145	131	119	109	99	90	82	74	67	60	54	48				
	18.23				1.4	1.5	1.6	1.7	1.7	1.8	1.8	1.8	1.8	1.7	1.7	1.6	1.5	1.4	1.2				
					1.7	1.8	1.9	1.9	1.9	1.9	1.9	1.8	1.7	1.6	1.4	1.1	0.9	0.6	0.2				
188-D1	7.59								150	137	125	114	104	94	86	77	70	63	56	50	45		
	17.98								1.8	1.9	1.9	2.0	2.0	2.0	2.0	2.0	1.9	1.8	1.7	1.5	1.3		
									2.1	2.1	2.1	2.1	2.0	2.0	1.8	1.7	1.4	1.1	0.8	0.4	0.0		
208-D1	6.48											105	96	87	79	72	65	58	52	47			
	17.73											2.2	2.2	2.2	2.2	2.2	2.1	1.9	1.8	1.6			
												2.2	2.1	2.0	1.7	1.4	1.1	0.7	0.2				
228-D1	5.57																88	80	73	66	60	54	49
	17.48																2.4	2.4	2.3	2.3	2.2	2.0	1.8
																	2.1	2.0	1.8	1.5	1.2	0.9	0.4

10DT32 + 2 — 2" Normal Weight Topping

Table of safe superimposed service load (psf) and cambers

Strand Pattern	e_e / e_c	42	44	46	48	50	52	54	56	58	60	62	64	66	68	70	72	74
108-D1	16.08	175	154	134	118	103	90	78	68	58	50							
	18.98	0.9	0.9	0.9	1.0	1.0	1.0	1.0	1.0	1.0	0.9							
		0.8	0.8	0.7	0.7	0.7	0.6	0.5	0.3	0.2	0.0							
128-D1	12.81		195	172	152	135	119	105	93	82	72	63						
	18.73		1.1	1.1	1.2	1.2	1.2	1.2	1.3	1.3	1.2	1.2						
			1.0	1.0	0.9	0.9	0.8	0.8	0.6	0.5	0.3	0.1						
148-D1	10.48				182	162	145	129	115	103	91	81	72	63				
	18.48				1.3	1.4	1.4	1.5	1.5	1.5	1.5	1.5	1.5	1.4				
					1.1	1.1	1.1	1.0	1.0	0.8	0.7	0.5	0.3	0.0				
168-D1	8.98					190	170	153	137	123	110	99	88	78	67			
	18.23					1.4	1.5	1.6	1.7	1.7	1.8	1.8	1.8	1.8	1.7			
						1.2	1.2	1.2	1.2	1.1	1.0	0.9	0.7	0.5	0.2			
188-D1	7.59									142	128	116	104	92	80	70	61	
	17.98									1.8	1.9	1.9	2.0	2.0	2.0	2.0	2.0	
										1.2	1.1	1.0	0.9	0.8	0.6	0.3	0.0	
208-D1	6.48													93	82	72	62	
	17.73													2.2	2.2	2.2	2.2	
														0.7	0.5	0.3	0.0	

Bold type — *Capacity governed by stresses, others governed by flexural strength*

Values below heavy line require release strengths higher than 3500 psi.

DOUBLE TEE

10'-0" x 32"
Lightweight Concrete

Strand Pattern Designation

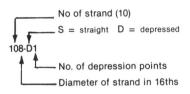

No of strand (10)
S = straight   D = depressed
108-D1
No. of depression points
Diameter of strand in 16ths

Safe loads shown include dead load of 10 psf for untopped members and 15 psf for topped members. Remainder is live load. Long-time cambers include superimposed dead load but do not include live load.

Key
128 — Safe superimposed service load, psf
2.2 — Estimated camber at erection, in.
2.6 — Estimated long-time camber, in.

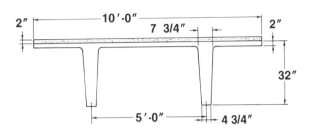

f'_c = 5000 psi
f_{pu} = 270,000 psi

Section Properties

	Untopped	Topped
A =	615 in.²	—
I =	59,720 in.⁴	77,118 in.⁴
Y_b =	21.98 in.	24.54 in.
Y_t =	10.02 in.	9.46 in.
Z_b =	2717 in.³	3142 in.³
Z_t =	5960 in.³	8152 in.³
wt =	491 plf	741 plf
	49 psf	74 psf
V/S =	1.69 in.	

10LDT32 — No Topping

Table of safe superimposed service load (psf) and cambers

Strand Pattern	e_e / e_c	54	56	58	60	62	64	66	68	70	72	74	76	78	80	82	84	86	88	90	92	94	96	98
128-D1	12.81 / 18.73	128	116	106	96	88	80	73	66	60	55	50	45	41	37									
		2.2	2.3	2.4	2.4	2.4	2.4	2.4	2.4	2.4	2.3	2.2	2.1	1.9	1.7									
		2.6	2.6	2.6	2.6	2.5	2.4	2.3	2.1	1.9	1.7	1.4	1.0	0.6	0.2									
148-D1	10.48 / 18.48	150	137	125	114	104	95	87	80	73	67	62	56	51	46	41	37							
		2.5	2.6	2.7	2.8	2.9	2.9	3.0	3.0	2.9	2.9	2.8	2.8	2.6	2.5	2.3	2.1							
		2.9	3.0	3.1	3.2	3.1	3.1	3.0	2.9	2.7	2.5	2.3	2.0	1.6	1.2	0.7	0.2							
168-D1	8.98 / 18.23	172	157	144	132	121	111	102	94	86	78	72	65	60	54	49	45	40	36					
		2.6	2.7	2.9	3.0	3.1	3.2	3.3	3.4	3.5	3.5	3.5	3.5	3.4	3.3	3.1	2.9	2.7	2.5					
		3.1	3.2	3.3	3.4	3.5	3.5	3.6	3.5	3.5	3.4	3.2	2.9	2.6	2.3	1.8	1.4	0.8	0.2					
188-D1	7.59 / 17.98							114	105	96	88	81	74	68	62	57	52	47	43	39	35			
								3.5	3.6	3.7	3.8	3.9	3.9	4.0	4.0	3.9	3.7	3.6	3.3	3.1	2.8			
								3.8	3.8	3.8	3.8	3.7	3.6	3.4	3.2	2.9	2.4	1.9	1.4	0.7	0.0			
208-D1	6.48 / 17.73													82	76	70	64	59	54	49	45	41	37	
														4.2	4.2	4.3	4.3	4.3	4.2	4.1	4.0	3.7	3.4	
														4.0	3.9	3.7	3.5	3.2	2.8	2.4	1.9	1.2	0.5	
228-D1	5.57 / 17.48																	60	55	50	46	42	38	35
																		4.6	4.5	4.5	4.3	4.2	4.0	3.6
																		3.4	3.1	2.6	2.1	1.6	0.9	0.0

10LDT32 + 2 — 2" Normal Weight Topping

Table of safe superimposed service load (psf) and cambers

Strand Pattern	e_e / e_c	42	44	46	48	50	52	54	56	58	60	62	64	66	68	70	72	74	76	78
108-D1	16.08 / 18.98	188	166	147	130	115	102	90	80	71	62	55	48							
		1.3	1.5	1.6	1.7	1.8	1.8	1.9	1.9	1.9	2.0	2.0	1.9							
		1.2	1.4	1.4	1.4	1.3	1.3	1.2	1.0	0.9	0.7	0.5	0.2							
128-D1	12.81 / 18.73			184	165	147	132	118	105	94	84	75	67	60	53					
				1.9	1.8	2.1	2.1	2.2	2.3	2.4	2.4	2.4	2.4	2.4	2.4					
				1.7	1.6	1.7	1.7	1.6	1.5	1.4	1.2	1.0	0.8	0.5	0.1					
148-D1	10.48 / 18.48				195	175	157	142	127	115	104	93	84	75	67	58	50			
					2.1	2.2	2.3	2.5	2.6	2.7	2.8	2.9	2.9	3.0	3.0	2.9	2.9			
					1.8	1.9	1.9	1.9	1.9	1.8	1.8	1.6	1.4	1.1	0.8	0.4	0.0			
168-D1	8.98 / 18.23							165	149	135	122	111	101	90	80	70	62	54		
								2.6	2.7	2.9	3.0	3.1	3.2	3.3	3.4	3.5	3.5	3.5		
								2.1	2.1	2.0	2.0	1.9	1.8	1.6	1.4	1.1	0.8	0.4		
188-D1	7.59 / 17.98													103	92	82	73	64	57	49
														3.5	3.6	3.7	3.8	3.9	3.9	4.0
														1.8	1.6	1.4	1.1	0.8	0.4	0.0
208-D1	6.48 / 17.73																66	59		
																	4.2	4.2		
																	0.8	0.4		

Bold type — *Capacity governed by stresses, others governed by flexural strength*

Values below heavy line require release strengths higher than 3500 psi.

SINGLE TEE

8'-0" x 36"
Normal Weight Concrete

Strand Pattern Designation

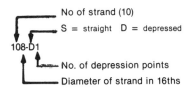

- No of strand (10)
- S = straight D = depressed
- 108-D1
- No. of depression points
- Diameter of strand in 16ths

Safe loads shown include dead load of 10 psf for untopped members and 15 psf for topped members. Remainder is live load. Long-time cambers include superimposed dead load but do not include live load.

Key
157 — Safe superimposed service load, psf
1.5 — Estimated camber at erection, in.
1.8 — Estimated long-time camber, in.

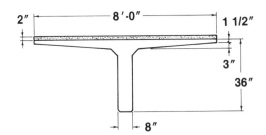

$f'_c = 5000$ psi
$f_{pu} = 270,000$ psi

Section Properties

	Untopped	Topped
A =	570 in.²	—
I =	68,917 in.⁴	83,212 in.⁴
Y_b =	26.01 in.	28.28 in.
Y_t =	9.99 in.	9.72 in.
Z_b =	2650 in.³	2942 in.³
Z_t =	6899 in.³	8561 in.³
wt =	594 plf	794 plf
	74 psf	99 psf
V/S =	2.16 in.	

8ST36

Table of safe superimposed service load (psf) and cambers — No Topping

Strand Pattern	e_e / e_c	56	58	60	62	64	66	68	70	72	74	76	78	80	82	84	86	88	90	92	94	96	98	100
128-D1	13.68	157	142	129	117	106	96	87	79	72	65	58	52	47	42									
	22.76	1.5	1.6	1.6	1.6	1.6	1.6	1.6	1.6	1.5	1.4	1.3	1.2	1.1	0.9									
		1.8	1.8	1.8	1.8	1.7	1.7	1.6	1.4	1.3	1.1	0.9	0.6	0.3	0.0									
148-D1	11.15	186	169	154	141	128	117	107	98	89	81	74	67	61	55	50	44							
	22.51	1.6	1.7	1.8	1.8	1.9	1.9	2.0	2.0	1.9	1.9	1.8	1.7	1.6	1.5	1.3	1.1							
		1.9	2.0	2.0	2.1	2.1	2.1	2.1	2.0	1.9	1.7	1.6	1.3	0.8	0.4	0.0								
168-D1	9.26							126	116	106	98	89	82	75	68	61	55	49	44					
	22.26							2.1	2.2	2.2	2.2	2.2	2.2	2.2	2.1	1.9	1.8	1.6	1.4					
								2.3	2.3	2.2	2.2	2.1	2.0	1.8	1.5	1.2	0.9	0.5	0.0					
188-D1	7.79												96	87	80	72	66	59	53	48				
	22.01												2.4	2.4	2.4	2.4	2.4	2.2	2.0	1.8				
													2.3	2.2	2.0	1.9	1.6	1.3	0.9	0.5				
208-D1	6.61																	69	63	57	51	46		
	21.76																	2.6	2.5	2.4	2.2	2.0		
																		1.8	1.6	1.2	0.8	0.4		
228-D1	5.65																				60	54	49	44
	21.51																				2.6	2.5	2.3	2.1
																					1.4	1.1	0.6	0.1

8ST36 + 2

Table of safe superimposed service load (psf) and cambers — 2" Normal Weight Topping

Strand Pattern	e_e / e_c	50	52	54	56	58	60	62	64	66	68	70	72	74	76	78	80	82
108-D1	17.61	165	147	130	116	102	90	80	70	61	53							
	23.01	1.2	1.2	1.3	1.3	1.3	1.3	1.3	1.3	1.3	1.2							
		1.0	1.0	1.0	0.9	0.9	0.8	0.6	0.5	0.3	0.1							
128-D1	13.68		185	165	148	133	119	106	95	85	75	66	58					
	22.76		1.4	1.5	1.5	1.6	1.6	1.6	1.6	1.6	1.6	1.6	1.5					
			1.2	1.2	1.2	1.2	1.1	1.0	0.9	0.7	0.6	0.3	0.1					
148-D1	11.15			179	162	146	131	119	107	96	86	77	68	58				
	22.51			1.6	1.7	1.8	1.8	1.9	1.9	2.0	2.0	1.9	1.9	1.8				
				1.3	1.3	1.3	1.3	1.2	1.1	1.0	0.9	0.6	0.4	0.1				
168-D1	9.26							116	105	95	84	74	64	56				
	22.26							2.1	2.2	2.2	2.2	2.2	2.2	2.2				
								1.2	1.1	0.9	0.8	0.5	0.3	0.0				
188-D1	7.79															79	69	61
	22.01															2.4	2.4	2.4
																0.6	0.3	0.0

Bold type — *Capacity governed by stresses, others governed by flexural strength*

Values below heavy line require release strengths higher than 3500 psi.

SINGLE TEE

8'-0" x 36"
Lightweight Concrete

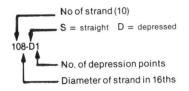

No of strand (10)

S = straight   D = depressed

108-D1

No. of depression points

Diameter of strand in 16ths

Safe loads shown include dead load of 10 psf for untopped members and 15 psf for topped members. Remainder is live load. Long-time cambers include superimposed dead load but do not include live load.

Key

132 — Safe superimposed service load, psf
2.8 — Estimated camber at erection, in.
3.3 — Estimated long-time camber, in.

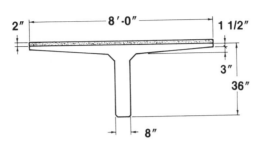

$f'_c = 5000$ psi
$f_{pu} = 270,000$ psi

Section Properties

	Untopped	Topped
A =	570 in.²	—
I =	68,917 in.⁴	88,260 in.⁴
Y_b =	26.01 in.	29.09 in.
Y_t =	9.99 in.	8.91 in.
Z_b =	2650 in.³	3034 in.³
Z_t =	6899 in.³	9906 in.³
wt =	455 plf	655 plf
	57 psf	82 psf
V/S =	2.16 in.	

8LST36
No Topping

Table of safe superimposed service load (psf) and cambers

Strand Pattern	e_e / e_c	62	64	66	68	70	72	74	76	78	80	82	84	86	88	90	92	94	96	98	100	102	104	106	108	110
128-D1	13.68	132	121	111	102	93	86	79	72	66	61	56	51	47	43	39	34									
	22.76	2.8	3.0	3.1	3.1	3.2	3.2	3.2	3.1	3.1	3.0	2.9	2.8	2.7	2.5	2.3	2.0									
		3.3	3.4	3.4	3.4	3.4	3.3	3.2	3.0	2.8	2.5	2.2	1.9	1.5	1.1	0.5	0.0									
148-D1	11.15		143	131	121	112	103	95	88	81	75	69	64	58	53	48	44	39	35							
	22.51		3.1	3.3	3.4	3.5	3.6	3.7	3.8	3.8	3.8	3.8	3.7	3.6	3.5	3.3	3.1	2.9	2.6							
			3.6	3.7	3.8	3.9	3.9	3.8	3.8	3.7	3.4	3.1	2.8	2.4	2.0	1.5	0.9	0.2								
168-D1	9.26									96	88	81	75	68	63	58	53	48	44	40	36					
	22.26									4.1	4.2	4.3	4.3	4.3	4.3	4.3	4.2	4.0	3.7	3.5	3.2					
										4.2	4.2	4.1	4.0	3.8	3.6	3.3	2.9	2.4	1.8	1.2	0.5					
188-D1	7.79															67	62	57	52	47	43	39	36			
	22.01															4.7	4.7	4.7	4.6	4.5	4.3	4.1	3.7			
																4.0	3.8	3.4	3.1	2.6	2.1	1.4	0.6			
208-D1	6.61																			51	46	43	39	35		
	21.76																			5.0	4.9	4.7	4.5	4.2		
																				3.1	2.6	2.0	1.3	0.5		
228-D1	5.65																								41	38
	21.51																								5.0	4.8
																									1.7	1.0

8LST36+2
2" Normal Weight Topping

Table of safe superimposed service load (psf) and cambers

Strand Pattern	e_e / e_c	50	52	54	56	58	60	62	64	66	68	70	72	74	76	78	80	82	84	86
108-D1	17.61	179	161	145	130	117	105	94	84	75	67	60	53	47						
	23.01	2.0	2.1	2.2	2.3	2.3	2.4	2.5	2.5	2.5	2.5	2.5	2.5	2.5						
		1.8	1.9	1.9	1.8	1.8	1.7	1.6	1.5	1.3	1.1	0.9	0.6	0.3						
128-D1	13.68		199	180	163	147	133	121	109	99	89	81	73	65	59	52				
	22.76		2.2	2.3	2.5	2.6	2.7	2.8	3.0	3.1	3.1	3.2	3.2	3.2	3.1	3.1				
			2.0	2.1	2.1	2.1	2.1	2.1	2.0	2.0	1.8	1.6	1.4	1.1	0.7	0.4				
148-D1	11.15							133	121	110	100	91	83	75	66	58	51			
	22.51							3.1	3.3	3.4	3.5	3.6	3.7	3.8	3.8	3.8	3.8			
								2.3	2.2	2.2	2.1	1.9	1.7	1.5	1.2	0.9	0.4			
168-D1	9.26															81	72	64	57	50
	22.26															4.1	4.2	4.3	4.3	4.3
																1.6	1.4	1.0	0.7	0.2

Bold type — *Capacity governed by stresses, others governed by flexural strength*

Values below heavy line require release strengths higher than 3500 psi.

SINGLE TEE

10'-0" x 48"

Strand Pattern Designation

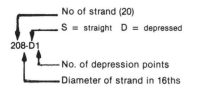

208-D1
- No of strand (20)
- S = straight   D = depressed
- No. of depression points
- Diameter of strand in 16ths

Safe loads shown include dead load of 10 psf for untopped members and 15 psf for topped members. Remainder is live load. Long-time cambers include superimposed dead load but do not include live load.

Key
- 126 — Safe superimposed service load, psf
- 1.5 — Estimated camber at erection, in.
- 1.7 — Estimated long-time camber, in.

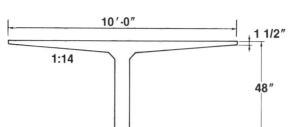

$f'_c = 5000$ psi
$f_{pu} = 270{,}000$ psi

Section Properties

A = 782 in.²
I = 168,968 in.⁴
Y_b = 36.64 in.
Y_t = 11.36
Z_b = 4612 in.³
Z_t = 14,873 in.³
wt = 815 plf
     82 psf
V/S = 2.33 in.

10ST48

Table of safe superimposed service load (psf) and cambers — Normal Weight Concrete — No Topping

Strand Pattern	e_e / e_c	68	70	72	74	76	78	80	82	84	86	88	90	92	94	96	98	100	102	104
148-D1	23.78 / 33.14	126	115	105	96	88	80	73	66	60	54	49	44	39						
		1.5	1.5	1.5	1.5	1.5	1.4	1.4	1.4	1.3	1.2	1.1	1.0	0.9						
		1.7	1.7	1.6	1.6	1.5	1.4	1.3	1.2	1.1	0.9	0.7	0.5	0.2						
168-D1	20.39 / 32.89	148	136	125	115	105	97	89	81	74	68	62	56	51	46	42				
		1.7	1.7	1.7	1.8	1.8	1.8	1.8	1.7	1.7	1.6	1.5	1.5	1.4	1.2	1.1				
		2.0	2.0	2.0	1.9	1.9	1.8	1.8	1.7	1.5	1.4	1.2	1.0	0.8	0.6	0.3				
188-D1	17.75 / 32.64	169	156	144	133	122	113	104	96	88	81	75	69	63	57	52	48	43		
		1.8	1.8	1.9	2.0	2.0	2.0	2.0	2.0	2.0	2.0	1.9	1.9	1.8	1.7	1.6	1.4	1.3		
		2.1	2.2	2.2	2.2	2.2	2.2	2.1	2.1	2.0	1.9	1.8	1.6	1.4	1.2	0.9	0.6	0.3		
208-D1	15.44 / 32.39							119	110	102	94	87	80	74	68	63	58	53	48	44
								2.2	2.2	2.3	2.3	2.3	2.3	2.2	2.1	2.0	1.9	1.8	1.6	1.4
								2.4	2.4	2.4	2.3	2.3	2.1	1.9	1.7	1.5	1.2	1.0	0.3	0.3
228-D1	13.73 / 32.14								99	92	85	79	73	67	62	57	53	48	**43**	
									2.5	2.5	2.5	2.5	2.5	2.4	2.3	2.2	2.1	2.0	1.8	1.6	
									2.5	2.5	2.4	2.3	2.1	1.9	1.6	1.3	1.0	0.6	0.2	
248-D1	12.14 / 31.89											83	77	72	66	61	**56**	**51**	**46**	**42**
												2.7	2.7	2.6	2.6	2.5	2.3	2.1	1.9	1.7
												2.5	2.3	2.1	1.9	1.7	1.3	0.9	0.5	0.0

10LST48

Table of safe superimposed service load (psf) and cambers — Lightweight Concrete — No Topping

Strand Pattern	e_e / e_c	76	78	80	82	84	86	88	90	92	94	96	98	100	102	104	106	108	110	112	114	116	118	120
148-D1	23.78 / 33.14	103	96	88	82	76	70	64	59	55	50	46	42	39	35	32								
		2.8	2.8	2.9	2.9	2.9	2.9	2.8	2.8	2.7	2.6	2.5	2.4	2.3	2.1	1.9								
		3.1	3.1	3.1	3.0	2.9	2.8	2.6	2.5	2.3	2.0	1.7	1.4	1.1	0.7	0.2								
168-D1	20.39 / 32.89	121	112	104	97	90	84	78	72	67	62	57	53	49	45	41	38	35						
		3.2	3.3	3.3	3.4	3.4	3.4	3.4	3.4	3.3	3.3	3.2	3.1	3.0	2.9	2.7	2.6	2.3						
		3.7	3.7	3.7	3.7	3.6	3.5	3.4	3.3	3.1	2.9	2.7	2.4	2.1	1.8	1.4	1.0	0.5						
188-D1	17.75 / 32.64		129	120	112	104	97	90	84	78	73	68	63	59	55	51	47	43	40	**36**				
			3.5	3.6	3.7	3.8	3.9	3.9	4.0	4.0	3.9	3.9	3.8	3.8	3.7	3.5	3.4	3.2	3.0	2.8				
			4.0	4.1	4.1	4.2	4.2	4.2	4.1	4.0	3.8	3.6	3.4	3.1	2.8	2.5	2.1	1.7	1.2	0.7				
208-D1	15.44 / 32.39								96	90	84	78	73	68	64	59	55	51	**47**	**43**	**39**	**36**		
									4.2	4.3	4.4	4.4	4.5	4.5	4.4	4.3	4.2	4.0	3.9	3.7	3.5	3.2		
									4.5	4.5	4.4	4.4	4.3	4.1	3.8	3.5	3.2	2.8	2.4	1.9	1.3	0.8		
228-D1	13.73 / 32.14													78	73	68	**63**	**58**	**54**	**50**	**46**	**42**	**39**	
														4.8	4.9	4.9	4.9	4.9	4.7	4.6	4.4	4.2	3.9	
														4.6	4.5	4.3	4.1	3.9	3.5	3.1	2.6	2.1	1.5	
248-D1	12.14 / 31.89																		**61**	**57**	**53**	**49**	**45**	**41**
																			5.2	5.2	5.1	5.0	4.9	4.6
																			4.3	4.0	3.7	3.3	2.8	2.2

Bold type — *Capacity governed by stresses, others governed by flexural strength*

Values below heavy line require release strengths higher than 3500 psi.

HOLLOW-CORE

4'-0" x 6"
Normal Weight Concrete

Strand Patterns

Producer may vary size and strength of strands. See "explanation of load tables"

Safe loads shown include dead load of 10 psf for untopped members and 15 psf for topped members. Remainder is live load. Long-time cambers include superimposed dead load but do not include live load.

Capacity of sections of other configurations are similar. For precise values, see local hollow-core manufacturer.

Key

263 — Safe superimposed service load, psf
0.1 — Estimated camber at erection, in.
0.2 — Estimated long-time camber, in.

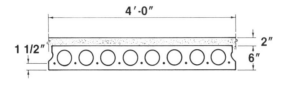

Section Properties

	Untopped	Topped
A =	187 in.²	—
I =	763 in.⁴	1640 in.⁴
Y_b =	3.00 in.	4.14 in.
Y_t =	3.00 in.	3.86 in.
Z_b =	254 in.³	396 in.³
Z_t =	254 in.³	425 in.³
b_w =	16.00 in.	16.00 in.
wt =	195 plf	295 plf
	49 psf	74 psf
V/S =	1.73 in.	

f'_c = 5000 psi
f'_{ci} = 3500 psi

4HC6

Table of safe superimposed service load (psf) and cambers — No Topping

Strand Designation Code	Span, ft.													
	12	13	14	15	16	17	18	19	20	21	22	23	24	25
30-S	263	218	183	154	131	112	95	82	70	60	51			
	0.1	0.1	0.1	0.2	0.1	0.1	0.1	0.1	0.1	0.0	0.0			
	0.2	0.2	0.2	0.2	0.1	0.1	0.1	0.0	−0.1	−0.2	−0.3			
40-S		262	231	205	181	156	135	117	100	87	75	64	55	47
		0.2	0.2	0.3	0.3	0.3	0.3	0.3	0.2	0.2	0.2	0.1	0.0	0.0
		0.3	0.3	0.3	0.3	0.3	0.2	0.2	0.2	0.1	0.0	−0.2	−0.3	−0.5
50-S			291	259	225	194	168	145	126	110	96	84	73	63
			0.3	0.4	0.4	0.4	0.4	0.4	0.4	0.4	0.4	0.4	0.3	0.2
			0.4	0.4	0.4	0.5	0.4	0.4	0.4	0.3	0.3	0.2	0.0	−0.1
60-S					266	230	200	174	152	134	118	103	91	80
					0.5	0.5	0.6	0.6	0.6	0.6	0.6	0.6	0.6	0.5
					0.6	0.6	0.6	0.6	0.6	0.6	0.6	0.5	0.4	0.3
70-S						266	232	203	178	157	139	123	109	97
						0.7	0.7	0.7	0.8	0.8	0.8	0.8	0.8	0.8
						0.8	0.8	0.9	0.9	0.9	0.9	0.8	0.7	0.7

4HC6 + 2

Table of safe superimposed service load (psf) and cambers — 2" Normal Weight Topping

Strand Designation Code	Span, ft.													
	14	15	16	17	18	19	20	21	22	23	24	25	26	27
30-S	257	217	184	150	122	97	77	59	43					
	0.1	0.2	0.1	0.1	0.1	0.1	0.1	0.0	0.0					
	0.1	0.1	0.0	0.0	−0.1	−0.2	−0.3	−0.4	−0.6					
40-S		289	248	207	172	143	118	96	77	61	47			
		0.3	0.3	0.3	0.3	0.3	0.2	0.2	0.2	0.1	0.0			
		0.2	0.2	0.1	0.1	0.0	−0.1	−0.2	−0.3	−0.5	−0.7			
50-S				263	222	187	158	133	111	91	75	60	47	
				0.4	0.4	0.4	0.4	0.4	0.4	0.4	0.3	0.2	0.2	
				0.3	0.3	0.2	0.1	0.0	−0.1	−0.2	−0.4	−0.6	−0.9	
60-S					272	232	198	169	144	122	103	86	71	57
					0.6	0.6	0.6	0.6	0.6	0.6	0.6	0.5	0.5	0.4
					0.4	0.4	0.4	0.3	0.2	0.1	−0.1	−0.3	−0.5	−0.8
70-S						277	239	206	178	153	131	112	95	79
						0.7	0.8	0.8	0.8	0.8	0.8	0.8	0.8	0.7
						0.6	0.6	0.5	0.4	0.3	0.2	0.0	−0.2	−0.4

Bold type — *Capacity governed by stresses, others governed by flexural or shear strength*

HOLLOW-CORE

4'-0" x 6"
Lightweight Concrete

Strand Patterns

Producer may vary size and strength of strands. See "explanation of load tables"

Safe loads shown include dead load of 10 psf for untopped members and 15 psf for topped members. Remainder is live load. Long-time cambers include superimposed dead load but do not include live load.

Capacity of sections of other configurations are similar. For precise values, see local hollow-core manufacturer.

Key

268 — Safe superimposed service load, psf
0.2 — Estimated camber at erection, in.
0.3 — Estimated long-time camber, in.

f'_c = 5000 psi
f'_{ci} = 3500 psi

Section Properties

	Untopped	Topped
A =	187 in.²	—
I =	763 in.⁴	1915 in.⁴
Y_b =	3.00 in.	4.49 in.
Y_t =	3.00 in.	3.51 in.
Z_b =	254 in.³	426 in.³
Z_t =	254 in.³	546 in.³
b_w =	16.00 in.	16.00 in.
wt =	149 plf	249 plf
	37 psf	62 psf
V/S =	1.73 in.	

4LHC6
No Topping

Table of safe superimposed service load (psf) and cambers

Strand Designation Code	Span, ft.													
	12	13	14	15	16	17	18	19	20	21	22	23	24	25
30-S	268	228	192	164	140	121	105	91	79	69	61	53	46	
	0.2	0.2	0.3	0.3	0.3	0.3	0.3	0.3	0.3	0.2	0.2	0.1	0.0	
	0.3	0.3	0.3	0.3	0.3	0.3	0.2	0.2	0.1	0.0	−0.1	−0.3	−0.5	
40-S		267	236	211	189	164	142	123	108	94	83	72	63	56
		0.4	0.4	0.4	0.5	0.5	0.5	0.5	0.5	0.5	0.5	0.5	0.4	0.3
		0.4	0.5	0.5	0.5	0.5	0.5	0.5	0.5	0.4	0.3	0.2	0.0	−0.1
50-S			296	265	228	198	173	151	133	117	103	91	81	72
			0.5	0.6	0.6	0.7	0.7	0.8	0.8	0.8	0.8	0.8	0.8	0.8
			0.7	0.7	0.8	0.8	0.8	0.8	0.8	0.8	0.8	0.7	0.6	0.4
60-S					268	233	204	179	158	140	124	110	98	88
					0.8	0.9	0.9	1.0	1.1	1.1	1.1	1.2	1.2	1.2
					1.0	1.1	1.1	1.2	1.2	1.2	1.2	1.2	1.1	1.0
70-S						267	234	207	183	162	145	129	116	104
						1.1	1.2	1.2	1.3	1.4	1.5	1.5	1.6	1.6
						1.3	1.4	1.5	1.6	1.6	1.6	1.6	1.6	1.6

4LHC6 + 2
2" Normal Weight Topping

Table of safe superimposed service load (psf) and cambers

Strand Designation Code	Span, ft.													
	14	15	16	17	18	19	20	21	22	23	24	25	26	27
30-S	267	226	193	165	142	118	97	78	62	48				
	0.3	0.3	0.3	0.3	0.3	0.3	0.3	0.2	0.2	0.1				
	0.2	0.2	0.1	0.1	0.0	−0.1	−0.2	−0.4	−0.6	−0.8				
40-S		293	263	227	196	165	139	116	97	79	64	51		
		0.4	0.5	0.5	0.5	0.5	0.5	0.5	0.5	0.5	0.4	0.3		
		0.4	0.3	0.3	0.3	0.2	0.1	0.0	−0.2	−0.4	−0.6	−0.9		
50-S				284	248	212	181	154	131	111	94	78	64	
				0.7	0.7	0.8	0.8	0.8	0.8	0.8	0.8	0.8	0.7	
				0.6	0.5	0.5	0.4	0.3	0.2	0.1	−0.1	−0.4	−0.7	
60-S					293	258	223	192	166	143	123	105	89	75
					0.9	1.0	1.1	1.1	1.1	1.2	1.2	1.2	1.2	1.1
					0.8	0.8	0.8	0.7	0.6	0.5	0.3	0.1	−0.1	−0.4
70-S						293	259	230	201	175	152	132	114	98
						1.2	1.3	1.4	1.5	1.5	1.6	1.6	1.6	1.6
						1.1	1.1	1.1	1.0	0.9	0.8	0.6	0.4	0.2

Bold type — *Capacity governed by stresses, others governed by flexural or shear strength*

HOLLOW-CORE

4'-0" x 8"
Normal Weight Concrete

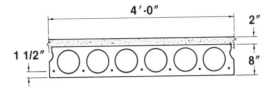

$f'_c = 5000$ psi
$f'_{ci} = 3500$ psi

Strand Patterns

Producer may vary size and strength of strands. See "explanation of load tables"

Safe loads shown include dead load of 10 psf for untopped members and 15 psf for topped members. Remainder is live load. Long-time cambers include superimposed dead load but do not include live load.

Capacity of sections of other configurations are similar. For precise values, see local hollow-core manufacturer.

Section Properties

	Untopped	Topped
A =	215 in.²	—
I =	1666 in.⁴	3071 in.⁴
Y_b =	4.00 in.	5.29 in.
Y_t =	4.00 in.	4.71 in.
Z_b =	416 in.³	580 in.³
Z_t =	416 in.³	652 in.³
b_w =	12.00 in.	12.00 in.
wt =	224 plf	323 plf
	56 psf	81 psf
V/S =	1.92 in.	

Key

284 — Safe superimposed service load, psf
0.1 — Estimated camber at erection, in.
0.2 — Estimated long-time camber, in.

4HC8
No Topping

Table of safe superimposed service load (psf) and cambers

Strand Designation Code	Span, ft.																			
	14	15	16	17	18	19	20	21	22	23	24	25	26	27	28	29	30	31	32	33
30-S	284	242	207	178	154	134	117	102	89	77	67	59	51							
	0.1	0.2	0.2	0.2	0.2	0.2	0.2	0.1	0.1	0.1	0.1	0.0	0.0							
	0.2	0.2	0.2	0.2	0.2	0.1	0.1	0.1	0.0	-0.1	-0.1	-0.3								
40-S				285	247	216	189	166	147	130	115	102	90	80	71	63	56	49		
				0.2	0.3	0.3	0.3	0.3	0.3	0.3	0.3	0.3	0.2	0.2	0.2	0.1	0.0	0.0		
				0.3	0.3	0.3	0.3	0.3	0.3	0.3	0.3	0.2	0.1	0.1	0.0	-0.1	-0.3	-0.4		
50-S				287	269	241	213	189	169	150	135	**120**	**107**	95	85	75	66	59	52	45
				0.4	0.4	0.4	0.4	0.5	0.5	0.5	0.5	0.5	0.4	0.4	0.4	0.3	0.3	0.2	0.1	0.0
				0.5	0.5	0.5	0.5	0.5	0.5	0.5	0.5	0.4	0.4	0.3	0.2	0.1	0.0	-0.2	-0.4	-0.6
60-S				296	275	260	244	224	**205**	183	163	146	131	117	105	94	84	76	67	60
				0.5	0.5	0.5	0.6	0.6	0.6	0.6	0.6	0.7	0.7	0.7	0.7	0.6	0.6	0.6	0.5	0.4
				0.6	0.6	0.7	0.7	0.7	0.8	0.8	0.8	0.7	0.7	0.7	0.6	0.5	0.4	0.3	0.1	
70-S					284	266	250	236	223	209	190	**172**	**155**	139	126	113	102	92	83	75
					0.6	0.7	0.7	0.8	0.8	0.8	0.9	0.9	0.9	0.9	0.9	0.9	0.9	0.9	0.8	0.8
					0.8	0.9	0.9	0.9	1.0	1.0	1.0	1.0	1.0	1.0	1.0	0.9	0.8	0.7	0.6	0.5

4HC8 + 2
2" Normal Weight Topping

Table of safe superimposed service load (psf) and cambers

Strand Designation Code	Span, ft.																	
	16	17	18	19	20	21	22	23	24	25	26	27	28	29	30	31	32	33
30-S	260	223	192	166	143	124	107	93	76	61	48							
	0.2	0.2	0.2	0.2	0.2	0.1	0.1	0.1	0.1	0.0	0.0							
	0.1	0.1	0.1	0.1	0.0	0.0	-0.1	-0.2	-0.3	-0.4	-0.5							
40-S			269	235	206	181	**158**	135	115	97	82	67	55	43				
			0.3	0.3	0.3	0.3	0.3	0.3	0.3	0.2	0.2	0.2	0.1	0.0				
			0.2	0.2	0.2	0.1	0.0	0.0	-0.1	-0.3	-0.4	-0.6	-0.8					
50-S				299	264	234	**205**	178	154	133	115	98	83	70	58	47		
				0.4	0.4	0.5	0.5	0.5	0.5	0.5	0.4	0.4	0.4	0.3	0.3	0.2		
				0.4	0.3	0.3	0.3	0.3	0.2	0.1	0.0	-0.1	-0.2	-0.4	-0.6	-0.8		
60-S						284	**251**	220	193	169	148	129	112	97	83	71	59	49
						0.6	0.6	0.6	0.7	0.7	0.7	0.7	0.6	0.6	0.6	0.5	0.5	0.4
						0.5	0.5	0.5	0.4	0.4	0.3	0.2	0.1	0.0	-0.2	-0.4	-0.6	-0.9
70-S						297	280	**263**	232	205	181	**160**	141	124	108	94	81	70
						0.8	0.8	0.8	0.9	0.9	0.9	0.9	0.9	0.9	0.9	0.9	0.8	0.8
						0.7	0.7	0.7	0.7	0.6	0.6	0.5	0.4	0.3	0.2	0.0	-0.2	-0.4

Bold type — *Capacity governed by stresses, others governed by flexural or shear strength*

HOLLOW-CORE

4'-0" x 8"
Lightweight Concrete

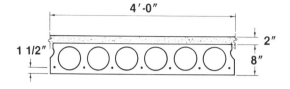

f'_c = 5000 psi
f'_{ci} = 3500 psi

Strand Patterns

Producer may vary size and strength of strands. See "explanation of load tables"

Safe loads shown include dead load of 10 psf for untopped members and 15 psf for topped members. Remainder is live load. Long-time cambers include superimposed dead load but do not include live load.

Capacity of sections of other configurations are similar. For precise values, see local hollow-core manufacturer.

Section Properties

	Untopped	Topped
A =	215 in.²	—
I =	1666 in.⁴	3529 in.⁴
Y_b =	4.00 in.	5.70 in.
Y_t =	4.00 in.	4.30 in.
Z_b =	416 in.³	619 in.³
Z_t =	416 in.³	821 in.³
b_w =	12.00 in.	12.00 in.
wt =	172 plf	271 plf
	43 psf	68 psf
V/S =	1.92 in.	

Key

295 — Safe superimposed service load, psf
0.2 — Estimated camber at erection, in.
0.3 — Estimated long-time camber, in.

4LHC8
No Topping

Table of safe superimposed service load (psf) and cambers

Strand Designation Code	Span, ft.																			
	14	15	16	17	18	19	20	21	22	23	24	25	26	27	28	29	30	31	32	33
30-S	295	253	218	189	165	145	127	112	99	88	78	69	62	55	49	43				
	0.2	0.3	0.3	0.3	0.3	0.3	0.3	0.3	0.3	0.3	0.3	0.3	0.2	0.2	0.1	0.0				
	0.3	0.3	0.3	0.3	0.3	0.3	0.3	0.3	0.3	0.2	0.1	0.1	0.0	−0.2	−0.3	−0.5				
40-S			291	258	226	200	177	157	140	126	113	101	91	82	74	66	59	52	47	41
			0.4	0.4	0.5	0.5	0.5	0.6	0.6	0.6	0.6	0.6	0.6	0.5	0.5	0.5	0.4	0.3	0.2	0.1
			0.5	0.5	0.5	0.5	0.6	0.6	0.6	0.6	0.6	0.6	0.5	0.5	0.1	−0.1	−0.3	−0.5	−0.7	
50-S				298	277	252	224	200	**178**	**160**	143	128	116	104	94	84	76	69	62	55
				0.6	0.6	0.7	0.7	0.8	0.8	0.8	0.8	0.9	0.9	0.9	0.9	0.9	0.8	0.8	0.7	0.7
				0.7	0.8	0.8	0.9	0.9	0.9	0.9	0.9	0.9	0.9	0.8	0.8	0.7	0.6	0.4	0.2	0.0
60-S					286	268	252	231	210	**189**	**170**	153	139	125	114	103	93	85	77	70
					0.8	0.9	0.9	1.0	1.1	1.1	1.2	1.2	1.3	1.3	1.3	1.3	1.3	1.3	1.3	1.2
					1.0	1.1	1.2	1.2	1.3	1.3	1.3	1.4	1.4	1.3	1.3	1.3	1.2	1.1	0.9	0.8
70-S					292	274	258	243	231	216	**197**	**178**	161	147	133	121	111	101	92	84
					1.0	1.1	1.1	1.2	1.3	1.4	1.5	1.5	1.6	1.6	1.7	1.7	1.8	1.8	1.8	1.8
					1.2	1.3	1.4	1.5	1.6	1.7	1.7	1.8	1.8	1.9	1.9	1.8	1.8	1.7	1.6	1.5

4LHC8 + 2
2" Normal Weight Topping

Table of safe superimposed service load (psf) and cambers

Strand Designation Code	Span, ft.																	
	16	17	18	19	20	21	22	23	24	25	26	27	28	29	30	31	32	33
30-S	271	234	203	176	154	135	118	104	91	79	**67**	**54**	44					
	0.3	0.3	0.3	0.3	0.3	0.3	0.3	0.3	0.3	0.3	0.2	0.2	0.1					
	0.2	0.2	0.2	0.2	0.1	0.1	0.0	−0.1	−0.2	−0.3	−0.5	−0.6	−0.9					
40-S		280	245	216	191	170	151	134	**117**	**101**	86	73	61	51				
		0.5	0.5	0.5	0.6	0.6	0.6	0.6	0.6	0.6	0.5	0.5	0.5	0.4				
		0.4	0.4	0.4	0.4	0.3	0.2	0.1	0.0	−0.2	−0.4	−0.6	−0.8					
50-S			275	244	218	195	175	**154**	**135**	118	103	89	76	65	55			
			0.7	0.8	0.8	0.8	0.9	0.9	0.9	0.9	0.9	0.9	0.8	0.8	0.7			
			0.6	0.6	0.6	0.6	0.5	0.5	0.4	0.3	0.1	−0.1	−0.3	−0.5	−0.8			
60-S				294	263	236	213	**191**	**169**	149	132	116	102	89	77	67		
				1.0	1.1	1.1	1.2	1.2	1.3	1.3	1.3	1.3	1.3	1.3	1.3	1.2		
				0.9	0.9	0.9	0.9	0.9	0.8	0.7	0.6	0.5	0.3	0.1	−0.1	−0.4		
70-S							288	273	248	225	**203**	**181**	161	144	128	113	100	88
							1.3	1.4	1.5	1.5	1.6	1.6	1.7	1.7	1.8	1.8	1.8	1.8
							1.2	1.2	1.3	1.2	1.2	1.2	1.1	1.0	0.9	0.7	0.5	0.3

Bold type — *Capacity governed by stresses, others governed by flexural or shear strength*

HOLLOW-CORE

4'-0" x 10"
Normal Weight Concrete

Strand Patterns

Producer may vary size and strength of strands. See "explanation of load tables"

Safe loads shown include dead load of 10 psf for untopped members and 15 psf for topped members. Remainder is live load. Long-time cambers include superimposed dead load but do not include live load.

Capacity of sections of other configurations are similar. For precise values, see local hollow-core manufacturer.

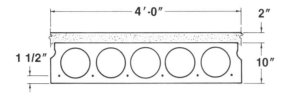

$f'_c = 5000$ psi
$f'_{ci} = 3500$ psi

Section Properties

	Untopped	Topped
A =	259 in.²	—
I =	3223 in.⁴	5328 in.⁴
Y_b =	5.00 in.	6.34 in.
Y_t =	5.00 in.	5.66 in.
Z_b =	645 in.³	840 in.³
Z_t =	645 in.³	941 in.³
b_w =	10.50 in.	10.50 in.
wt =	270 plf	370 plf
	68 psf	93 psf
V/S =	2.23 in.	

Key

227 — Safe superimposed service load, psf
0.2 — Estimated camber at erection, in.
0.3 — Estimated long-time camber, in.

4HC10 — No Topping

Table of safe superimposed service load (psf) and cambers

Strand Designation Code	20	21	22	23	24	25	26	27	28	29	30	31	32	33	34	35	36	37	38	39	40	41	42
40-S	227	201	178	158	141	126	112	100	89	80	71	63	56										
	0.2	0.2	0.3	0.3	0.2	0.2	0.2	0.2	0.2	0.1	0.1	0.1	0.0										
	0.3	0.3	0.3	0.3	0.2	0.2	0.2	0.1	0.1	0.0	−0.1	−0.2	−0.3										
50-S	274	254	232	207	186	167	150	136	122	110	100	90	81	73	66	59	52						
	0.3	0.4	0.4	0.4	0.4	0.4	0.4	0.4	0.4	0.4	0.3	0.3	0.2	0.2	0.1	0.1	0.0						
	0.4	0.4	0.4	0.4	0.4	0.4	0.4	0.4	0.3	0.3	0.2	0.1	0.0	−0.1	−0.2	−0.4	−0.5						
60-S	280	263	249	233	222	207	187	170	154	140	127	116	105	**95**	**85**	**77**	**69**	**61**	**55**	**49**			
	0.4	0.5	0.5	0.5	0.5	0.5	0.6	0.6	0.6	0.6	0.6	0.5	0.5	0.5	0.4	0.4	0.3	0.2	0.1	0.0			
	0.5	0.6	0.6	0.6	0.6	0.6	0.6	0.6	0.6	0.6	0.5	0.5	0.4	0.3	0.2	0.1	−0.1	−0.3	−0.4	−0.7			
70-S	289	269	255	242	228	217	205	194	181	**168**	**153**	**139**	**126**	**115**	**104**	94	85	77	70	63	56	50	45
	0.5	0.6	0.6	0.6	0.7	0.7	0.7	0.7	0.8	0.8	0.8	0.8	0.8	0.7	0.7	0.7	0.6	0.5	0.5	0.4	0.3	0.2	0.0
	0.7	0.7	0.8	0.8	0.8	0.8	0.9	0.9	0.9	0.8	0.8	0.8	0.7	0.7	0.6	0.5	0.3	0.2	0.0	−0.1	−0.4	−0.6	−0.9
80-S	295	278	261	248	234	223	211	203	193	183	168	156	**147**	**135**	**123**	**112**	**102**	**93**	**85**	**77**	**70**	**63**	**57**
	0.6	0.7	0.7	0.8	0.8	0.9	0.9	0.9	1.0	1.0	1.0	1.0	1.0	1.0	1.0	1.0	0.9	0.9	0.8	0.8	0.7	0.6	0.5
	0.8	0.9	0.9	1.0	1.0	1.1	1.1	1.1	1.1	1.1	1.1	1.1	1.1	1.0	1.0	0.9	0.8	0.7	0.5	0.4	0.2	0.0	−0.3

4HC10 + 2 — 2" Normal Weight Topping

Table of safe superimposed service load (psf) and cambers

Strand Designation Code	20	21	22	23	24	25	26	27	28	29	30	31	32	33	34	35	36	37	38	39	40
40-S	266	234	207	183	162	144	127	113	100	**85**	**72**	**60**	**49**								
	0.2	0.2	0.3	0.3	0.2	0.2	0.2	0.2	0.2	0.1	0.1	0.1	0.0								
	0.2	0.2	0.2	0.1	0.1	0.0	0.0	−0.1	−0.2	−0.3	−0.4	−0.5	−0.7								
50-S		270	241	215	192	172	154	**137**	**119**	**103**	**89**	**76**	**65**	54	44						
		0.4	0.4	0.4	0.4	0.4	0.4	0.4	0.4	0.3	0.3	0.2	0.2	0.1	0.1						
		0.3	0.3	0.3	0.2	0.2	0.1	0.1	0.0	−0.1	−0.2	−0.4	−0.5	−0.7	−0.9						
60-S			297	280	263	239	215	194	**173**	**153**	**135**	**119**	**104**	**91**	**78**	**67**	**57**	**47**			
			0.5	0.5	0.5	0.5	0.6	0.6	0.6	0.6	0.5	0.5	0.5	0.5	0.4	0.4	0.3	0.2			
			0.5	0.5	0.4	0.4	0.4	0.4	0.3	0.2	0.2	0.1	0.0	−0.2	−0.3	−0.5	−0.7	−0.9			
70-S				286	272	256	244	230	**209**	**187**	**167**	**148**	**132**	**117**	**103**	**90**	79	68	58		
				0.6	0.7	0.7	0.7	0.7	0.8	0.8	0.8	0.8	0.8	0.7	0.7	0.7	0.6	0.5	0.5		
				0.6	0.6	0.6	0.6	0.6	0.5	0.5	0.4	0.4	0.3	0.2	0.0	−0.1	−0.3	−0.5	−0.7		
80-S				295	278	265	250	236	226	215	**198**	**178**	**160**	**143**	**128**	**114**	**101**	**89**	**78**	**68**	**59**
				0.8	0.8	0.9	0.9	0.9	1.0	1.0	1.0	1.0	1.0	1.0	1.0	1.0	0.9	0.9	0.8	0.8	0.7
				0.8	0.8	0.8	0.8	0.8	0.8	0.8	0.7	0.6	0.6	0.5	0.4	0.2	0.1	−0.1	−0.3	−0.5	−0.8

Bold type — *Capacity governed by stresses, others governed by flexural or shear strength*

HOLLOW-CORE

4'-0" x 10"
Lightweight Concrete

Strand Patterns

Producer may vary size and strength of strands. See "explanation of load tables"

Safe loads shown include dead load of 10 psf for untopped members and 15 psf for topped members. Remainder is live load. Long-time cambers include superimposed dead load but do not include live load.

Capacity of sections of other configurations are similar. For precise values, see local hollow-core manufacturer.

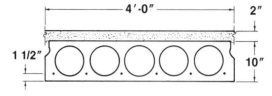

f'_c = 5000 psi
f'_{ci} = 3500 psi

Section Properties

		Untopped	Topped
A	=	259 in.²	—
I	=	3223 in.⁴	5328 in.⁴
Y_b	=	5.00 in.	6.34 in.
Y_t	=	5.00 in.	5.66 in.
Z_b	=	645 in.³	840 in.³
Z_t	=	645 in.³	941 in.³
b_w	=	10.50 in.	10.50 in.
wt	=	207 plf	307 plf
		52 psf	77 psf
V/S	=	2.23 in.	

Key

240 — Safe superimposed service load, psf
0.4 — Estimated camber at erection, in.
0.5 — Estimated long-time camber, in.

4LHC10

Table of safe superimposed service load (psf) and cambers

No Topping

Strand Designation Code	Span, ft.																						
	20	21	22	23	24	25	26	27	28	29	30	31	32	33	34	35	36	37	38	39	40	41	42
40-S	240	214	191	171	154	139	125	113	102	93	84	76	69	62	56	51	46						
	0.4	0.4	0.5	0.5	0.5	0.5	0.5	0.5	0.5	0.5	0.4	0.4	0.3	0.3	0.2	0.1	0.0						
	0.5	0.5	0.5	0.5	0.5	0.5	0.5	0.4	0.4	0.3	0.3	0.2	0.1	−0.1	−0.2	−0.4	−0.6						
50-S	284	267	244	220	199	180	163	149	135	123	113	103	94	86	**78**	71	64	58	52	47	42		
	0.6	0.6	0.6	0.7	0.7	0.7	0.7	0.8	0.8	0.8	0.8	0.7	0.7	0.7	0.6	0.6	0.5	0.4	0.3	0.2	0.1		
	0.7	0.7	0.8	0.8	0.8	0.8	0.8	0.8	0.8	0.8	0.7	0.6	0.6	0.5	0.3	0.2	0.0	−0.1	−0.4	−0.6	−0.9		
60-S	293	273	259	246	232	215	200	183	167	152	139	127	116	106	97	88	81	73	67	61	55	50	45
	0.7	0.8	0.8	0.9	0.9	0.9	1.0	1.0	1.1	1.1	1.1	1.1	1.1	1.1	1.1	1.0	1.0	0.9	0.9	0.8	0.7	0.6	0.4
	0.9	1.0	1.0	1.1	1.1	1.1	1.2	1.2	1.2	1.2	1.2	1.1	1.1	1.0	0.9	0.8	0.7	0.5	0.4	0.2	−0.1	−0.3	−0.6
70-S	299	282	265	252	241	227	218	204	187	177	162	148	136	**125**	**115**	**105**	97	89	81	75	68	63	57
	0.9	0.9	1.0	1.1	1.1	1.2	1.2	1.3	1.3	1.4	1.4	1.5	1.5	1.5	1.5	1.5	1.5	1.4	1.4	1.4	1.3	1.2	1.1
	1.1	1.2	1.3	1.3	1.4	1.4	1.5	1.5	1.6	1.6	1.6	1.6	1.6	1.6	1.5	1.4	1.3	1.2	1.1	0.9	0.7	0.5	0.2
80-S		288	274	258	247	233	224	213	202	193	181	166	155	**144**	**133**	122	113	104	96	88	82	75	69
		1.1	1.2	1.2	1.3	1.4	1.5	1.6	1.6	1.7	1.8	1.8	1.9	1.9	1.9	1.9	2.0	2.0	1.9	1.9	1.9	1.8	1.7
		1.4	1.5	1.6	1.7	1.8	1.8	1.9	2.0	2.0	2.1	2.1	2.1	2.1	2.1	2.0	2.0	1.9	1.8	1.7	1.5	1.3	1.1

4LHC10+2

Table of safe superimposed service load (psf) and cambers

2" Normal Weight Topping

Strand Designation Code	Span, ft.																				
	20	21	22	23	24	25	26	27	28	29	30	31	32	33	34	35	36	37	38	39	40
40-S	279	247	220	196	175	157	140	126	113	101	88	76	65	55	46						
	0.4	0.4	0.5	0.5	0.5	0.5	0.5	0.5	0.5	0.5	0.4	0.4	0.3	0.3	0.2						
	0.4	0.4	0.3	0.3	0.3	0.2	0.2	0.1	0.0	−0.1	−0.2	−0.4	−0.5	−0.7	−0.9						
50-S		283	254	228	205	185	167	**150**	133	118	104	92	80	70	60	51					
		0.6	0.7	0.7	0.7	0.7	0.8	0.8	0.8	0.8	0.7	0.7	0.7	0.6	0.6	0.5					
		0.6	0.6	0.5	0.5	0.5	0.4	0.4	0.3	0.2	0.1	−0.1	−0.2	−0.4	−0.6	−0.9					
60-S			290	276	252	228	206	**185**	**166**	148	132	118	105	93	82	72	63	55			
			0.9	0.9	0.9	1.0	1.0	1.1	1.1	1.1	1.1	1.1	1.1	1.1	1.0	1.0	0.9	0.9			
			0.8	0.8	0.8	0.8	0.8	0.7	0.7	0.6	0.5	0.4	0.3	0.1	−0.1	−0.3	−0.5	−0.8			
70-S				296	282	269	254	243	**220**	198	179	161	145	130	117	105	93	83	74	65	57
				1.1	1.1	1.2	1.2	1.2	1.3	1.3	1.4	1.4	1.5	1.5	1.5	1.5	1.5	1.4	1.4	1.4	1.3
				1.0	1.1	1.1	1.1	1.1	1.1	1.1	1.0	0.9	0.9	0.7	0.6	0.5	0.3	0.1	−0.2	−0.4	−0.7
80-S					288	275	260	249	236	228	**209**	189	171	155	140	127	115	103	93	83	74
					1.3	1.4	1.5	1.6	1.6	1.7	1.8	1.8	1.9	1.9	1.9	1.9	2.0	2.0	1.9	1.9	1.9
					1.3	1.4	1.4	1.4	1.4	1.4	1.4	1.4	1.3	1.2	1.1	1.0	0.9	0.7	0.5	0.2	0.0

Bold type — *Capacity governed by stresses, others governed by flexural or shear strength*

HOLLOW-CORE

4'-0" x 12"
Normal Weight Concrete

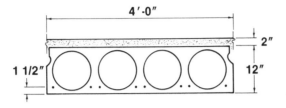

Section Properties

		Untopped	Topped
A	=	262 in.²	—
I	=	4949 in.⁴	7811 in.⁴
Y_b	=	6.00 in.	7.55 in.
Y_t	=	6.00 in.	6.45 in.
Z_b	=	825 in.³	1035 in.³
Z_t	=	825 in.³	1211 in.³
b_w	=	8.00 in.	8.00 in.
wt	=	273 plf	373 plf
		68 psf	93 psf
V/S	=	2.18 in.	

Strand Patterns

Producer may vary size and strength of strands. See "explanation of load tables"

Safe loads shown include dead load of 10 psf for untopped members and 15 psf for topped members. Remainder is live load. Long-time cambers include superimposed dead load but do not include live load.

Capacity of sections of other configurations are similar. For precise values, see local hollow-core manufacturer.

$$f'_c = 5000 \text{ psi}$$
$$f'_{ci} = 3500 \text{ psi}$$

Key

124 — Safe superimposed service load, psf
0.3 — Estimated camber at erection, in.
0.2 — Estimated long-time camber, in.

4HC12

No Topping

Table of safe superimposed service load (psf) and cambers

Strand Designation Code	Span, ft.																						
	28	29	30	31	32	33	34	35	36	37	38	39	40	41	42	43	44	45	46	47	48	49	50
40-S	124	112	101	91	82	74	67	60	54														
	0.3	0.2	0.2	0.2	0.2	0.1	0.1	0.1	0.0														
	0.2	0.2	0.2	0.1	0.0	0.0	−0.1	−0.2	−0.3														
50-S	166	152	138	126	115	105	95	87	79	72	66	59	54										
	0.4	0.4	0.4	0.4	0.4	0.4	0.3	0.3	0.3	0.2	0.2	0.1	0.0										
	0.5	0.4	0.4	0.4	0.3	0.3	0.2	0.1	0.0	−0.1	−0.2	−0.3	−0.5										
60-S	173	167	159	151	144	134	123	113	104	95	88	80	73	66	60	54	48						
	0.6	0.6	0.6	0.6	0.6	0.6	0.6	0.6	0.5	0.5	0.5	0.4	0.3	0.3	0.2	0.1	0.0						
	0.7	0.7	0.7	0.6	0.6	0.6	0.5	0.5	0.4	0.3	0.2	0.1	0.0	−0.2	−0.3	−0.5	−0.7						
70-S	179	173	165	157	150	143	136	130	122	113	**107**	98	90	82	75	69	63	57	51	46			
	0.7	0.7	0.8	0.8	0.8	0.8	0.8	0.8	0.8	0.8	0.8	0.7	0.7	0.6	0.6	0.5	0.4	0.3	0.2	0.1			
	0.9	0.9	0.9	0.9	0.9	0.9	0.9	0.8	0.8	0.7	0.6	0.5	0.4	0.3	0.2	0.0	−0.2	−0.4	−0.6	−0.8			
80-S	185	179	171	163	156	149	142	136	131	125	120	112	105	97	**91**	**84**	77	71	65	59	54	49	44
	0.9	0.9	1.0	1.0	1.0	1.0	1.1	1.1	1.1	1.1	1.1	1.0	1.0	1.0	0.9	0.9	0.8	0.7	0.6	0.5	0.4	0.3	0.1
	1.1	1.1	1.2	1.2	1.2	1.2	1.2	1.2	1.1	1.1	1.0	1.0	0.9	0.8	0.7	0.5	0.4	0.2	0.0	−0.2	−0.4	−0.7	−0.9

4HC12 + 2

2" Normal Weight Topping

Table of safe superimposed service load (psf) and cambers

Strand Designation Code	Span, ft.																						
	23	24	25	26	27	28	29	30	31	32	33	34	35	36	37	38	39	40	41	42	43	44	45
40-S	236	210	188	168	151	135	121	108	96	86	76	**67**	**57**	**47**									
	0.3	0.3	0.3	0.3	0.3	0.3	0.2	0.2	0.2	0.2	0.1	0.1	0.1	0.0									
	0.2	0.2	0.2	0.1	0.1	0.1	0.0	−0.1	−0.1	−0.2	−0.3	−0.4	−0.6	−0.7									
50-S	247	234	219	209	196	182	165	149	135	122	111	**97**	**85**	74	64	55	46						
	0.4	0.4	0.4	0.4	0.4	0.4	0.4	0.4	0.4	0.4	0.4	0.3	0.3	0.3	0.2	0.2	0.1						
	0.3	0.3	0.3	0.3	0.3	0.3	0.2	0.2	0.1	0.0	0.0	−0.1	−0.2	−0.4	−0.5	−0.7	−0.8						
60-S	256	240	228	215	202	194	183	174	167	157	**143**	**128**	**114**	101	90	79	69	60	51				
	0.5	0.5	0.5	0.5	0.6	0.6	0.6	0.6	0.6	0.6	0.6	0.6	0.6	0.5	0.5	0.5	0.4	0.3	0.3				
	0.5	0.5	0.5	0.5	0.5	0.5	0.4	0.4	0.4	0.3	0.2	0.2	0.1	0.0	−0.2	−0.3	−0.5	−0.6	−0.8				
70-S	262	249	234	221	211	200	189	180	173	165	157	149	142	**128**	**115**	103	92	82	72	63	55		
	0.6	0.6	0.6	0.7	0.7	0.7	0.8	0.8	0.8	0.8	0.8	0.8	0.8	0.8	0.8	0.8	0.7	0.7	0.6	0.6	0.5		
	0.6	0.6	0.6	0.7	0.7	0.7	0.6	0.6	0.6	0.6	0.5	0.5	0.4	0.3	0.2	0.1	−0.1	−0.2	−0.4	−0.6	−0.8		
80-S	271	255	240	230	217	206	195	186	176	171	163	155	148	141	135	**127**	**115**	103	93	83	74	65	57
	0.7	0.7	0.8	0.8	0.9	0.9	0.9	1.0	1.0	1.0	1.0	1.1	1.1	1.1	1.1	1.1	1.0	1.0	1.0	0.9	0.9	0.8	0.7
	0.7	0.8	0.8	0.8	0.8	0.9	0.9	0.9	0.8	0.8	0.8	0.7	0.7	0.6	0.5	0.4	0.3	0.2	0.0	−0.1	−0.3	−0.6	−0.8

Bold type — *Capacity governed by stresses, others governed by flexural or shear strength*

HOLLOW-CORE

4'-0" x 12"
Lightweight Concrete

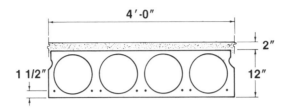

f'_c = 5000 psi
f'_{ci} = 3500 psi

Section Properties

	Untopped	Topped
A =	262 in.²	—
I =	4949 in.⁴	8800 in.⁴
Y_b =	6.00 in.	8.08 in.
Y_t =	6.00 in.	5.92 in.
Z_b =	825 in.³	1089 in.³
Z_t =	825 in.³	1486 in.³
b_w =	8.00 in.	8.00 in.
wt =	209 plf	309 plf
	52 psf	77 psf
V/S =	2.18 in.	

Key

137 — Safe superimposed service load, psf
0.5 — Estimated camber at erection, in.
0.5 — Estimated long-time camber, in.

4LHC12

No Topping

Table of safe superimposed service load (psf) and cambers

Strand Designation Code	Span, ft. 28	29	30	31	32	33	34	35	36	37	38	39	40	41	42	43	44	45	46	47	48	49	50
40-S	137	125	114	104	95	87	80	73	67	61	56	51	46	42									
	0.5	0.5	0.5	0.5	0.5	0.5	0.5	0.4	0.4	0.3	0.3	0.2	0.1	0.0									
	0.5	0.5	0.5	0.5	0.4	0.3	0.3	0.2	0.1	-0.1	-0.2	-0.4	-0.5	-0.7									
50-S	180	165	151	139	128	118	109	100	92	85	79	73	67	62	57	52	47	43					
	0.8	0.8	0.8	0.8	0.8	0.8	0.8	0.8	0.8	0.8	0.7	0.7	0.7	0.6	0.5	0.4	0.4	0.2	0.1				
	0.9	0.9	0.9	0.9	0.8	0.8	0.7	0.7	0.6	0.5	0.4	0.3	0.1	0.0	-0.2	-0.4	-0.6	-0.9					
60-S	186	177	169	164	157	147	136	126	117	**108**	99	92	85	78	72	66	61	56	51	47	43		
	1.0	1.0	1.1	1.1	1.1	1.1	1.2	1.2	1.2	1.2	1.1	1.1	1.1	1.0	1.0	0.9	0.8	0.7	0.6	0.5	0.4		
	1.2	1.2	1.2	1.3	1.3	1.3	1.2	1.2	1.2	1.1	1.0	0.9	0.8	0.7	0.5	0.4	0.2	-0.1	-0.3	-0.6	-0.9		
70-S	192	183	175	167	163	156	149	140	135	126	118	**109**	101	94	87	81	75	69	64	59	54	50	46
	1.2	1.3	1.4	1.4	1.4	1.5	1.5	1.5	1.6	1.6	1.6	1.6	1.6	1.6	1.5	1.5	1.4	1.4	1.3	1.2	1.1	1.0	0.8
	1.5	1.6	1.6	1.7	1.7	1.7	1.7	1.7	1.7	1.7	1.6	1.6	1.5	1.4	1.3	1.1	1.0	0.8	0.6	0.4	0.1	-0.2	-0.5
80-S	198	189	181	173	166	159	152	149	144	135	130	122	115	107	**102**	95	88	82	76	71	66	61	57
	1.5	1.6	1.6	1.7	1.8	1.8	1.9	1.9	2.0	2.0	2.0	2.1	2.1	2.1	2.1	2.1	2.0	2.0	1.9	1.9	1.8	1.7	1.6
	1.9	1.9	2.0	2.1	2.1	2.2	2.2	2.2	2.2	2.2	2.2	2.2	2.2	2.1	2.0	1.9	1.8	1.6	1.5	1.3	1.1	0.8	0.5

4LHC12+2

2" Normal Weight Topping

Table of safe superimposed service load (psf) and cambers

Strand Designation Code	Span, ft. 25	26	27	28	29	30	31	32	33	34	35	36	37	38	39	40	41	42	43	44	45	46	47
40-S	201	181	164	148	134	121	110	99	89	81	73	65	**58**	49									
	0.5	0.5	0.5	0.5	0.5	0.5	0.5	0.5	0.5	0.5	0.4	0.4	0.3	0.3									
	0.4	0.3	0.3	0.3	0.2	0.2	0.1	0.0	-0.1	-0.2	-0.4	-0.5	-0.7	-0.9									
50-S	232	219	210	196	178	163	148	135	124	113	103	94	**83**	74	65	57							
	0.7	0.7	0.7	0.8	0.8	0.8	0.8	0.8	0.8	0.8	0.8	0.8	0.7	0.7	0.7	0.6							
	0.6	0.6	0.6	0.6	0.5	0.5	0.5	0.4	0.3	0.2	0.1	0.0	-0.2	-0.3	-0.5	-0.8							
60-S	238	225	216	204	197	187	177	169	157	144	133	**121**	109	98	88	79	70	62	54				
	0.9	0.9	1.0	1.0	1.0	1.1	1.1	1.1	1.1	1.2	1.2	1.2	1.2	1.1	1.1	1.0	1.0	0.9					
	0.8	0.9	0.9	0.9	0.9	0.8	0.8	0.8	0.7	0.6	0.6	0.5	0.3	0.2	0.0	-0.1	-0.4	-0.6	-0.8				
70-S	244	234	222	210	203	193	183	175	167	162	155	148	**135**	123	111	101	91	82	73	65	58		
	1.1	1.1	1.2	1.2	1.3	1.4	1.4	1.4	1.5	1.5	1.5	1.6	1.6	1.6	1.6	1.6	1.6	1.5	1.5	1.4	1.4		
	1.1	1.1	1.2	1.2	1.2	1.2	1.2	1.1	1.1	1.1	1.0	0.9	0.9	0.7	0.6	0.5	0.3	0.1	-0.2	-0.4	-0.7		
80-S	253	240	228	216	206	199	189	181	173	165	158	151	148	142	**134**	123	112	102	92	84	76	68	61
	1.3	1.3	1.4	1.5	1.6	1.6	1.7	1.8	1.8	1.9	1.9	2.0	2.0	2.0	2.1	2.1	2.1	2.1	2.1	2.0	2.0	1.9	1.9
	1.3	1.4	1.4	1.5	1.5	1.5	1.5	1.5	1.5	1.5	1.5	1.4	1.4	1.3	1.2	1.1	0.9	0.7	0.5	0.3	0.1	-0.2	-0.5

Bold type — *Capacity governed by stresses, others governed by flexural or shear strength*

HOLLOW-CORE SLABS

Fig. 2.4.1 Section properties — normal weight concrete

Trade name: Dy-Core
Licensing Organization: Dyform Engineering Limited, Vancouver, British Columbia

4' - 0'' x 6''

Area (sq. in.)	y_b (in.)	I (in.^{4})	Weight (psf)
178	3	750	46

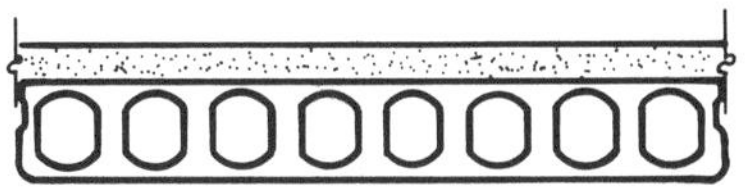

4' - 0'' x 6'' with 2'' Topping

y_b (in.)	Mom. of Inertia (in.^{4})	Weight (psf)
4.18	1614	71

4' - 0'' x 8''

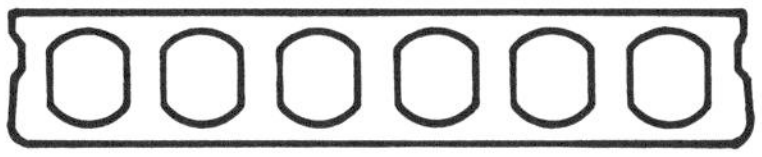

Area (sq. in.)	y_b (in.)	I (in.^{4})	Weight (psf)
193	4	1600	50

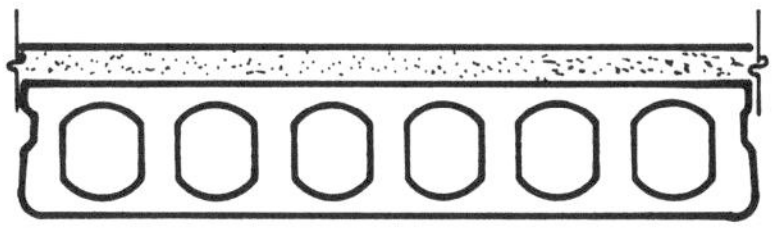

4' - 0'' x 8'' with 2'' Topping

y_b (in.)	Mom. of Inertia (in.^{4})	Weight (psf)
5.39	2967	75

4' - 0'' x 12''

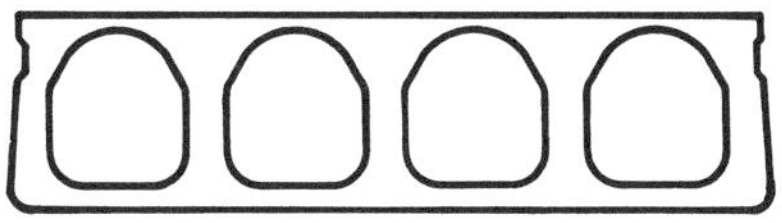

Area (sq. in.)	y_b (in.)	I (in.^{4})	Weight (psf)
265	6.28	4900	69

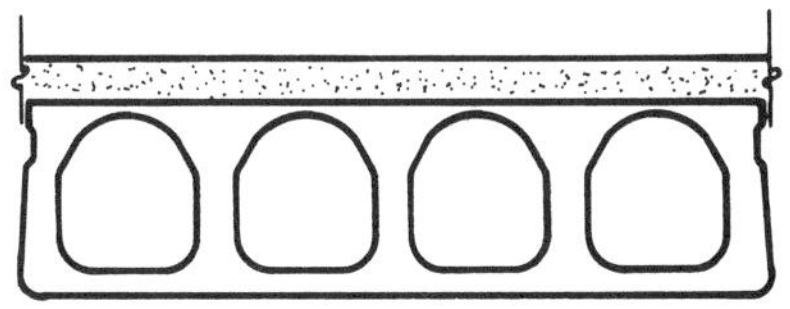

4' - 0'' x 12'' with 2'' Topping

y_b (in.)	Mom. of Inertia (in.^{4})	Weight (psf)
7.75	7547	94

Note: All sections not available from all producers. Check availability with local manufacturers.

HOLLOW-CORE SLABS

Fig. 2.4.2 Section properties — normal weight concrete

Dynaspan

Trade name: Dynaspan®

Equipment Manufacturers: Hastings Dynamold Corp., Hastings, Nebraska

4' - 0'' x 4''

4' - 0'' x 4'' with 2'' Topping

Area (sq in.)	y_b (in.)	I (in.⁴)	Weight (psf)	y_b (in.)	Mom. of Inertia (in.⁴)	Weight (psf)
133	2.00	235	35	3.08	689	60

4' - 0'' x 6''

4' - 0'' x 6'' with 2'' Topping

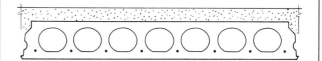

Area (sq in.)	y_b (in.)	I (in.⁴)	Weight (psf)	y_b (in.)	Mom. of Inertia (in.⁴)	Weight (psf)
165	3.02	706	43	4.25	1543	68

4' - 0'' x 8''

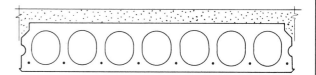

4' - 0'' x 8'' with 2'' Topping

Area (sq in.)	y_b (in.)	I (in.⁴)	Weight (psf)	y_b (in.)	Mom. of Inertia (in.⁴)	Weight (psf)
233	3.93	1731	61	5.16	3205	86

4' - 0'' x 10''

4' - 0'' x 10'' with 2'' Topping

Area (sq in.)	y_b (in.)	I (in.⁴)	Weight (psf)	y_b (in.)	Mom. of Inertia (in.⁴)	Weight (psf)
260	4.91	3145	68	6.26	5314	93

Note: All sections not available from all producers. Check availability with local manufacturers.

HOLLOW-CORE SLABS

Fig. 2.4.2 Section properties — normal weight concrete

Trade name: Dynaspan®
Equipment Manufacturers: Hastings Dynamold Corp., Hastings, Nebraska

8' - 0" x 6"		8' - 0" x 6" with 2" Topping	

Area (sq in.)	y_b (in.)	I (in.⁴)	Weight (psf)	y_b (in.)	Mom. of Inertia (in.⁴)	Weight (psf)
338	3.05	1445	44	4.26	3106	69

8' - 0" x 8"		8' - 0" x 8" with 2" Topping	

Area (sq in.)	y_b (in.)	I (in.⁴)	Weight (psf)	y_b (in.)	Mom. of Inertia (in.⁴)	Weight (psf)
470	3.96	3525	61	5.17	6444	86

8' - 0" x 10"		8' - 0" x 10" with 2" Topping	

Area (sq in.)	y_b (in.)	I (in.⁴)	Weight (psf)	y_b (in.)	Mom. of Inertia (in.⁴)	Weight (psf)
532	4.96	6422	69	6.28	10,712	94

8' - 0" x 12"		8' - 0" x 12" with 2" Topping	

Area (sq in.)	y_b (in.)	I (in.⁴)	Weight (psf)	y_b (in.)	Mom. of Inertia (in.⁴)	Weight (psf)
615	5.95	10,505	80	7.32	16,507	105

Note: All sections not available from all producers. Check availability with local manufacturers.

HOLLOW-CORE SLABS

Fig. 2.4.3 Section properties — normal weight concrete

Flexicore

Trade name: Flexicore®

Licensing Organization: The Flexicore Co. Inc., Dayton, Ohio

1' - 4'' x 6''

1' - 4'' x 6'' with 2'' Topping

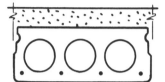

Area (sq in.)	y_b (in.)	I (in.^{4})	Weight (psf)	y_b (in.)	Mom. of Inertia (in.^{4})	Weight (psf)
55	3.00	243	43	4.23	523	68

2' - 0'' x 6''

2' - 0'' x 6'' with 2'' Topping

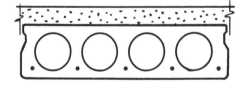

Area (sq in.)	y_b (in.)	I (in.^{4})	Weight (psf)	y_b (in.)	Mom. of Inertia (in.^{4})	Weight (psf)
86	3.00	366	45	4.20	793	70

1' - 4'' x 8''

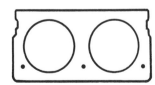

1' - 4'' x 8'' with 2'' Topping

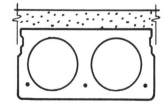

Area (sq in.)	y_b (in.)	I (in.^{4})	Weight (psf)	y_b (in.)	Mom. of Inertia (in.^{4})	Weight (psf)
73	4.00	560	57	5.26	1028	82

Note: All sections not available from all producers. Check availability with local manufacturers.

HOLLOW-CORE SLABS

Fig. 2.4.3 Section properties — normal weight concrete

Trade name: Flexicore®
Licensing Organization: The Flexicore Co. Inc., Dayton, Ohio

2' - 0'' x 8''

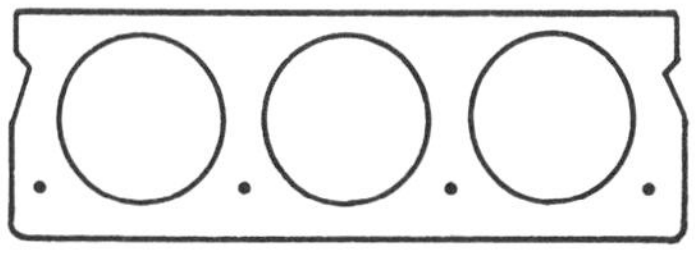

Area (sq in.)	y_b (in.)	I (in.⁴)	Weight (psf)
110	4.00	843	57

2' - 0'' x 8'' with 2'' Topping

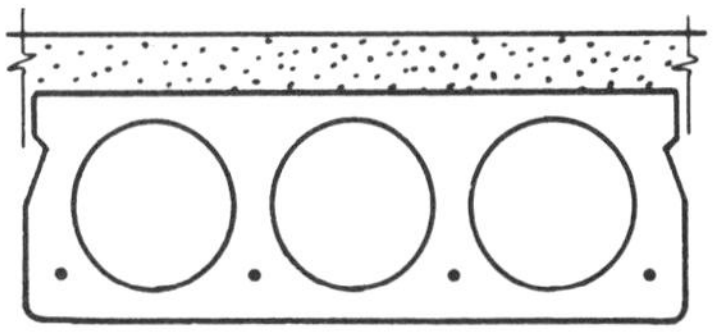

y_b (in.)	Mom. of Inertia (in.⁴)	Weight (psf)
5.26	1547	82

1' - 8'' x 10''

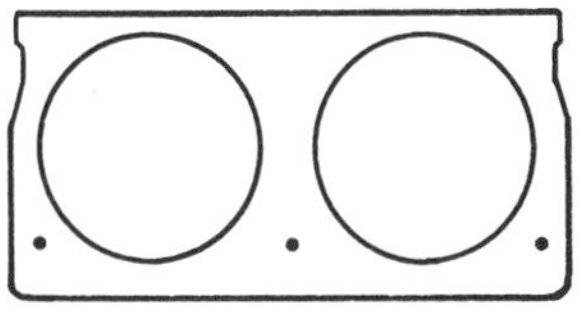

Area (sq in.)	y_b (in.)	I (in.⁴)	Weight (psf)
98	5.00	1254	61

1' - 8'' x 10'' with 2'' Topping

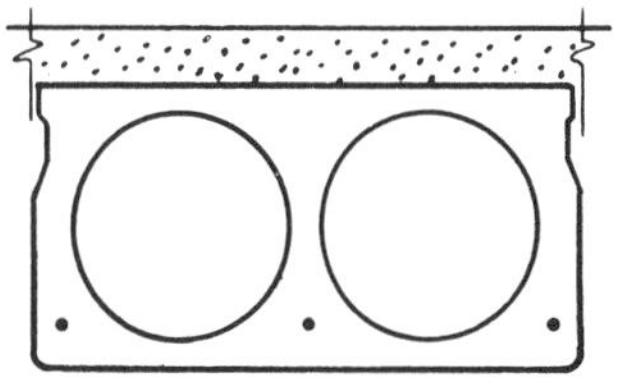

y_b (in.)	Mom. of Inertia (in.⁴)	Weight (psf)
6.43	2109	86

2' - 0'' x 10''

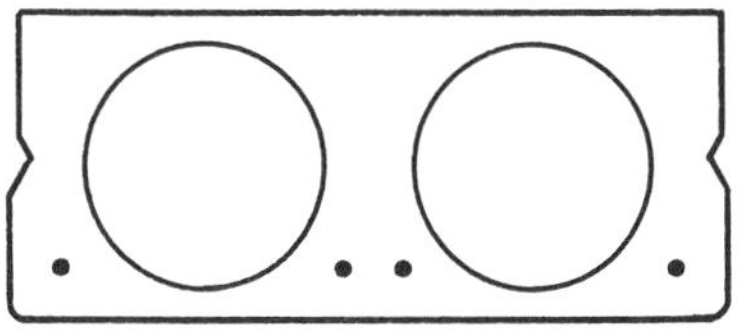

Area (sq in.)	y_b (in.)	I (in.⁴)	Weight (psf)
138	5.00	1587	72

2' - 0'' x 10'' with 2'' Topping

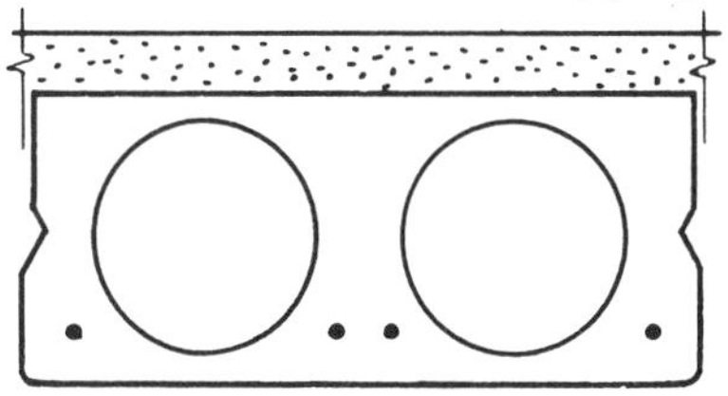

y_b (in.)	Mom. of Inertia (in.⁴)	Weight (psf)
6.27	2651	97

2' - 0'' x 12''

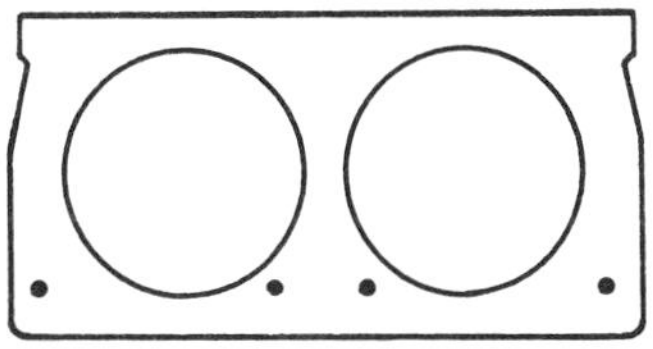

Area (sq in.)	y_b (in.)	I (in.⁴)	Weight (psf)
141	6.00	2595	73

2' - 0'' x 12'' with 2'' Topping

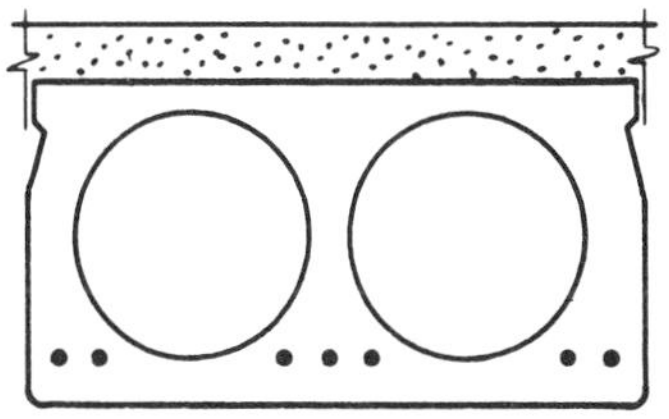

y_b (in.)	Mom. of Inertia (in.⁴)	Weight (psf)
7.46	4049	98

Note: All sections not available from all producers. Check availability with local manufacturers.

HOLLOW-CORE SLABS

Fig. 2.4.4   Section properties — normal weight concrete

Trade name: Spancrete®

Licensing Organization: Spancrete Industries, Inc., Waukesha, Wisconsin

3' - 4'' x 4''

3' - 4'' x 4'' with 2'' Topping

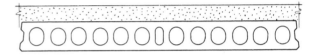

Area (sq in.)	y_b (in.)	I (in.^{4})	Weight (psf)	y_b (in.)	Mom. of Inertia (in.^{4})	Weight (psf)
119	2.01	199	37	3.20	659	62

3' - 4'' x 6''

3' - 4'' x 6'' with 2'' Topping

Area (sq in.)	y_b (in.)	I (in.^{4})	Weight (psf)	y_b (in.)	Mom. of Inertia (in.^{4})	Weight (psf)
160	2.92	635	50	4.28	1585	75

Note: All sections not available from all producers. Check availability with local manufacturers.

HOLLOW-CORE SLABS

Fig. 2.4.4 Section properties — normal weight concrete

Trade name: Spancrete®

Licensing Organization: Spancrete Industries, Inc., Waukesha, Wisconsin

3' - 4'' x 8''

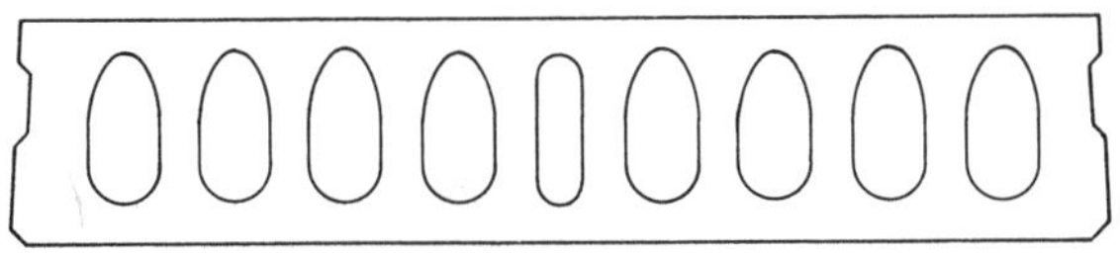

3' - 4'' x 8'' with 2'' Topping

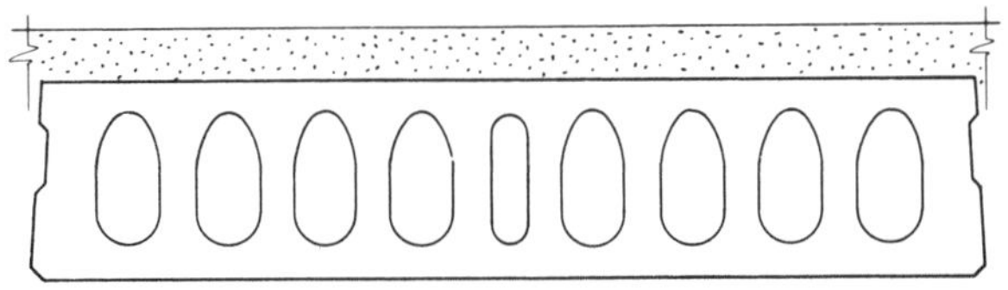

Area (sq in.)	y_b (in.)	I (in.^{4})	Weight (psf)	y_b (in.)	Mom. of Inertia (in.^{4})	Weight (psf)
218	3.98	1515	68	5.33	3000	93

3' - 4'' x 10''

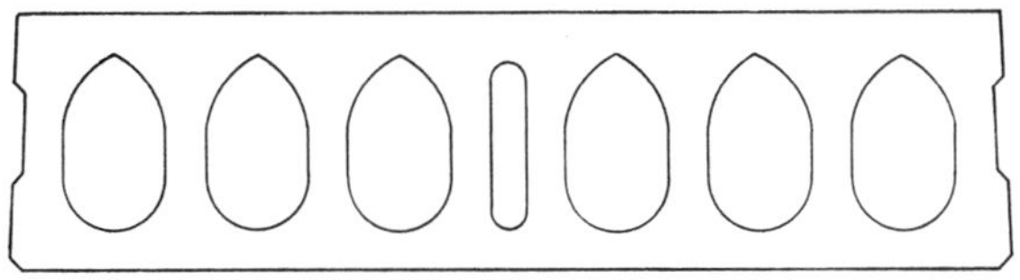

3' - 4'' x 10'' with 2'' Topping

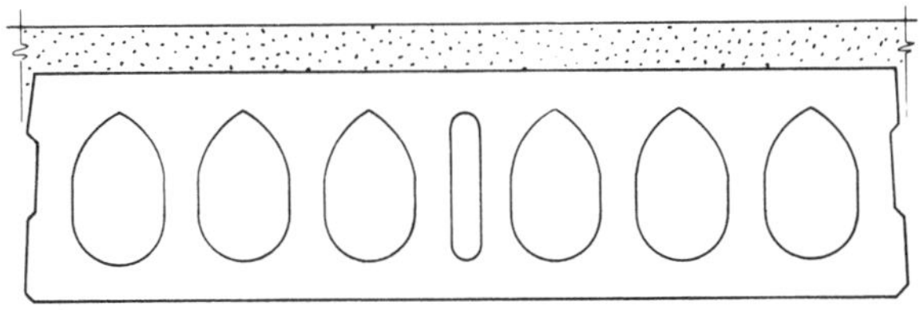

Area (sq in.)	y_b (in.)	I (in.^{4})	Weight (psf)	y_b (in.)	Mom. of Inertia (in.^{4})	Weight (psf)
257	5.19	2933	80	6.56	5015	105

3' - 4'' x 12''

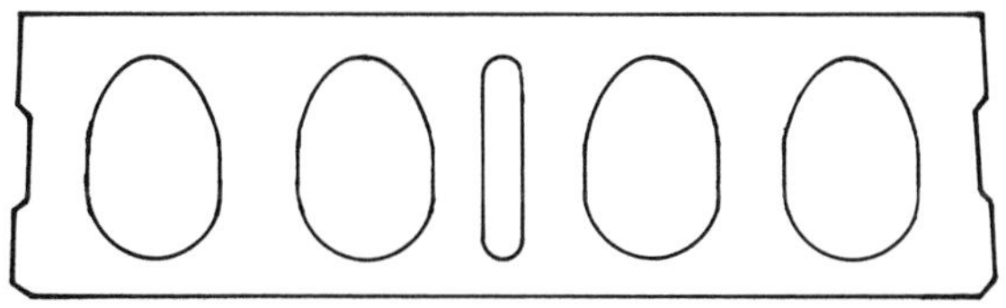

3' - 4'' x 12'' with 2'' Topping

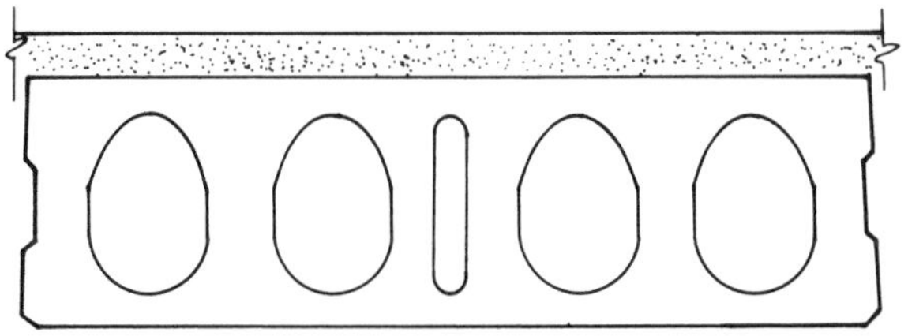

Area (sq in.)	y_b (in.)	I (in.^{4})	Weight (psf)	y_b (in.)	Mom. of Inertia (in.^{4})	Weight (psf)
325	6.20	4981	102	7.54	7976	127

Note: All sections not available from all producers. Check availability with local manufacturers.

HOLLOW-CORE SLABS

Fig. 2.4.5 Section properties

Trade name: Span-Deck®

Licensing Organization: Span-Deck Incorporated, Franklin, Tennessee

4' - 0'' x 8'' Textured Soffit

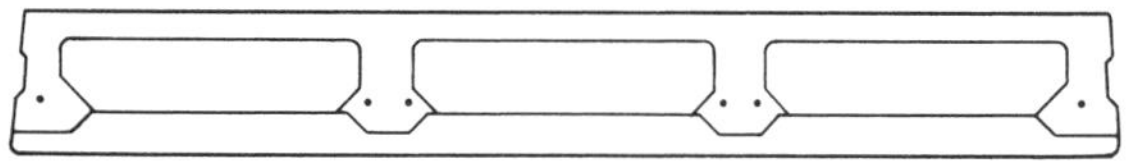

4' - 0'' x 8'' Textured Soffit with 2'' Topping

Area (sq in.)	y_b (in.)	I (in.⁴)	Weight (psf)	y_b (in.)	Mom. of Inertia (in.⁴)	Weight (psf)
144*	4.70	1033*	45	6.16	1965	70

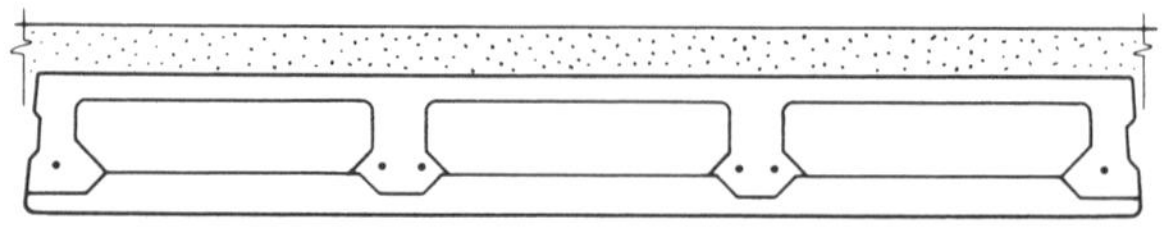

4' - 0'' x 8'' Smooth Soffit

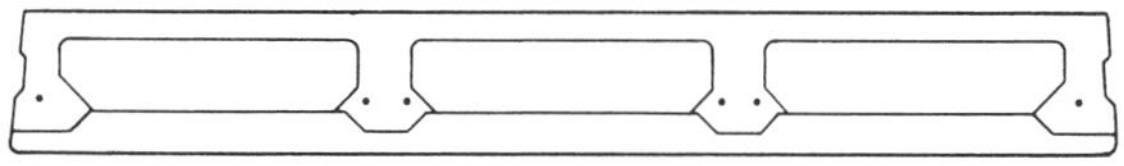

4' - 0'' x 8'' Smooth Soffit with 2'' Topping

Area (sq in.)	y_b (in.)	I (in.⁴)	Weight (psf)	y_b (in.)	Mom. of Inertia (in.⁴)	Weight (psf)
173	3.90	1527	45	5.43	2904	70

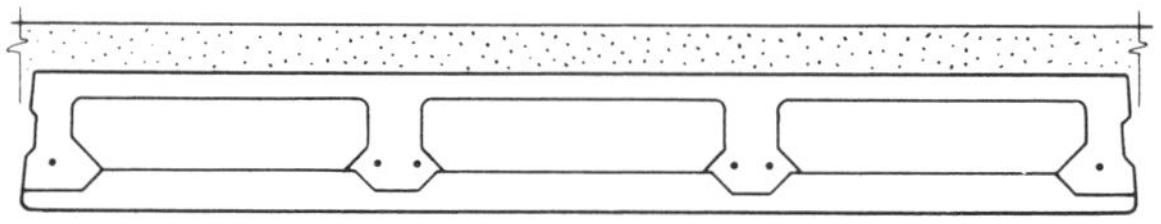

4' - 0'' x 12'' Textured Soffit

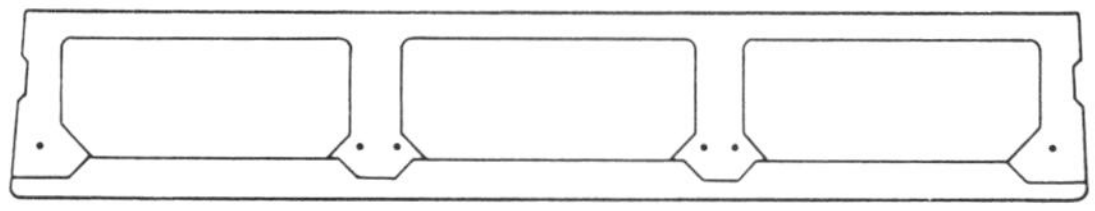

4' - 0'' x 12'' Textured Soffit with 2'' Topping

Area (sq in.)	y_b (in.)	I (in.⁴)	Weight (psf)	y_b (in.)	Mom. of Inertia (in.⁴)	Weight (psf)
177	6.90	3170	54	8.70	5143	79

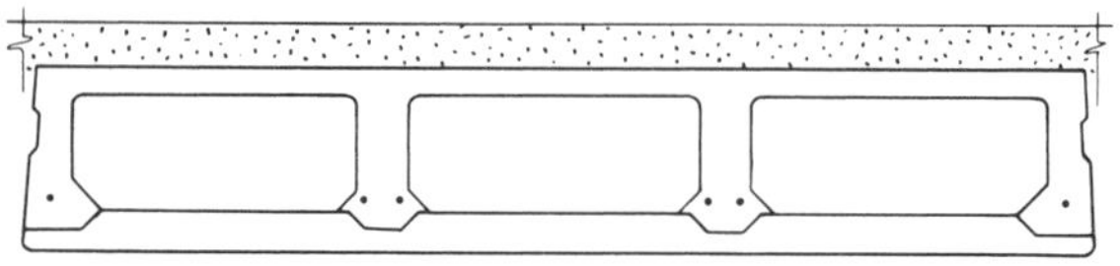

4' - 0'' x 12'' Smooth Soffit

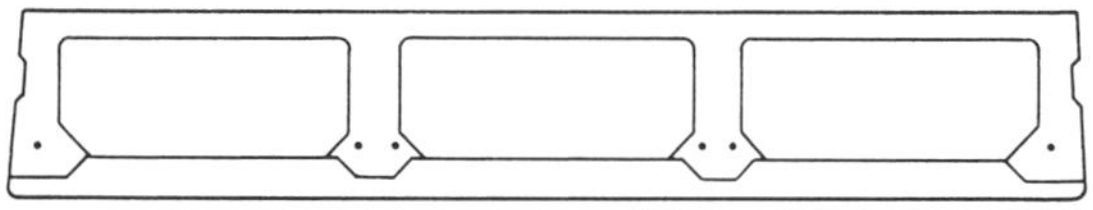

4' - 0'' x 12'' Smooth Soffit with 2'' Topping

Area (sq in.)	y_b (in.)	I (in.⁴)	Weight (psf)	y_b (in.)	Mom. of Inertia (in.⁴)	Weight (psf)
205	5.83	4186	53	7.74	7016	78

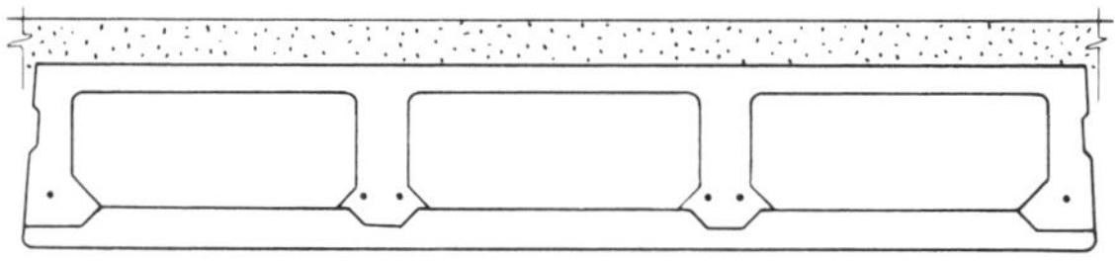

*Section properties for lightweight soffit sections are calculated using a method in which the bottom soffit is transformed into an effective width having the same properties of the top concrete.

Effective width = $\dfrac{E_c \text{ (soffit)}}{E_c \text{ (top)}}$ (Actual width) = 19.7 in. for 4' - 0'' wide sections, 39.5 in. for 8' - 0'' wide sections.

Note: All sections not available from all producers. Check availability with local manufacturers.

6'' and 10'' depths are of similar configuration.

HOLLOW-CORE SLABS

Fig. 2.4.5 Section properties

Trade name: Span-Deck®
Licensing Organization: Span-Deck, Incorporated, Franklin, Tennessee

8' - 0" x 8" Textured Soffit

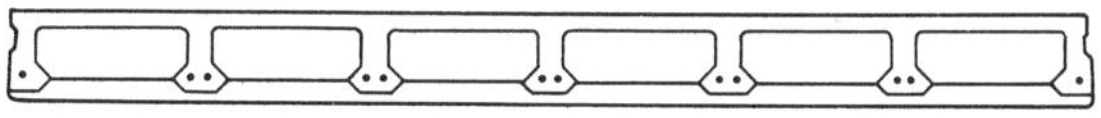

8' - 0" x 8" Textured Soffit with 2" Topping

Area (sq in.)	y_b (in.)	I (in.^{4})	Weight (psf)	y_b (in.)	Mom. of Inertia (in.^{4})	Weight (psf)
284	4.71	2088	45	6.18	3934	70

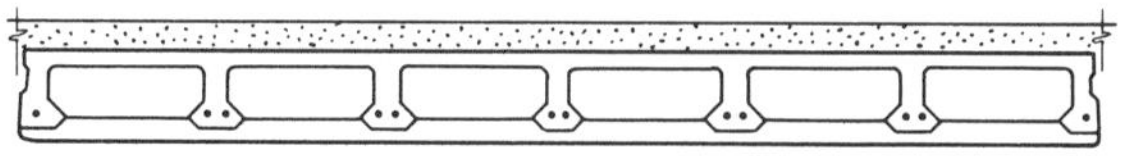

8' - 0" x 8" Smooth Soffit

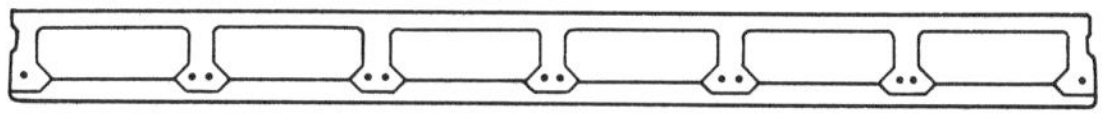

8' - 0" x 8" Smooth Soffit with 2" Topping

Area (sq in.)	y_b (in.)	I (in.^{4})	Weight (psf)	y_b (in.)	Mom. of Inertia (in.^{4})	Weight (psf)
349	3.90	3063	45	5.42	5825	70

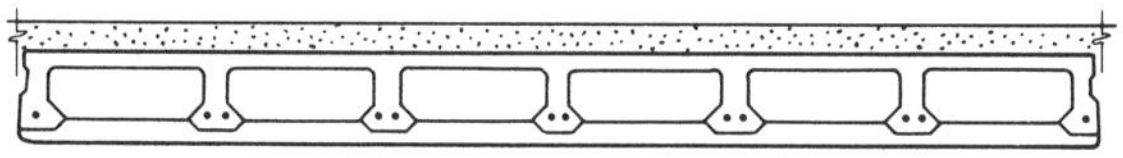

8' - 0" x 12" Textured Soffit

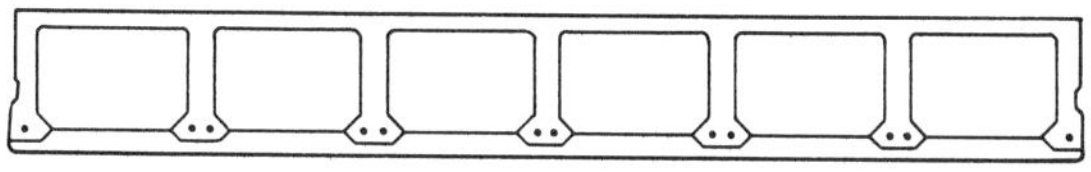

8' - 0" x 12" Textured Soffit with 2" Topping

Area (sq in.)	y_b (in.)	I (in.^{4})	Weight (psf)	y_b (in.)	Mom. of Inertia (in.^{4})	Weight (psf)
350	6.84	6135	54	8.68	10,145	79

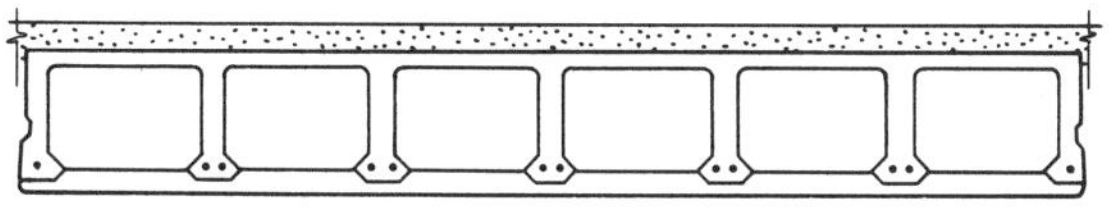

8' - 0" x 12" Smooth Soffit

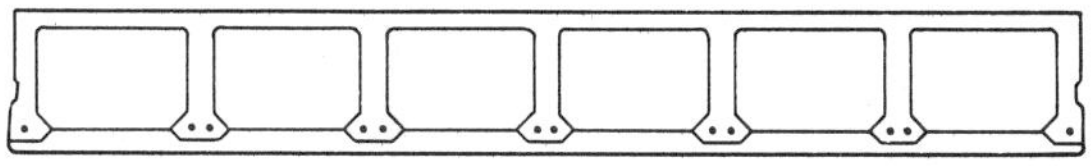

8' - 0" x 12" Smooth Soffit with 2" Topping

Area (sq in.)	y_b (in.)	I (in.^{4})	Weight (psf)	y_b (in.)	Mom. of Inertia (in.^{4})	Weight (psf)
415	5.82	8419	54	7.71	14,113	79

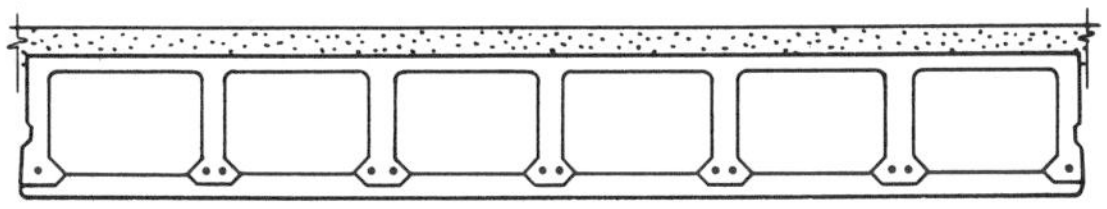

Note: All sections not available from all producers. Check availability with local manufacturers.
6" and 10" depths are of similar configuration.

HOLLOW-CORE SLABS

Fig. 2.4.6   Section properties — normal weight concrete

Trade name: Spiroll,® Corefloor
Licensing Organization: Spiroll Corporation Limited, Winnipeg, Manitoba

4' - 0'' x 4''

4' - 0'' x 4'' with 2'' Topping

Area (sq in.)	y_b (in.)	I (in.^{4})	Weight (psf)	y_b (in.)	Mom. of Inertia (in.^{4})	Weight (psf)
154	2.00	247	40	2.98	723	65

4' - 0'' x 6''

4' - 0'' x 6'' with 2'' Topping

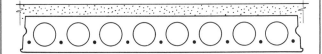

Area (sq in.)	y_b (in.)	I (in.^{4})	Weight (psf)	y_b (in.)	Mom. of Inertia (in.^{4})	Weight (psf)
188	3.00	764	49	4.13	1641	74

4' - 0'' x 8''

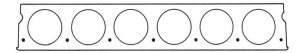

4' - 0'' x 8'' with 2'' Topping

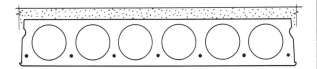

Area (sq in.)	y_b (in.)	I (in.^{4})	Weight (psf)	y_b (in.)	Mom. of Inertia (in.^{4})	Weight (psf)
214	4.00	1666	56	5.29	3070	81

Note: All sections not available from all producers. Check availability with local manufacturers.

HOLLOW-CORE SLABS

Fig. 2.4.6 Section properties — normal weight concrete

Trade name: Spiroll,® Corefloor

Licensing Organization: Spiroll Corporation Limited, Winnipeg, Manitoba

4' - 0'' x 10''

4' - 0'' x 10'' with 2'' Topping

Area (sq in.)	y_b (in.)	I (in.^{4})	Weight (psf)	y_b (in.)	Mom. of Inertia (in.^{4})	Weight (psf)
259	5.00	3223	67	6.34	5328	92

4' - 0'' x 12''

4' x 0'' x 12'' with 2'' Topping

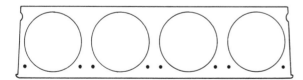
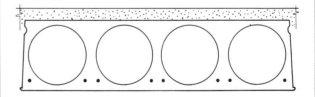

Area (sq in.)	y_b (in.)	I (in.^{4})	Weight (psf)	y_b (in.)	Mom. of Inertia (in.^{4})	Weight (psf)
289	6.00	5272	75	7.43	8195	100

Note: All sections not available from all producers. Check availability with local manufacturers.

SOLID FLAT SLAB

4″ Thick
Normal Weight Concrete

Strand Patterns

Producer may vary size and strength of strands. See "explanation of load tables"

Safe loads shown include dead load of 10 psf for untopped members and 15 psf for topped members. Remainder is live load. Long-time cambers include superimposed dead load but do not include live load.

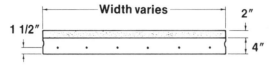

Width varies — 2″

1 1/2″ | 4″

$f'_c = 5000$ psi
$f'_{ci} = 3500$ psi

Section Properties per foot of width

		Untopped	Topped
A	=	48.00 in.²	—
I	=	64.00 in.⁴	190.75 in.⁴
Y_b	=	2.00 in.	2.84 in.
Y_t	=	2.00 in.	3.16 in.
Z_b	=	32.00 in.³	67.18 in.³
Z_t	=	32.00 in.³	60.38 in.³
b_w	=	12.00 in.	12.00 in.
wt	=	50 psf	75 psf
V/S	=	2.00 in.	

Key

168 — Safe superimposed service load, psf
0.0 — Estimated camber at erection, in.
0.0 — Estimated long-time camber, in.

FS4

Table of safe superimposed service load (psf) and cambers

No Topping

Strand Designation Code	Span, ft.							
	10	11	12	13	14	15	16	17
30-S	**168** 0.0 0.0	**130** 0.0 0.0	**101** 0.0 −0.1	79 0.0 −0.1				
40-S	195 0.1 0.1	**163** 0.1 0.1	**131** 0.1 0.0	**104** 0.0 0.0	83 0.0 −0.1			
50-S	245 0.1 0.1	200 0.1 0.1	**160** 0.1 0.1	**129** 0.1 0.1	**105** 0.1 0.0	85 0.0 −0.1	68 0.0 −0.3	
60-S	291 0.2 0.2	234 0.2 0.2	190 0.2 0.2	**154** 0.2 0.2	**126** 0.2 0.1	**103** 0.1 0.0	85 0.1 −0.1	69 0.0 −0.3

FS4 + 2

Table of safe superimposed service load (psf) and cambers

2″ Normal Weight Topping

Strand Designation Code	Span, ft.							
	10	11	12	13	14	15	16	17
30-S	**300** 0.0 0.0	**220** 0.0 −0.1	**160** 0.0 −0.1	**113** 0.0 −0.2				
40-S		**294** 0.1 0.0	**222** 0.1 −0.1	**166** 0.0 −0.1	**122** 0.0 −0.3			
50-S		284 0.1 0.0	219 0.1 0.0	**167** 0.1 −0.1	**125** 0.0 −0.3	91 0.0 −0.5		
60-S		271 0.2 0.0	212 0.2 0.0	165 0.1 −0.2	**126** 0.1 −0.3	93 0.0 −0.5		

Bold type — *Capacity governed by stresses, others governed by flexural or shear strength*

SOLID FLAT SLAB

4″ Thick
Lightweight Concrete

Strand Patterns

Producer may vary size and
strength of strands. See
"explanation of load tables"

Safe loads shown include dead load
of 10 psf for untopped members and
15 psf for topped members. Remainder
is live load. Long-time cambers include
superimposed dead load but do not include live load.

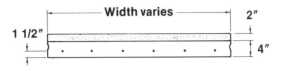

$f'_c = 5000$ psi
$f'_{ci} = 3500$ psi

Section Properties per foot of width		
	Untopped	Topped
A =	48.00 in.²	—
I =	64.00 in.⁴	231.25 in.⁴
Y_b =	2.00 in.	3.10 in.
Y_t =	2.00 in.	2.90 in.
Z_b =	32.00 in.³	74.60 in.³
Z_t =	32.00 in.³	79.75 in.³
b_w =	12.00 in.	12.00 in.
wt =	38 psf	63 psf
V/S =	2.00 in.	

Key

175 — Safe superimposed service load, psf
0.1 — Estimated camber at erection, in.
0.1 — Estimated long-time camber, in.

LFS4

No Topping

Table of safe superimposed service load (psf) and cambers

Strand Designation Code	Span, ft.							
	10	11	12	13	14	15	16	17
30-S	**175** 0.1 0.1	**138** 0.1 0.1	**110** 0.1 0.0	**88** 0.0 −0.1				
40-S	**201** 0.2 0.2	**169** 0.2 0.2	**138** 0.2 0.1	**112** 0.2 0.1	91 0.1 0.0	75 0.1 −0.1	61 0.0 −0.3	
50-S	**250** 0.2 0.3	**205** 0.2 0.3	**166** 0.3 0.3	**136** 0.3 0.2	112 0.3 0.2	93 0.2 0.1	77 0.2 −0.1	64 0.1 −0.3
60-S	**297** 0.3 0.4	**239** 0.3 0.4	**195** 0.4 0.4	**160** 0.4 0.4	133 0.4 0.3	111 0.4 0.3	93 0.3 0.2	78 0.3 0.0

LFS4+2

2″ Normal Weight Topping

Table of safe superimposed service load (psf) and cambers

Strand Designation Code	Span, ft.							
	11	12	13	14	15	16	17	18
30-S	**263** 0.1 0.0	**197** 0.1 −0.1	**146** 0.0 −0.2	**106** 0.0 −0.4				
40-S		**263** 0.2 0.0	**203** 0.2 −0.1	**154** 0.1 −0.2	**115** 0.1 −0.4	84 0.0 −0.6		
50-S			**259** 0.3 0.1	**203** 0.3 −0.1	**158** 0.2 −0.2	**121** 0.2 −0.4	90 0.1 −0.7	
60-S				**251** 0.4 0.1	**200** 0.4 0.0	**158** 0.3 −0.2	**123** 0.3 −0.4	94 0.2 −0.7

Bold type — *Capacity governed by stresses, others governed by flexural or shear strength*

SOLID FLAT SLAB

6″ Thick
Normal Weight Concrete

Strand Pattern Designation

Strand Patterns

Producer may vary size and strength of strands. See "explanation of load tables"

Safe loads shown include dead load of 10 psf for untopped members and 15 psf for topped members. Remainder is live load. Long-time cambers include superimposed dead load but do not include live load.

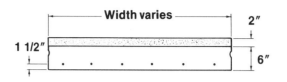

$f'_c = 5000$ psi
$f'_{ci} = 3500$ psi

Section Properties per foot of width

	Untopped	Topped
A =	72.00 in.²	—
I =	216.00 in.⁴	458.50 in.⁴
Y_b =	3.00 in.	3.82 in.
Y_t =	3.00 in.	4.18 in.
Z_b =	72.00 in.³	120.03 in.³
Z_t =	72.00 in.³	109.70 in.³
b_w =	12.00 in.	12.00 in.
wt =	75 psf	100 psf
V/S =	3.00 in.	

Key

282 — Safe superimposed service load, psf
0.1 — Estimated camber at erection, in.
0.1 — Estimated long-time camber, in.

FS6

No Topping

Table of safe superimposed service load (psf) and cambers

Strand Designation Code	Span, ft.														
	11	12	13	14	15	16	17	18	19	20	21	22	23	24	25
30-S	282	241	196	161	133	109	90	72							
	0.1	0.1	0.1	0.1	0.1	0.0	0.0	0.0							
	0.1	0.1	0.1	0.1	0.0	0.0	−0.1	−0.2							
40-S		279	240	209	177	147	122	100	82	67					
		0.1	0.2	0.2	0.2	0.1	0.1	0.1	0.0	0.0					
		0.2	0.2	0.2	0.1	0.1	0.1	0.0	−0.1	−0.2					
50-S			261	218	182	153	128	107	90	74	61				
			0.2	0.2	0.2	0.2	0.2	0.2	0.1	0.1	0.0				
			0.3	0.3	0.3	0.2	0.2	0.1	0.0	−0.1	−0.2				
60-S					258	218	184	156	133	112	95	80	67	55	
					0.3	0.3	0.3	0.3	0.3	0.3	0.3	0.2	0.1	0.0	
					0.4	0.4	0.4	0.4	0.3	0.2	0.1	0.0	−0.1	−0.3	
70-S					299	253	216	184	158	135	116	99	84	71	60
					0.4	0.4	0.5	0.5	0.5	0.5	0.4	0.4	0.3	0.3	0.2
					0.5	0.5	0.5	0.5	0.5	0.4	0.4	0.3	0.1	0.0	−0.2

FS6 + 2

2″ Normal Weight Topping

Table of safe superimposed service load (psf) and cambers

Strand Designation Code	Span, ft.												
	13	14	15	16	17	18	19	20	21	22	23	24	25
30-S	286	236	186	144	108	78							
	0.1	0.1	0.1	0.0	0.0	0.0							
	0.0	0.0	0.0	−0.1	−0.2	−0.3							
40-S		254	203	161	125	96	70						
		0.2	0.1	0.1	0.1	0.0	0.0						
		0.1	0.0	0.0	−0.1	−0.2	−0.4						
50-S			262	213	172	138	108	82	60				
			0.2	0.2	0.2	0.2	0.1	0.1	0.0				
			0.2	0.1	0.0	−0.1	−0.2	−0.3	−0.5				
60-S			266	219	180	146	117	92	70	50			
			0.3	0.3	0.3	0.3	0.3	0.2	0.1	0.0			
			0.2	0.2	0.1	0.0	−0.1	−0.3	−0.5	−0.7			
70-S				266	221	184	151	123	98	77	58		
				0.5	0.5	0.5	0.4	0.4	0.3	0.3	0.2		
				0.3	0.3	0.2	0.1	0.0	−0.2	−0.4	−0.7		

Bold type — *Capacity governed by stresses, others governed by flexural or shear strength*

SOLID FLAT SLAB

6″ Thick
Lightweight Concrete

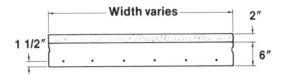

Strand Patterns

Producer may vary size and strength of strands. See "explanation of load tables"

Safe loads shown include dead load of 10 psf for untopped members and 15 psf for topped members. Remainder is live load. Long-time cambers include superimposed dead load but do not include live load.

$f'_c = 5000$ psi
$f'_{ci} = 3500$ psi

Section Properties per foot of width

	Untopped	Topped
A =	72.00 in.²	—
I =	216.00 in.⁴	545.25 in.⁴
Y_b =	3.00 in.	4.11 in.
Y_t =	3.00 in.	3.89 in.
Z_b =	72.00 in.³	132.68 in.³
Z_t =	72.00 in.³	140.18 in.³
b_w =	12.00 in.	12.00 in.
wt =	58 psf	83 psf
V/S =	3.00 in.	

Key

291 — Safe superimposed service load, psf
0.1 — Estimated camber at erection, in.
0.2 — Estimated long-time camber, in.

LFS6 — No Topping

Table of safe superimposed service load (psf) and cambers

Strand Designation Code	11	12	13	14	15	16	17	18	19	20	21	22	23	24	25
30-S	291	252	211	176	147	124	104	86	72						
	0.1	0.2	0.2	0.2	0.2	0.2	0.1	0.1	0.0						
	0.2	0.2	0.2	0.1	0.1	0.1	0.0	0.0	-0.1						
40-S		287	250	220	189	159	134	113	96	81	68	57	47		
		0.2	0.3	0.3	0.3	0.3	0.3	0.3	0.3	0.2	0.1	0.1	0.0		
		0.3	0.3	0.3	0.3	0.3	0.3	0.2	0.1	0.0	-0.1	-0.3	-0.5		
50-S				270	228	193	164	140	120	103	88	75	64	54	45
				0.4	0.4	0.5	0.5	0.5	0.5	0.4	0.4	0.3	0.3	0.2	0.0
				0.5	0.5	0.5	0.5	0.5	0.4	0.4	0.3	0.1	-0.1	-0.3	-0.5
60-S				266	227	195	167	144	125	108	93	80	69	59	
				0.6	0.6	0.6	0.7	0.7	0.7	0.7	0.6	0.6	0.6	0.5	0.4
				0.7	0.7	0.7	0.7	0.7	0.7	0.6	0.5	0.4	0.2	0.0	
70-S					261	225	194	169	147	128	111	97	84	73	
					0.8	0.8	0.9	0.9	0.9	0.9	0.9	0.9	0.8	0.8	
					0.9	1.0	1.0	1.0	1.0	1.0	0.9	0.8	0.6	0.5	

LFS6 + 2 — 2″ Normal Weight Topping

Table of safe superimposed service load (psf) and cambers

Strand Designation Code	14	15	16	17	18	19	20	21	22	23	24	25	26
30-S	250	210	176	145	113	86	63						
	0.2	0.2	0.2	0.1	0.1	0.0	0.0						
	0.1	0.0	-0.1	-0.2	-0.4	-0.5							
40-S		277	246	201	163	131	103	79	59	41			
		0.3	0.3	0.3	0.3	0.3	0.2	0.1	0.1	0.0			
		0.2	0.2	0.1	0.0	-0.1	-0.2	-0.4	-0.6	-0.9			
50-S			257	213	175	143	116	92	71	53			
			0.5	0.5	0.5	0.4	0.4	0.3	0.3	0.2			
			0.3	0.3	0.2	0.0	-0.1	-0.3	-0.5	-0.8			
60-S				262	220	184	152	125	102	81	63		
				0.7	0.7	0.7	0.7	0.6	0.6	0.5	0.4		
				0.5	0.4	0.3	0.2	0.1	-0.1	-0.4	-0.7		
70-S					265	224	189	159	132	109	89	70	
					0.9	0.9	0.9	0.9	0.9	0.8	0.8	0.7	
					0.7	0.6	0.5	0.4	0.2	0.0	-0.2	-0.6	

Bold type — *Capacity governed by stresses, others governed by flexural or shear strength*

SOLID FLAT SLAB

8″ Thick
Normal Weight Concrete

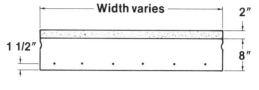

Section Properties per foot of width

	Untopped	Topped
A =	96.00 in.²	—
I =	512.00 in.⁴	907.50 in.⁴
Y_b =	4.00 in.	4.81 in.
Y_t =	4.00 in.	5.19 in.
Z_b =	128.00 in.³	188.68 in.³
Z_t =	128.00 in.³	174.85 in.³
b_w =	12.00 in.	12.00 in.
wt =	100 psf	125 psf
V/S =	4.00 in.	

Strand Patterns

Producer may vary size and strength of strands. See "explanation of load tables"

Safe loads shown include dead load of 10 psf for untopped members and 15 psf for topped members. Remainder is live load. Long-time cambers include superimposed dead load but do not include live load.

f'_c = 5000 psi
f'_{ci} = 3500 psi

Key

248 — Safe superimposed service load, psf
0.1 — Estimated camber at erection, in.
0.1 — Estimated long-time camber, in.

FS8

No Topping

Table of safe superimposed service load (psf) and cambers

Strand Designation Code	Span, ft.															
	14	15	16	17	18	19	20	21	22	23	24	25	26	27	28	29
30-S	248	206	171	142	118	98	80									
	0.1	0.1	0.1	0.1	0.0	0.0	0.0									
	0.1	0.1	0.1	0.0	0.0	−0.1	−0.1									
40-S		280	248	211	179	153	130	109	90	74						
		0.1	0.1	0.1	0.1	0.1	0.1	0.0	0.0	0.0						
		0.2	0.2	0.1	0.1	0.1	0.0	0.0	−0.1	−0.2						
50-S				267	227	194	165	140	119	100	84	70				
				0.2	0.2	0.2	0.2	0.2	0.1	0.1	0.0	0.0				
				0.2	0.2	0.2	0.2	0.2	0.1	0.1	0.0	−0.1	−0.2			
60-S					270	232	200	172	148	127	108	92	77	65		
					0.3	0.3	0.3	0.3	0.3	0.2	0.2	0.2	0.1	0.0		
					0.4	0.3	0.3	0.3	0.3	0.2	0.1	0.0	−0.1	−0.3		
70-S						271	235	204	177	153	132	114	98	84	71	59
						0.4	0.4	0.4	0.4	0.4	0.4	0.3	0.3	0.2	0.1	0.0
						0.5	0.5	0.5	0.4	0.4	0.3	0.2	0.1	0.0	−0.1	−0.3

FS8 + 2

2″ Normal Weight Topping

Table of safe superimposed service load (psf) and cambers

Strand Designation Code	Span, ft.												
	17	18	19	20	21	22	23	24	25	26	27	28	29
40-S	273	232	192	155	123	96	72						
	0.1	0.1	0.1	0.1	0.0	0.0	0.0						
	0.1	0.0	0.0	−0.1	−0.1	−0.2	−0.3						
50-S		298	249	206	170	139	111	87	66				
		0.2	0.2	0.2	0.2	0.1	0.1	0.0	0.0				
		0.2	0.1	0.1	0.0	−0.1	−0.2	−0.3	−0.4				
60-S			258	217	181	150	123	99	77	58			
			0.3	0.3	0.3	0.2	0.2	0.2	0.1	0.0			
			0.2	0.2	0.1	0.0	−0.1	−0.2	−0.4	−0.5			
70-S				263	223	189	158	131	108	86	67	50	
				0.4	0.4	0.4	0.4	0.3	0.3	0.2	0.1	0.0	
				0.3	0.3	0.2	0.1	0.0	−0.1	−0.3	−0.5	−0.7	

Bold type — *Capacity governed by stresses, others governed by flexural or shear strength*

SOLID FLAT SLAB

8″ Thick
Lightweight Concrete

Section Properties per foot of width

	Untopped	Topped
A =	96.00 in.²	—
I =	512.00 in.⁴	1058.50 in.⁴
Y_b =	4.00 in.	5.12 in.
Y_t =	4.00 in.	4.88 in.
Z_b =	128.00 in.³	206.75 in.³
Z_t =	128.00 in.³	216.90 in.³
b_w =	12.00 in.	12.00 in.
wt =	77 psf	102 psf
V/S =	4.00 in.	

Strand Patterns

Producer may vary size and strength of strands. See "explanation of load tables"

Safe loads shown include dead load of 10 psf for untopped members and 15 psf for topped members. Remainder is live load. Long-time cambers include superimposed dead load but do not include live load.

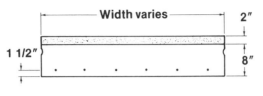

Width varies — 2″
1 1/2″ — 8″

f'_c = 5000 psi
f'_{ci} = 3500 psi

Key

267 — Safe superimposed service load, psf
0.1 — Estimated camber at erection, in.
0.2 — Estimated long-time camber, in.

LFS8

No Topping

Table of safe superimposed service load (psf) and cambers

Strand Designation Code	Span, ft.																			
	14	15	16	17	18	19	20	21	22	23	24	25	26	27	28	29	30	31	32	33
30-S	267	225	190	161	137	117	100	85	72											
	0.1	0.1	0.1	0.1	0.1	0.1	0.1	0.0	0.0											
	0.2	0.2	0.2	0.1	0.1	0.1	0.0	-0.1	-0.2											
40-S		294	263	230	198	172	148	127	109	93	79	67								
		0.2	0.3	0.3	0.3	0.3	0.2	0.2	0.2	0.2	0.1	0.0								
		0.3	0.3	0.3	0.3	0.3	0.2	0.2	0.1	0.0	-0.1	-0.2								
50-S				281	242	210	182	158	137	119	103	89	76	65	55					
				0.4	0.4	0.4	0.4	0.4	0.4	0.4	0.3	0.3	0.2	0.1	0.0					
				0.5	0.5	0.5	0.4	0.4	0.4	0.3	0.2	0.1	0.0	-0.2	-0.4					
60-S						284	247	215	188	164	144	126	110	96	83	72	62	53	45	
						0.5	0.6	0.6	0.6	0.6	0.6	0.6	0.5	0.5	0.4	0.4	0.3	0.1	0.0	
						0.7	0.7	0.7	0.7	0.6	0.6	0.5	0.5	0.3	0.2	0.0	-0.2	-0.4	-0.7	
70-S						284	249	218	192	169	149	132	116	102	89	78	68	59	50	43
						0.7	0.7	0.8	0.8	0.8	0.8	0.8	0.8	0.7	0.7	0.6	0.5	0.4	0.3	0.1
						0.9	0.9	0.9	0.9	0.9	0.9	0.8	0.7	0.6	0.5	0.3	0.1	-0.1	-0.4	-0.7

LFS8 + 2

2″ Normal Weight Topping

Table of safe superimposed service load (psf) and cambers

Strand Designation Code	Span, ft.														
	17	18	19	20	21	22	23	24	25	26	27	28	29	30	31
30-S															
40-S	292	252	217	188	163	136	110	88	68						
	0.3	0.3	0.3	0.2	0.2	0.2	0.2	0.1	0.0						
	0.2	0.2	0.1	0.1	0.0	-0.1	-0.2	-0.3	-0.5						
50-S		282	247	214	181	151	125	103	83	65	49				
		0.4	0.4	0.4	0.4	0.4	0.3	0.3	0.2	0.1	0.0				
		0.3	0.3	0.2	0.2	0.1	-0.1	-0.2	-0.4	-0.6	-0.8				
60-S			263	225	192	163	137	115	94	76	60				
			0.6	0.6	0.6	0.6	0.5	0.5	0.4	0.4	0.3				
			0.5	0.4	0.3	0.2	0.1	0.0	-0.2	-0.4	-0.7				
70-S				270	233	201	172	147	124	104	86	69	54		
				0.8	0.8	0.8	0.8	0.8	0.7	0.7	0.6	0.5	0.4		
				0.6	0.6	0.5	0.4	0.3	0.2	0.0	-0.3	-0.5	-0.8		

Bold type — *Capacity governed by stresses, others governed by flexural or shear strength*

RECTANGULAR BEAMS

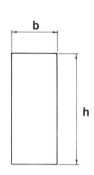

$f'_c = 5000$ psi

$f_{pu} = 270{,}000$ psi

Section Properties

Designation	b (in.)	h (in.)	A (in.²)	I (in.⁴)	Y_b (in.)	Z (in.³)	wt (plf)
12RB16	12	16	192	4096	8.00	512	200
12RB20	12	20	240	8000	10.00	800	250
12RB24	12	24	288	13,824	12.00	1152	300
12RB28	12	28	336	21,952	14.00	1568	350
12RB32	12	32	384	32,768	16.00	2048	400
12RB36	12	36	432	46,656	18.00	2592	450
16RB24	16	24	384	18,432	12.00	1536	400
16RB28	16	28	448	29,269	14.00	2091	467
16RB32	16	32	512	43,691	16.00	2731	533
16RB36	16	36	576	62,208	18.00	3456	600
16RB40	16	40	640	85,333	20.00	4267	667

Key

3246 — Safe superimposed service load, plf
0.4 — Estimated camber at erection, in.
0.1 — Estimated long-time camber, in.

Safe loads shown include 50% dead load and 50% live load. 800 psi top tension has been allowed, therefore additional top reinforcement is required.

Table of safe superimposed service load (plf) and cambers

Desig-nation	No. Strand	e	16	18	20	22	24	26	28	30	32	34	36	38	40	42	44	46	48	50
12RB16	5	5.67	3246 0.4 0.1	2527 0.5 0.1	2012 0.6 0.2	1632 0.7 0.2	1342 0.7 0.2	1117 0.8 0.2												
12RB20	8	6.60	5816 0.3 0.1	4548 0.4 0.1	3641 0.5 0.1	2970 0.6 0.2	2459 0.7 0.2	2062 0.8 0.2	1747 0.9 0.2	1493 1.0 0.2	1285 1.1 0.2	1112 1.2 0.2								
12RB24	10	7.76	8585 0.3 0.1	6726 0.4 0.1	5397 0.4 0.1	4413 0.5 0.1	3665 0.6 0.1	3083 0.7 0.2	2621 0.8 0.2	2248 0.9 0.2	1940 1.0 0.2	1684 1.1 0.2	1470 1.2 0.2	1288 1.3 0.2	1133 1.3 0.2	1000 1.4 0.1				
12RB28	12	8.89		9074 0.3 0.1	7290 0.4 0.1	5970 0.5 0.1	4966 0.5 0.2	4184 0.6 0.2	3564 0.7 0.2	3064 0.8 0.2	2655 0.9 0.2	2316 1.0 0.2	2031 1.1 0.2	1791 1.1 0.2	1585 1.2 0.2	1409 1.3 0.2	1255 1.4 0.2	1122 1.4 0.1	1002 1.5 0.1	
12RB32	13	10.48			9584 0.3 0.1	7858 0.4 0.1	6545 0.5 0.1	5524 0.5 0.1	4713 0.6 0.2	4059 0.7 0.2	3524 0.8 0.2	3080 0.9 0.2	2708 0.9 0.2	2394 1.0 0.2	2125 1.1 0.2	1894 1.2 0.2	1694 1.2 0.2	1519 1.3 0.1	1365 1.4 0.1	1230 1.4 0.1
12RB36	15	11.64					8450 0.4 0.1	7140 0.5 0.1	6100 0.6 0.1	5261 0.6 0.2	4575 0.7 0.2	4006 0.8 0.2	3530 0.9 0.2	3123 0.9 0.2	2775 1.0 0.2	2475 1.1 0.2	2215 1.2 0.2	1989 1.2 0.2	1790 1.3 0.2	1614 1.4 0.1
16RB24	13	7.86		8847 0.4 0.1	7098 0.4 0.1	5803 0.5 0.1	4819 0.6 0.2	4052 0.7 0.2	3444 0.8 0.2	2954 0.9 0.2	2552 1.0 0.2	2220 1.1 0.2	1941 1.2 0.1	1705 1.2 0.1	1503 1.3 0.1	1330 1.4 0.1	1180 1.5 0.0			
16RB28	13	8.89			9720 0.4 0.1	7959 0.5 0.1	6621 0.5 0.1	5579 0.6 0.2	4752 0.7 0.2	4086 0.8 0.2	3540 0.9 0.2	3087 1.0 0.2	2708 1.1 0.2	2388 1.1 0.2	2114 1.2 0.2	1878 1.3 0.2	1674 1.4 0.2	1496 1.4 0.1	1335 1.5 0.1	1194 1.5 0.1
16RB32	18	10.29					8808 0.5 0.1	7434 0.5 0.1	6343 0.6 0.2	5464 0.7 0.2	4744 0.8 0.2	4147 0.9 0.2	3647 0.9 0.2	3224 1.0 0.2	2863 1.1 0.2	2549 1.2 0.2	2275 1.3 0.2	2036 1.3 0.2	1827 1.4 0.1	1642 1.5 0.1
16RB36	20	11.64					9519 0.5 0.1	8133 0.6 0.1	7015 0.6 0.2	6100 0.7 0.2	5342 0.8 0.2	4706 0.9 0.2	4165 0.9 0.2	3700 1.0 0.2	3300 1.1 0.2	2954 1.2 0.2	2651 1.2 0.2	2386 1.3 0.2	2152 1.4 0.1	
16RB40	22	13.00							8647 0.6 0.1	7527 0.6 0.1	6599 0.7 0.1	5821 0.8 0.2	5163 0.9 0.2	4601 0.9 0.2	4117 1.0 0.2	3698 1.1 0.2	3332 1.1 0.2	3011 1.2 0.1	2728 1.3 0.1	

L-SHAPED BEAMS

Normal Weight Concrete

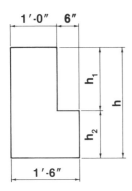

1'-0" 6"

f'_c = 5000 psi

f_{pu} = 270,000 psi

1'-6"

Section Properties								
Designation	h (in.)	h_1/h_2 (in.)	A (in.²)	I (in.⁴)	y_b (in.)	Z_b (in.³)	Z_t (in.³)	wt (plf)
18LB20	20	12/8	288	9696	9.00	1077	882	300
18LB24	24	12/12	360	16,762	10.80	1552	1270	375
18LB28	28	16/12	408	26,611	12.59	2114	1727	425
18LB32	32	20/12	456	39,695	14.42	2753	2258	475
18LB36	36	24/12	504	56,407	16.29	3463	2862	525
18LB40	40	24/16	576	77,568	18.00	4309	3526	600
18LB44	44	28/16	624	103,153	19.85	5197	4271	650
18LB48	48	32/16	672	133,705	21.71	6159	5086	700
18LB52	52	36/16	720	169,613	23.60	7187	5972	750
18LB56	56	40/16	768	211,264	25.50	8285	6927	800
18LB60	60	44/16	816	259,046	27.41	9451	7949	850

Key

6486 — Safe superimposed service load, plf
0.3 — Estimated camber at erection, in.
0.1 — Estimated long-time camber, in.

Safe loads shown include 50% dead load and 50% live load. 800 psi top tension has been allowed, therefore additional top reinforcement is required.

Table of safe superimposed service load (plf) and cambers

Desig-nation	No. Strand	e	16	18	20	22	24	26	28	30	32	34	36	38	40	42	44	46	48	50
18LB20	9	6.26	6486 0.3 0.1	5068 0.4 0.1	4053 0.5 0.1	3303 0.6 0.2	2732 0.7 0.2	2288 0.8 0.2	1935 0.9 0.2	1650 0.9 0.2	1414 1.0 0.2	1218 1.1 0.2	1054 1.2 0.1							
18LB24	10	7.67	9179 0.3 0.1	7182 0.3 0.1	5753 0.4 0.1	4696 0.5 0.1	3891 0.5 0.1	3266 0.6 0.1	2769 0.6 0.1	2369 0.7 0.1	2041 0.8 0.1	1769 0.8 0.1	1541 1.0 0.1	1349 1.0 0.1	1184 1.1 0.0					
18LB28	12	8.93			8039 0.3 0.1	6578 0.4 0.1	5466 0.5 0.1	4600 0.6 0.1	3914 0.6 0.2	3360 0.7 0.2	2906 0.8 0.2	2531 0.8 0.2	2216 0.9 0.1	1949 1.0 0.1	1722 1.0 0.1	1524 1.1 0.1	1351 1.1 0.0	1200 1.2 0.0		
18LB32	14	10.22				8814 0.4 0.1	7331 0.4 0.1	6176 0.5 0.1	5260 0.6 0.2	4521 0.7 0.2	3916 0.7 0.2	3414 0.8 0.2	2994 0.9 0.2	2639 0.9 0.2	2335 1.0 0.2	2074 1.1 0.2	1847 1.1 0.1	1650 1.2 0.1	1476 1.2 0.1	1323 1.3 0.0
18LB36	16	11.52					9358 0.4 0.1	7903 0.5 0.1	6744 0.5 0.1	5807 0.6 0.1	5040 0.7 0.2	4405 0.7 0.2	3872 0.8 0.2	3422 0.9 0.2	3037 0.9 0.2	2706 1.0 0.2	2419 1.1 0.2	2168 1.1 0.1	1948 1.2 0.1	1755 1.2 0.1
18LB40	18	12.52						9693 0.4 0.1	8284 0.5 0.1	7146 0.5 0.2	6215 0.6 0.2	5443 0.7 0.2	4797 0.7 0.2	4250 0.8 0.2	3783 0.9 0.2	3380 0.9 0.2	3026 1.0 0.2	2718 1.0 0.2	2447 1.1 0.2	2208 1.1 0.2
18LB44	19	14.19							8729 0.5 0.1	7601 0.6 0.2	6666 0.6 0.2	5883 0.7 0.2	5219 0.7 0.2	4653 0.8 0.2	4166 0.9 0.2	3743 0.9 0.2	3370 1.0 0.2	3042 1.0 0.2	2752 1.1 0.2	
18LB48	21	15.48								9166 0.5 0.1	8048 0.6 0.1	7110 0.6 0.1	6313 0.7 0.2	5629 0.8 0.2	5041 0.8 0.2	4531 0.9 0.2	4086 0.9 0.2	3695 1.0 0.2	3351 1.0 0.2	
18LB52	23	16.78									9538 0.5 0.2	8427 0.6 0.2	7486 0.7 0.2	6683 0.7 0.2	5992 0.8 0.2	5393 0.8 0.2	4871 0.9 0.2	4412 0.9 0.2	4007 1.0 0.2	
18LB56	25	18.07										9842 0.6 0.2	8752 0.6 0.2	7820 0.7 0.2	7019 0.7 0.2	6324 0.8 0.2	5718 0.8 0.2	5186 0.9 0.2	4717 1.0 0.2	
18LB60	27	19.36											9026 0.6 0.2	8116 0.7 0.2	7326 0.7 0.2	6630 0.8 0.2	6020 0.9 0.2	5481 0.9 0.2		

Span, ft.

INVERTED TEE BEAMS

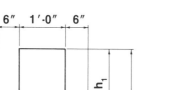

6" 1'-0" 6"

h_1 / h / h_2

2'-0"

$f'_c = 5000$ psi

$f_{pu} = 270,000$ psi

Section Properties

Designation	h (in.)	h_1/h_2 (in.)	A (in.²)	I (in.⁴)	y_b (in.)	Z_b (in.³)	Z_t (in.³)	wt (plf)
24IT20	20	12/8	336	10,981	8.29	1325	938	350
24IT24	24	12/12	432	19,008	10.00	1901	1358	450
24IT28	28	16/12	480	30,131	11.60	2598	1837	500
24IT32	32	20/12	528	44,969	13.27	3388	2401	550
24IT36	36	24/12	576	63,936	15.00	4262	3045	600
24IT40	40	24/16	672	87,845	16.57	5301	3749	700
24IT44	44	28/16	720	116,877	18.27	6397	4542	750
24IT48	48	32/16	768	151,552	20.00	7578	5413	800
24IT52	52	36/16	816	192,275	21.76	8836	6358	850
24IT56	56	40/16	864	239,445	23.56	10,163	7381	900
24IT60	60	44/16	912	293,460	25.37	11,567	8474	950

Safe loads shown include 50% dead load and 50% live load. 800 psi top tension has been allowed, therefore additional top reinforcement is required.

Key

6888 — Safe superimposed service load, plf
0.3 — Estimated camber at erection, in.
0.1 — Estimated long-time camber, in.

Table of safe superimposed service load (plf) and cambers

Desig-nation	No. Strand	e	16	18	20	22	24	26	28	30	32	34	36	38	40	42	44	46	48	50
24IT20	9	6.20	6888 0.3 0.1	5376 0.3 0.1	4294 0.4 0.1	3494 0.5 0.1	2886 0.6 0.1	2412 0.6 0.1	2033 0.7 0.1	1726 0.8 0.1	1474 0.8 0.0	1266 0.9 0.0								
24IT24	11	7.17	9759 0.2 0.1	7625 0.3 0.1	6099 0.3 0.1	4970 0.4 0.1	4111 0.5 0.1	3443 0.5 0.1	2913 0.6 0.1	2485 0.7 0.1	2135 0.7 0.0	1845 0.8 0.0	1601 0.8 0.0							
24IT28	13	8.44			8505 0.3 0.1	6951 0.4 0.1	5768 0.4 0.1	4848 0.5 0.1	4118 0.6 0.1	3529 0.6 0.1	3047 0.7 0.1	2648 0.7 0.1	2313 0.8 0.1	2030 0.8 0.0	1786 0.9 0.0					
24IT32	15	9.77				9248 0.3 0.1	7691 0.4 0.1	6480 0.5 0.1	5519 0.5 0.1	4744 0.6 0.1	4109 0.6 0.1	3583 0.7 0.1	3138 0.8 0.1	2760 0.8 0.1	2437 0.9 0.1	2159 0.9 0.1	1919 1.0 0.0	1709 1.0 0.0		
24IT36	16	11.50					9879 0.4 0.1	8337 0.4 0.1	7114 0.5 0.1	6127 0.5 0.1	5320 0.6 0.1	4644 0.6 0.1	4077 0.7 0.1	3598 0.7 0.1	3189 0.8 0.1	2836 0.9 0.1	2531 0.9 0.1	2265 0.9 0.0	2031 1.0 0.0	1825 1.0 0.0
24IT40	19	12.02					8675 0.4 0.1	7475 0.5 0.1	6494 0.5 0.1	5680 0.6 0.1	4998 0.6 0.1	4421 0.7 0.1	3928 0.7 0.1	3504 0.8 0.1	3137 0.8 0.1	2816 0.9 0.1	2535 0.9 0.0	2286 1.0 0.0		
24IT44	20	13.73							9300 0.4 0.1	8083 0.5 0.1	7075 0.5 0.1	6230 0.6 0.1	5514 0.6 0.1	4903 0.7 0.1	4378 0.7 0.1	3922 0.8 0.1	3525 0.8 0.1	3176 0.9 0.1	2868 0.9 0.0	
24IT48	22	15.08								9723 0.5 0.1	8522 0.5 0.1	7515 0.6 0.1	6663 0.6 0.1	5935 0.7 0.1	5309 0.7 0.1	4766 0.8 0.1	4293 0.8 0.1	3877 0.9 0.1	3510 0.9 0.1	
24IT52	24	16.44									8917 0.5 0.1	7916 0.6 0.1	7061 0.6 0.1	6326 0.7 0.1	5688 0.7 0.1	5132 0.8 0.1	4644 0.8 0.1	4213 0.9 0.1		
24IT56	26	17.82										9279 0.6 0.1	8287 0.6 0.1	7433 0.7 0.1	6692 0.7 0.1	6046 0.8 0.1	5480 0.8 0.1	4979 0.9 0.1		
24IT60	28	19.18											9597 0.6 0.1	8616 0.6 0.1	7766 0.7 0.1	7025 0.7 0.1	6374 0.8 0.2	5800 0.8 0.1		

AASHTO GIRDERS

Normal Weight Concrete

Section Properties						
Designation	A (in.²)	I (in.⁴)	y_b (in.)	Z_b (in.³)	Z_t (in.³)	wt (plf)
Type II	369	50,979	15.83	3220	2527	384
Type III	560	125,390	20.27	6186	5070	583
Type IV	789	260,741	24.73	10,544	8908	822

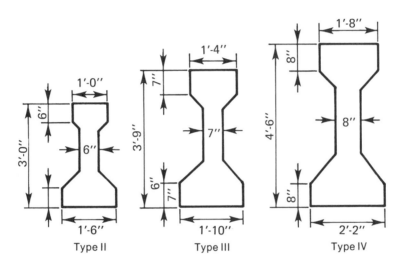

Type II   Type III   Type IV

f'_c = 5000 psi
f_{pu} = 270,000 psi

Key
5221 — Safe superimposed service load, plf
0.6 — Estimated camber at erection, in.
0.1 — Estimated long-time camber, in.

Safe loads shown include 50% dead load and 50% live load. 800 psi top tension has been allowed, therefore additional top reinforcement is required.

Table of safe superimposed service load (plf) and cambers

Desig-nation	No. Strand	e_e / e_c	Span, ft.																	
			30	32	34	36	38	40	42	44	46	48	50	52	54	56	58	60	62	64
Type II	14	7.83 13.12	5221 0.6 0.1	4547 0.6 0.1	3988 0.7 0.1	3520 0.8 0.1	3123 0.9 0.1	2785 0.9 0.1	2494 1.0 0.1	2241 1.1 0.1	2021 1.2 0.1	1826 1.2 0.1	1653 1.3 0.1							
Type III	22	10.27 17.50		9292 0.5 0.1	8171 0.6 0.1	7231 0.7 0.1	6436 0.7 0.2	5757 0.8 0.2	5173 0.9 0.2	4667 0.9 0.2	4225 1.0 0.2	3837 1.1 0.2	3495 1.2 0.2	3192 1.2 0.2	2921 1.3 0.2	2679 1.4 0.2	2462 1.5 0.2	2266 1.5 0.2	2089 1.6 0.1	1928 1.6 0.1
Type IV	32	2.86 21.26						9848 0.5 0	8856 0.6 0	7996 0.6 0	9246 0.7 0	6588 0.7 −0.1	6007 0.8 −0.1	5492 0.9 −0.1	5033 0.9 −0.1	4622 1.0 −0.1	4253 1.0 −0.1	3920 1.0 −0.2	3619 1.1 −0.2	3346 1.1 −0.2

PRECAST, PRESTRESSED COLUMNS

Fig. 2.6.1 Design strength interaction curves for precast, prestressed concrete columns

Criteria

1. Minimum prestress = 225 psi
2. All strand assumed 1/2 in. diameter, f_{pu} = 270 ksi
3. Curves shown for partial development of strand near member end, where $f_{ps} \approx f_{se}$
4. Horizontal portion of curve is the maximum for tied columns = $0.80\phi P_o$
5. ϕ = 0.9 for ϕP_n = 0
   = 0.7 for $\phi P_n \geqslant 0.10 f'_c A_g$

   Varies from 0.9 to 0.7 for points between

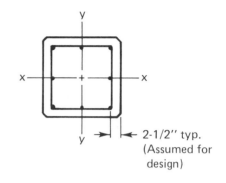

2-1/2'' typ.
(Assumed for design)

Use of curves

1. Enter at left with applied factored axial load, P_u
2. Enter at bottom with applied magnified factored moment, δM_u
3. Intersection point must be to the left of curve indicating required concrete strength.

Notation

ϕP_n = Design axial strength
ϕM_n = Design flexural strength
ϕP_o = Design axial strength at zero ecentricity
A_g = Gross area of the column
δ = Moment magnifier (Sect. 10.11, ACI 318-77)

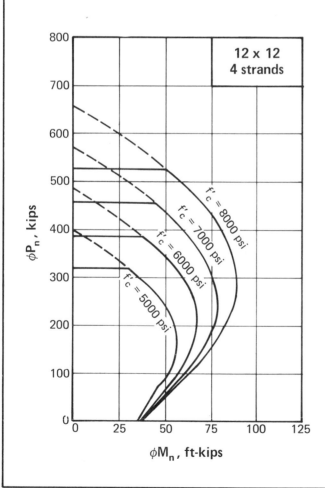

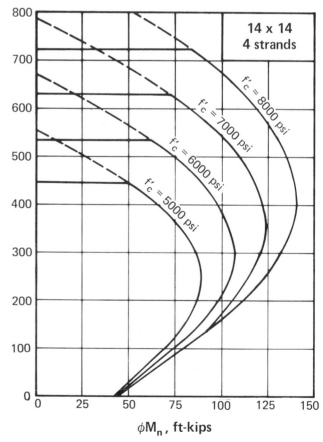

PRECAST, PRESTRESSED COLUMNS

Fig. 2.6.1 (cont.) Design strength interaction curves for precast, prestressed concrete columns

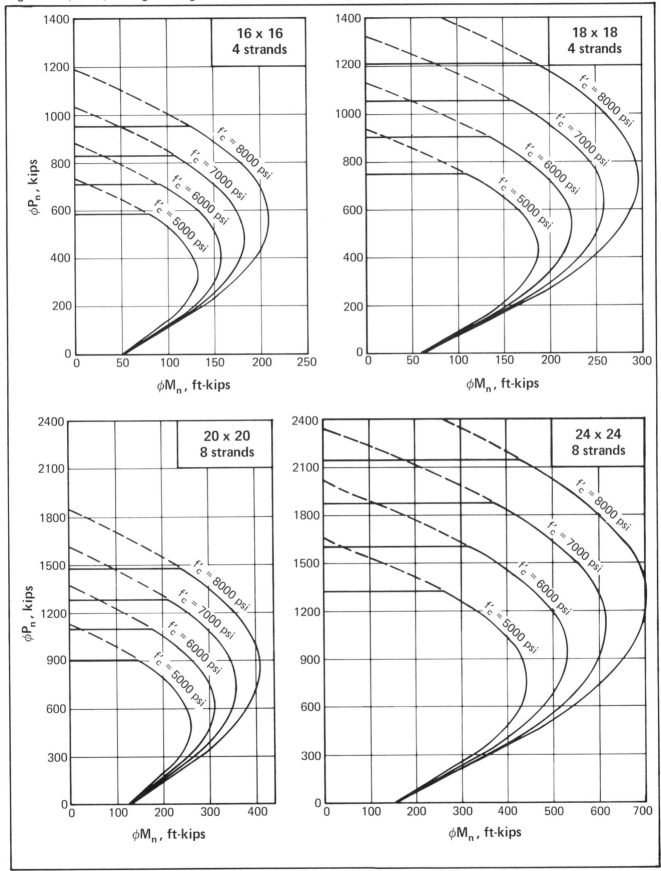

PRECAST, REINFORCED COLUMNS

Fig. 2.6.2 Design strength interaction curves for precast, reinforced concrete columns

Criteria

1. Concrete f'_c = 5000 psi
2. Reinforcement f_y = 60,000 psi
3. Curves shown for full development of reinforcement
4. Horizontal portion of curve is the maximum for tied columns = $0.80\phi P_o$.
5. ϕ = 0.9 for $\phi P_n = 0$
   = 0.7 for $\phi P_n \geq 0.10\, f'_c A_g$

   Varies from 0.9 to 0.7 for points between

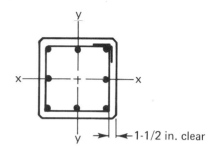

Use of curves

1. Enter at left with applied factored axial load, P_u
2. Enter at bottom with applied magnified factored moment, δM_u
3. Intersection point must be to the left of curve indicating required reinforcement.

Notation

ϕP_n = Design axial strength
ϕM_n = Design flexural strength
ϕP_o = Design axial strength at zero eccentricity
A_g = Gross area of the column
δ = Moment magnifier (Sect. 10.11, ACI 318-77)

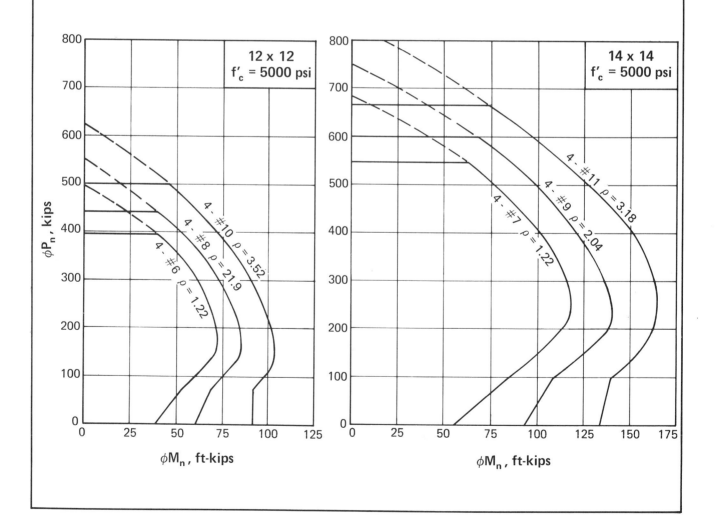

PCI Design Handbook

PRECAST, REINFORCED COLUMNS

Fig. 2.6.2 (Cont.) Design strength interaction curves for precast, reinforced concrete columns

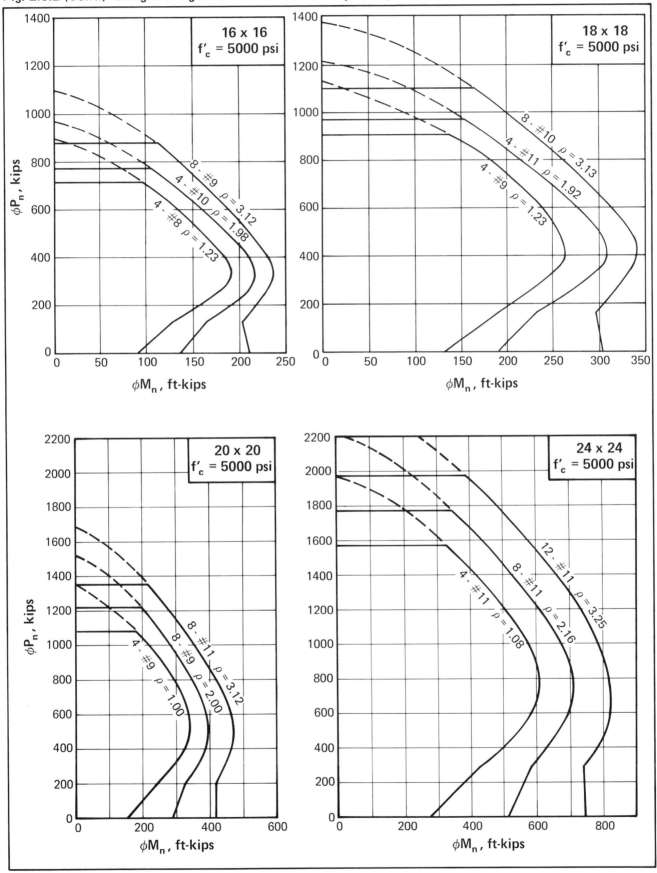

DOUBLE TEE WALL PANELS

Fig. 2.6.3 Partial interaction curve for prestressed double tee wall panels

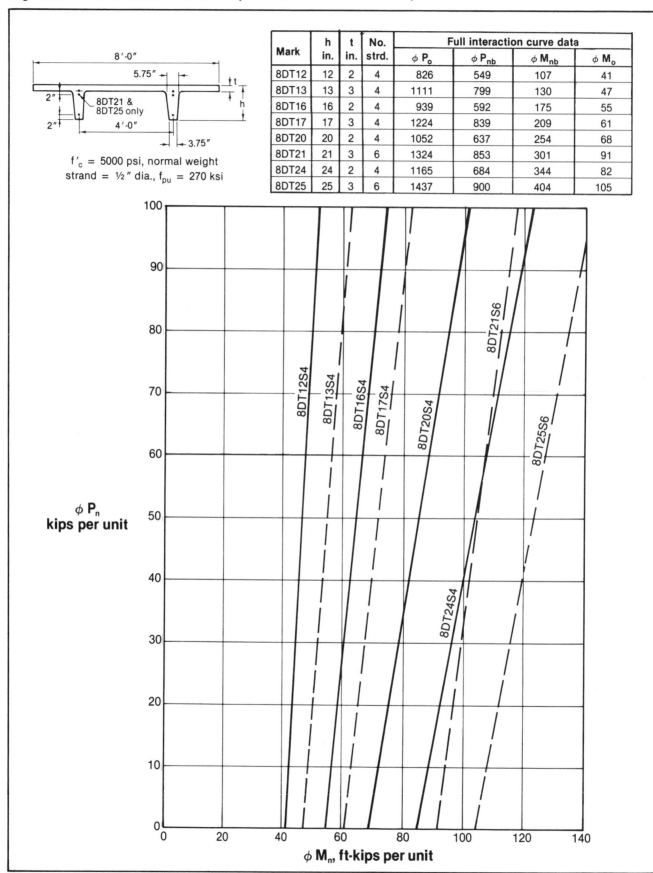

8'-0"

5.75"

2"

8DT21 &
8DT25 only

2"

4'-0"

3.75"

t

h

f'_c = 5000 psi, normal weight
strand = ½" dia., f_{pu} = 270 ksi

Mark	h in.	t in.	No. strd.	Full interaction curve data			
				ϕP_o	ϕP_{nb}	ϕM_{nb}	ϕM_o
8DT12	12	2	4	826	549	107	41
8DT13	13	3	4	1111	799	130	47
8DT16	16	2	4	939	592	175	55
8DT17	17	3	4	1224	839	209	61
8DT20	20	2	4	1052	637	254	68
8DT21	21	3	6	1324	853	301	91
8DT24	24	2	4	1165	684	344	82
8DT25	25	3	6	1437	900	404	105

ϕP_n
kips per unit

8DT12S4

8DT13S4

8DT16S4

8DT17S4

8DT20S4

8DT21S6

8DT25S6

8DT24S4

ϕM_n, ft-kips per unit

HOLLOW-CORE WALL PANELS

Fig. 2.6.4 Partial interaction curve for prestressed hollow-core wall panels

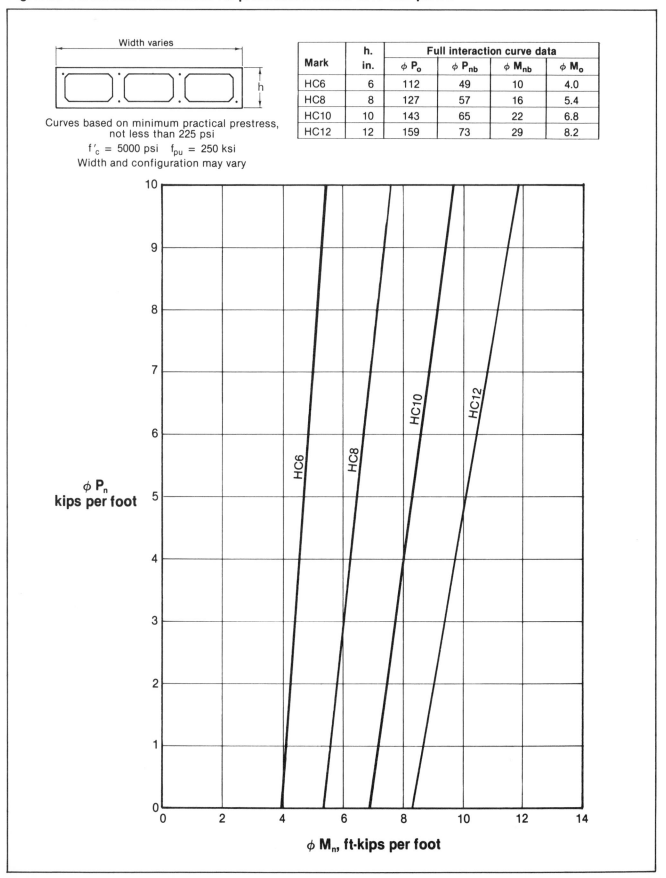

Width varies

h

Curves based on minimum practical prestress,
not less than 225 psi

f'_c = 5000 psi f_{pu} = 250 ksi

Width and configuration may vary

Mark	h. in.	Full interaction curve data			
		ϕP_o	ϕP_{nb}	ϕM_{nb}	ϕM_o
HC6	6	112	49	10	4.0
HC8	8	127	57	16	5.4
HC10	10	143	65	22	6.8
HC12	12	159	73	29	8.2

ϕP_n
kips per foot

HC6   HC8   HC10   HC12

ϕM_n, ft-kips per foot

PRECAST, PRESTRESSED SOLID WALL PANELS

Fig. 2.6.5 Partial interaction curve for prestressed solid wall panels

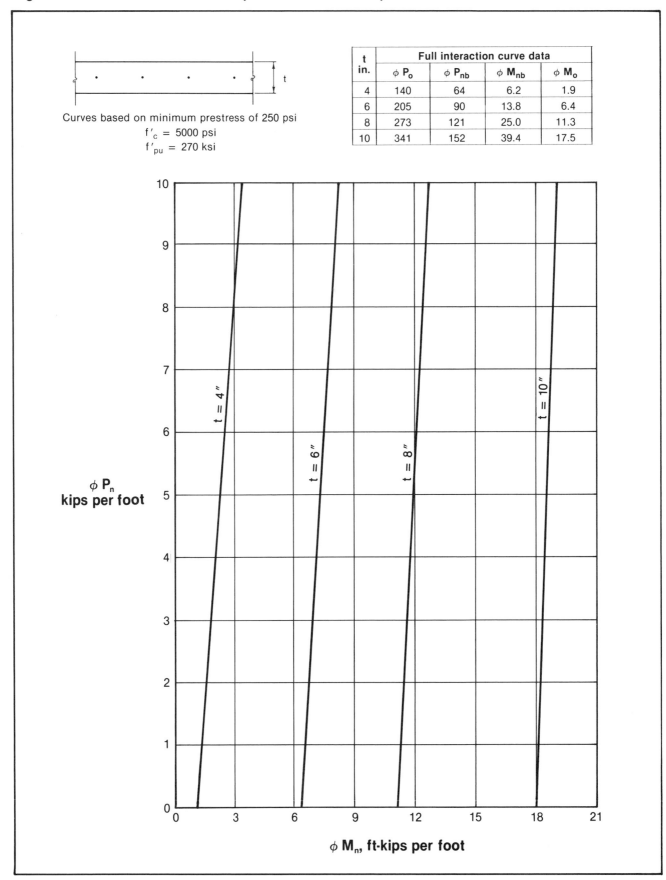

Curves based on minimum prestress of 250 psi
$f'_c = 5000$ psi
$f'_{pu} = 270$ ksi

t in.	Full interaction curve data			
	$\phi\,P_o$	$\phi\,P_{nb}$	$\phi\,M_{nb}$	$\phi\,M_o$
4	140	64	6.2	1.9
6	205	90	13.8	6.4
8	273	121	25.0	11.3
10	341	152	39.4	17.5

$\phi\,P_n$
kips per foot

$\phi\,M_n$, ft·kips per foot

PRECAST, REINFORCED SOLID WALL PANELS

Fig. 2.6.6 Partial interaction curve for precast, reinforced concrete wall panels

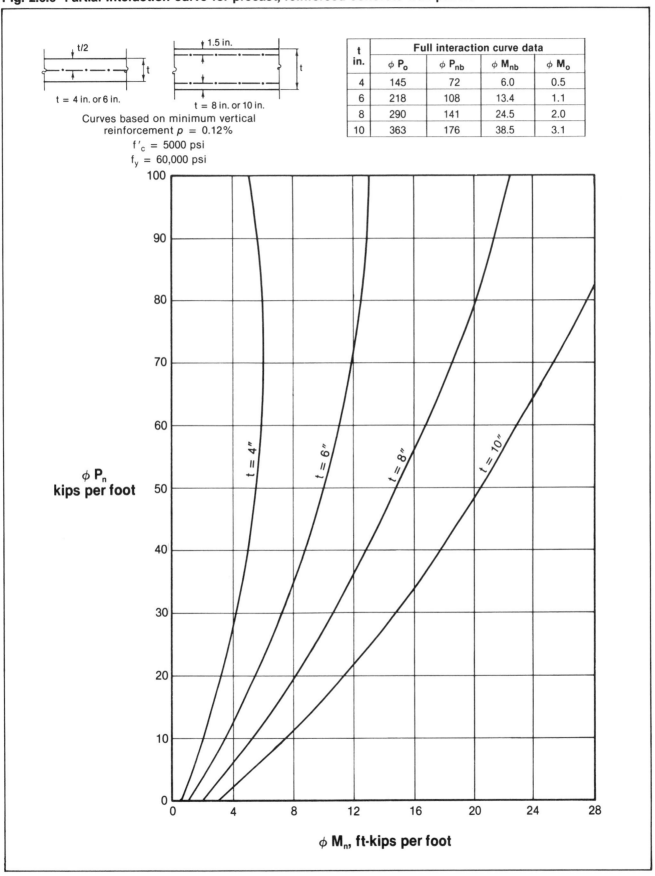

Curves based on minimum vertical
reinforcement $p = 0.12\%$
$f'_c = 5000$ psi
$f_y = 60{,}000$ psi

$t = 4$ in. or 6 in.

$t = 8$ in. or 10 in.

t in.	Full interaction curve data			
	ϕP_o	ϕP_{nb}	ϕM_{nb}	ϕM_o
4	145	72	6.0	0.5
6	218	108	13.4	1.1
8	290	141	24.5	2.0
10	363	176	38.5	3.1

ϕP_n **kips per foot**

$t = 4''$ $t = 6''$ $t = 8''$ $t = 10''$

ϕM_n, **ft-kips per foot**

PILES

Table 2.7.1 Section properties and allowable loads of prestressed concrete piles

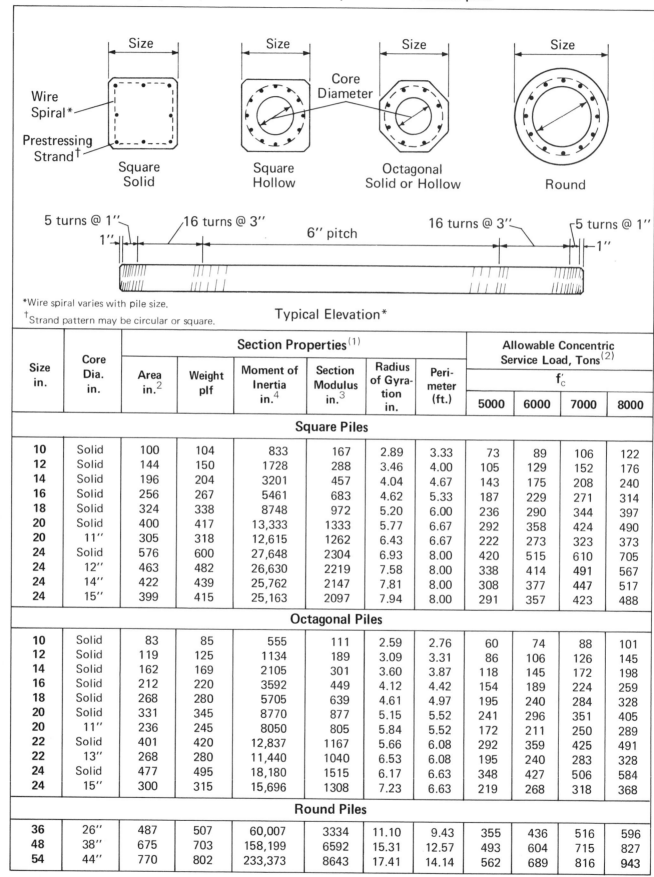

*Wire spiral varies with pile size.
†Strand pattern may be circular or square.

Typical Elevation*

Size in.	Core Dia. in.	Section Properties[1]						Allowable Concentric Service Load, Tons[2]			
		Area in.^{2}	Weight plf	Moment of Inertia in.^{4}	Section Modulus in.^{3}	Radius of Gyration in.	Perimeter (ft.)	f'_c			
								5000	6000	7000	8000
Square Piles											
10	Solid	100	104	833	167	2.89	3.33	73	89	106	122
12	Solid	144	150	1728	288	3.46	4.00	105	129	152	176
14	Solid	196	204	3201	457	4.04	4.67	143	175	208	240
16	Solid	256	267	5461	683	4.62	5.33	187	229	271	314
18	Solid	324	338	8748	972	5.20	6.00	236	290	344	397
20	Solid	400	417	13,333	1333	5.77	6.67	292	358	424	490
20	11″	305	318	12,615	1262	6.43	6.67	222	273	323	373
24	Solid	576	600	27,648	2304	6.93	8.00	420	515	610	705
24	12″	463	482	26,630	2219	7.58	8.00	338	414	491	567
24	14″	422	439	25,762	2147	7.81	8.00	308	377	447	517
24	15″	399	415	25,163	2097	7.94	8.00	291	357	423	488
Octagonal Piles											
10	Solid	83	85	555	111	2.59	2.76	60	74	88	101
12	Solid	119	125	1134	189	3.09	3.31	86	106	126	145
14	Solid	162	169	2105	301	3.60	3.87	118	145	172	198
16	Solid	212	220	3592	449	4.12	4.42	154	189	224	259
18	Solid	268	280	5705	639	4.61	4.97	195	240	284	328
20	Solid	331	345	8770	877	5.15	5.52	241	296	351	405
20	11″	236	245	8050	805	5.84	5.52	172	211	250	289
22	Solid	401	420	12,837	1167	5.66	6.08	292	359	425	491
22	13″	268	280	11,440	1040	6.53	6.08	195	240	283	328
24	Solid	477	495	18,180	1515	6.17	6.63	348	427	506	584
24	15″	300	315	15,696	1308	7.23	6.63	219	268	318	368
Round Piles											
36	26″	487	507	60,007	3334	11.10	9.43	355	436	516	596
48	38″	675	703	158,199	6592	15.31	12.57	493	604	715	827
54	44″	770	802	233,373	8643	17.41	14.14	562	689	816	943

(1) Form dimensions may vary with producers, with corresponding variations in section properties.
(2) Allowable loads based on N = A_c (0.33 f'_c - 0.27 f_{pc}); f_{pc} = 700 psi; Check local producer for available concrete strengths.

PCI Design Handbook

SHEET PILES

Table 2.7.2 Section properties and allowable moments of prestressed sheet piles

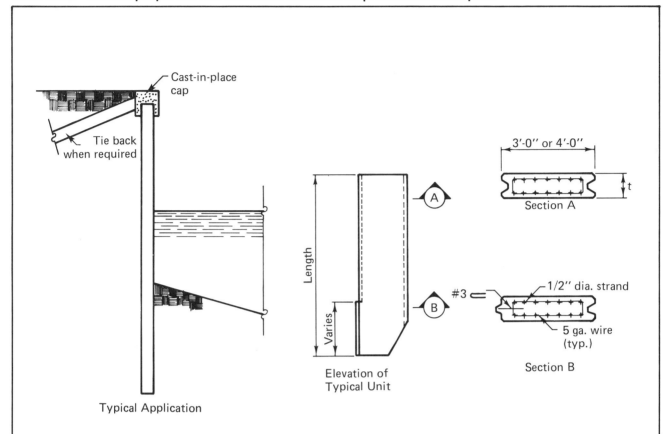

Thickness t in	Section Properties per Foot of Width				Maximum Allowable Service Load Moment[2] ft-kips per foot	
	Area in²	Weight[1] psf	Moment of Inertia in⁴	Section Modulus in³	f'_c = 5000 psi	f'_c = 6000 psi
6[3]	72	75	216	72	6.0	7.2
8[3]	96	100	512	128	10.6	12.8
10	120	125	1000	200	16.6	20.0
12	144	150	1728	288	24.0	28.8
16	192	200	4096	512	42.7	51.2
18	216	225	5832	648	54.0	64.8
20	240	250	8000	800	66.7	80.0
24	288	300	13,824	1152	96.0	115.2

(1) Normal weight concrete
(2) Based on zero tension and maximum $0.4f'_c$ compression
(3) Strand can be placed in a single layer in thin sections. Where site conditions require it, strand may be placed eccentrically.

PART 3
DESIGN OF PRECAST, PRESTRESSED CONCRETE COMPONENTS

DESIGN OF PRECAST, PRESTRESSED CONCRETE COMPONENTS

3.1 General

3.1.1 Notation

A = cross sectional area

A_{cr} = area of crack interface

A_{cs} = area of horizontal shear ties

A_{comp} = cross-sectional area of the equivalent rectangular stress block

A_ℓ = total area of longitudinal reinforcement to resist torsion

A_{ps} = area of prestressed reinforcement

A_s = area of non-prestressed tension reinforcement

A'_s = area of compression reinforcement

A_t = area of one leg of closed stirrup

A_{top} = effective area of cast-in-place composite topping

A_v = area of shear reinforcement

a = depth of equivalent rectangular stress block

b = width of compression face of member

b_v = width of interface surface in a composite member

b_w = web width

C = coefficient as defined in section used (with subscripts)

C = compressive force

C_c = compressive force capacity of composite topping

C_f = coefficient = $bd^2/12{,}000$

c = distance from extreme compression fiber to neutral axis

d = distance from extreme compression fiber to centroid of tension reinforcement

d' = distance from extreme compression fiber to centroid of compression reinforcement

d_b = nominal diameter of reinforcing bar or prestressing strand

E = modulus of elasticity

E_c = modulus of elasticity of concrete

E_{ci} = modulus of elasticity of concrete at time of initial prestress

E_s = modulus of elasticity of steel

e = eccentricity of design load or prestress force parallel to axis measured from the centroid of the section

e' = distance between c.g. of strand at end and c.g. of strand at lowest point, $e_c - e_e$

e_c = eccentricity of prestress force from the centroid of the section at the center of the span

e_e = eccentricity of prestress force from the centroid of the section at the end of the span

F = force as defined in section used (with subscripts)

F_h = horizontal shear force

f_b = stress in the bottom fiber of the cross section

f_c = unit stress in concrete

f'_c = specified compressive strength of concrete

f'_{cc} = specified compressive strength of composite topping

f_{cds} = concrete compressive stress at center of gravity of prestressing force due to all permanent (dead) loads not used in computing f_{cr}.

f'_{ci} = compressive strength of concrete at time of initial prestress

f_{cr} = concrete stress at center of gravity of prestressing force immediately after transfer

f_{ct} = splitting tensile strength of lightweight concrete

f_d = stress due to service dead load

f_e = total load stress in excess of f_r

f_ℓ = stress due to service live load

f_{pc} = average compressive stress in concrete due to effective prestress force only (after all losses)

f_{pe} = compressive stress in concrete due to prestress only after all losses, at the extreme fiber of a section at which tensile stresses are caused by applied loads

f_{ps}	=	stress in prestressed reinforcement at nominal strength	M_u	=	applied factored moment at a section
f_{pu}	=	ultimate strength of prestressing steel	M_1	=	smaller factored end moment on a compression member, positive if bent in single curvature, negative if double curvature
f_{py}	=	specified yield strength of prestressing steel			
f_r	=	modulus of rupture of concrete	M_2	=	larger factored end moment, always positive
f_s, f'_s	=	stress in non-prestressed reinforcement			
f_{se}	=	effective stress in prestressing steel after losses	N	=	unfactored axial load
			n	=	modular ratio E_s/E_c
f_t	=	stress in the top fiber of the cross section	P	=	prestress force after losses
			P_i	=	initial prestress force
$f_{t\ell}$	=	final calculated stress in the member	P_n	=	axial load nominal strength of a compression member at given eccentricity
f_y	=	specified yield strength of non-prestressed reinforcement			
			P_{nb}	=	axial load nominal strength under balanced conditions
h	=	overall depth of member			
I	=	moment of inertia	P_o	=	prestress force at transfer
I_{cr}	=	moment of inertia of cracked section transformed to concrete	P_o	=	axial load nominal strength of a compression member with zero eccentricity
I_e	=	effective moment of inertia for computation of deflection			
			P_u	=	factored axial load
I_g	=	moment of inertia of gross section	$R_{t'}$, R_v	=	coefficients used in torsion design
$\overline{j}_u$	=	$j_u (f_{ps}/f_{pu})$	r	=	radius of gyration
j_u	=	for resisting lever arm, used in $j_u d$	s	=	shear or torsion reinforcement spacing in a direction perpendicular to the longitudinal reinforcement
K_u	=	a coefficient = $\phi M_n (12,000)/bd^2$			
k	=	effective length factor for compression members			
ℓ	=	span length	T	=	tensile force
ℓ_d	=	development length	TL	=	total prestress loss
ℓ_u	=	unsupported length of a compression member	T_u	=	factored torsional moment on a section
ℓ_{vh}	=	horizontal shear length as defined in Fig. 3.3.5	t	=	thickness (used for various parts of members with subscripts)
M	=	service load moment	V_c	=	nominal shear strength provided by the concrete
M_a	=	total moment at the section			
M_{cr}	=	cracking moment	V_d	=	dead load shear (unfactored)
M_d	=	moment due to service dead load (unfactored)	V_i	=	factored shear force at section due to externally applied loads occurring simultaneously with M_{max}.
M_ℓ	=	moment due to service live load (unfactored)			
			V_ℓ	=	live load shear (unfactored)
M_{max}	=	maximum factored moment at section due to externally applied loads	V_{nh}, V_{nv}	=	nominal shear strength of the connection in the horizontal and vertical directions, respectively
M_n	=	nominal moment strength of a section			
M_{nb}	=	nominal moment strength under balanced condition	V_p	=	vertical component of the effective prestress force at the section considered
M_o	=	nominal moment strength of a compressive member with zero axial load			
			V/S	=	volume-surface ratio
M_{sd}	=	moment due to superimposed dead load (unfactored)	V_u	=	factored shear force at section
M_{top}	=	moment due to topping (unfactored)	V_{uh}, V_{uv}	=	applied factored shear loads in horizontal and vertical directions, respectively

v_c	=	nominal shear stress carried by concrete
v'_c	=	nominal shear stress carried by concrete if no torsion is present
v_{ci}	=	shear stress at diagonal cracking due to all design loads when such cracking is the result of combined shear and moment
v_{cw}	=	shear stress at diagonal cracking due to all design loads when such cracking is the result of excessive principal tensile stresses in the web
v_{max}	=	maximum factored shear stress on a section
v_{tc}	=	nominal torsional stress carried by the concrete
v'_{tc}	=	nominal torsional stress carried by concrete if no shear is present
v_{tu}	=	factored torsional stress
v_u	=	factored shear stress
w	=	unfactored load per unit length of beam or per unit area of slab
w_d	=	unfactored dead load per unit length
w_ℓ	=	unfactored live load per unit of length
w_{sd}	=	dead load due to superimposed loading
$w_{t\ell}$	=	unfactored total load per unit of length $= w_d + w_\ell$
x	=	distance from support to point being investigated
x, x_1	=	shorter side of component rectangle and closed stirrup, respectively (torsion design)
y, y_1	=	longer side of component rectangle and closed stirrup, respectively (torsion design)
y'	=	distance from top to c.g. of A_{comp}
y_b	=	distance from bottom fiber to center of gravity of the section
y_t	=	distance from top fiber to center of gravity of the section
Z	=	section modulus
Z_b	=	section modulus with respect to the bottom fiber of a cross section
Z_t	=	section modulus with respect to the top fiber of a cross section
Δ	=	deflection (with subscripts)
α_t, β	=	coefficients used in torsion design
β_1	=	factor defined in Sect. 3.2.1.3

γ_t	=	prestress factor used in torsion design
ϵ_{ps}	=	strain in prestressing steel corresponding to f_{ps}
ϵ_{sa}	=	strain in prestressing steel caused by external loads $= \epsilon_{ps} - \epsilon_{se}$
ϵ_{se}	=	strain in prestressing steel after losses
λ	=	a conversion factor for shear in lightweight concrete
μ	=	shear-friction coefficient
μ_e	=	effective shear-friction coefficient
ρ_p	=	A_{ps}/bd = ratio of prestressed reinforcement
ϕ	=	strength reduction factor
ω_p	=	$\rho_p \, f_{ps}/f'_c$
$\overline{\omega}_p$	=	$\rho_p \, f_{pu}/f'_c$

3.1.2 Introduction

This part of the Handbook provides a summary of design theory and procedures used in the design of precast, prestressed members. Designs are based on the provisions of the "Building Code Requirements for Reinforced Concrete (ACI 318-77)", of the American Concrete Institute (often referred to as "the Code" in this Handbook). Design procedures will vary among experienced designers. The procedures described in this part are typical, but not mandatory.

Rarely will the load tables in Part 2 of this Handbook provide all the design data necessary. In most cases, the engineer will select a standard section, with the detailed design calculations furnished by the staff or consulting engineer of the local prestressed concrete producer. Under some circumstances, the engineer of record will furnish complete designs. When complete designs are furnished, some economy may be realized if the producers are permitted to suggest modifications to better fit their own production procedures.

3.2 Flexure

Design for flexure in accordance with the Code requires that prestressed concrete members be checked for flexural stresses at transfer and service loads, and for flexural design strength.

3.2.1 Strength Design

Strength design is based on the solution of the equations of equilibrium, normally using the rectangular stress block in accordance with Sect. 10.2.7 of the Code (see Fig. 3.2.1). The steel stress at nominal strength (f_{ps}) can be determined by strain compatibility or by the approximate equation given in the Code (Eq. 18-3).

Fig. 3.2.1 Strength design relationships

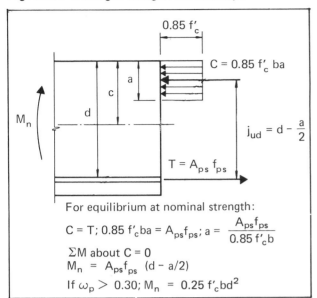

For equilibrium at nominal strength:

$C = T$; $0.85 f'_c ba = A_{ps}f_{ps}$; $a = \dfrac{A_{ps}f_{ps}}{0.85 f'_c b}$

ΣM about $C = 0$

$M_n = A_{ps}f_{ps} (d - a/2)$

If $\omega_p > 0.30$; $M_n = 0.25 f'_c bd^2$

The designer will normally choose a section and reinforcement and then determine if it has the capacity to meet the design requirements.

Critical section:

Because of the Code limitation on end stresses at release (see Sect. 3.2.2), prestressing strands are often draped or depressed, producing a varying effective depth, d, along the member length. For this reason, and when non-uniform loading is applied, it may be necessary to check the applied factored moment (M_u) and moment strength (ϕM_n) at points other than midspan. This is illustrated in Fig. 3.2.2. For uniform loads with straight strands, it is sufficient to check the midspan condition. For uniform loads with single point depressed strands, the critical point is usually near 0.4ℓ, and it is common practice to check only at midspan or 0.4ℓ.

3.2.1.1 Analysis Using Approximate Equation

Eq. 18-3 of the Code is a convenient, conservative method of determining the value of f_{ps}, the only unknown in solving for the design capacity as illustrated in Fig. 3.2.1. Such a solution is illustrated in Example 3.2.1. Eq. 18-3 is:

$$f_{ps} = f_{pu} \left(1 - 0.5\rho_p \frac{f_{pu}}{f'_c}\right)$$

Design aids:

Tables 3.9.1 through 3.9.4 are provided to assist in the strength design of flexural members using the approximate equation in the Code. Table 3.9.1 yields a coefficient, K_u, used to determine flexural design strength. The factor $\phi = 0.90$ is included in the table. It should be noted that whenever $\overline{\omega}_p$ exceeds 0.368, ω_p exceeds 0.30, and the flexural design strength is controlled by compression as stated in Section 18.8 (ACI 318-77). Although no design aids are provided for this situation, the flexural strength can easily be computed using the expression $\phi M_n = \phi (0.25 f'_c bd^2)$* for rectangular sections or flanged sections in which the neutral axis lies within the flange.

---

*ACI 318-77 has introduced a new system of notation to be used in strength design of reinforced concrete. Previously, the subscript "u" denoted either the applied factored forces (M_u, V_u, P_u, etc.) or the design strength (termed "ultimate strength" prior to 1971). In the 1977 edition of the Code, the subscript "u" denotes only the applied factored forces. The subscript "n" denotes the "nominal strength." The "design strength" (or "ultimate strength") is the nominal strength multiplied by the strength reduction factor, ϕ, for example ϕM_n, ϕV_n, ϕP_n. Thus, the design of a member or component requires that $M_u \leqslant \phi M_n$, $V_u \leqslant \phi V_n$, $P_u \leqslant \phi P_n$.

In presenting equations for strength design, this leads to the apparent algebraic redundancy of having the term ϕ on both sides of the equation. The Committee on Design Handbook decided, however, that to avoid the inadvertent neglect of the ϕ-factor, the equations should be in terms of "design strength" rather than "nominal strength."

Fig. 3.2.2 Critical sections for flexural design

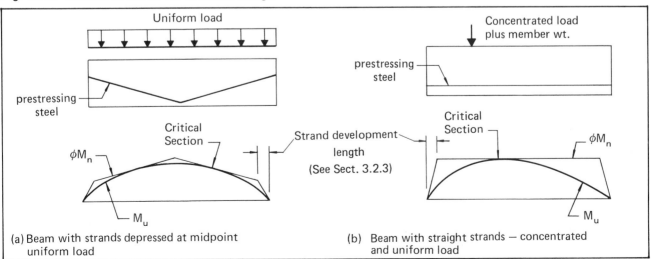

(a) Beam with strands depressed at midpoint uniform load

(b) Beam with straight strands — concentrated and uniform load

Table 3.9.2 is an extension of Table 3.9.1 used to determine the location of the neutral axis. In a flanged section in which the neutral axis falls outside the flange, Table 3.9.1 is not valid and the flexural strength is determined as shown in Example 3.2.6. Table 3.9.2 includes values of f_{ps}/f_{pu} for comparison with values obtained from stress-strain compatibility (Figs. 3.9.5, 3.9.6, and 3.9.7).

Table 3.9.3 presents the same data as that in Table 3.9.1 but in a form more suitable for determination of the steel area required to resist a given factored moment.

Table 3.9.4 furnishes coefficients to select b and d for a given factored moment and K_u, or to determine the flexural strength for a given b and d with K_u known.

Use of these design aids is shown in Examples 3.2.1 through 3.2.4.

Example 3.2.1 — Determing flexural strength by use of approximate equation or by use of Table 3.9.1.

Given:

3'-4'' x 8'' hollow core

Concrete:

f'_c = 5000 psi
Normal weight

Prestressing steel:

10 - 3/8'' diameter 250K strand
A_{ps} = 10 x 0.080 = 0.800 sq in.

Problem:

Find design flexural strength, ϕM_n

Solution:

By use of Code Equation 18-3:

Determine $\rho_p = \dfrac{A_{ps}}{bd} = \dfrac{0.800}{(40)(7)} = 0.00286$

$f_{ps} = f_{pu}\left(1 - 0.5\rho_p \dfrac{f_{pu}}{f'_c}\right)$

$= 250\left[1 - 0.5(0.00286)\left(\dfrac{250}{5}\right)\right]$

$= 232.1$ ksi

$a = \dfrac{A_{ps}f_{ps}}{0.85\,f'_c b} = \dfrac{0.800(232.1)}{0.85(5)(40)} = 1.09$ in.

$\phi M_n = \phi A_{ps} f_{ps}(d - a/2)$

$= 0.90(0.800)(232.1)\left(7 - \dfrac{1.09}{2}\right)$

$= 1079$ in-kips $= 89.9$ ft-kips

To use Table 3.9.1, determine

$\overline{\omega}_p = \dfrac{A_{ps}f_{pu}}{bd\,f'_c} = \dfrac{0.800(250)}{(40)(7)(5)} = 0.143$

Table 3.9.1 yields the following values of K_u for f'_c = 5000 psi:

for $\overline{\omega}_p$ = 0.14, K_u = 540

for $\overline{\omega}_p$ = 0.15, K_u = 573

Interpolating for $\overline{\omega}_p$ = 0.143 yields K_u = 550

The design flexural strength thus becomes

$\phi M_n = \dfrac{K_u bd^2}{12000} = \dfrac{550(40)(7)^2}{12000} = 89.8$ ft-kips

Example 3.2.2 — Use of Table 3.9.2 Coefficients for determining flexural design strength

Given:

PCI standard double tee
8DT24 + 2

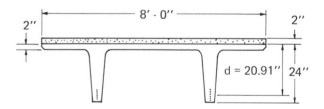

Concrete:

Precast: f'_c = 5000 psi
Topping: f'_c = 3000 psi
Normal weight

Prestressing steel:

12 - 7/16'' diameter 270K strand (6 in each stem)
A_{ps} = 12 x 0.115 = 1.38 sq in.

Problem:

Find the design flexural strength for both the precast section and the composite section.

Solution:

A. *Precast section:*

To use Table 3.9.2 first determine $\overline{\omega}_p$ for the precast section:

$$\overline{\omega}_p = \frac{A_{ps} f_{pu}}{bd f'_c} = \frac{1.38}{8 \times 12 \times 20.91} \times \frac{270}{5}$$
$$= 0.037$$

For $f'_c = 5000$ psi and $\overline{\omega}_p = 0.037$, Table 3.9.2 yields (by interpolation):

$c/d = 0.054$
$f_{ps}/f_{pu} = 0.982$
$\overline{j}_u = 0.960$
$c = 0.054 \times d$
$\quad = 0.054 \times 20.91 = 1.13$ in. < 2.0 in.

This shows that the neutral axis lies within the flange, and that the section can be analyzed like a rectangular one as follows:

$\phi M_n = \phi A_{ps} f_{pu} \overline{j}_u d$
$\quad = 0.9 \times 1.38 \times 270 \times 0.960 \times 20.91/12$
$\quad = 561$ ft-kips
$f_{ps} = 0.982 f_{pu}$
$\quad = 0.982 \times 270 = 265$ ksi

B. Composite section:

Similar to the procedure outlined above, first determine $\overline{\omega}_p$ with modified values for f'_c and d:

$$\overline{\omega}_p = \frac{A_{ps} f_{pu}}{bd f'_c} = \frac{1.38}{8 \times 12 \times 22.91} \times \frac{270}{3}$$
$$= 0.056$$

For $f'_c = 3000$ psi, and $\overline{\omega}_p = 0.056$, Table 3.9.2 yields (by interpolation):

$c/d = 0.075$
$f_{ps}/f_{pu} = 0.972$
$\overline{j}_u = 0.941$
$c = 0.075 \times d$
$\quad = 0.075 \times 22.91 = 1.72$ in. < 4.0 in.

The neutral axis lies within the flange.

$\phi M_n = \phi A_{ps} f_{pu} \overline{j}_u d$
$\quad = 0.9 \times 1.38 \times 270 \times 0.941 \times 22.91/12$
$\quad = 602$ ft-kips
$f_{ps} = 0.972 \times f_{pu}$
$\quad = 0.972 \times 270 = 262$ ksi

Example 3.2.3 — Use of Table 3.9.3 Coefficients for determination of prestressing steel requirements — bonded prestressing steel

Given:

PCI standard rectangular beam 16RB24

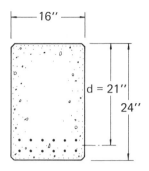

Concrete:

$f'_c = 6000$ psi
Normal weight

Applied factored moment, M_u, under dead and live load = 600 ft-kips.

Problem:

Find the required amount of prestressing steel based on $f_{pu} = 270$ ksi.

Solution:

To use Table 3.9.3, determine:

$$K_u = \frac{M_u (12,000)}{bd^2} \quad \text{(for } M_u \leqslant \phi M_n)$$
$$= \frac{600 \times 12,000}{16 \times 21^2} = 1020$$

Table 3.9.3 yields for this value, and for $f'_c = 6000$ psi:

$\overline{\omega}_p = 0.247$

The required amount of prestressing steel, thereby, becomes:

$A_{ps} = \overline{\omega}_p bd f'_c/f_{pu}$
$\quad = 0.247 \times 16 \times 21 \times 6/270$
$\quad = 1.84$ sq in.

Provide 12 - 1/2" diameter 270K strand

Example 3.2.4 — Use of Table 3.9.4 Coefficients for resisting moments of rectangular and T-sections

Given:

PCI standard rectangular beam 16RB24
Applied factored moment
$M_u = 600$ ft-kips

Problem:

Find the value, $K_u = \dfrac{M_u}{C_f}$, for use with other design aids to determine the required prestressing steel (see Example 3.2.3).

Solution:

To find C_f, enter Table 3.9.4 with b = 16 in. and d = 21 in. The latter value requires interpolation between d = 20 in. and d = 22 in. Linear interpolation is accurate enough and yields C_f = 0.589.

$$K_u = \frac{M_u}{C_f} = \frac{600}{0.589} = 1019$$

3.2.1.2 Analysis using Strain Compatibility

Figs. 3.9.5, 3.9.6, and 3.9.7 represent a plot of the stress induced in the steel as a ratio of its ultimate strength based on stress-strain compatibility. It can be seen from Fig. 3.9.5 for bonded strand that for high values of $C\overline{\omega}_p$, 5 to 15 percent savings in steel can be achieved using this method as compared to Eq. 18-3 of the Code. After obtaining f_{ps} from the above figures, the design flexural strength is determined using the expression $\phi M_n = \phi A_{ps}d(1 - 0.59 \omega_p)$ or $\phi M_n = \phi A_{ps} f_{ps} (d - a/2)$ as shown in Example 3.2.5.

Example 3.2.5 — Use of Fig. 3.9.5 Values of f_{ps} by stress-strain relationship — bonded strand

Given:

PCI standard inverted tee beam 24IT36

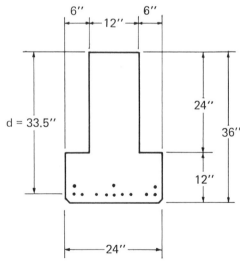

Concrete:

f'_c = 5000 psi
Normal weight

Prestressing steel:

12-1/2″ diameter 270K strand
A_{ps} = 12 x 0.153 = 1.84 sq in.

Problem:

Find the design flexural strength, ϕM_n

Solution:

Determine $C\overline{\omega}_p$ for the section:

$$C\overline{\omega}_p = C \frac{A_{ps} f_{pu}}{bd f'_c}$$

$$= 1.06 \times \frac{1.84}{12 \times 33.50} \times \frac{270}{5} = 0.262$$

Entering Fig. 3.9.5 with this parameter and an assumed effective stress of f_{se} = 150 ksi gives a value of:

$$f_{ps}/f_{pu} = 0.93 \text{ or } f_{ps} = 0.93 \times 270 = 251 \text{ ksi}$$

(Note: This value is higher than the value obtained by using Eq. 18-3 (ACI 318-77). From Fig. 3.9.5, f_{ps}/f_{pu} by Eq. 18-3 equals 0.87 or f_{ps} = 0.87 x 270 = 235 ksi).

Determine the design flexural strength:

$$\phi M_n = \phi A_{ps} f_{ps} (d - a/2)$$

$$a = \frac{A_{ps} f_{ps}}{0.85 f'_c b}$$

$$= \frac{1.84 \times 251}{0.85 \times 5 \times 12} = 9.06 \text{ in.}$$

$$\phi M_n = 0.90 \times 1.84 \times 251 (33.50 - 9.06/2)/12$$
$$= 1003 \text{ ft-kips}$$

(Note: Using f_{ps} = 235 ksi by Eq. 18-3; a = 8.48 in. and ϕM_n = 949 ft-kips)

3.2.1.3 Flanged Members with Neutral Axis Below Flange

The expression $\phi M_n = \phi A_{ps} f_{ps} (d - a/2)$ applies only to rectangular sections and flanged sections in which the compression block (dimension a in Fig. 3.2.1) lies within the flange. Dimension a = $\beta_1 c$. Values of β_1 (see Code Sect. 10.2.7) are shown as follows:

f'_c, psi	β_1
3000	0.85
4000	0.85
5000	0.80
6000	0.75
7000	0.70
8000 and higher	0.65

If the compression block is deeper than the flange, fundamental strain compatibility analysis may be required to calculate the design moment capacity. This procedure is illustrated in Example 3.2.6.

Example 3.2.6 — Flexural design strength of flanged section.

Given:

Inverted tee beam with 2 in. composite topping as shown

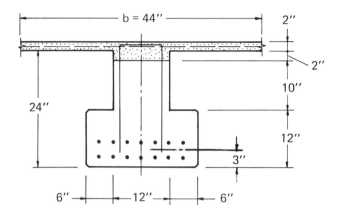

Concrete

f'_c (precast) = 5000 psi

f'_c (topping) = 3000 psi

Prestressing steel:

14-1/2″ diameter 270K strands

A_{ps} = 14 x 0.153 = 2.142 sq in.

Problem:

Find design flexural strength, ϕM_n.

Solution:

Determine effective flange width, b

From Sect. 8.10.2 of the Code, overhanging width = 8 times thickness.

b = b_w + 2 x 8t = 12 + 2(8)(2) = 44 in.

d = 26 − 3 = 23 in.

See strain diagram below:

$$\frac{\epsilon_{sa} + 0.003}{0.003} = \frac{d}{c}$$

$$\epsilon_{sa} = \frac{0.003d}{c} - 0.003 = \frac{0.069}{c} - 0.003$$

Assume 22% loss and E_s = 27,500 ksi

$$f_{se} = 0.78\,(0.7 \times 270) = 147.4 \text{ ksi}$$

$$\epsilon_{se} = \frac{f_{se}}{E_s} = \frac{147.4}{27,500} = 0.0054 \text{ in./in.}$$

$$\epsilon_{ps} = \epsilon_{sa} + \epsilon_{se} = \frac{0.069}{c} - 0.003 + 0.0054$$

$$= \frac{0.069}{c} + 0.0024$$

Try c = 11 in.

$$\epsilon_{ps} = \frac{0.069}{11} + 0.0024 = 0.0087 \text{ in./in.}$$

From stress-strain curve, Fig. 8.2.5

f_{ps} = 234 ksi

a_3 (for 5000 psi portion) = 0.80(11) − 4 = 4.8″

C_1 = 0.85 (3)(2)(44)	=	224.4 kips
C_2 = 0.85 (3)(2)(12)	=	61.2 kips
		285.6
C_3 = 0.85 (5)(4.8)(12)	=	244.8
		530.4 kips

T = 2.142(234) = 501.2 < 530.4

Try c = 10.5 in.

$$\epsilon_{ps} = \frac{0.069}{10.5} + 0.0024 = 0.0090$$

f_{ps} = 238 ksi

a_3 = 0.80(10.5) − 4 = 4.40″

$C_1 + C_2$ =		285.6 kips
C_3 = 0.85(5)(4.40)(12) =		224.4
		510.0 kips

$$T = 2.142(238) = 509.8 \approx 510.0 \text{ OK}$$

$$M_n = 224.4(23 - 1) + 61.2(23 - 3)$$
$$\qquad + 224.4(23 - 4 - \frac{4.40}{2})$$
$$\qquad = 4937 + 1224 + 3770 = 9931 \text{ in.-kips}$$

$$\phi M_n = 0.9(9931) = 8938 \text{ in.-kips}$$
$$\qquad = 744.8 \text{ ft-kips}$$

Check for tension or compression control (See ACI 318-77, Sect. 18.8.1)

Steel required to develop web:

$$A_{pw} = \frac{(b_w/b)C_1 + C_2 + C_3}{f_{ps}}$$
$$\qquad = \frac{61.2 + 61.2 + 224.4}{238} = 1.457 \text{ sq in.}$$

$$\omega_{pw} = \frac{A_{pw}\,f_{ps}}{b_w\,d\,f'_c} \quad \text{(use ave } f'_c = 4050 \text{ psi)}$$
$$\qquad = \frac{1.457(238)}{12(23)(4.05)} \; 0.31 > 0.30$$

Compression controls. Calculate moment strength on the basis of the compression moment couple. Or modify design with higher strength topping concrete or added nonprestressed reinforcement.

Note: See Sect. 3.3.4 for tie requirements for horizontal shear at interface.

When the concrete strength is the same throughout the member, a simpler approach may be used. This is illustrated in Sect. 10.3.1 of "Commentary on Building Code Requirements for Reinforced Concrete (ACI 318-77)."

3.2.2 Service Load Stresses

Unlike concrete reinforced with non-prestressed steel, the ACI Code requires that service load stresses be checked at critical points, in addition to the design strength of the member. Code limitations on the service load stresses are summarized as follows (see Code Sects. 18.4 and 18.5):

Concrete:

1. At release (transfer) of prestress, before time-dependent losses:
   a. Compression $0.60\ f'_{ci}$
   b. Tension (except at ends) $3\sqrt{f'_{ci}}$
   c. Tension at ends of simply supported members $6\sqrt{f'_{ci}}$

2. Under service loads:
   a. Compression $0.45\ f'_c$
   b. Tension in precompressed

tensile zone when deflections are calculated based on gross section $6\sqrt{f'_c}$

  c. Tension in precompressed tensile zone when deflections are calculated based on bilinear relationships (see Sect. 3.4.2) $12\sqrt{f'_c}$

Prestressing steel:

  a. Tension due to tendon jacking force: $0.80\ f_{pu}$ or $0.94\ f_{py}$
  b. Tension immediately after prestress transfer: $0.70\ f_{pu}$

It is common practice in the precast, prestressed concrete industry to follow the above recommendations with the following clarifications.

Tension in precompressed tensile zone at service loads:

Hollow-core and solid flat slabs — $6\sqrt{f'_c}$

Stemmed deck members and beams* — $12\sqrt{f'_c}$

Initial stress in steel due to jacking force — $0.7\ f_{pu}$

These values should not be exceeded without consulting the product manufacturer.

Calculations of stresses at critical points follow classical straight line theory as illustrated in Fig. 3.2.3.

Critical sections:

For stresses immediately after transfer, the most critical section is usually near the end of the member, although in members with single-point depressed strands, the release stresses at midspan may also be critical and should be checked. The actual critical end stress is at the point where the prestressing force has been completely transferred to the concrete, usually assumed to be 50 strand diameters from the end. For convenience, using hand calculation, it is normal practice to calculate the stress at the end (assuming full transfer), and check the transfer point only if necessary to meet Code requirements.

If release stresses are higher than allowed by Code, it may be necessary to either increase the specified release strength, or provide supplemental tensile reinforcement. In short span, heavily loaded members, such as beams, it is usually more practical to reinforce for the release tension. This is illustrated in Example 3.2.9.

Under uniform service loads, the critical section is at midspan for members with straight

---

*Many prestressed concrete designers prefer to hold beam tension below the cracking level. For most standard beam sections, the maximum capacity is controlled by strength rather than stresses.

Fig. 3.2.3 Calculation of service load stresses

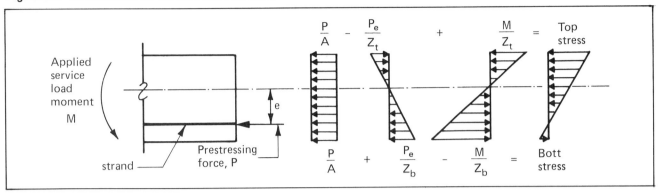

strands and near 0.4ℓ from the end (ℓ = span) for members with strand depressed at midpoint. The exact critical point can be determined by a detailed analysis, but designing for critical stresses at the midspan and 0.4ℓ points will usually determine the capacity within 1 or 2 percent.

Strand profiles other than straight or single-point depressed are uncommon in building products and should be avoided unless absolutely necessary. Straight strands are used in flat deck members and beams and either straight or depressed are used in stemmed deck members.

For unusual loading conditions, such as concentrated loads, other sections may be critical. In these cases it is common to estimate where the critical section may be, and check two or three points near that estimated position.

Composite members:

It is usually more economical to place cast-in-place composite topping without shoring the member, especially for deck members. This means that the weight of the topping must be carried by the precast member alone. Additional superimposed dead and live loads are carried by the composite section.

The following examples illustrate a tabular form of superimposing the stresses caused by the prestress force and the dead and live load moments.

Sign Convention:

The customary sign convention used in the design of precast, prestressed concrete members for service load stresses is positive (+) for compression and negative (−) for tension. This convention is used throughout this Handbook.

Example 3.2.7 — Calculation of critical stresses — straight strands
Given:

Span = 36'-0"

Select 4HC12 + 2

Superimposed dead load = 20 psf = 80 plf
Superimposed live load = 50 psf = 200 plf

Precast concrete:
    f'_c = 6000 psi
    f'_{ci} = 4000 psi
    E_c = 4700 ksi
    Normal weight

Topping concrete:
    f'_c = 4000 psi
    E_c = 3800 ksi
    Normal weight

Prestressing steel:
    5-1/2 in. dia. 270K strands
    A_{ps} = 5 x 0.153 = 0.765 sq in.
    Straight strands

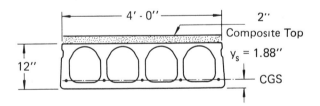

Section properties:

Non-composite		Composite
A	= 265 sq in.	361 sq in.
I	= 4771 in.⁴	7209 in.⁴
y_b	= 6.67 in.	8.10 in.
y_t	= 5.33 in.	5.90 in.
Z_b	= 715.3 in.³	890.0 in.³
Z_t	= 895.1 in.³	1221.9 in.³
wt	= 276 plf	376 plf
e	= 4.79 in.	

Problem:

Find critical service load stresses

Solution:

Prestress force:

P_i = 0.765 (0.7 × 270) = 145 kips

P_o = (Assume 10% initial loss)
= 0.90 (145) = 130 kips

P = (Assume 22% total loss)
= 0.78 (145) = 113 kips

Midspan service load moments:

M_d = 0.276 (36)² (12/8) = 537 in.-kips
M_{top} = 0.100 (36)² (12/8) = 194 in.-kips

M_{sd} = 0.080 (36)² (12/8) = 156 in.-kips
M_ℓ = 0.200 (36)² (12/8) = 389 in.-kips

Allow $6\sqrt{f'_c}$ tension at service load

Load	Support at Release $P = P_o$		Midspan at Release $P = P_o$		Midspan at Service load $P = P$	
	f_b	f_t	f_b	f_t	f_b	f_t
P/A	+ 491	+ 491	+ 491	+ 491	+ 426	+ 426
Pe/Z	+ 871	− 696	+ 871	− 696	+757	− 605
M_d/Z			− 751	+ 600	− 751	+ 600
M_{top}/Z					− 271	+ 217
M_{sd}/Z^c					− 175	+ 128
M_ℓ/Z^c					− 437	+ 318
Stresses	+ 1362	− 205	+611	+ 395	− 451	+ 1084
Allowable Stresses	0.6 f'$_{ci}$	6 $\sqrt{f'_{ci}}$	0.6 f'$_{ci}$	0.6 f'$_{ci}$	6 $\sqrt{f'_c}$	0.45 f'$_c$
	2400	− 379	2400	2400	− 465	2700
	OK	OK	OK	OK	OK	OK

Example 3.2.8 — Calculation of critical stresses — single point depressed strand

Given:

Span = 70 ft
Superimposed dead load = 10 psf = 80 plf
Superimposed live load = 35 psf = 280 plf

Select 8DT24 as shown

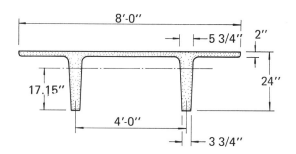

Concrete:

f'_c = 5000 psi
f'_{ci} = 3500 psi
Normal weight

Prestressing steel

14-1/2'' dia. 270K strands
A_{ps} = 14 × 0.153 = 2.142 sq in.

Section properties

A = 401 sq in.
I = 20,985 in.⁴
y_b = 17.15 in.
y_t = 6.85 in.
Z_b = 1224 in.³
Z_t = 3063 in.³
wt = 418 plf = 52 psf

Eccentricities, single point depression

e_e = 4.29 in., e_c = 13.65 in.
e @ 0.4ℓ = 11.78 in.
e' = 13.65 − 4.29 = 9.36 in.

Problem:

Find critical service load stresses

Solution:

Prestress force:

P_i = 2.142 (0.7 × 270) = 405 kips
P_o = (Assume 10% initial loss)
= 0.90 (405) = 364 kips
P = (Assume 24% total loss)
= 0.76 (405) = 308 kips

Service load moments:
at midspan:

$M_d = 0.418(70)^2 (12/8) = 3072$ in.-kips
$M_{sd} = 0.080(70)^2 (12/8) = 588$ in.-kips
$M_\ell = 0.280(70)^2 (12/8) = 2058$ in.-kips
at 0.4ℓ:
$M_d = 3072 \times 0.96 = 2949$ in.-kips
$M_{sd} = 588 \times 0.96 = 564$ in.-kips

$M_\ell = 2058 \times 0.96 = 1976$ in.-kips
Allow $12\sqrt{f'_c}$ final tension

In this example, a release strength of $f'_{ci} = \dfrac{2457}{0.6} = 4095$ psi should be provided. Also deflection should be checked.

Load	Support at Release $P = P_o$		Midspan at Release $P = P_o$		0.4ℓ at Service load $P = P$	
	f_b	f_t	f_b	f_t	f_b	f_t
P/A	+ 908	+908	+ 908	+ 908	+ 768	+ 768
Pe/Z	+1276	−510	+4059	− 1622	+ 2964	− 1185
M_d/Z			− 2510	+ 1003	− 2409	+ 963
M_{sd}/Z					− 461	+ 184
M_ℓ/Z					− 1614	+ 645
Stresses	+ 2184	+ 398	+ 2457	+ 289	− 752	+ 1375
Allowable Stresses	$0.60\,f'_{ci}$ +2100	$0.60\,f'_{ci}$ +2100	$0.60\,f'_{ci}$ 2100	$0.60\,f'_{ci}$ 2100	$12\sqrt{f'_c}$ − 848	$0.45\,f'_c$ 2250
	HIGH	OK	HIGH	OK	OK	OK

Example 3.2.9 — Use of Fig. 3.9.8 — Tensile force to be resisted by top reinforcement

Given:

Span = 24 ft
24IT26 as shown

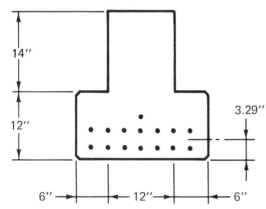

Concrete:
$f'_c = 6000$ psi
$f'_{ci} = 4000$ psi

Prestressing steel
15-1/2'' 270K strands
$A_{ps} = 15 (0.153) = 2.295$ sq in.

Section properties:
A = 456 sq in.
I = 24,132 in.[4]
$y_b = 10.79$ in.

$y_t = 15.21$ in.
$Z_b = 2237$ in.[3]
$Z_t = 1587$ in.[3]
wt = 475 plf
e = 7.5 in.

Problem:

Find critical service load stresses

Solution:

Prestress force:
$P_i = 2.295 (189) = 434$ kips
P_o = (Assume 10% initial loss)
    = 0.90 (434) = 391 kips

Moment due to member weight:
at midspan
$M_d = 0.475 (24)^2 (12/8) = 410$ in.-kips
at 50 strand diameters (2.08 ft) (transfer point)

$M_d = \dfrac{wx}{2}(\ell - x) = \dfrac{0.475 (2.08)}{2}$

(24 − 2.08) (12) = 130 in.-kips

(stresses are tabulated on p. 3—14)

Since the tensile stress is high, reinforcement is required to resist the total tensile force.

Load	Transfer point at Release $P = P_o$		Midspan at Release $P = P_o$	
	f_b	f_t	f_b	f_t
P/A	+ 857	+ 857	+ 857	+ 857
Pe/Z	+ 1311	− 1848	+ 1311	− 1848
M_d/Z	− 58	+ 82	− 183	+ 258
Stresses	+ 2110	− 909	+ 1985	− 733
Allowable Stresses	0.6 f'_{ci}	6 $\sqrt{f'_{ci}}$	0.6 f'_{ci}	3 $\sqrt{f'_{ci}}$
	2400	− 379	2400	− 190
	OK	HIGH	OK	HIGH

$$c = \frac{f_t}{f_t - f_b}(h) = \frac{-909}{-909 - 2110} \quad (26)$$

$$= 7.83 \text{ in.}$$

$$T = \frac{c\,f_t b}{2} = \frac{7.83\,(909)\,(12)}{2} = 42,700 \text{ lb}$$

Alternatively, Fig. 3.9.8 may be used.

Enter the chart at left with

f_t = 909 psi, proceed to right to f_b

= 2110 psi, then up to bh = 12 x 26

= 312 sq in., and right again to value of T

= 42.5 kips

Similarly, the tension at midspan can be read as 30.5 kips.

The Commentary to the Code, Sect. 18.4.1(b) and (c), recommends that reinforcement be proportioned to resist this tensile force at a stress of 0.6f_y, but not more than 30 ksi. Using reinforcement with f_y = 60 ksi:

0.6 (60) = 36 ksi, use 30 ksi.

$$A_s \text{ (end)} = \frac{42.5}{30} = 1.42 \text{ sq in.}$$

$$A_s \text{ (midspan)} = \frac{30.5}{30} = 1.02 \text{ sq in.}$$

3.2.3 Prestress Transfer and Strand Development

In a pretensioned member, the prestress force is transferred to the concrete by bond. The length required to accomplish this transfer is called the "transfer length," and is assumed by the Code to be approximately 50 times the nominal diameter of the strand.

However, the length required to develop the full design strength of the strand is much longer, and is specified in the Code in Sect. 12.10 by the equation:

$$\ell_d = (f_{ps} - 2/3\,f_{se})\,d_b$$

In the Commentary to the Code, the variation of strand stress along the development length is shown as in Fig. 3.2.4.

For convenience, this curve may be approximated by straight lines. Also, to be consistent with the 50 diameter transfer length specified in other Code sections, the value of f_{se}, the stress which must be transferred is assumed to be 150 ksi.

Fig. 3.9.9 is a curve plotted according to the above assumptions, and can be used as a design aid as illustrated in Example 3.2.10.

Fig. 3.2.4 Variation of steel stress with development length

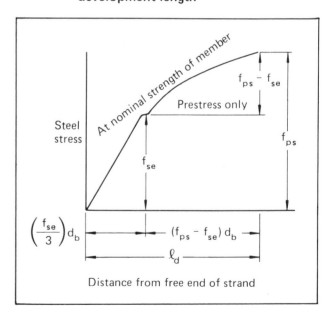

Example 3.2.10 − Use of Fig. 3.9.9 − Design stress for underdeveloped strand.

Given:

Span = 12 ft

4HC8 as shown

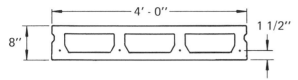

Concrete:

f'_c = 5000 psi

Normal weight

Prestressing steel

4-1/2" diameter 270K strands

A_{ps} = 4 x 0.153 = 0.612 sq in.

If the strand is fully developed (see Fig. 3.9.5):

$$C\bar{\omega}_p = C \frac{A_{ps} f_{pu}}{bd f'_c} = \frac{1.06 (0.612)(270)}{48 (6.5)(5)}$$

$$= 0.11$$

From Fig. 3.9.5 with f_{se} = 150 ksi

f_{ps}/f_{pu} = 0.97

f_{ps} = 0.97 (270) = 262 ksi

The maximum development length =

$\ell/2$ = 12 x 12/2 = 72 in.

From Fig. 3.9.9, the maximum f_{ps} = 244 ksi

This value should be used to calculate the design strength (ϕM_n) of the member.

3.2.4 Partial Prestressing

Partial prestressing, that is, using a combination of prestressed and non-prestressed steel is a legitimate alternative to employing only prestressing steel and can in many instances provide an economical solution. Although many aspects of partial prestressing are still in the developmental stage, guidance into this method of design can be obtained from the article "Design of Partially Prestressed Concrete Flexural Members" by Saad Moustafa, published in the May-June 1977 *PCI Journal.* In addition, the Readers Comments to this article, published in the May-June 1978 *PCI Journal,* are useful.

3.3. Shear

For flat deck members (hollow-core and solid slabs) and others proven by test, no shear reinforcement is required if the factored shear force, V_u, does not exceed the design shear strength of the concrete, ϕV_c. For other members, the minimum shear reinforcement is usually adequate. (Note: If V_u is less than $\phi V_c/2$, no shear reinforcement is required.)

3.3.1 Use of Code Equations

Shear design of prestressed concrete members is covered in ACI 318-77 by Equations 11-10 through 11-13. These equations are expressed in terms of forces. However, it is generally more convenient to work shear design problems using unit stresses. The factored shear stress can be expressed by:

$$v_u = \frac{V_u}{\phi b_w d} \qquad \text{(Eq. 3.3-1)}$$

where ϕ = 0.85.

Equations 11-10 through 11-13 of ACI 318-77 can be expressed in terms of unit stresses by Eqs. 3.3-2 through 3.3-5, respectively:

$$v_c = 0.6\sqrt{f'_c} + 700\frac{V_u d}{M_u} \qquad \text{(Eq. 3.3-2)}$$

$$\frac{V_u d}{M_u} \leqslant 1$$

$$v_{ci} = 0.6\sqrt{f'_c} + \frac{V_d + \dfrac{V_i M_{cr}}{M_{max}}}{b_w d} \qquad \text{(Eq. 3.3-3)}$$

$$M_{cr} = \left(\frac{I}{y_b}\right)(6\sqrt{f'_c} + f_{pe} - f_d) \qquad \text{(Eq. 3.3-4)}$$

$$v_{cw} = 3.5\sqrt{f'_c} + 0.3f_{pc} + \frac{V_p}{b_w d} \qquad \text{(Eq. 3.3-5)}$$

An option is given in the Code for calculating the value of v_c. Either Eq. 3.3-2 or the lesser of Eqs. 3.3-3 or 3.3-5 may be used. The Code places certain upper and lower limits on the use of these equations, which are shown in Fig. 3.3.1.

The value of d in Eq. 3.3-2 must always be the distance from the extreme compression fiber to the centroid of the reinforcement. In all other equations, d need not be less than $0.8h$.

In unusual cases, such as members which carry heavy concentrated loads, or short spans with high superimposed loads, it may be necessary to construct a shear resistance diagram (v_c) and superimpose upon that a unit shear (v_u) diagram. The procedure is illustrated in Fig. 3.3.1.

The steps for constructing the shear resistance diagram are as follows:

1. Draw a horizontal line at a value of $2\sqrt{f'_c}$. (Note: The Code requires that this minimum be reduced to $1.7\sqrt{f'_c}$ when the stress in the strand after all losses is less than $0.4 f_{pu}$. For precast, prestressed members the value will always be above $0.4 f_{pu}$.)

2. Construct the curved portion of the diagram. For this, either Eq. 3.3-3 or, more conserva-

Fig. 3.3.1 Shear design

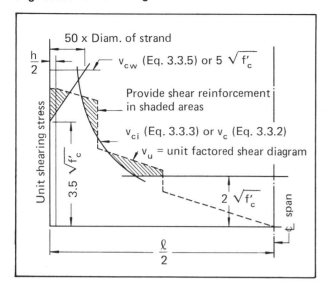

Fig. 3.3.2 Diagrams for Example 3.3.1

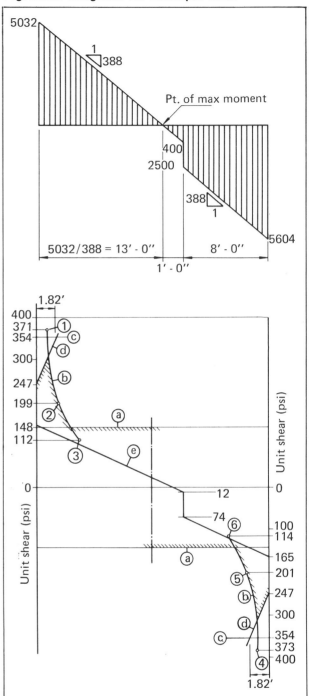

tively, Eq. 3.3-2 may be used. Usually it is adequate to find 3 points on the curve. For uniformly loaded simple span members, Fig. 3.9.10 may be used to quickly determine the points on the curve.

3. Draw the upper limit line, v_{cw} from Eq. 3.3-5 if Eq. 3.3-3 has been used in step 2, or $5\sqrt{f'_c}$ if Eq. 3.3-2 has been used.

4. The diagonal line at the upper left of Fig. 3.3.1 delineates the upper limit of the shear resistance diagram in the prestress transfer zone. This line starts at a value of $3.5\sqrt{f'_c}$ at the end of the member, and intersects the v_{cw} line or $5\sqrt{f'_c}$ line at 50 strand diameters from the end of the member.

Example 3.3.1 – Construction of applied and resisting design shear diagrams

Given:

2HC8 with span and loadings shown

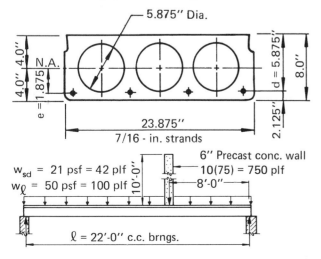

Section properties:

$A = 110$ sq in.

$I = 843$ in.^{4}

$y_b = 4.0$ in.

$b_w = 6.25$ in.

$wt = 57$ psf $= 114$ plf

Concrete:

$f'_c = 5000$ psi

Normal weight

Solution:

1. Determine factored loads

   Uniform dead $\quad = 1.4\,(42 + 114) = 218$ plf

   Uniform live $\quad\quad = 1.7\,(100) = 170$ plf

   Concentrated dead $= 1.4\,(2 \times 750) = 2100$ lbs

2. Construct shear diagram as shown in Fig. 3.3.2.

3. Construct the shear resistance diagram as described in previous section.

   a. Construct line at $2\sqrt{f'_c} = 141$ psi

   b. Construct v_c line by Eq. 3.3-2,

   $$v_c = 0.6\sqrt{f'_c} + 700\,\frac{V_u d}{M_u}$$

   $$= 42.4 + 4112.5\,\frac{V_u}{M_u}$$

   At 1', 2', and 4' from each end

   $$V_u \text{ (left)} = 5032 - 388x$$

   $$M_u \text{ (left)} = \left(5032x - \frac{388x^2}{2}\right)12$$

   $$V_u \text{ (right)} = 5604 - 388x$$

   $$M_u \text{ (right)} = \left(5604x - \frac{388x^2}{2}\right)12$$

Point	x	V_u	M_u	$4112.5\,\dfrac{V_u}{M_u}$	v_c
1	1	4644	58056	329.0	371.4
2	2	4256	111456	157.0	199.4
3	4	3480	204288	70.1	112.5
4	1	5216	64920	330.4	372.8
5	2	4828	125184	158.6	201.0
6	4	4052	231744	71.9	114.3

   c. Construct upper limit line at $5\sqrt{f'_c} = 354$ psi

   d. Construct diagonal line at transfer zone from $3.5\sqrt{f'_c} = 247$ psi at end of member to 354 psi at $50d_b = 50(7/16) = 21.9'' = 1.82'$

   e. Construct v_u diagram by dividing V_u points by $\phi b_w d = 0.85(6.25)\,(6.4) = 34$
   (Note: The Code permits the use of 0.8h in lieu of d in this equation.)

   $5032/34 = 148$

   $400/34 \quad = \quad 12$

   $2500/34 = \quad 74$

   $5604/34 = 165$

It is apparent from construction of these diagrams that no shear reinforcement is required.

3.3.2 Design using Design Aids

Figs. 3.9.10 through 3.9.15 are design aids to assist in determining the shear strength of precast prestressed members.

Lightweight Concrete:

Figs. 3.9.10, 3.9.14, and 3.9.15 employ a coefficient, $\lambda = \dfrac{f_{ct}/6.7}{\sqrt{f'_c}}$ for use with lightweight concrete. For normal weight concrete, λ is equal to 1.0. If the value of f_{ct} is not known, $\lambda = 0.85$ for sand-lightweight concrete, and 0.75 for all light-weight concrete. Figs. 3.9.11 to 3.9.13 provide separate charts for normal weight and lightweight concrete. In these charts, it is assumed that f_{ct} is not known and the material is sand-lightweight.

Fig. 3.9.10 is useful for finding the points on the v_c curve, when the procedure illustrated in Example 3.3.1 is used.

Example 3.3.2 — Use of Fig. 3.9.10 Concrete shear strength by Eq. 11-10 (ACI 318-77)

Given:

PCI standard single tee 10LST48

Simple span, uniformly loaded

2-1/2 in. topping

Concrete:

Precast:　$f'_c = 5000$ psi
　　　　　Sand-lightweight
　　　　　$f_{ct} = 470$ psi

Topping:　$f'_c = 4000$ psi
　　　　　Normal weight

Span = 80 ft

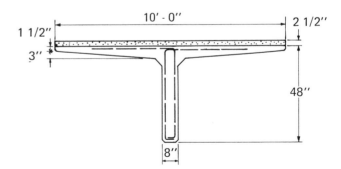

Problem:

Determine v_c at $x = 14$ ft from the support face

Solution:

$x/\ell = 14/80 = 0.175$

d (at that section) = 38 in.

$\ell/d = 80 \times 12/38 = 25$

To use Fig. 3.9.10, determine the parameters:

$$\lambda = (f_{ct}/6.7)/\sqrt{f'_c}$$
$$= (470/6.7)/\sqrt{5000} = 0.99$$
$$\lambda^2 f'_c = 0.98 \times 5000 = 4900 \text{ psi}$$

Enter Fig. 3.9.10, as the dashed arrows show, at $x/\ell = 0.175$ and proceed to the right to $\ell/d = 25$, and then downward to $\lambda^2 f'_c = 4900$. The value, v_c, can be read off the right margin as:

$$v_c = 170 \text{ psi}$$

Figs. 3.9.11 through 3.9.13 allow the graphical solution of Eq. 3.3-2 (Eq. 11-10 of ACI 318-77) for simple spans with uniform loads. They may also be used for other loadings with small error, provided the majority of the load is uniform. The charts are shown for $f'_c = 5000$ psi. However, Eq. 3.3-2 is relatively insensitive to f'_c so they can be used for strength of 4000 to 6000 psi with an error of less than 10%.

Example 3.3.3 — Use of Figs. 3.9.11 through 3.9.13 — Graphical solution of Eq. 11-10 (ACI 318-77)

Given:

PCI standard single tee 8LST36
2-1/2 in. topping

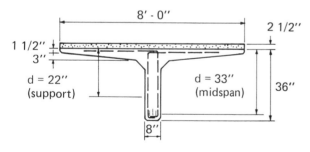

Concrete:

Precast: $f'_c = 5000$ psi
Sand-lightweight

Topping: $f'_c = 4000$ psi
Normal weight

Prestress:

13-1/2'' diameter 270K strands
$A_{ps} = 13(0.153) = 1.99$ sq in.
Single depression at midspan

Span = 70 ft
Dead load, $w_d = 723$ lbs per ft

Live load, $w_\ell = 600$ lbs per ft
Shear reinforcement $f_y = 40,000$ psi

Problem:

Find the maximum value of excess unit shear, $v_u - v_c$, along the span in compliance with Eq. 11-10 (ACI 318-77).

Solution (see Fig. 3.3.3):

To use Figs. 3.9.11 through 3.9.13, first determine the drape pattern.

The strands in this example are draped 33 - 22 = 11 inches, which is equal to $d/3$. This is defined as a shallow drape and Fig. 3.9.12 applies.

The parameters needed for use of Fig. 3.9.12 are determined as follows:

$$\ell/d = 70 \times 12/(33 + 2.5) = 23.7$$

for determining v_u, $d = 0.8h = 0.8 (38.5) = 30.8$ in.

$$v_u \text{ at support} = \frac{(1.4\ w_d + 1.7\ w_\ell)\ \ell/2}{\phi\ b_w\ d}$$

$$= \frac{(1.4 \times 723 + 1.7 \times 600) \times 70/2}{0.85 \times 8 \times 30.8}$$

$$= 340 \text{ psi}$$

The graphical solution (Fig. 3.3.3) follows these steps:

a. Draw a diagonal line from $v_u = 340$ psi at support to $v_u = 0$ at midspan

b. Draw diagonal line from $v_c = 3.5 \sqrt{f'_c}$ at support to $5 \sqrt{f'_c}$ at 50 strand diameters to meet requirements of Sect. 11.4.3 of the Code. 50 (0.5) = 25'' = 0.03ℓ

c. Draw a curved line at $\ell/d = 23.7$

d. Draw a vertical line at $d/2$ from support

e. Design shear reinforcement for shaded area. (See Example 3.3.7)

Fig. 3.9.14 is used to obtain the design shear strength at $d/2$ from the support by Code Eq. 11-13. It is conservative to assume that the excess shear $(v_u - v_c)$ varies linearly from a maximum at the support to zero at midspan for a uniformly loaded member.

Example 3.3.4 — Use of Fig. 3.9.14 Concrete shear strength at support by Eq. 11-13 (ACI 318-77)

Given:

14 in. deep double tee as shown
Span = 38 ft

Fig. 3.3.3 Solution of Example 3.3.3

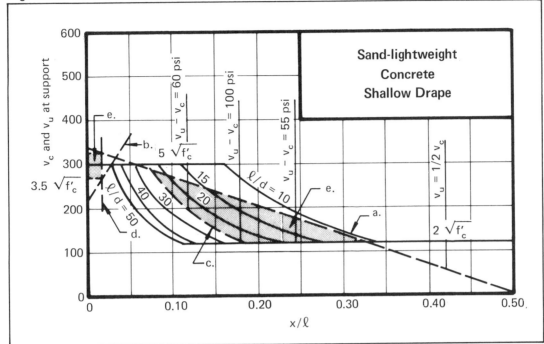

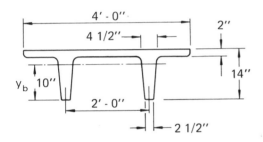

Concrete:

Precast: f'_c = 5000 psi
Normal weight

A = 180 sq in.

Prestressing steel:

6-7/16" diameter 270 K strands

Prestress force after losses, P = 108 kips

Single depression at midspan

e at support = 4.4 in. (d = 8.4 in.)

e at midspan = 7.5 in.

Problem:

Determine the nominal shear strength, V_{cw}, at the support by Eq. 11-13 (ACI 318-77).

Solution:

(Unit shears will be used for convenience.)

The parameters required to use Fig. 3.9.14 are determined as follows:

d = 0.8 h = 0.8 x 14 = 11.2 in.

(Note: d in Eq. 11-13 is the distance from the ex-

treme compression fiber to the centroid of the prestressing tendons or 0.8h, whichever is greater.)

e' = 7.5 – 4.4 = 3.1 in.

b_w = 2 x (2.5 + 4.5)/2 = 7.0 in.
(average width of stems)

$$\frac{e' A}{\ell b_w d} = \frac{3.1 \times 180}{38.0 \times 12 \times 7.0 \times 11.2} = 0.016$$

P/A = 108 x 1000/180 = 600 psi

Enter Fig. 3.9.14, as indicated by the arrow, at $e'A/\ell\, b_w d$ = 0.016. Proceed horizontally to P/A = 600 and vertically to the $\lambda^2 f'_c$ = 5000 line.

Then read at the right margin:

v_{cw} = 445 psi

If w_d = 72 psf and w_ℓ = 50 psf, then V_u = (1.4 x 72 + 1.7 x 50) x 4 x 38/2 = 14,100 lb and v_u = 14,100/(0.85 x 7.0 x 11.2) = 212 psi. Since v_u is less than one-half of v_c (in this case v_{cw}) stirrups are not required by Eq. 11-13.

Fig. 3.9.15 provides an alternate solution for v_{cw} in Eq. 11-13 (ACI 318-77) on the basis of computed principal tensile stress of $4\lambda \sqrt{f'_c}$ as provided in Sect. 11.4.2.2 (ACI 318-77).

Example 3.3.5 – Use of Fig. 3.9.15 – v_{cw} corresponding to principal tensile stress = $4\,\lambda\sqrt{f'_c}$

Given:

Same section as Example 3.3.4

Problem:

Is it advantageous to use the alternate value of V_{cw}, as stipulated in Section 11.4.2.2 (ACI 318-77), which results in a principal tensile stress of $4\lambda\sqrt{f'_c}$?

Solution:

(Unit shears will be used for convenience.)

To find the shear stress, v_{cw}, which results in a principal tensile stress of $4\lambda\sqrt{f'_c} = 4 \times \sqrt{5000}$ = 283 psi, compute the normal stress at the center of gravity of the section or at the junction of the web and the flange when the centroidal axis is in the flange.

In this example, the center of gravity is below the intersection of the web and the flange and the normal stress, hence, amounts to:

$$f_c = P/A = 108 \times 1000/180 = 600 \text{ psi}$$

(Note that the general form of the equation for normal stress is $f_c = P/A \pm Pe/Z \pm M/Z$. At the center of gravity, the last two terms of the equation drop out since $c = 0$ in the calculation $Z = I_g/c$). Enter Fig. 3.9.15 with the above value of $4\lambda\sqrt{f'_c}$ = 283 psi and proceed horizontally to the curve of f_c = 600 psi. Proceeding vertically down from the intersection, yields a value of:

$$v_{cw} = 500 \text{ psi}$$

This value is larger than the 445 psi obtained by Eq. 11-13. It is therefore, advantageous to use v_{cw} = 500 psi in $v_u - v_{cw}$ when computing the required shear reinforcement.

3.3.3 Shear Reinforcement

Shear reinforcement is required in all prestressed concrete members except as noted in Section 11.5.5 (ACI 318-77). The minimum area required by the ACI Code is determined using Eq. 11-14, $A_v = 50 b_w s/f_y$, or, alternatively, using Eq. 11-15, $A_v = \dfrac{A_{ps}}{80}\dfrac{f_{pu}}{f_y}\dfrac{s}{d}\sqrt{\dfrac{d}{b_w}}$.

Fig. 3.9.16 is a graphical solution for the minimum shear reinforcement by Eq. 11-15.

Example 3.3.6 — Use of Fig. 3.9.16 — Minimum shear reinforcement by Eq. 11-15 (ACI 318-77)

Given:

Single tee of Example 3.3.3

Stirrups

#3, 2-leg (f_y = 40 ksi)

A_v per stirrup = 2 x 0.11 = 0.22 sq in.

Problem:

Determine the amount of minimum shear reinforcement required by Eq. 11-15 of the Code.

Solution:

Determine $b_w d$ = 8 (0.8 x 38.5) = 246 sq in.

Enter Fig. 3.9.16 at left with A_{ps} = 1.99 sq in., proceed right to the line f_y grade 40 and f_{pu} = 270 ksi, then up to $b_w d$ = 246 and right to read A_v = 0.13 sq. in. per ft.

Required stirrup spacing = $\dfrac{12 \times 0.22}{0.13}$ = 20.3 in., say 20 in.

Maximum spacing (Sect. 11.5.4 of the Code) = 0.75h or 24 in.

0.75 (38.5) = 29 in.

Shear reinforcement requirements are defined in ACI 318-77 by Eq. 11-17. This equation can be rewritten in terms of unit shear stresses as:

$$A_v = \frac{(v_u - v_c)b_w s}{f_y} \qquad \text{(Eq. 3.3-6)}$$

Table 3.9.17 is used to design shear reinforcement by Eq. 3.3-6 for a given excess shear. Stirrup size, strength or spacing can be varied. Welded wire fabric may also be used for shear reinforcement in accordance with Section 11.5.1 (ACI 318-77).

Example 3.3.7 — Use of Table 3.9.17 — Shear Reinforcement

Given:

Single tee of examples 3.3.3 and 3.3.6

Problem:

Determine the required spacing for #3, two-leg stirrups (f_y = 40 ksi)

Solution:

From Table 3.9.17, the minimum stirrup spacing of 20 in. will resist $(v_u - v_c) b_w$ of 440 lb/in. or $v_u - v_c$ = 440/8 = 55 psi. By scaling from the plot in Fig. 3.3.3, 55 psi is at about 0.25ℓ = 17.5 ft from the end. The maximum value of $v_u - v_c$ can be scaled as about 100 psi. Therefore, $(v_u - v_c) b_w$ = 800. From Table 3.9.17, this requires a spacing of 11 in. Also note that from about 0.11ℓ (about 8 ft) to the support, minimum spacing is adequate. From about 0.42ℓ to midspan, $v_u < 1/2 v_c$ and no stirrups are required.

3.3.4 Horizontal Shear Transfer in Composite Members

In order for a precast, prestressed member with topping to behave compositely, full transfer of the horizontal shear forces must be assured at the interface of the precast member and the cast-in-place topping. The procedure recommended in this section is based on Section 17.5.5 of ACI 318-77.

The horizontal shear force, F_h, which must be resisted is the total force in the topping; compression in positive moment regions, and tension in negative moment regions, as shown in Fig. 3.3.4.

In a composite member which has an interface surface that is intentionally roughened, but does not have horizontal shear ties, F_h should not ex-

ceed $40\phi b_v \ell_{vh}$, where b_v is the width of the interface surface and ℓ_{vh} is the horizontal shear length as defined in Fig. 3.3.5. (Note: this is an average value over the length of the member that will result in a maximum ϕV_{nh} of approximately $80 b_v d$ for a uniformly loaded member in accordance with Section 17.5.4 of ACI 318-77.)

The area of horizontal shear ties required in length ℓ_{vh} (see Fig. 3.3.5) may be calculated by:

$$A_{cs} = \frac{F_h}{\phi \mu_e f_y} \qquad \text{(Eq. 3.3-7)}$$

where:

A_{cs} = area of horizontal shear ties, sq in.

F_h = horizontal shear force, lb

f_y = yield strength of horizontal shear ties, psi

Fig. 3.3.4 Horizontal shear in composite section

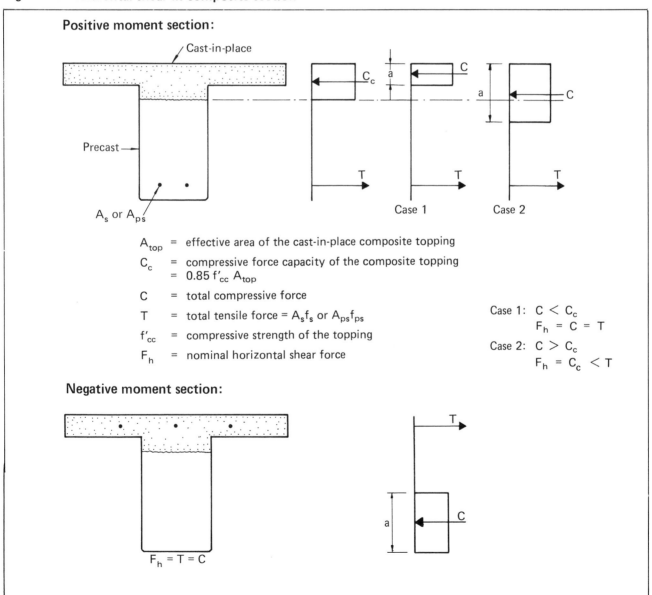

Positive moment section:

A_{top} = effective area of the cast-in-place composite topping

C_c = compressive force capacity of the composite topping
     = $0.85 f'_{cc} A_{top}$

C = total compressive force

T = total tensile force = $A_s f_s$ or $A_{ps} f_{ps}$

f'_{cc} = compressive strength of the topping

F_h = nominal horizontal shear force

Case 1: $C < C_c$
        $F_h = C = T$

Case 2: $C > C_c$
        $F_h = C_c < T$

Negative moment section:

$F_h = T = C$

Fig. 3.3.5 Horizontal shear length

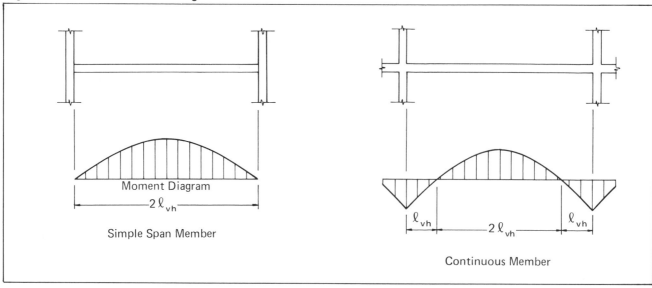

Moment Diagram

$2\ell_{vh}$

Simple Span Member

ℓ_{vh} — $2\ell_{vh}$ — ℓ_{vh}

Continuous Member

μ_e = effective shear-friction coefficient as defined in Sect. 5.6

$$= \frac{1000 \lambda^2 A_{cr} \mu}{V_u}$$

ϕ = 0.85

For composite members, $\mu = 1.0$ and $A_{cr} = b_v \ell_{vh}$, thus:

$$\mu_e = \frac{1000 \lambda^2 b_v \ell_{vh}}{F_h} \qquad \text{(Eq. 3.3-8)}$$

The value of F_h is limited to:

$$F_h \text{ (max)} = 0.25 f'_c b_v \ell_{vh} \leqslant 1000 b_v \ell_{vh} \qquad \text{(Eq. 3.3-9)}$$

where f'_c is the lesser compressive strength of the precast member or the composite topping.

When ties are provided, the minimum should be:

$$A_{cs} \text{ (min)} = \frac{120 b_v \ell_{vh}}{f_y} \qquad \text{(Eq. 3.3-10)}$$

where f_y is in psi, unless the amount provided is at least one-third more than that required by calculation.

Section 17.6.1 of ACI 318-77 also requires that ties, when required, be spaced no more than four times the least dimension of the supported element, and meet the minimum shear reinforcement requirements of Section 11.5.5.3.

$$A_{cs} \text{ (min)} = \frac{50 b_v \ell_{vh}}{f_y} \qquad \text{(Eq. 3.3-11)}$$

Example 3.3.8: Horizontal shear design for composite beam

Given:

Inverted tee beam with 2 in. composite topping
(See Example 3.2.6)
Beam length = 20'-0''

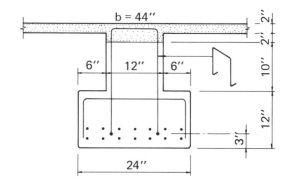

b = 44''

6'' 12'' 6''

2'' 2'' 10'' 12'' 3''

24''

Concrete:

f'_c (precast) = 5000 psi
f'_{cc} (topping) = 3000 psi

Prestressing steel:

14-1/2'' diameter 270K strands
$A_{ps} = 14 \times 0.153 = 2.142$ sq in.

Tie steel: $f_y = 40{,}000$ psi

Problem:

Determine the tie requirements to transfer horizontal shear force.

Solution:

b_v = 12 in.

$\ell_{vh} = \dfrac{20 \times 12}{2} = 120$ in.

$A_{top} = 2 \times 44 + 2 \times 12 = 112$ sq in.

$C_c = 0.85\, f'_{cc}\, A_{top} = 0.85(3)(112)$
$ = 285.6$ kips

$f_{ps} = 238$ ksi (see Ex. 3.2.6)

$A_{ps}f_{ps} = 2.142(238) = 509.8$ kips > 285.6

Therefore, $F_h = 285.6$ kips

$40\,\phi\, b_v \ell_{vh} = 40(0.85)(12)(120)/1000$
$\phantom{40\,\phi\, b_v \ell_{vh}} = 49.0 < 285.6$

Therefore, ties are required.

$\lambda = 1.0$ (normal weight concrete)

$\mu_e = \dfrac{1000\,\lambda^2\, b_v \ell_{vh}}{F_h} = \dfrac{1000(1.0)(12)(120)}{285,600}$

$ = 5.04$

$A_{cs} = \dfrac{F_h}{\phi \mu_e f_y} = \dfrac{285.6}{0.85(5.04)(40)}$

$\phantom{A_{cs}} = 1.67$ sq in.

Check minimum requirements:

1. A_{cs} (min) $= \dfrac{120\, b_v \ell_{vh}}{f_y} = \dfrac{120(12)(120)}{40,000}$

$\phantom{1. A_{cs} (min)} = 4.32$ sq in.

or $1.33 \times 1.67 = 2.22$ sq in. (controls)

2. A_{cs} (min) $= \dfrac{50\, b_v \ell_{vh}}{f_y} = \dfrac{50(12)(120)}{40,000}$

$\phantom{2. A_{cs} (min)} = 1.80$ sq in.

Use #3 ties, area per tie = $2 \times 0.11 = 0.22$ sq in.

Maximum tie spacing = $4 \times 4 = 16$ in.

$s = \dfrac{120 \times 0.22}{2.22} = 11.9$ in.

Use #3 ties at 12 in. o.c.

3.4 Camber and Deflection

Most precast, prestressed concrete flexural members will have a net positive (upward) camber at the time of transfer of prestress, caused by the eccentricity of the prestressing force. This camber may increase or decrease with time, depending on the stress distribution across the member under sustained loads. Camber tolerances are suggested in Part 7 of this Handbook.

Limitations on instantaneous deflections and time-dependent cambers and deflections are specified in the ACI Code. Table 9.5(b) of the Code is reprinted below for reference.

The following sections contain suggested methods for computing cambers and deflections. There are many inherent variables that affect camber and deflection, such as concrete mix, storage method, time of release of prestress, time of erection and placement of superimposed loads, rela-

Table 9.5(b) — Maximum permissible computed deflections

Type of member	Deflection to be considered	Deflection limitation
Flat roofs not supporting or attached to nonstructural elements likely to be damaged by large deflections	Immediate deflection due to live load	$\dfrac{\ell^*}{180}$
Floors not supporting or attached to nonstructural elements likely to be damaged by large deflections	Immediate deflection due to live load	$\dfrac{\ell^*}{360}$
Roof or floor construction supporting or attached to nonstructural elements likely to be damaged by large deflections	That part of the total deflection occurring after attachment of nonstructural elements (sum of the long-time deflection due to all sustained loads and the immediate deflection due to any additional live load)‡	$\dfrac{\ell^\dagger}{480}$
Roof or floor construction supporting or attached to nonstructural elements not likely to be damaged by large deflections		$\dfrac{\ell^\S}{240}$

* Limit not intended to safeguard against ponding. Ponding should be checked by suitable calculations of deflection, including added deflections due to ponded water, and considering long-time effects of all sustained loads, camber, construction tolerances, and reliability of provisions for drainage.

† Limit may be exceeded if adequate measures are taken to prevent damage to supported or attached elements.

‡ Long-time deflection shall be determined in accordance with Section 9.5.2.5 or 9.5.4.2 but may be reduced by amount of deflection calculated to occur before attachment of nonstructural elements. This amount shall be determined on basis of accepted engineering data relating to time-deflection characteristics of members similar to those being considered.

§ But not greater than tolerance provided for nonstructural elements. Limit may be exceeded if camber is provided so that total deflection minus camber does not exceed limit.

tive humidity, etc. *Because of this, calculated long-time values should never be considered any better than estimates.* Non-structural components attached to members which could be affected by camber variations, such as partitions or folding doors, should be placed with adequate allowance for error. Calculation of topping quantities should also recognize the imprecision of camber calculations.

It should also be recognized that camber of precast, prestressed members is a result of the placement of the strands needed to resist the design moments and service load stresses. It is not practical to alter the forms of the members to produce a desired camber. Therefore, cambers should not be specified, but their inherent existence should be recognized.

3.4.1 Initial Camber

Initial camber can be calculated using conventional moment-area equations. Fig. 3.9.18 has equations for the camber caused by prestress force for the most common strand patterns used in precast, prestressed members. Figs. 8.1.3 and 8.1.4 provide deflection equations for typical loading conditions and more general camber equations.

Example 3.4.1 — Calculation of initial camber

Given:

8DT24 of Example 3.2.8

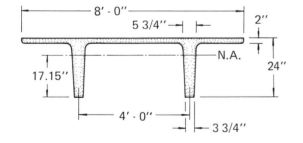

Section Properties

A	=	401 in.^{2}
I	=	20,985 in.^{4}
y_b	=	17.15 in.
y_t	=	6.85 in.
Z_b	=	1224 in.^{3}
Z_t	=	3063 in.^{3}
wt	=	418 plf
		52 psf

Concrete:

f'_c = 5000 psi

Normal weight (150 pcf)

$$E_c = 33w\sqrt{w f'_c} = 33\,(150)\sqrt{(150)\,(5000)}$$
$$= 4287 \text{ ksi}$$

f'_{ci} = 3500 psi

$$E_{ci} = 33\,(150)\sqrt{(150)\,(3500)} = 3587 \text{ ksi}$$

(Note: The values of E_c and E_{ci} could also be read from Fig. 8.2.2.)

Problem:

Find the initial camber at time of transfer of prestress

Solution:

The prestress force at transfer and strand eccentricities are calculated in Example 3.2.8 and are shown in the illustration below.

Calculate the upward component using equations given in Fig. 3.9.18.

$$\Delta\uparrow = \frac{P_o\, e_e\, \ell^2}{8\, E_{ci}\, I} + \frac{P_o\, e'\, \ell^2}{12\, E_{ci}\, I}$$

$$= \frac{364\,(3.72)\,(70\times 12)^2}{8\,(3587)\,(20,985)} +$$

$$\frac{(364)\,(9.93)\,(70\times 12)^2}{(12)\,(3587)\,(20,985)}$$

$$= 1.59 + 2.82 = 4.41 \text{ in.}\uparrow$$

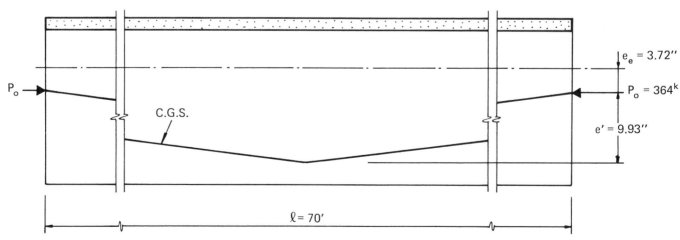

Deduct deflection caused by weight of member:

$$\Delta \downarrow = \frac{5 \, w\ell^4}{384 \, E_{ci} \, I}$$

$$= \frac{5\left(\frac{0.418}{12}\right) (70 \times 12)^4}{384 \ (3587) \ (20,985)} = 3.00 \text{ in.} \downarrow$$

Net camber at release = 4.41↑ – 3.00↓
= 1.41 in.↑

3.4.2 Elastic Deflections

Calculation of instantaneous deflections caused by superimposed service loads follow classical methods of mechanics. Design equations for various load conditions are given in Chapter 8 of this Handbook. If the bottom tension in a simple span member does not exceed the modulus of rupture, the deflection is calculated using the uncracked moment of inertia of the section. The modulus of rupture of concrete is defined in Chapter 9 of the Code as:

$$f_r = 7.5 \, \lambda \sqrt{f'_c} \qquad \text{(Eq. 3.4-1)}$$

(See Sect. 3.3.2 for definition of λ)

Fig. 3.4.1 Bilinear moment-deflection relationship

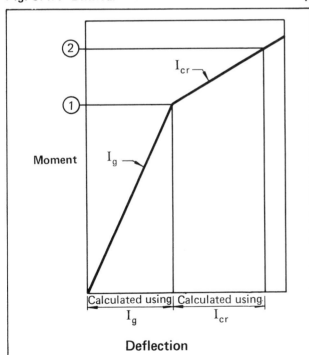

1. Moment which results in bottom tension of f_r

2. Moment corresponding to final bottom tension

(Note: Prestress effects not included in illustration.)

3.4.2.1 Bilinear Behavior

Section 18.4.2 of the Code requires that "bilinear moment-deflection relationships" be used to calculate instantaneous deflections when the bottom tension exceeds $6\sqrt{f'_c}$. This means that the deflection before the member has cracked is calculated using the gross (uncracked) moment of inertia (I_g) and the *additional* deflection after cracking is calculated using the moment of inertia of the cracked section. This is illustrated graphically in Fig. 3.4.1.

In lieu of a more exact analysis, the empirical relationship:

$$I_{cr} = nA_{ps}d^2 \, (1 - \sqrt{\rho_p}) \qquad \text{(Eq. 3.4-2)*}$$

may be used to determine the cracked moment of inertia. Table 3.9.19 gives coefficients for use in solving this equation.

Example 3.4.2 – Deflection calculation using bilinear moment-deflection relationships

Given:

8DT24 of Examples 3.2.8 and 3.4.1

Problem:

Determine the total instantaneous deflection caused by the specified uniform live load.

Solution:

Determine $f_r = 7.5\sqrt{f'_c} = 530$ psi

From Example 3.2.8, the final tensile stress is 782 psi, which is more than 530 psi, so the bilinear behavior must be considered.

Determine I_{cr} from Table 3.9.19

A_{ps} = 2.142 sq in. (See Ex. 3.2.8)
d at midspan = $e_c + y_t$ = 13.65 + 6.85
= 20.5 in.†

$$\rho_p = \frac{A_{ps}}{bd} = \frac{2.142}{(96)(20.5)} = 0.00109$$

C = 0.0067

I_{cr} = Cbd^3 = 0.0067 (96)(20.5)³
= 5541 in.⁴

---

*"Allowable Tensile Stresses for Prestressed Concrete," *PCI Journal,* Feb, 1970.

†It is within the precision of the calculation method and observed behavior to use midspan d and to calculate the deflection at midspan, although the maximum tensile stress in this case is assumed at 0.4 ℓ.

Determine the portion of the live load that would result in a bottom tension of 530 psi.

782 − 530 = 252 psi

The tension caused by live load alone is 1614 psi, therefore, the portion of the live load that would result in a bottom tension of 530 psi is:

$$\frac{1614 - 252}{1614} (0.280) = 0.236 \text{ kips/ft}$$

and

$$\Delta_g = \frac{5 \, w\ell^4}{384 \, E_c I_g} = \frac{5 \left(\dfrac{0.236}{12}\right) (70 \times 12)^4}{384 \, (4287)(20{,}985)}$$

$$= 1.42 \text{ in.}$$

$$\Delta_{cr} = \frac{5 \left(\dfrac{0.044}{12}\right) (70 \times 12)^4}{384 \, (4287)(5541)} = 1.00 \text{ in.}$$

Total deflection = 1.42 + 1.00 = 2.42 in.

3.4.2.2 Effective Moment of Inertia

The Code allows an alternative to the method of calculation described in the previous section. An effective moment of inertia, I_e, can be determined and the deflection then calculated by substituting I_e for I_g in the deflection calculation.

The equation for effective moment of inertia is:

$$I_e = \left(\frac{M_{cr}}{M_a}\right)^3 I_g + \left[1 - \left(\frac{M_{cr}}{M_a}\right)^3\right] I_{cr}$$

(Eq. 3.4-3)

The difference between the bilinear method and the I_e method is illustrated in Fig. 3.4.2.

The use of I_e with prestressed concrete members is described in a paper by Branson.* The value of M_{cr}/M_a for use in determining live load deflections can be expressed as:

$$\frac{M_{cr}}{M_a} = 1 - \left(\frac{f_{t\ell} - f_r}{f_\ell}\right)$$

(Eq. 3.4-4)

where $f_{t\ell}$ = final calculated total stress in the member

f_ℓ = calculated stress due to live load

Example 3.4.3 — Deflection calculation using effective moment of inertia

Given:

Same section and loading conditions of Example 3.4.2

---

*Branson, D. E., "The Deformation of Noncomposite and Composite Prestressed Concrete Members" *Deflections of Concrete Structures,* SP-43, American Concrete Institute.

Fig. 3.4.2 Effective moment of inertia

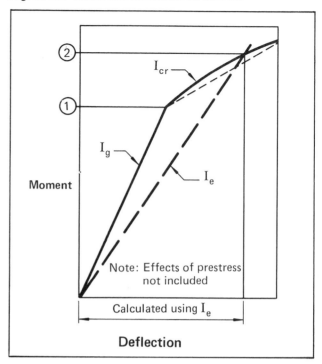

Problem:

Determine the deflection caused by live load using the I_e method.

Solution:

From the table of stresses in Example 3.2.8:

$f_{t\ell}$ = 782 psi (tension)
f_ℓ = 1614 psi (tension)
f_r = $7.5\sqrt{f'_c}$ = 530 psi (tension)

$$\frac{M_{cr}}{M_a} = 1 - \left(\frac{782 - 530}{1614}\right) = 0.844$$

$$\left(\frac{M_{cr}}{M_a}\right)^3 = (0.844)^3 = 0.601$$

$$1 - \left(\frac{M_{cr}}{M_a}\right)^3 = 1 - 0.601 = 0.399$$

$$I_e = 0.601 \, (20{,}985) + 0.399 \, (5541)$$
$$= 14{,}823 \text{ in.}^4$$

I_e can also be found using Fig. 3.9.20

$$f_e = 782 - 530 = 252 \text{ psi}$$

$$\frac{f_e}{f_\ell} = \frac{252}{1614} = 0.16$$

$$\frac{I_{cr}}{I_g} = \frac{5541}{20{,}985} = 0.26$$

Follow arrows on the chart

$$\frac{I_e}{I_g} = 0.70$$

$$I_e = 0.70 \ (20{,}985) = 14{,}690 \ \text{in.}^4$$

$$\Delta_\ell = \frac{5 \ w\ell^4}{384 \ E_c \ I_e} = \frac{5 \left(\dfrac{0.280}{12}\right)(70 \times 12)^4}{384 \ (4287) \ (14{,}823)} = 2.38 \ \text{in.}$$

3.4.3 Long-Time Camber/Deflection

ACI 318-77 provides a convenient equation for estimating the additional long-time deflection of non-prestressed reinforced concrete members (Section 9.5.2.5):

$$[2 - 1.2 \ (A'_s/A_s)] \geqslant 0.6, \text{ where}$$

A'_s is the compressive reinforcement and A_s is the tensile reinforcement. No such convenient guide is given for prestressed concrete.

The determination of long-time cambers and deflections in precast, prestressed members is somewhat more complex because of (1) the effect of prestress and the loss of prestress over time, (2) the strength gain of concrete after release of prestress, and because (3) the camber or deflection is important not only at the "initial" and "final" stages, but also at erection, which occurs at some intermediate stage, usually from 30 to 60 days after casting.

It has been customary in the design of precast, prestressed concrete to estimate the camber of a member after a period of time by multiplying the initial calculated camber by some factor, usually based on the experience of the designer. To properly use these "multipliers," the upward and downward components of the initial calculated camber should be separated in order to take into account the effects of loss of prestress, which only affect the upward component.

Table 3.4.1 provides suggested multipliers which can be used as a guide in estimating long-time cambers and deflections for typical members, i.e., those members which are within the span-depth ratios recommended in this Handbook. Derivation of these multipliers is contained in a paper by Martin.[*]

Long-time effects can be substantially reduced by adding non-prestressed reinforcement in prestressed concrete members. The reduction effects proposed by Shaikh and Branson[†] can be applied to the approximate multipliers of Table 3.4.1 as follows:

$$C_2 = \frac{C_1 + A_s/A_{ps}}{1 + A_s/A_{ps}}$$

where
C_1 = multiplier from Table 3.4.1
C_2 = revised multiplier
A_s = area of non-prestressed reinforcement
A_{ps} = area of prestressed steel

[*]Martin, L. D. "A Rational Method for Estimating Camber and Deflection of Precast, Prestressed Concrete Members" *PCI Journal*, Jan-Feb, 1977.

[†]Shaikh, A. F., and Branson, D. E., "Non-Tensioned Steel in Prestressed Concrete Beams," *PCI Journal*, Feb, 1970.

Table 3.4.1   Suggested multipliers to be used as a guide in estimating long-time cambers and deflections for typical members

		Without Composite Topping	With Composite Topping
	At erection:		
(1)	Deflection (downward) component — apply to the elastic deflection due to the member weight at release of prestress	1.85	1.85
(2)	Camber (upward) component — apply to the elastic camber due to prestress at the time of release of prestress	1.80	1.80
	Final:		
(3)	Deflection (downward) component — apply to the elastic deflection due to the member weight at release of prestress	2.70	2.40
(4)	Camber (upward) component — apply to the elastic camber due to prestress at the time of release of prestress	2.45	2.20
(5)	Deflection (downward) — apply to elastic deflection due to superimposed dead load only	3.00	3.00
(6)	Deflection (downward) — apply to elastic deflection caused by the composite topping	—	2.30

	(1) Release	Multiplier	(2) Erection	Multiplier	(3) Final
Prestress	4.41 ↑	1.80 x (1)	7.94 ↑	2.45 x (1)	10.80 ↑
w_d	3.00 ↓	1.85 x (1)	5.55 ↓	2.7 x (1)	8.10 ↓
	1.41 ↑		2.39 ↑		2.70 ↑
w_{sd}			0.48 ↓	3.0 x (2)	1.44 ↓
			1.91 ↑		1.26 ↑
w_ℓ					2.38 ↓
					1.12 ↓

Example 3.4.4 — Use of multipliers for determining long-time cambers and deflections.

Given:

8DT24 of Examples 3.2.8, 3.4.1, 3.4.2 and 3.4.3. Non-structural elements are attached, but not likely to be damaged by deflections (light fixtures, etc.)

Problem:

Estimate the camber and deflection and determine if it meets the requirements of Table 9.5(b) of the Code.

Solution:

Calculate the instantaneous deflections caused by the superimposed dead and live loads.

$$\Delta_d = \frac{5 \, w\ell^4}{384 \, E_c I} = \frac{5 \left(\dfrac{0.080}{12} \right) (70 \times 12)^4}{384 \ (4287)(20{,}985)}$$

$$= 0.48 \text{ in.} \downarrow$$

$\Delta_\ell = 2.38$ in.↓ (see Example 3.4.3)

For convenience, a tabular format is used (above).

The estimated critical cambers and deflections would then be:

At erection of the member after
w_{sd} is applied   = 1.91 in.
"Final" long-time camber   = 1.26 in.

The deflection limitation of Table 9.5(b) for the above condition is $\ell/240$.

$$(70 \times 12)/240 = 3.50 \text{ in.}$$

Total deflection occurring after attachment of non-structural elements:

$$\Delta_{t\ell} = (1.91 - 1.26) + 2.38$$
$$= 3.03 \text{ in.} < 3.50 \text{ in. OK}$$

3.5 Loss of Prestress

Loss of prestress is the reduction of tensile stress in prestressing tendons due to shortening of the concrete around the tendons, relaxation of stress within the tendons and external factors which reduce the total initial force before it is applied to the concrete. ACI 318 identifies the following sources of loss of prestress:

(a) Anchorage seating loss

(b) Elastic shortening of concrete

(c) Creep of concrete

(d) Shrinkage of concrete

(e) Relaxation of tendon stress

(f) Friction loss due to intended or unintended curvature in post-tensioning tendons.

Accurate determination of losses is more important in some prestressed concrete members than in others. Losses have no effect on the ultimate strength of a flexural member unless the tendons are unbonded or if the final stress after losses is less than $0.50 \, f_{pu}$. Underestimation or overestimation of losses can affect service conditions such as camber, deflection and cracking.

3.5.1 Sources of Stress Loss

Anchorage seating loss and friction

These two sources of loss are mechanical. They represent the difference between the tension applied to the tendon by the jacking unit and the initial tension available for application to the concrete by the tendon. Their magnitude can be determined with reasonable accuracy and, in many cases, they are fully or partially compensated for by overjacking.

Elastic shortening of concrete

The concrete around the tendons shortens as the prestressing force is applied to it. Those tendons which are already bonded to the concrete shorten with it.

Shrinkage of concrete

Loss of stress in the tendon due to shrinkage of the concrete surrounding it is proportional to that part of the shrinkage that takes place after the transfer of prestress force to the concrete.

Creep of concrete and relaxation of tendons

Losses due to creep of concrete and relaxation of tendons complicate stress loss calculations. The rate of loss due to each of these factors changes when the stress level changes and the stress level is changing constantly throughout the life of the structure. Therefore, the rates of loss due to creep and relaxation are constantly changing.

3.5.2 Range of Values for Total Losses

Total loss of prestress in typical members will range from about 35,000 to 50,000 psi for normal weight concrete members, and from about 40,000 to 55,000 psi for sand-lightweight members.

The value of 35,000 psi appeared in the ACI-ASCE Committee 423 report* in 1958 and has been widely used with satisfactory results in typical members since then. Many designers prefer to assign a fixed value within the ranges above for their designs of typical members. For example, the load tables in this Handbook use values of 22 percent for normal weight and 25 percent for sand-lightweight concrete (41,500 psi and 47,250 psi respectively for 270K strand initially tensioned to 0.7 f_{pu}). Use of such fixed values will normally have satisfactory results for members that are within the span-depth ratios, prestress levels and loading conditions of the sections shown in the tables. For members with high span-depth ratios and prestress levels, or with heavy sustained loads, a more detailed analysis may be warranted. For critical members, the use of low-relaxation strand (see Eqs. 3.5-1 through 3.5-4) can significantly reduce losses.

3.5.3 PCI Committee Calculation Method

"Recommendations for Estimating Prestress Losses," prepared by the PCI Committee on Prestress Losses, appears on pages 44 to 75 of the *PCI Journal* for July/August 1975. Reader's Comment on the Recommendations appears on pages 108 to 126 of the *PCI Journal* for March/April 1976. That report contains all the information needed to make a complete stress loss analysis. It presents a "General Method" and a "Simplified Method."

The General Method considers a number of

time intervals during the life of the structure. It establishes the stress levels existing in the tendons and the concrete at the beginning of each interval, computes the loss due to each factor during the interval, and determines the stress levels in tendons and concrete at the end of the interval. This method can be included in a computer program for analysis of prestressed concrete members without too much trouble but it involves a great deal of work when applied to individual members by long-hand procedures.

The Simplified Method, which is illustrated by Examples 3.5.1 and 3.5.2 is based on the General Method but eliminates most of the mathematics.

Using the Simplified Method, stress loss is determined by computing the value of f_{cr} and f_{cds} and substituting them in the appropriate Eqs. 3.5-1 through 3.5-4. These equations are used to compute total loss, TL, in kips per sq in. Total loss is the sum of losses due to shrinkage, elastic shortening and creep of concrete plus loss due to relaxation of tendons.

For normal weight concrete:

Using stress-relieved strand:

$$TL = 33.0 + 13.8\, f_{cr} - 4.5\, f_{cds} \qquad \text{(Eq. 3.5-1)}$$

Using low-relaxation strand:

$$TL = 19.8 + 16.3\, f_{cr} - 5.4\, f_{cds} \qquad \text{(Eq. 3.5-2)}$$

For sand-lightweight concrete:

Using stress-relieved strand:

$$TL = 31.2 + 16.8\, f_{cr} - 3.8\, f_{cds} \qquad \text{(Eq. 3.5-3)}$$

Using low-relaxation strand:

$$TL = 17.5 + 20.4\, f_{cr} - 4.8\, f_{cds} \qquad \text{(Eq. 3.5-4)}$$

The empirical Eqs. 3.5-1 through 3.5-4 were developed by setting up the General Method on a computer, processing a number of examples for each type of concrete and strand and plotting curves from the results. They apply to pretensioned members only. For typical members it was found that the only variable that is not included in the equations but could make an appreciable difference in the net result is volume/surface ratio. A correction factor is applied for that:

V/S ratio, in.	1.0	2.0	3.0	4.0
Adjustment, percent	+3.2	0	−3.8	−7.6

Example: For V/S = 3.0 reduce losses by 3.8%.

Eqs. 3.5-1 through 3.5-4 are based on the initial tension, after reduction for anchor slip, normally used in pretensioned members, i.e.,

---

*ACI-ASCE Committee 423 "Tentative Recommendation for Prestressed Concrete," *ACI Journal, Proceedings*, V. 54, No. 7, Jan. 1958, pp. 545-578.

0.7 f_{pu} for stress-relieved strand and 0.75 f_{pu} for low-relaxation strand. Use of a higher or lower initial tension will result in an appreciable change in net losses especially in the case of stress-relieved tendons.

Use of Eqs. 3.5-1 through 3.5-4 requires calculation of the stresses f_{cr} and f_{cds}.

$$f_{cr} = \frac{P_o}{A} + \frac{P_o e^2}{I} - \frac{M_d e}{I} \qquad \text{(Eq. 3.5-5)}$$

$$f_{cds} = \frac{M_{sd} e}{I} \qquad \text{(Eq. 3.6-6)}$$

where

A	=	area of the precast section
e	=	eccentricity of the strand at the critical section
f_{cds}	=	concrete stress at C.G. of strand at the critical section caused by sustained loads not included in the calculation of f_{cr}.
f_{cr}	=	concrete compressive stress at C.G. of strand at critical section immediately after transfer.
I	=	moment of inertia of the section
M_d	=	moment due to weight of the member
M_{sd}	=	moment due to all sustained loads *except* the member weight
P_o	=	prestress force at transfer, after initial loss. It is within reasonable accuracy to assume 10% initial loss for stress relieved strand and 7.5% initial loss for low-relaxation strand.

3.5.4 Critical Locations

Computations for stress losses due to elastic shortening and creep of concrete are based on the compressive stress in the concrete at the center of gravity (cgs) of the tendons.

For bonded tendons, stress losses are computed at that point on the span where flexural tensile stresses are most critical. In members with straight, parabolic or approximately parabolic tendons this is usually mid-span. In members with tendons deflected at mid-span only the critical point is generally near the 0.4 point of the span. Since the tendons are bonded, only the stresses at the critical point need be considered. Stresses or stress changes at other points along the member do not affect the stresses or stress losses at the critical point.

Example 3.5.1 — Loss of prestress — stress-relieved strand

Given:

10LDT 32 + 2 as shown

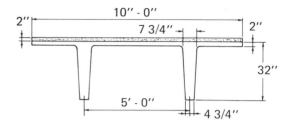

Span = 70 ft

No superimposed dead load except topping

Section properties (untopped):

A = 615 sq in.

I = 59,720 in.^{4}

Z_b = 2717 in.^{3}

V/S = 615/364 = 1.69 in.

wt = 491 plf

wt of topping = 250 plf

Concrete:

Precast: f'_c = 5000 psi

f'_{ci} = 3500 psi

Sand-lightweight

Topping: Normal weight

Prestressing steel:

12-1/2″ dia. 270K stress-relieved strands

A_{ps} = 12 x 0.153 = 1.836 sq in.

Depressed at mid-span

e_e = 12.81 in.

e_c = 18.73 in.

Problem:

Determine total loss of prestress by PCI Committee simplified method.

Solution:

For depressed strand, critical section is at 0.4ℓ. Determine moments, eccentricity, and prestress force.

$$M @ 0.4\ell = \frac{wx}{2}(\ell - x) = \frac{w(0.4\ell)}{2}(\ell - 0.4\ell)$$

$$= 0.12 \, w\ell^2$$

$$M_d = 0.12 \, (0.491)(70)^2 = 289 \text{ ft-kips}$$

$$M_{sd} = 0.12 \, (0.250)(70)^2 = 147 \text{ ft-kips}$$

e at $0.4\ell = 12.81 + 0.8 (18.73 - 12.81)$
$= 17.55$ in.

$P_i = 0.7 A_{ps} f_{pu} = 0.7 (1.836)(270)$
$= 347.0$ kips

$P_o = 0.9 P_i = 0.9 (347) = 312.3$ kips

Determine f_{cr} and f_{cds}

$$f_{cr} = \frac{P_o}{A} + \frac{P_o e^2}{I} - \frac{M_d e}{I}$$

$$= \frac{312.3}{615} + \frac{312.3 (17.55)^2}{59,720}$$

$$- \frac{289 (12)(17.55)}{59,720}$$

$$= 0.508 + 1.611 - 1.019 = 1.10 \text{ ksi}$$

$$f_{cds} = \frac{M_{sd} e}{I} = \frac{147 (12)(17.55)}{59,720} = 0.518 \text{ ksi}$$

For sand-lightweight concrete and stress-relieved strand, use Eq. 3.5-3 for TL.

$$\text{TL} = 31.2 + 16.8 f_{cr} - 3.8 f_{cds}$$

$$= 31.2 + 16.8 (1.10) - 3.8 (0.518)$$

$$= 31.2 + 18.5 - 2.0 = 47.7 \text{ ksi}$$

Adjust for V/S ratio (interpolate between 1.0 and 2.0)

Adjustment factor $= 3.2 - 0.69 (3.2 - 0)$
$= +0.99\%$

(+ means additional loss)

$$\frac{0.99}{100} (47.7) = 0.5 \text{ ksi}$$

Final loss is $47.7 + 0.5 = 48.2$ ksi or 25.5% of 189 ksi

Final prestress force:

$$P = 347 - 48.2 (1.836) = 258.5 \text{ kips}$$

Example 3.5.2 — Loss of prestress — low-relaxation strand

Given:

Same as Example 3.5.1, except use low-relaxation strands.

Solution:

Calculations are same up to the determination of P_i.

$$P_i = 0.75 A_{ps} f_{pu} = 0.75 (1.836)(270)$$
$$= 371.8 \text{ kips}$$

$$P_o = 0.925 P_i = 0.925 (371.8) = 343.9 \text{ kips}$$

$$f_{cr} = \frac{343.9}{615} + \frac{343.9 (17.55)^2}{59,720}$$

$$- \frac{289 (12)(17.55)}{59,720}$$

$$= 0.559 + 1.774 - 1.019 = 1.314 \text{ ksi}$$

$$f_{cds} = 0.518 \text{ ksi (from Example 3.5.1)}$$

For sand-lightweight concrete and low-relaxation strand, use Eq. 3.5-4.

$$\text{TL} = 17.5 + 20.4 f_{cr} - 4.8 f_{cds}$$

$$= 17.5 + 20.4 (1.314) - 4.8 (0.518)$$

$$= 17.5 + 26.8 - 2.5 = 41.8 \text{ ksi}$$

V/S Adjustment:

$$\frac{0.99}{100} (41.8) = 0.4 \text{ ksi}$$

Final loss $= 41.8 + 0.4 = 42.2$ ksi or 20.8%

$$P = 371.8 - 42.2 (1.836) = 294.3 \text{ kips}$$

3.6 Compression Members

Precast and prestressed concrete columns and load-bearing wall panels are usually proportioned on the basis of strength design. Stresses under service conditions, particularly during handling and erection (especially wall panels) must also be considered. The procedures in this section are based on Chapter 10 of the Code and on the recommendations of the PCI Committee on Pre-stressed Concrete Columns."* (Referred to in this section as "the Recommended Practice.")

3.6.1 Strength Design of Precast, Reinforced Concrete Compression Members

The capacity of a reinforced concrete compression member with eccentric loads is most easily determined by constructing a capacity interaction curve. Points on this curve are calculated using the compatibility of strains and solving the equations of equilibrium as prescribed in Chapter 10 of the Code. Solution of these equations is illustrated in Fig. 3.6.1.

Interaction curves for typical square columns are given in Part 2 of this handbook. Tables for a wider range of reinforced concrete columns are provided in the CRSI Handbook.[†]

---

*"Recommended Practice for the Design of Prestressed Concrete Columns and Bearing Walls," *PCI Journal,* Nov-Dec, 1976.

†"CRSI Handbook" Third edition, 1978. Concrete Reinforcing Steel Institute, Chicago, IL.

Construction of the interaction curve usually follows these steps:

Step 1. Determine P_o [$M_n = 0$ — see Fig. 3.6.1 (c)]

Step 2. Determine P_{nb} and M_{nb} at balance point [see Fig. 3.6.1(d)]

Step 3. Determine M_o. This can be done by using design aids in the ACI Design Handbook (ACI Publication SP-17), by trial and error using Fig. 3.6.1(a), or, conservatively, by basic strength equations neglecting the reinforcement above the neutral axis. In many cases it is adequate to select a point near the value of M_o by the procedures in Steps 4 through 6.

Step 4. Select a value "a" or "c" between the points determined above.

Step 5. Determine the value of A_{comp} from the geometry of the section (shaded portions in Fig. 3.6.1).

Step 6. Solve the equations in Fig. 3.6.1(a) or (b).

Strength reduction factors (ϕ): The Code prescribes that the ϕ factors for compression members is 0.7, except that it can vary from 0.7 at a point where $\phi P_n = 0.10 f'_c A_g$ to 0.9 where $\phi P_n = 0$. This ϕ variation is accomplished on the interaction curve by first constructing the curve with $\phi = 0.7$. A straight line is then drawn from the point on that curve where $\phi P_n = 0.10 f'_c A_g$ to a point on the $\phi P_n = 0$ line corresponding to ϕM_o calculated with $\phi = 0.9$.

Maximum design load: ACI 318-77 specifies that the maximum applied design load is $0.80 \phi P_o$ for tied columns and $0.85 \phi P_o$ for spiral columns. This has the same effect as specifying a minimum eccentricity as was done in previous codes, except that it is dependent on the strength of the column, rather than on its dimensions.

Example 3.6.1 — Construction of interaction curve for a precast, reinforced concrete column

Given:

Column cross-section shown

Concrete:

f'_c = 5000 psi

Reinforcement:

Grade 60

f_y = 60,000 psi

E_s = 29,000 ksi

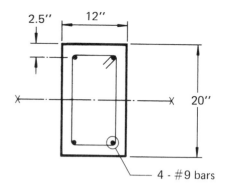

4 - #9 bars

Problem:

Construct interaction curve for bending about x-x axis

Solution:

Determine following parameters:

β_1 = 0.85 – 0.05 = 0.80

d = 20 – 2.5 = 17.5 in.

d' = 2.5 in.

y_t = 10 in.

$0.85 f'_c = 0.85(5) = 4.25$ ksi

A_g = 12 x 20 = 240 sq in.

$A_s = A'_s = 2.00$ sq in.

Step 1 — Determine P_o from Fig. 3.6.1(c)

$P_o = 0.85 f'_c (A_g - A'_s - A_s) + (A'_s + A_s) f_y$

$\phi P_o = 0.70 [4.25(240 - 4) + (4)(60)]$
      = 870 kips

Step 2 — Determine P_{nb} and M_{nb} from Fig. 3.6.1(d)

$$c = \frac{0.003d}{0.003 + f_y/E_s} = \frac{0.003(17.5)}{0.003 + 60/29,000}$$

= 10.36 in.

$$f'_s = 29,000 \left[\frac{0.003}{10.36} (10.36 - 2.5) \right]$$

= 66.0 > 60

therefore $f'_s = f_y = 60$ ksi

A_{comp} = ab = β_1 c b = 0.80(10.36)(12)
       = 99.5 sq in.

$$y' = \frac{a}{2} = \frac{0.80(10.36)}{2} = 4.14 \text{ in.}$$

P_{nb} = (99.5 – 2)4.25 + 2(60) – 2(60)
     = 414.4 kips

ϕP_{nb} = 0.70(414.4) = 290 kips

Fig. 3.6.1 Equilibrium equations for reinforced concrete compression members

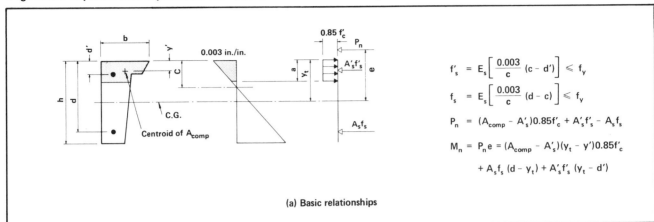

$$f'_s = E_s\left[\frac{0.003}{c}(c - d')\right] \leq f_y$$

$$f_s = E_s\left[\frac{0.003}{c}(d - c)\right] \leq f_y$$

$$P_n = (A_{comp} - A'_s)0.85f'_c + A'_s f'_s - A_s f_s$$

$$M_n = P_n e = (A_{comp} - A'_s)(y_t - y')0.85f'_c$$
$$+ A_s f_s (d - y_t) + A'_s f'_s (y_t - d')$$

(a) Basic relationships

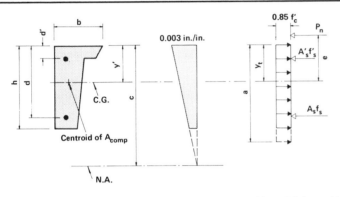

$$A_{comp} \approx A_g \text{ if } a > h$$

$$f'_s = E_s\left[\frac{0.003}{c}(c - d')\right] \leq f_y$$

$$f_s = E_s\left[\frac{0.003}{c}(c - d)\right] \leq f_y$$

$$P_n = (A_g - A'_s - A_s)0.85f'_c + A'_s f'_s + A_s f_s$$

$$M_n = P_n e = A'_s f'_s (y_t - d') - A_s f_s (d - y_t)$$

(b) Special case with Neutral Axis outside of the section

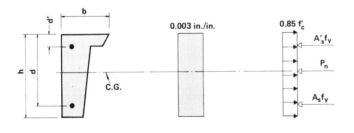

$$P_o = 0.85f'_c (A_g - A'_s - A_s) + (A'_s + A_s) f_y$$

$$M_n = 0$$

(c) Special Case when $M_n = 0$ $P_n = P_o$

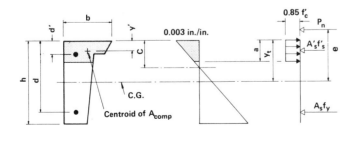

$$c = \frac{0.003d}{0.003 + f_y/E_s}$$

$$f'_s = E_s\left[\frac{0.003}{c}(c - d')\right] \leq f_y$$

$$P_{nb} = (A_{comp} - A'_s)0.85f'_c + A'_s f'_s - A_s f_y$$

$$M_{nb} = P_{nb} e = (A_{comp} - A'_s)(y_t - y')0.85f'_c$$
$$+ A_s f_y (d - y_t) + A'_s f'_s (y_t - d')$$

(d) Special Case at Balance Point

$$M_{nb} = (97.5)(10 - 4.14)(4.25)$$
$$+ 2.0(60)(17.5 - 10)$$
$$+ 2.0(60)(10 - 2.5)$$

$$\phi M_{nb} = 0.70(2428 + 900 + 900)$$
$$= 2960 \text{ in-kips} = 247 \text{ ft-kips}$$

Step 3 — Determine M_o — Use conservative solution neglecting compressive reinforcement

$$a = \frac{A_s f_y}{0.85 \, f'_c \, b} = \frac{2.0(60)}{4.25(12)} = 2.35 \text{ in.}$$

$$M_o = A_s f_y \left(d - \frac{a}{2}\right) = (2.0)(60)(17.5 - 2.35/2)$$
$$= 1959 \text{ in.-kips}$$

For $\phi = 0.7$, $\phi M_o = 1371$ in.-kips = 114 ft-kips
(This point is found for curve projection)

For $\phi = 0.9$, $\phi M_o = 1763$ in.-kips = 147 ft kips

To determine intermediate points on the curve:

Step 4 — set $a = 6$ in., $c = \dfrac{6}{0.80} = 7.5$ in.

Step 5 — $A_{comp} = 6(12) = 72$ sq in.

Step 6 — Use Fig. 3.6.1(a)

$$f'_s = 29,000 \left[\frac{0.003}{7.5}(7.5 - 2.5) \right]$$
$$= 58.0 \text{ ksi} < f_y$$

$$f_s = 29,000 \left[\frac{0.003}{7.5}(17.5 - 7.5) \right]$$
$$= 116 \text{ ksi} > f_y$$

Use $f_s = f_y = 60$ ksi

$$P_n = (72 - 2)4.25 + 2.0(58) - 2.0(60)$$
$$= 293.5 \text{ kips}$$

$$\phi P_n = 0.7(293.5) = 205 \text{ kips}$$

$$\phi M_n = 0.70 \, [(72 - 2)(10 - 3)\, 4.25$$
$$+ 2.0(60)(17.5 - 10)$$
$$+ 2.0(58)(10 - 2.5)]$$
$$= 0.70(2082.5 + 900 + 870)$$
$$= 2697 \text{ in.-kips} = 225 \text{ ft-kips}$$

(Note: Steps 4 to 6 can be repeated for as many points as desired.)

A plot of these points is shown as Fig. 3.6.2.

Also calculate:

Maximum design load $= 0.80 \, \phi P_o$
$= 0.80(870) = 696$ kips

Transition point for $\phi = 0.7$ to $\phi = 0.9$
$= 0.10 \, f'_c A_g = 0.10(5)(240) = 120$ kips

Fig. 3.6.2 Interaction curve for Example 3.6.1

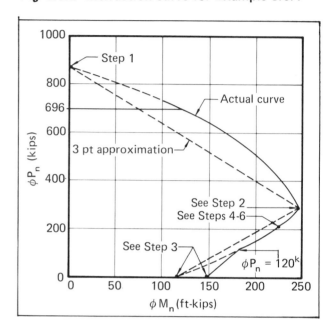

3.6.2 Strength Design of Precast, Prestressed Concrete Compression Members

Construction of the interaction curve for prestressed compression members differs from that for reinforced members in that the yield point for prestressing steel is not well defined, and the stress-strain curve is non-linear over a broad range (see Fig. 8.2.5). Therefore, the balance point is usually determined by a series of trials. The strain compatibility and equilibrium equations for prestressed compression members is illustrated in Fig. 3.6.3.

ACI 318-77 waives the minimum vertical reinforcement requirements for compression members if the concrete is prestressed to at least an average of 225 psi after all losses. In addition, the Recommended Practice permits the elimination of column ties, if the nominal capacity is multiplied by 0.85. Interaction curves for typical prestressed square columns and wall panels are provided in Part 2.

Construction of the interaction curve usually follows these steps:

Step 1. Determine P_o [$M_n = 0$ — see Fig. 3.6.3 (c)]

Step 2. Determine M_o. This is normally done by neglecting the steel above the neutral axis and determining the moment capa-

Fig. 3.6.3 Strain compatibility relationships for prestressed concrete compression members

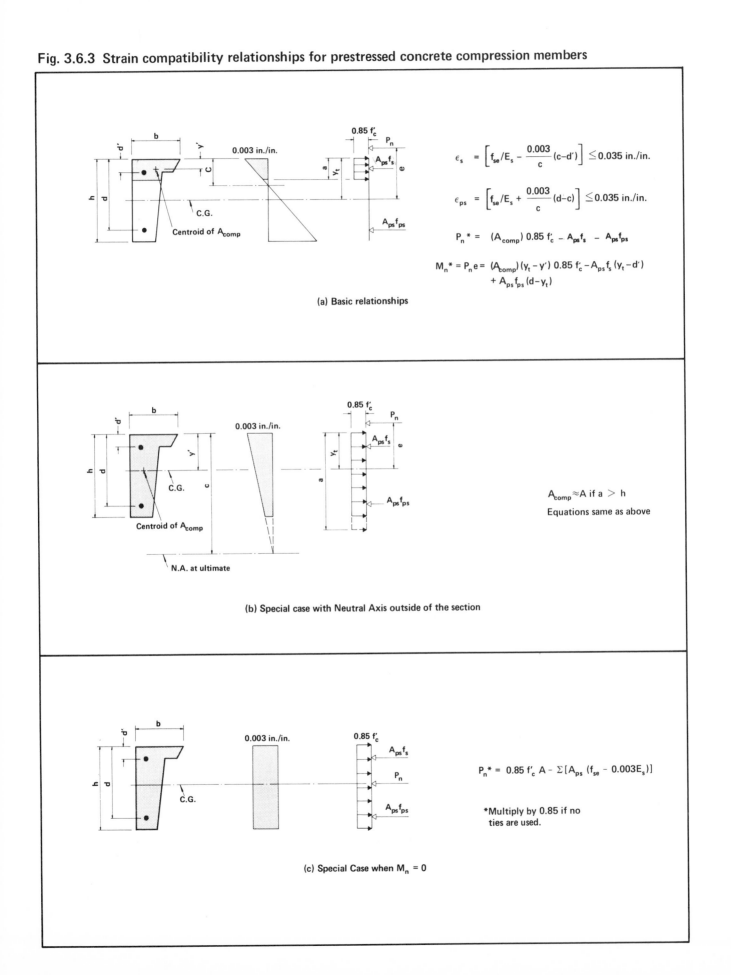

$$\epsilon_s = \left[f_{se}/E_s - \frac{0.003}{c}(c-d') \right] \leq 0.035 \text{ in./in.}$$

$$\epsilon_{ps} = \left[f_{se}/E_s + \frac{0.003}{c}(d-c) \right] \leq 0.035 \text{ in./in.}$$

$$P_n{}^* = (A_{comp}) \, 0.85 \, f'_c - A_{ps} f_s - A_{ps} f_{ps}$$

$$M_n{}^* = P_n e = (A_{comp})(y_t - y') \, 0.85 \, f'_c - A_{ps} f_s (y_t - d') + A_{ps} f_{ps} (d - y_t)$$

(a) Basic relationships

$A_{comp} \approx A$ if $a > h$

Equations same as above

(b) Special case with Neutral Axis outside of the section

$$P_n{}^* = 0.85 \, f'_c \, A - \Sigma [A_{ps}(f_{se} - 0.003 E_s)]$$

*Multiply by 0.85 if no ties are used.

(c) Special Case when $M_n = 0$

city by one of the methods described in Section 3.2.1.

For each additional point on the curve follow Steps 3 through 8.

Step 3. Select a value "a" or "c" for each point on the interaction curve. Determine the corresponding "a" or "c" from the equation $a = \beta_1 c$.

Step 4. Determine the value of A_{comp} from the geometry of the section (shaded portion in Fig. 3.6.3). (Note: Since the strand area is usually quite small, it is normally neglected.)

Step 5. Determine the strain in the strand caused by the prestressing, $\epsilon_{es} = f_{se}/E_s$.

Step 6. Determine the strain in the strand caused by external loading by similar triangles as shown in Fig. 3.6.3(a). The total strain is then the prestressing strain ± the strain caused by external loading.

Step 7. Determine the stress in the strand from the stress-strain curve (Fig. 8.2.5).

Step 8. Calculate the values of P_n and M_n for each point selected by statics as shown in Fig. 3.6.3(a).

Capacity reduction factors, ϕ, and maximum design load requirements are the same as for reinforced columns.

Development length: If applied factored moments occur near the ends of members where the strand is not fully developed, appropriate reductions in the values of f_{ps} should be taken, as described in Sect. 3.2.3, or the prestressing strands can be supplemented by mild steel that is anchored to end plates, or otherwise developed.

The interaction curves in Part 2 are based on a maximum value of $f_{ps} = f_{se}$, which is equivalent to a development length equal to the assumed transfer length. The required area of end reinforcement can be determined by matching interaction curves, or can be approximated by the following equation if the bar locations approximately match the strand locations:

$$A_s = \frac{A_{ps} f_{se}}{f_y}$$

where

A_s = required area of bars

f_{se} = strand stress after losses

f_y = yield strength of bars

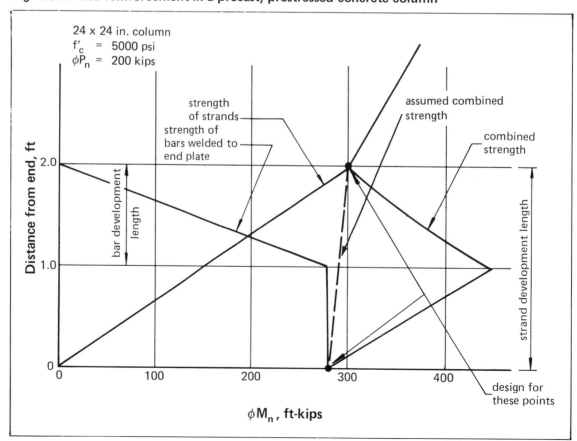

Fig. 3.6.4 End reinforcement in a precast, prestressed concrete column

The effects of adding end reinforcement to a 24 x 24 in. prestressed concrete column, thus improving moment capacity in the end 2 ft, are shown in Fig. 3.6.4.

Example 3.6.2 — Calculation of interaction points for prestressed concrete compression members

Given:

Hollow-core wall panel shown

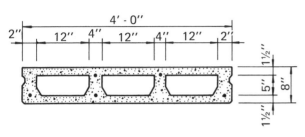

Concrete:

f'_c = 6000 psi

A = 204 sq in.

Prestressing steel:

f_{pu} = 270 ksi

E_s = 27,500 ksi

f_{se} = 150 ksi

5-3/8" dia., 270K strands

A_{ps} (bott) = 3(0.085) = 0.255 sq in.

A_{ps} (top) = 2(0.085) = 0.170 sq in.

Problem:

Calculate a point on the design interaction curve for a = 2"

Solution:

Step 1: β_1 = 0.85 - 2(.05) = 0.75

a = 2"   c = $\dfrac{2}{0.75}$ = 2.67 in.

Step 2: A_{comp} = 48(1.5) + 12(2 - 1.5)

= 78 sq in.

$y' = \dfrac{48(1.5)(1.5/2) + 12(0.5)(1.5 + 0.5/2)}{78}$

= 0.83 in.

Step 3: $\dfrac{f_{se}}{E_s} = \dfrac{150}{27,500}$ = 0.00545 in./in.

Step 4: From Fig. 3.6.3(a)

ϵ_s = 0.00545 - $\dfrac{0.003}{2.67}$ (2.67 - 1.5)

= 0.00545 - 0.00131 = 0.00414 in./in.

From Fig. 8.2.5, this strain is on the linear part of the curve:

f_s = $\epsilon_s E_s$ = 0.00414(27,500) = 114 ksi

ϵ_{ps} = 0.00545 + $\dfrac{0.003}{2.67}$ (6.5 - 2.67)

= 0.00545 + 0.00430 = 0.00975

From Fig. 8.2.5, f_{ps} = 245 ksi

From Fig. 3.6.3(a)

P_n = (A_{comp}) 0.85f'_c - $A_{ps}f_s$ - $A_{ps}f_{ps}$

= 78 (0.85)(6) - 0.170(114) - 0.255(245)

= 397.8 - 19.4 - 62.5 = 315.9 kips

ϕP_n = 0.7(315.9) = 221.1 kips

M_n = 397.8(4 - 0.83) - 19.4(4 - 1.5)

+ 62.5(6.5 - 4)

= 1261.0 - 48.5 + 156.3 = 1368.8 in.-kips

= 114 ft-kips

ϕM_n = 0.7(1368.8) = 958.2 in.-kips

= 79.8 ft-kips

Since no lateral ties are used in this member, the values should be multiplied by 0.85.

ϕP_n = 0.85 (221.1) = 187.9 kips

ϕM_n = 0.85 (79.8) = 67.8 ft-kips

Note that this is for *fully developed* strand. If the capacity at a point near the end of the transfer zone is desired, then $f_{ps} \leq f_{se}$ = 150 ksi

ϕP_n = 0.85 (0.7) [397.8 - 19.4 - 0.255 (150)]

= 202.4 kips

ϕM_n = 0.85 (0.7) [1261.0 - 48.5

+ 0.255(150)(6.5 - 4)]

= 778.3 in.-kips = 64.9 ft-kips

Note: In prestressed wall panels, the effects of unsymmetrical prestress should also be investigated.

3.6.3 Slenderness Effects

Sections 10.10 and 10.11 of ACI 318-77 contain provisions for evaluating slenderness effects (buckling) of columns. Use of these provisions is described in Part 4 of this Handbook. Additional recommendations are given in the Recommended Practice, and in the *PCI Manual for Structural Design of Architectural Precast Concrete.**

---

*Available from Prestressed Concrete Institute

3.6.4 Service Load Stresses

There are no limitations in ACI 318-77 on service load stresses in compression members subject to bending. The Recommended Practice suggests tension limitations as shown in Table 3.6.1.

3.6.5 Effective Width of Wall Panels

The Recommended Practice specifies the following for effective width (the portion of the wall assumed to resist the loads):

Flat wall panels (solid or hollow-core): length of loaded area plus six times the thickness of the wall on either side.

Ribbed wall panels: Width of the rib at the flange plus six times the flange thickness on either side.

In no case should the effective width exceed the center-to-center distance between loads.

These limitations may be waived if the design load is less than $P_u = 0.035 f'_c$ A where A is the total area of the effective section. In that case, design may be based on the full section.

3.6.6 Insulated Load-Bearing Wall Panels

Insulated wall panels, or sandwich panels of the type shown in Fig. 3.6.6, are usually designed assuming only one wythe as structurally effective, unless it can be shown by experience, test, or calculation that both wythes are fully or partially effective. The method of transmitting the superimposed loads to the panel (corbels, for example) should be carefully detailed to assure that the non-structural elements do not carry loads they are not designed for. Detailed design procedures are given in the *PCI Manual for the Design of Architectural Precast Concrete.*

Fig. 3.6.5 Effective width of wall panels

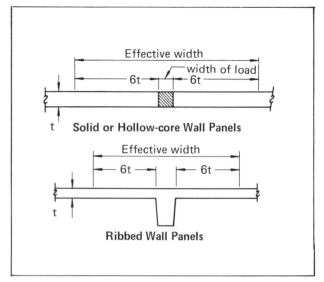

Fig. 3.6.6 Typical precast concrete load-bearing insulated wall panels

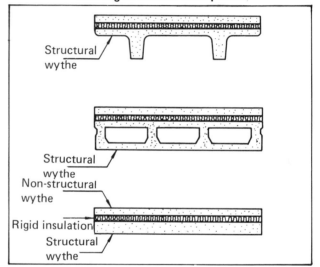

Table 3.6.1   Suggested tension limitations for compression members with bending

		Short columns or walls [1]	Medium columns or walls [2]	Long columns or walls [3]
(a)	Tension (including slenderness effects) for members having a cross-sectional dimension in the direction of bending greater than or equal to 5 in.	$6\sqrt{f'_c}$	$3\sqrt{f'_c}$	$0^{[4]}$
(b)	Tension for members having a cross-sectional dimension in the direction of bending less than 5 in.	$6\sqrt{f'_c}$	$1.6\sqrt{f'_c}$	$0^{[4]}$
(c)	Tension for members not exposed to freezing temperatures or corrosive environment which contain bonded prestressed or non-prestressed reinforcement located so as to control cracking.	$12\sqrt{f'_c}$	$6\sqrt{f'_c}$	$0^{[4]}$

(1)   Braced members, $k\ell_u/r < 34-12M_1/M_2$. Unbraced members $k\ell_u/r < 22$.

(2)   Members with $k\ell_u/r$ greater than in (1), but less than 100

(3)   Members with $k\ell_u/r > 100$ (braced members only)

(4)   If it can be shown that tensile stresses under transportation, handling and construction loads will at no time exceed $3\sqrt{f'_c}$, this may be increased to $1.67\sqrt{f'_c}$

3.6.7 Piles

3.6.7.1 General

The pile designs considered here are based upon structural capacity alone. The ability of the soil to carry these loads must be established by load tests or evaluated by soils engineers.

In the following design procedure for pretensioned concrete piles, load capacity is limited by the service load stresses. An overall factor of safety based on the nominal strength ($\phi = 1.0$) of the section is computed, and limits suggested for various loading conditions. Stresses caused by transporting, handling and driving should also be considered. Experience has shown that the frictional and bearing resistance of the soil will control the design more often than the strength and service load stresses on the pile.

The values used in sample calculations are based on a concrete strength of 6,000 psi. In many areas, higher concrete strengths have been effectively used in prestressed pile design. Engineers should check with local prestressed concrete pile producers to determine what concrete strengths are available in their areas as well as sizes and types, i.e., square, octagonal, or round.

The values in the pile load table (Table 2.7.1) may be modified to fit the actual service conditions and to maintain reasonable safety factors consistent with the character of the applied loads and how and where the piles are to be used.

3.6.7.2 Strengths under Direct Loads

(a) Nominal strength

It is assumed that the concrete stress at failure of a concrete pile will be $0.85 f'_c$. The amount of prestress remaining in the tendons must be deducted. Assuming an ultimate concrete strain of 0.003 it can be shown that only about 60% of the effective prestress, f_{pc}, is left in the member when it reaches its nominal strength. Thus, if a 6000 psi concrete pile is prestressed to an effective prestress of 700 psi, the nominal strength can be computed as:

$$P_n = (0.85 f'_c - 0.60 f_{pc}) A$$

$$P_n = (0.85 \times 6000 - 0.60 \times 700) A = 4680A$$

Where A = gross sectional area of the concrete in square inches, and P_n is the concentric nominal strength in pounds.

(b) Service load

For a concentric load on a short column pile, a factor of safety of between 2.0 and 3.0 is usually adequate.

Based upon a recent study by the Portland Cement Association, which has been accepted by most current building codes, the formula for service loads on concentrically loaded short column prestressed piles is:

$$N = (0.33 f'_c - 0.27 f_{pc}) A$$

Thus, for 6,000 psi concrete and 700 psi prestress

$$N = (0.33 \times 6,000 - 0.27 \times 700) A = 1790 A$$

If the service load is 1790 A, the overall safety factor for the pile as given in Section (a) will be $\frac{4,680}{1,790} = 2.61$ which is considered quite adequate.

For a pile with $f'_c = 6,000$ psi and an effective prestress of 1,200 psi, the corresponding overall safety factor will be approximately the same, 2.65. The value of $0.2 f'_c$ is considered to be about the desirable upper limit for the prestressing force, and a value of 700 psi is recommended as the desirable lower limit for all piles over about 40 ft in length. The overall safety factors for this range of prestress fall within the acceptable limits as indicated above.

(c) Unsupported length.

It is suggested that service loads based on the short column value of $(0.33 f'_c - 0.27 f_{pc}) A$ be limited to values of h/r up to 60

where

h = unsupported length of the pile
r = radius of gyration

For piles considered fully fixed at one end and hinged at the other end, it is suggested that h be taken as 0.7 of the length between hinge and assumed point of fixity. For piles fully fixed at both ends, h may be taken as 0.5 of the length between the assumed points of fixity.

3.6.7.3 Moment Resisting Capacities

(a) Service load stresses

(1) Allowing no tension. Under this criterion, if the effective prestress is $f_{pc} = P/A$, the moment capacity is $M = f_{pc} I/c$, where c = distance from extreme fiber to neutral axis. This criterion of "no tension" is much too conservative for prestressed piles under normal conditions. Zero tension will result in an overall safety factor of about 3 or more, which is greater than required for bending and greater than the safety factors used in steel or reinforced concrete design.

(2) The modulus of rupture for 6,000 psi concrete is generally taken to be $7.5 \sqrt{f'_c}$ or ap-

proximately 600 psi. Allowing tension up to about 50% of the modulus of rupture, the allowable tension for normal bending can be taken as $4\sqrt{f_c'}$ or 300 psi. If the prestress is 700 psi, the total stress available for bending is 1,000 psi and the moment capacity is $M = 1000\,I/c$.

This usually gives a safety factor of 2.5 or more. For earthquake and other transient loads this is conservative and the allowable tension may be increased to 600 psi. This gives a moment capacity of $M = 1300\,I/c$.

(b) Strength design

The nominal moment capacity of a pretensioned pile can be computed by the ultimate strength of the tendons multiplied by a proper lever arm. As an approximation, the total ultimate strength in all the tendons can be used. The lever arm is approximately 0.37 t for solid square piles and 0.32 t for solid circular and octagonal piles. For hollow piles, the lever arm will be a little longer, approximately 0.38 t for square piles and 0.34 t for circular and octagonal piles.

Thus, the approximate nominal moment strength is given by:

$M_n = 0.37\,t\,A_{ps}\,f_{pu}$ for solid square piles

$M_n = 0.32\,t\,A_{ps}\,f_{pu}$ for solid circular and octagonal piles

$M_n = 0.38\,t\,A_{ps}\,f_{pu}$ for hollow square piles

$M_n = 0.34\,t\,A_{ps}\,f_{pu}$ for hollow circular and octagonal piles

where A_{ps} = total steel area of all the tendons in square inches, f_{pu} = ultimate strength of the tendons in psi, and t = diameter or thickness of the pile in inches.

(c) Allowable service load moment

In general, the allowable moment should be based on an allowable tensile stress based on the modulus of rupture as in (a) and checked by the nominal strength moment as in (b) to insure a factor of safety of 2 for normal loading. For wind, earthquake, or other short-time loads a safety factor of 1.5 is considered adequate. In corrosive conditions the Engineer should make an evaluation to determine whether to reduce or eliminate the allowable tension.

3.6.7.4 Combined Moment and Direct Load

(a) Service load stresses

By the elastic theory, the existence of direct load delays the cracking of the concrete piles and thereby increases the moment carrying capacity. For example, if the prestress in the concrete is 700 psi, an external load induces 400 psi, and the allowable tensile stress is 300 psi, the total fiber stress available for bending moment is 1400 psi, and the moment capacity is:

$$M = 1400\,I/c$$

700　　　　400　　　　− 1400　　　　− 300

+ 　　　　+ 　　　　=

+ 1400　　　　2500

Prestress　　Direct Load　　Bending　　Working Stress

$$\frac{P}{A} + \frac{N}{A} \pm \frac{M}{I/c} = \begin{array}{l} \text{tens.} \leqslant 4\sqrt{f_c'} \\ \text{comp.} \leqslant .45\,f_c' \end{array}$$

Piles controlled by compression should also be checked against the interaction formula:

$$\frac{N}{N'} + \frac{M}{M'} \leqslant 1$$

where N' and M' are the allowable axial loads and moments, respectively.

(b) Strength design

The nominal moment capacity of a prestressed pile is reduced by the presence of external direct load, because the area of the compression zone is increased and the available lever arm for the resisting steel is correspondingly reduced.

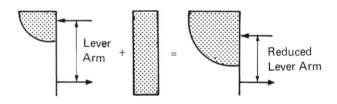

It can be roughly estimated that, for an external load producing 700 psi in the concrete, the nominal moment capacity is given by:

$M_n = 0.29\,t\,A_{ps}\,f_{pu}$ for solid square piles

$M_n = 0.25\,t\,A_{ps}\,f_{pu}$ for solid round and octagonal piles

For hollow piles the lever arm is a little longer, 0.30 t and 0.26 t respectively.

At the pile head, where combined moment and direct load may be critical, the transfer in which the prestress increases from zero to full value (about 50 tendon diameters) must be considered.

(c) Allowable service load moment

With the presence of external direct load, the

moment capacity of the piles should be first computed by the elastic theory as in (a), and the nominal strength checked as in (b) to insure a safety factor of 2. For wind, earthquake, or other short-time loads, a safety factor of 1.5 is considered adequate. In corrosive conditions the Engineer should make an evaluation to determine whether to reduce or eliminate the allowable tension.

Example 3.6.3 — Bearing pile

For a 12" square solid pile, compute the allowable load and moments.

Given:

A = 144 sq in.

I = 1,728 in.^{4}

I/c = 288 in.^{3}

Prestress with 6 - 7/16" diameter 270K strands.

A_{ps} = 0.69 sq in.

f_{pu} = 270,000 psi or 31,000 lb. per strand

f'_c = 6000 psi

If the strands are stressed initially to 0.7 f_{pu} and total losses are assumed to be 22%, effective prestress = 0.7 x 0.78 x 31,000 = 16,900 lb/strand

$$f_{pc} = \frac{6 \times 16,900}{144} = 705 \text{ psi}$$

(a) Direct Load

Using the formula $N = (0.33 f'_c - 0.27 f_{pc}) A$

$N = 1,790 A$

Allowable N = 1,790 x 144 = 258 kips or 129 tons.

Nominal strength is given by:

$P_n = (0.85 f'_c - 0.60 f_{pc}) A$
= (0.85 x 6,000 - 0.60 x 705) 144 = 673 kips.

Thus, factor of safety $\frac{673}{258}$ = 2.61

(b) Moment Capacity

For an allowable tension of 300 psi:
$M = f_c \, I/c$
= (300 + 705) 288/1000 = 289 in.-kips

Nominal moment strength:

M_n = 0.37 t A_{ps} f_{pu} = 0.37 x 12 x 0.69 x 270
= 827 in.-kips

Thus, factor of safety = $\frac{827}{289}$ = 2.86, which is higher than necessary. Thus, for transient loads, the allowable tension could be increased beyond 300 psi.

(c) Allowable Unsupported Length

This will vary with the load. The maximum h/r for which the short column formula should be used is 60. From pile properties in Table 2.7.1, r = 3.46 in. for a 12-in. square pile. Thus, for a sustained load of 258 kips, and assuming pile fully fixed at both ends, the allowable unsupported length

$$\ell_u = \frac{60r}{0.5} = 415 \text{ in.} = 34.6 \text{ ft}$$

Example 3.6.4 — Sheet pile

For a 12" x 36" sheet pile, compute allowable moment.

Given:

A = 432 sq in.

I = 5184 in.^{4}

I/c = 864 in.^{3} (per pile)

Prestress with 20 - 1/2" 270 K strands

A_{ps} = 3.06 sq in.

f_{pu} = 270,000 psi or 41,300 lb per strand

f'_c = 6000 psi

Effective prestress = 0.7 x 0.78 x 41,300
= 22,560 lb/strand

$$f_{pc} = \frac{20 \times 22,560}{432} = 1045 \text{ psi}$$

Moment Capacity

(a) Service loads

For allowable tension of 300 psi:

Allowable M = (300 + 1045) x 864/1000
= 1160 in.-kips

or $\frac{1160}{3}$ = 386 in.-kips per foot of wall.

(b) Strength design

Nominal moment strength, M_n = 0.37 t A_{ps} f_{pu}
= (0.37 x 12 x 3.06 x 270) = 3670 in.-kips

indicating a factor of safety of $\frac{3670}{1160}$ = 3.1,

which is more than sufficient.

Thus, the above allowable tension of 300 psi may be increased for transient loads, except in corrosive conditions.

3.7 Torsion

ACI 318-77 does not include torsion design provisions for prestressed concrete. The provisions in the Code for non-prestressed concrete will usually be conservative if applied to prestressed concrete. The design procedure outlined in this section is a modified version of the method proposed by Zia and McGee,* which is an extension of the ACI provision for torsion in non-prestressed concrete. The basic approach consists of determining the concrete contribution to the torsional strength of a member, taking into account the effects of the prestressing force and the torsion-shear interaction, and then proportioning reinforcement for the remaining strength. (Note: Equations in ACI 318-77 are expressed in terms of forces. However, it is generally more convenient to work torsion problems using stresses. This is the procedure used in this section.)

3.7.1 Design Procedure

The step-by-step design procedure may be summarized as follows:

Step 1: Determine the factored torsional stress:

$$v_{tu} = \frac{3 T_u}{\phi \, \Sigma x^2 y} \qquad \text{(Eq. 3.7-1)}$$

where:

v_{tu} = factored torsional stress, psi

T_u = factored torque on the member, in.-lb

x = shorter side of a component rectangle, in.

y = longer side of the corresponding rectangle, in.

ϕ = 0.85

T-sections, L-sections or similar members are divided into component rectangles such that the quantity $\Sigma x^2 y$ is a maximum.

For a box section with a wall thickness, t, not less than b/10 nor more than b/4:

$$v_{tu} = \frac{b}{4t} \; \frac{T_u}{\phi b^2 h} \qquad \text{(Eq. 3.7-2)}$$

where:

b = overall width of the member

h = overall depth of the member

t = wall thickness

Where applicable, the factored shear stress, v_u, should also be determined (see Sect. 3.3).

Torsion may be neglected if v_{tu} is less than v_{tu} (min):

$$v_{tu} \text{ (min)} = 1.5 \, \lambda \sqrt{f'_c} \; \gamma_t \qquad \text{(Eq. 3.7-3)}$$

where:

f'_c = concrete compressive strength, psi

λ = 1.0 for normal weight concrete

    = $(f_{ct}/6.7) \sqrt{f'_c}$ for sand-lightweight or lightweight concrete

If f_{ct} is unknown:

λ = 0.85 for sand-lightweight concrete

    = 0.75 for all-lightweight concrete

f_{ct} = splitting tensile strength of concrete, psi

γ_t = $\sqrt{1 + 10 \, f_{pc}/f'_c}$

f_{pc} = average prestress on the section, psi

Fig. 3.9.21 gives v_{tu} (min) for various concrete strengths and average prestress values.

Step 2: If v_{tu} exceeds v_{tu} (min), torsion design is required. However, to prevent over-reinforcing the section, v_{tu} and v_u must not exceed their maximum permissible values:

$$v_{tu} \text{ (max)} = \frac{10 K_T \, \lambda \sqrt{f'_c}}{\sqrt{1 + \left(\dfrac{K_T \, v_u}{v_{tu}}\right)^2}} \qquad \text{(Eq. 3.7-4)}$$

$$v_u \text{ (max)} = \frac{10 \, \lambda \sqrt{f'_c}}{\sqrt{1 + \left(\dfrac{v_{tu}}{K_T \, v_u}\right)^2}} \qquad \text{(Eq. 3.7-5)}$$

where:

$$K_T = \gamma_t \, (1.4 - 4 \, f_{pc}/3 \, f'_c) \qquad \text{(Eq. 3.7-6)}$$

Fig. 3.9.22 gives values of v_{tu} (max) and Fig. 3.9.23 gives values of v_u (max).

Step 3: If $v_{tu} < v_{tu}$ (max) and $v_u < v_u$ (max), the

---

*Zia, Paul and McGee, W. Denis "Torsion Design of Prestressed Concrete" *PCI Journal,* March-April, 1974.

As of this writing, the ACI Torsion Committee was receiving recommendations for torsion in prestressed members. The modification of the Zia-McGee method is based on those recommendations. The final ACI version may be somewhat different than shown here, but it is anticipated that the effect of the differences will be minor.

torsion and shear stresses carried by the concrete are:

$$v_{tc} = R_t \, v'_{tc} \qquad \text{(Eq. 3.7-7)}$$

$$v_c = R_v \, v'_c \qquad \text{(Eq. 3.7-8)}$$

where:

$$R_t = \frac{1}{\sqrt{1 + \left(\dfrac{v_u}{\beta \, v_{tu}}\right)^2}} \qquad \text{(Eq. 3.7-9)}$$

$$R_v = \frac{1}{\sqrt{1 + \left(\dfrac{\beta \, v_{tu}}{v_u}\right)^2}} \qquad \text{(Eq. 3.7-10)}$$

$$v'_{tc} = 6\lambda\sqrt{f'_c} \; (\gamma_t - 0.60) \qquad \text{(Eq. 3.7-11)}$$

$$v'_c = V_c/b_w d \quad \text{(see Sect. 3.3)}$$

$$\beta = v'_c / 2 \, v'_{tc}$$

Values of R_t and R_v are given in Figs. 3.9.24 and 3.9.25.

Step 4: Closed stirrups spaced not more than 12 in. or $(x_1 + y_1)/4$ must be provided to carry the torsional stress in excess of that carried by the concrete:

$$A_t = \frac{(v_{tu} - v_{tc})s \, \Sigma x^2 y}{\alpha_t \, x_1 \, y_1 \, f_y} \qquad \text{(Eq. 3.7-12)}$$

where:

A_t = the required area of *one* leg of stirrup, sq in.

x_1 = shorter leg of the stirrup, in.

y_1 = longer leg of the stirrup, in.

s = stirrup spacing, in.

$\alpha_t = 2 + y_1/x_1 \leqslant 4.5$

f_y = yield strength of the stirrup, psi

The required stirrups for torsion are in addition to the web reinforcement required for shear.

If v_{tu} is greater than that calculated by Eq. 3.7-3, a minimum amount of shear and torsion reinforcement should be provided to ensure a reasonable amount of ductility. The minimum area of closed stirrups which should be provided is:

$$A_v + 2 A_t = \frac{50 \, b_w s \, \gamma_t^2}{f_y} \qquad \text{(Eq. 3.7-13)}$$

where b_w is the web width of the beam

Step 5: To resist the longitudinal component of the diagonal tension induced by torsion, longitudinal reinforcement equal in volume to that of stirrups for torsion should be provided in addition to that required by flexure. Thus:

$$A_\ell = \frac{2A_t \, (x_1 + y_1) \, f_y}{s \, f_{y\ell}} \qquad \text{(Eq. 3.7-14)}$$

where:

$f_{y\ell}$ = yield strength of the longitudinal reinforcement.

This design procedure assumes that the member is properly connected to a rigid support. The design of the connections to achieve this condition is critical. Also, the effects of torsion on the overall stability of the frame should be investigated. Guides for these design considerations may be found in Parts 4 and 5.

Example 3.7.1 – Torsion of a prestressed concrete member

Given:

Typical precast, prestressed concrete spandrel beam shown in Fig. 3.7.1.

D. L. of deck = 89.5 psf

L. L. = 50 psf

Beam properties:

A = 696 sq in.

wt = 725 plf

f'_c = 5000 psi, normal weight

f_y = 40 ksi

Prestressing:

6-1/2″ dia., 270K strands

A_{ps} = 6 x 0.153 = 0.918 sq in.

d = 69 in.

Find:

Required torsion reinforcement for the spandrel beam.

Solution:

1) Calculate V_u and T_u

Determine factored loads:

Fig. 3.7.1 Structure of Example 3.7.1

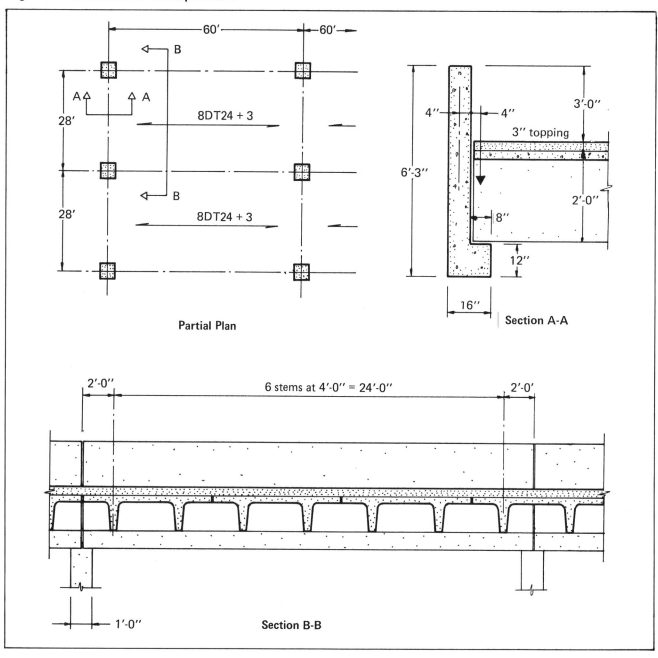

Partial Plan

Section A-A

Section B-B

D. L. of beam = 1.4 x 0.725 = 1.02 kips/ft

D. L. of deck = 1.4 x 0.0895 x 60/2 x 4
= 15.04 kips per stem

L. L.   = 1.7 x 0.050 x 30 x 4
= 10.2 kips per stem

V_u at support

= 1.02 x 14 + 7 (15.04 + 10.2) x 1/2

= 102.6 kips

Center of support to h/2 = 0.5 + 6.25/2
= 3.625 ft

V_u at h/2

= 102.6 – 15.04 – 10.2 – (1.02 x 3.625)

= 73.7 kips

T_u at support (assume torsion arm = 8 in.)

= (15.04 + 10.2) x 7 x 1/2 x 8 in.

= 707 in.-kips

T_u at h/2

= 707 – (15.04 + 10.2) x 8 in. = 505 in.-kips

2) Calculate shear stress:

$v_u = V_u/\phi\, b_w d = 73.7/(0.85 \times 8 \times 69)$
     $= 0.157$ ksi

3) Calculate torsion stress:

For maximum $\Sigma x^2 y$; consider the beam web ($x = 8''$, $y = 63''$) and bottom flange ($x = 12''$, $y = 16''$) as the component rectangles:

$\Sigma x^2 y = 8^2 \times 63 + 12^2 \times 16 = 6336$ in.³

$v_{tu} = \dfrac{3\,T_u}{\phi\,\Sigma x^2 y} = \dfrac{3\,(505)}{0.85\,(6336)} = 0.281$ ksi

4) Check for torsion design requirement:

Effective prestress (assume 22% loss):

$P = 0.918\,(0.7 \times 270)(0.78) = 135.3$ kips

$f_{pc} = P/A = 135.3/696 = 0.194$ ksi

From Fig. 3.9.21

v_{tu} (min) $= 125$ psi < 281 psi

Therefore torsion design is required.

5) Check v_{tu} (max) and v_u (max):

$f_{pc}/f'_c = 0.194/5 = 0.039$

$v_{tu}/v_u = 281/157 = 1.79$

From Fig. 3.9.22:

v_{tu} (max) $= 12\sqrt{f'_c} = 849$ psi > 281 OK

From Fig. 3.9.23:

v_u (max) $= 6.5\sqrt{f'_c} = 460$ psi > 157 OK

6) Calculate v_{tc} and v_c:

$\gamma_t = \sqrt{1 + 10\,f_{pc}/f'_c} = \sqrt{1 + 0.39}$
     $= 1.18$

$v'_{tc} = 6\lambda\sqrt{f'_c}\,(\gamma_t - 0.60) = 246$ psi

v'_c (see Fig. 3.9.11) $= 5\sqrt{f'_c} = 354$ psi

$v'_{tc}/v'_c = 246/354 = 0.69$

From Fig. 3.9.24, $R_t = 0.79$

$v_{tc} = R_t\, v'_{tc} = 0.79\,(246) = 194$ psi

From Fig. 3.9.25, $R_v = 0.54$

$v_c = R_v\, v'_c = 0.54\,(354) = 191$ psi

7) Calculate web reinforcement for torsion:

Assume 1.5 in. from surface to center of reinforcement.

$x_1 = 8 - (2 \times 1.5) = 5$ in.

$y_1 = 75 - (2 \times 1.5) = 72$ in.

$\alpha_t = 2 + y_1/x_1 = 16.4 > 4.5$

use $\alpha_t = 4.5$

$\dfrac{A_t}{s} = \dfrac{(v_{tu} - v_{tc})\,\Sigma x^2 y}{\alpha_t\, x_1\, y_1\, f_y}$

$= \dfrac{(281 - 194)(6336)}{4.5\,(5)(72)(40,000)}$

$= 0.0085$ sq in./in.

max $s = \dfrac{x_1 + y_1}{4} = \dfrac{5 + 72}{4} = 19.25'' > 12''$

use 12 in. (max.)

Use #3 closed stirrups at 12 in. o.c.

Fig. 3.7.2 Reinforcement for beam of Example 3.7.1

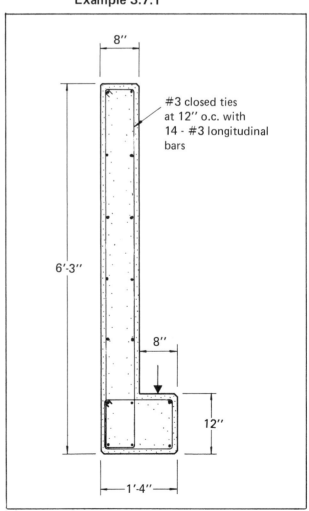

#3 closed ties at 12'' o.c. with 14 - #3 longitudinal bars

8''

6'-3''

8''

12''

1'-4''

$$\frac{A_t}{s} = \frac{0.11}{12} = 0.0092 \text{ sq in./in. OK}$$

8) Web reinforcement for shear:

$v_u = 157$ psi, $v_c = 191$ psi $> v_u$

but $v_u > v_c/2$ so minimum web reinforcement is required

$$A_v + 2 A_t = \frac{50 \, b_w \, s \, \gamma_t^2}{f_y}$$

$$= \frac{50 \, (8)(12)(1.18)^2}{40,000} = 0.167 \text{ sq in.}$$

But $2 A_t = 0.22$ sq in. > 0.167 sq in., so web reinforcement provided for torsion is more than adequate for shear plus torsion.

9) Longitudinal torsion reinforcement:

$$A_\ell = \frac{2 A_t (x_1 + y_1) f_y}{s \, f_{y\ell}}$$

$$= \frac{2 (0.11)(5 + 72)(40,000)}{12 \, (40,000)} = 1.41 \text{ sq in.}$$

Use 14 #3 bars distributed around the perimeter of the stirrups (see Fig. 3.7.2)

$A_\ell = 1.54$ sq in.

10) Reinforcement in the ledge may be designed as shown in Sect. 5.10.

3.8 Special Design Situations

3.8.1 Cantilevers

The most effective way to design cantilevered members will depend on the type of product, method of production, span conditions and section properties of the member. The designer is advised to consult with local producers on the most effective method.

Many producers prefer to design cantilevers as reinforced concrete members, using deformed reinforcing bars or short pieces of unstressed strand to provide the negative moment resistance. When using strand which is unstressed, the stress under factored loads should be limited to 60,000 psi. Pretensioned strands are sometimes used to improve performance characteristics of cantilevers, however design, production and quality control should be quite rigorous to assure proper performance.

It is suggested that service load tension be limited to 100 psi (including prestress) when pre-

stressed strands are used and to the cracking tension ($6 \sqrt{f'_c}$) when non-prestressed steel is used for negative moment resistance.

Strand development may be a problem in cantilevers, and it is suggested that when prestressed strand is used, the design stress, f_{ps}, be limited to one-half that indicated by Fig. 3.9.9.

3.8.2 Concentrated Load Distribution

Concentrated loads on deck members can be distributed to adjacent members by the shear strength of either grout keys (hollow-core slabs) or weld plates (flanged members and solid slabs). If the grout keys or weld plates are also used to transfer diaphragm shears (see Sect. 4.5), the strength of the connection should be tested by:

$$\frac{V_{uh}}{\phi V_{nh}} + \frac{V_{uv}}{\phi V_{nv}} \leqslant 1 \qquad \text{(Eq. 3.8-1)}$$

where

V_{uh}, V_{uv} = applied factored loads in the horizontal and vertical directions, respectively

V_{nh}, V_{nv} = nominal strength of the connection in the horizontal and vertical directions, respectively

ϕ = 0.85

Former ACI Standard 711-58 permitted distribution of concentrated loads uniformly over three units on each side of the loaded unit, but never over a greater total width than 0.4 of the clear span distance. Concentrated loads near a support may not distribute in this manner.

Example 3.8.1 — Distribution of concentrated load

Given:

A partition wall supported by an untopped hollow-core floor system as shown in Fig. 3.8.1.

Grout strength = 3000 psi

Grout key depth = 3 in.

Factored horizontal diaphragm load = 6.0 kips

Find:

Load on floor slabs in addition to floor live load, and determine if grout key is adequate to distribute the load.

Solution:

Load can be distributed to 3 units each side of loaded unit (7 units) or 0.4 times the clear span

Fig. 3.8.1   Example 3.8.1

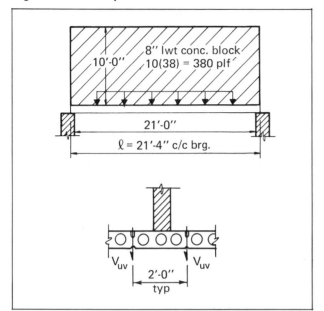

7 x 2 ft = 14 ft

0.4 x 21 = 8.4 ft

380/8.4 = 45.2 psf to be distributed to adjacent slabs in addition to other superimposed loads.

$$V_{uv} = 1.4\,[380 - 45.2\,(2)]\,/2 = 203 \text{ plf}$$

$$V_{uh} = 6000/21.33 = 281 \text{ plf}$$

Assume shear strength of grout = $2\sqrt{f'_c}$ = 110 psi

$$\phi V_{nv} = (0.85)\,110\,(3)(12) = 3366 \text{ plf}$$

$$\phi V_{nh} = (0.85)\,40 \text{ psi} \times 3 \times 12 = 1224 \text{ plf}$$
(see Sect. 4.5)

$$\frac{V_{uh}}{\phi V_{nh}} + \frac{V_{uv}}{\phi V_{nv}} = \frac{281}{1224} + \frac{203}{3366} = 0.29 < 1.0 \text{ OK}$$

3.8.3 Openings in Critical Areas

It is frequently necessary to provide openings in deck units to accommodate ducts, stairs or other items required for the building function. Examples of such openings, and suggested limitations are shown in Fig. 6.4.4.

If such openings are large and occur at or near the critical sections for moment or shear, the reduced section should be considered in the design. In the case of flanged members, the opening will normally not interfere with the strand placement. In these cases, the effect on the design is to reduce the compression block width and the section modulus. (See Sects. 3.2.1 and 3.2.2). In some cases it may be desirable to provide compression reinforcement to compensate for the reduced compression concrete.

When the opening reduces the strand area, as frequently occurs in hollow-core slabs, the reduced moment and shear capacity in the area should be compared with the applied moments and shears, taking full account of the strand development on each side of the opening. (See Sect. 3.2.3).

Quite often, an opening is the same width as one or more deck slabs, and steel headers are used to support the cut-off slabs. These headers are supported by the adjacent slabs and the design for the reaction is similar to that described in Sect. 3.8.2.

FLEXURE

Table 3.9.1 Coefficients, K_u, for determining flexural design strength — bonded prestressing steel

Procedure:

1. Determine $\overline{\omega}_p = \dfrac{A_{ps}}{bd}\dfrac{f_{pu}}{f'_c}$

2. Find K_u from table

3. Determine $\phi M_n = K_u \dfrac{bd^2}{12,000}$ (ft–kips)

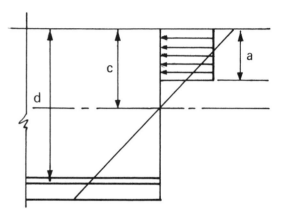

Note: Values below dashed line indicate

$$\omega_p = \frac{A_{ps}}{bd}\frac{f_{ps}}{f'_c} > 0.30$$

Values of K_u

f'_c	$\overline{\omega}_p$	.00	.01	.02	.03	.04	.05	.06	.07	.08	.09
3000 psi	0.0	0	26	52	78	103	127	151	175	197	220
	0.1	242	263	284	304	324	343	362	381	399	417
	0.2	434	451	467	483	499	514	529	543	557	571
	0.3	584	597	610	623	635	646	658	669	680	690
4000 psi	0.0	0	35	70	104	137	170	202	233	263	293
	0.1	322	351	379	406	432	458	483	508	532	556
	0.2	579	601	623	644	665	685	705	724	743	762
	0.3	779	797	814	830	846	862	877	892	906	920
5000 psi	0.0	0	44	88	130	172	213	252	291	329	367
	0.1	403	439	473	507	540	573	604	635	665	695
	0.2	723	751	779	805	831	857	882	906	929	952
	0.3	974	996	1017	1038	1058	1078	1097	1115	1133	1151
6000 psi	0.0	0	53	105	156	206	255	303	350	395	440
	0.1	484	526	568	609	649	687	725	762	799	834
	0.2	868	902	935	967	998	1028	1058	1087	1115	1143
	0.3	1169	1195	1221	1246	1270	1293	1316	1338	1360	1381
7000 psi	0.0	0	62	123	182	241	298	354	408	461	514
	0.1	564	614	663	710	757	802	846	890	932	973
	0.2	1013	1052	1091	1128	1164	1200	1234	1268	1301	1333
	0.3	1364	1395	1424	1453	1481	1509	1535	1561	1586	1611
8000 psi	0.0	0	71	140	209	275	340	404	466	527	587
	0.1	645	702	758	812	865	917	967	1017	1065	1112
	0.2	1158	1203	1246	1289	1331	1371	1411	1449	1487	1524
	0.3	1559	1594	1628	1661	1693	1724	1755	1784	1813	1841

PCI Design Handbook

FLEXURE

Table 3.9.2 Coefficients for determining flexural design strength — bonded prestressing steel

Procedure:

1. Determine $\overline{\omega}_p = \dfrac{A_{ps}}{bd}\dfrac{f_{pu}}{f'_c}$

2. Enter table with $\overline{\omega}_p$ and find moment capacity

   $\phi\, M_n = \phi\, A_{ps}\, f_{pu}\, \overline{j}_u\, d$

3. If needed for tee-sections, enter with $\overline{\omega}_p$ and find c and f_{ps}

   $f_{ps}/f_{pu} = 1.0 - 0.5\,\overline{\omega}_p \qquad \overline{j}_u = j_u\,\dfrac{f_{ps}}{f_{pu}} = (1.0 - a/2d)\,\dfrac{f_{ps}}{f_{pu}}$

Distance to Neutral Axis: Values of c/d

f'_c	$\overline{\omega}_p$	.00	.01	.02	.03	.04	.05	.06	.07	.08	.09
3000 psi	0.0	0.000	0.014	0.027	0.041	0.054	0.067	0.081	0.093	0.106	0.119
	0.1	0.131	0.144	0.156	0.168	0.180	0.192	0.204	0.215	0.227	0.238
	0.2	0.249	0.260	0.271	0.282	0.292	0.303	0.313	0.323	0.333	0.343
	0.3	0.353	0.363	0.372	0.381	0.391	0.400	0.409	0.417	0.426	0.435
4000 psi	0.0	0.000	0.014	0.027	0.041	0.054	0.067	0.081	0.093	0.106	0.119
	0.1	0.131	0.144	0.156	0.168	0.180	0.192	0.204	0.215	0.227	0.238
	0.2	0.249	0.260	0.271	0.282	0.292	0.303	0.313	0.323	0.333	0.343
	0.3	0.353	0.363	0.372	0.381	0.391	0.400	0.409	0.417	0.426	0.435
5000 psi	0.0	0.000	0.015	0.029	0.043	0.058	0.072	0.086	0.099	0.113	0.126
	0.1	0.140	0.153	0.166	0.179	0.191	0.204	0.216	0.229	0.241	0.253
	0.2	0.265	0.276	0.288	0.299	0.311	0.322	0.333	0.343	0.354	0.365
	0.3	0.375	0.385	0.395	0.405	0.415	0.425	0.434	0.443	0.453	0.462
6000 psi	0.0	0.000	0.016	0.031	0.046	0.061	0.076	0.091	0.106	0.120	0.135
	0.1	0.149	0.163	0.177	0.191	0.204	0.218	0.231	0.244	0.257	0.270
	0.2	0.282	0.295	0.307	0.319	0.331	0.343	0.355	0.366	0.378	0.389
	0.3	0.400	0.411	0.422	0.432	0.443	0.453	0.463	0.473	0.483	0.492
7000 psi	0.0	0.000	0.017	0.033	0.050	0.066	0.082	0.098	0.114	0.129	0.144
	0.1	0.160	0.175	0.190	0.204	0.219	0.233	0.247	0.261	0.275	0.289
	0.2	0.303	0.316	0.329	0.342	0.355	0.368	0.380	0.393	0.405	0.417
	0.3	0.429	0.440	0.452	0.463	0.474	0.485	0.496	0.507	0.517	0.528
8000 psi	0.0	0.000	0.018	0.036	0.053	0.071	0.088	0.105	0.122	0.139	0.156
	0.1	0.172	0.188	0.204	0.220	0.236	0.251	0.266	0.282	0.296	0.311
	0.2	0.326	0.340	0.354	0.368	0.382	0.396	0.409	0.423	0.436	0.449
	0.3	0.462	0.474	0.487	0.499	0.511	0.523	0.534	0.546	0.557	0.568

Values f_{ps}/f_{pu}

	$\overline{\omega}_p$	.00	.01	.02	.03	.04	.05	.06	.07	.08	.09
	0.0	1.000	0.995	0.990	0.985	0.980	0.975	0.970	0.965	0.960	0.955
	0.1	0.950	0.945	0.940	0.935	0.930	0.925	0.920	0.915	0.910	0.905
	0.2	0.900	0.895	0.890	0.885	0.880	0.875	0.870	0.865	0.860	0.855
	0.3	0.850	0.845	0.840	0.835	0.830	0.825	0.820	0.815	0.810	0.805

Values $\overline{j}_u$

	$\overline{\omega}_p$	.00	.01	.02	.03	.04	.05	.06	.07	.08	.09
	0.0	1.000	0.989	0.978	0.968	0.957	0.947	0.937	0.927	0.917	0.907
	0.1	0.897	0.887	0.878	0.868	0.859	0.850	0.840	0.831	0.822	0.813
	0.2	0.805	0.796	0.787	0.779	0.771	0.762	0.754	0.746	0.738	0.730
	0.3	0.723	0.715	0.707	0.700	0.692	0.685	0.678	0.670	0.663	0.656

FLEXURE

Table 3.9.3 Coefficients for determination of prestressing steel requirements — bonded prestressing steel

Procedure

1. Determine $K_u = \dfrac{M_u \, (12,000)}{bd^2}$
   (for $M_u \leqslant \phi M_n$)

2. Find $\overline{\omega_p}$ from table

3. Determine $A_{ps} = \overline{\omega_p} \, bd \dfrac{f'_c}{f_{pu}}$

M_u = Applied factored moment, ft-kips
b = Width of section, in.
d = Effective depth to steel, in.
A_{ps} = Required area of prestressing steel, sq. in.
f_{pu} = Ultimate strength of prestressing steel, psi

Note: Values below dashed line indicate $\omega_p = \dfrac{A_{ps}}{bd} \dfrac{f_{ps}}{f'_c} > 0.30$

	K_u	\multicolumn{10}{c}{Values of $\overline{\omega_p}$}
		0	**10**	**20**	**30**	**40**	**50**	**60**	**70**	**80**	**90**
f'_c = 3000 psi	0	0.000	0.004	0.007	0.011	0.015	0.019	0.023	0.027	0.031	0.035
	100	0.039	0.043	0.047	0.051	0.055	0.059	0.064	0.068	0.072	0.076
	200	0.081	0.085	0.090	0.094	0.099	0.104	0.108	0.113	0.118	0.123
	300	0.128	0.133	0.138	0.143	0.148	0.153	0.158	0.164	0.169	0.175
	400	0.180	0.186	0.192	0.197	0.203	0.209	0.215	0.221	0.228	0.234
	500	0.241	0.247	0.254	0.261	0.267	0.274	0.282	0.289	0.296	0.304
	600	0.312	0.319	0.328	0.336	0.344	0.353	0.362	0.371	0.380	0.389
f'_c = 4000 psi	0	0.000	0.003	0.006	0.008	0.011	0.014	0.017	0.020	0.023	0.026
	100	0.029	0.032	0.035	0.038	0.041	0.044	0.047	0.050	0.053	0.056
	200	0.059	0.062	0.066	0.069	0.072	0.075	0.079	0.082	0.085	0.089
	300	0.092	0.096	0.099	0.102	0.106	0.110	0.113	0.117	0.120	0.124
	400	0.128	0.131	0.135	0.139	0.143	0.147	0.151	0.154	0.158	0.162
	500	0.166	0.171	0.175	0.179	0.183	0.187	0.192	0.196	0.200	0.205
	600	0.209	0.214	0.218	0.223	0.228	0.232	0.237	0.242	0.247	0.252
	700	0.257	0.262	0.267	0.273	0.278	0.283	0.289	0.294	0.300	0.306
	800	0.312	0.317	0.323	0.330	0.336	0.342	0.348	0.355	0.362	0.368
	900	0.375	0.382	0.389	0.397	0.404	0.412	0.419	0.427	0.436	0.444
f'_c = 5000 psi	0	0.000	0.002	0.004	0.007	0.009	0.011	0.014	0.016	0.018	0.020
	100	0.023	0.025	0.027	0.030	0.032	0.035	0.037	0.039	0.042	0.044
	200	0.047	0.049	0.052	0.054	0.057	0.059	0.062	0.064	0.067	0.070
	300	0.072	0.075	0.077	0.080	0.083	0.085	0.088	0.091	0.094	0.096
	400	0.099	0.102	0.105	0.107	0.110	0.113	0.116	0.119	0.122	0.125
	500	0.128	0.131	0.134	0.137	0.140	0.143	0.146	0.149	0.152	0.155
	600	0.158	0.162	0.165	0.168	0.171	0.175	0.178	0.181	0.185	0.188
	700	0.192	0.195	0.199	0.202	0.206	0.209	0.213	0.217	0.220	0.224
	800	0.228	0.232	0.235	0.239	0.243	0.247	0.251	0.255	0.259	0.263
	900	0.267	0.272	0.276	0.280	0.284	0.289	0.293	0.298	0.302	0.307
	1000	0.312	0.316	0.321	0.326	0.331	0.336	0.341	0.346	0.351	0.356
	1100	0.362	0.367	0.372	0.378	0.384	0.389	0.395	0.401	0.407	0.413

PCI Design Handbook

FLEXURE

Table 3.9.3 (cont.) Coefficients for determination of prestressing steel requirements — bonded prestressing steel

		Values of $\overline{\omega}_p$								
K_u	0	10	20	30	40	50	60	70	80	90
$f'_c = 6000$ psi										
0	0.000	0.002	0.004	0.006	0.007	0.009	0.011	0.013	0.015	0.017
100	0.019	0.021	0.023	0.025	0.027	0.029	0.031	0.033	0.035	0.037
200	0.039	0.041	0.043	0.045	0.047	0.049	0.051	0.053	0.055	0.057
300	0.059	0.061	0.064	0.066	0.068	0.070	0.072	0.074	0.076	0.079
400	0.081	0.083	0.085	0.088	0.090	0.092	0.094	0.097	0.099	0.101
500	0.104	0.106	0.108	0.111	0.113	0.116	0.118	0.120	0.123	0.125
600	0.128	0.130	0.133	0.135	0.138	0.140	0.143	0.145	0.148	0.151
700	0.153	0.156	0.158	0.161	0.164	0.166	0.169	0.172	0.175	0.177
800	0.180	0.183	0.186	0.189	0.192	0.195	0.197	0.200	0.203	0.206
900	0.209	0.212	0.215	0.218	0.221	0.225	0.228	0.231	0.234	0.237
1000	0.241	0.244	0.247	0.250	0.254	0.257	0.261	0.264	0.267	0.271
1100	0.274	0.278	0.282	0.285	0.289	0.293	0.296	0.300	0.304	0.308
1200	0.312	0.316	0.319	0.323	0.328	0.332	0.336	0.340	0.344	0.348
1300	0.353	0.367	0.362	0.366	0.371	0.375	0.380	0.385	0.389	0.394
$f'_c = 7000$ psi										
0	0.000	0.002	0.003	0.005	0.006	0.008	0.010	0.011	0.013	0.015
100	0.016	0.018	0.019	0.021	0.023	0.024	0.026	0.028	0.030	0.031
200	0.033	0.035	0.036	0.038	0.040	0.042	0.043	0.045	0.047	0.049
300	0.050	0.052	0.054	0.056	0.057	0.059	0.061	0.063	0.065	0.067
400	0.068	0.070	0.072	0.074	0.076	0.078	0.080	0.082	0.083	0.085
500	0.087	0.089	0.091	0.093	0.095	0.097	0.099	0.101	0.103	0.105
600	0.107	0.109	0.111	0.113	0.115	0.117	0.119	0.121	0.123	0.126
700	0.128	0.130	0.132	0.134	0.136	0.138	0.141	0.143	0.145	0.147
800	0.149	0.152	0.154	0.156	0.158	0.161	0.163	0.165	0.168	0.170
900	0.172	0.175	0.177	0.179	0.182	0.184	0.187	0.189	0.192	0.194
1000	0.197	0.199	0.202	0.204	0.207	0.209	0.212	0.214	0.217	0.220
1100	0.222	0.225	0.228	0.230	0.233	0.236	0.239	0.241	0.244	0.247
1200	0.250	0.253	0.256	0.259	0.262	0.264	0.267	0.270	0.273	0.276
1300	0.280	0.283	0.286	0.289	0.292	0.295	0.298	0.302	0.305	0.308
1400	0.312	0.315	0.318	0.322	0.325	0.329	0.332	0.336	0.339	0.343
1500	0.347	0.350	0.354	0.358	0.362	0.365	0.369	0.373	0.377	0.381
$f'_c = 8000$ psi										
0	0.000	0.001	0.003	0.004	0.006	0.007	0.008	0.010	0.011	0.013
100	0.014	0.016	0.017	0.018	0.020	0.021	0.023	0.024	0.026	0.027
200	0.029	0.030	0.032	0.033	0.035	0.036	0.038	0.039	0.041	0.042
300	0.044	0.045	0.047	0.048	0.050	0.051	0.053	0.055	0.056	0.058
400	0.059	0.061	0.062	0.064	0.066	0.067	0.069	0.070	0.072	0.074
500	0.075	0.077	0.079	0.080	0.082	0.084	0.085	0.087	0.089	0.090
600	0.092	0.094	0.096	0.097	0.099	0.101	0.102	0.104	0.106	0.108
700	0.110	0.111	0.113	0.115	0.117	0.119	0.120	0.122	0.124	0.126
800	0.128	0.130	0.131	0.133	0.135	0.137	0.139	0.141	0.143	0.145
900	0.147	0.149	0.151	0.153	0.154	0.156	0.158	0.160	0.162	0.164
1000	0.166	0.169	0.171	0.173	0.175	0.177	0.179	0.181	0.183	0.185
1100	0.187	0.189	0.192	0.194	0.196	0.198	0.200	0.203	0.205	0.207
1200	0.209	0.212	0.214	0.216	0.218	0.221	0.223	0.225	0.228	0.230
1300	0.232	0.235	0.237	0.240	0.242	0.245	0.247	0.250	0.252	0.255
1400	0.257	0.260	0.262	0.265	0.267	0.270	0.273	0.275	0.278	0.281
1500	0.283	0.286	0.289	0.292	0.294	0.297	0.300	0.303	0.306	0.309
1600	0.312	0.315	0.317	0.320	0.323	0.327	0.330	0.333	0.336	0.339
1700	0.342	0.345	0.348	0.352	0.355	0.358	0.362	0.365	0.368	0.372

FLEXURE

Table 3.9.4 Coefficients for resisting moments of rectangular and T-sections

Values of $C_f = \dfrac{bd^2}{12{,}000}$

a. Enter table with known values of $C_f = \dfrac{M_u}{K_u}$
   Select b and d (in.)

or b. Enter table with known value of b and d
   Compute resisting moment in concrete:
   $\phi M_n = K_u \times C_f$ (ft-kips)

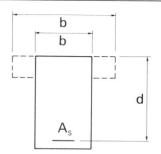

d	\multicolumn{14}{c}{b: Width of compressive area}
	4	8	12	16	20	24	28	32	36	40	48	60	72	96
10	0.033	0.067	0.100	0.133	0.167	0.200	0.233	0.267	0.300	0.333	0.400	0.500	0.600	0.800
11	0.040	0.081	0.121	0.161	0.202	0.242	0.282	0.323	0.363	0.403	0.484	0.605	0.726	0.968
12	0.048	0.096	0.144	0.192	0.240	0.288	0.336	0.384	0.432	0.480	0.576	0.720	0.864	1.152
13	0.056	0.113	0.169	0.225	0.282	0.338	0.394	0.451	0.507	0.563	0.676	0.845	1.014	1.352
14	0.065	0.131	0.196	0.261	0.327	0.392	0.457	0.523	0.588	0.653	0.784	0.980	1.176	1.568
15	0.075	0.150	0.225	0.300	0.375	0.450	0.525	0.600	0.675	0.750	0.900	1.125	1.350	1.800
16	0.085	0.171	0.256	0.341	0.427	0.512	0.597	0.683	0.768	0.853	1.024	1.280	1.536	2.048
17	0.096	0.193	0.289	0.385	0.482	0.578	0.674	0.771	0.867	0.963	1.156	1.445	1.734	2.312
18	0.108	0.216	0.324	0.432	0.540	0.648	0.756	0.864	0.972	1.080	1.296	1.620	1.944	2.592
19	0.120	0.241	0.361	0.481	0.602	0.722	0.842	0.963	1.083	1.203	1.444	1.805	2.166	2.888
20		0.267	0.400	0.533	0.667	0.800	0.933	1.067	1.200	1.333	1.600	2.000	2.400	3.200
22		0.323	0.484	0.645	0.807	0.968	1.129	1.291	1.452	1.613	1.936	2.420	2.904	3.872
24		0.384	0.576	0.768	0.960	1.152	1.344	1.536	1.728	1.920	2.304	2.880	3.456	4.608
26		0.451	0.676	0.901	1.127	1.352	1.577	1.803	2.028	2.253	2.704	3.380	4.056	5.408
28		0.523	0.784	1.045	1.307	1.568	1.829	2.091	2.352	2.613	3.136	3.920	4.704	6.272
30		0.600	0.900	1.200	1.500	1.800	2.100	2.400	2.700	3.000	3.600	4.500	5.400	7.200
32		0.683	1.024	1.365	1.707	2.048	2.389	2.731	3.072	3.413	4.096	5.120	6.144	8.192
34		0.771	1.156	1.541	1.927	2.312	2.697	3.083	3.468	3.853	4.624	5.780	6.936	9.248
36		0.864	1.296	1.728	2.160	2.592	3.024	3.456	3.888	4.320	5.184	6.480	7.776	10.368
38		0.963	1.444	1.925	2.407	2.888	3.369	3.851	4.332	4.813	5.776	7.220	8.664	11.552
40			1.600	2.133	2.667	3.200	3.733	4.267	4.800	5.333	6.400	8.000	9.600	12.800
42			1.764	2.352	2.940	3.528	4.116	4.704	5.292	5.880	7.056	8.820	10.584	14.112
44			1.936	2.581	3.227	3.872	4.517	5.163	5.808	6.453	7.744	9.680	11.616	15.488
46			2.116	2.821	3.527	4.232	4.937	5.643	6.348	7.053	8.464	10.580	12.696	16.928
48			2.304	3.072	3.840	4.608	5.376	6.144	6.912	7.680	9.216	11.520	13.824	18.432
50			2.500	3.333	4.167	5.000	5.833	6.667	7.500	8.333	10.000	12.500	15.000	20.000
52			2.704	3.605	4.507	5.408	6.309	7.211	8.112	9.013	10.816	13.520	16.224	21.632
54			2.916	3.888	4.860	5.832	6.804	7.776	8.748	9.720	11.664	14.580	17.496	23.328
56			3.136	4.181	5.227	6.272	7.317	8.363	9.408	10.453	12.544	15.680	18.816	25.088
58			3.364	4.485	5.607	6.728	7.849	8.971	10.092	11.213	13.456	16.820	20.184	26.912
60				4.800	6.000	7.200	8.400	9.600	10.800	12.000	14.400	18.000	21.600	28.800
62				5.125	6.407	7.688	8.969	10.251	11.532	12.813	15.376	19.220	23.064	30.752
64				5.461	6.827	8.192	9.557	10.923	12.288	13.653	16.384	20.480	24.576	32.768
66				5.808	7.260	8.712	10.164	11.616	13.068	14.520	17.424	21.780	26.136	34.848
68				6.165	7.707	9.248	10.789	12.331	13.872	15.413	18.496	23.120	27.744	36.992
70				6.533	8.167	9.800	11.433	13.067	14.700	16.333	19.600	24.500	29.400	39.200
72				6.912	8.640	10.368	12.096	13.824	15.552	17.280	20.736	25.920	31.104	41.472
74				7.301	9.127	10.952	12.777	14.603	16.428	18.253	21.904	27.380	32.856	43.808
76				7.701	9.627	11.552	13.477	15.403	17.328	19.253	23.104	28.880	34.656	46.208
78				8.112	10.140	12.168	14.196	16.224	18.252	20.280	24.336	30.420	36.504	48.672
80					10.667	12.800	14.933	17.067	19.200	21.333	25.600	32.000	38.400	51.200
82					11.207	13.448	15.689	17.931	20.172	22.413	26.896	33.620	40.344	53.792
84					11.760	14.112	16.464	18.816	21.168	23.520	28.224	35.280	42.336	56.448
86					12.327	14.792	17.257	19.723	22.188	24.653	29.584	36.980	44.376	59.168
88					12.907	15.488	18.069	20.651	23.232	25.813	30.976	38.720	46.464	61.952
90					13.500	16.200	18.900	21.600	24.300	27.000	32.400	40.500	48.600	64.800
92					14.107	16.928	19.749	22.571	25.392	28.213	33.856	42.320	50.784	67.712
94					14.727	17.672	20.617	23.563	26.508	29.453	35.344	44.180	53.016	70.688
96					15.360	18.432	21.504	24.576	27.648	30.720	36.864	46.080	55.296	73.728
98					16.007	19.208	22.409	25.611	28.812	32.013	38.416	48.020	57.624	76.832

FLEXURE

Fig. 3.9.5 Values of f_{ps} by stress-strain relationship — bonded strand

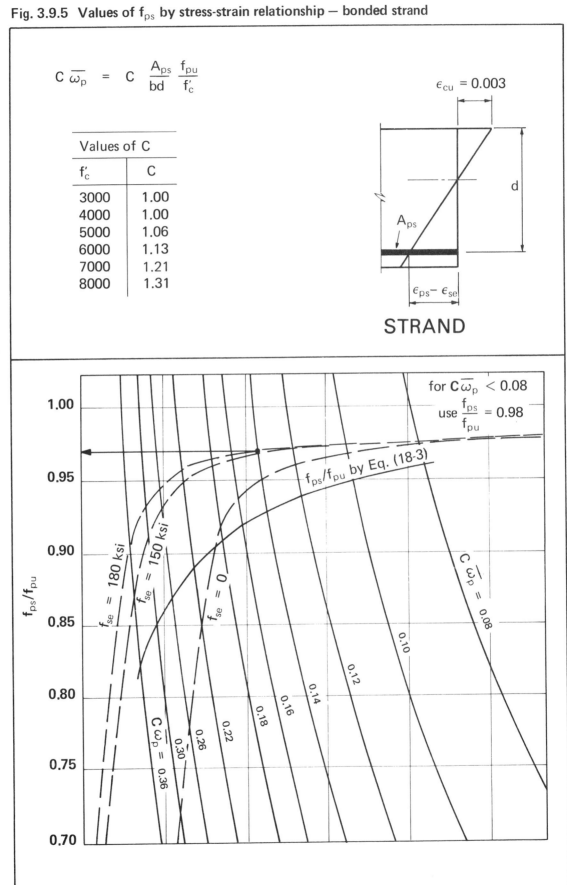

$$C \, \overline{\omega}_p = C \, \frac{A_{ps}}{bd} \, \frac{f_{pu}}{f'_c}$$

Values of C

f'_c	C
3000	1.00
4000	1.00
5000	1.06
6000	1.13
7000	1.21
8000	1.31

$\epsilon_{cu} = 0.003$

$\epsilon_{ps} - \epsilon_{se}$

STRAND

for $C\overline{\omega}_p < 0.08$ use $\dfrac{f_{ps}}{f_{pu}} = 0.98$

f_{ps}/f_{pu} by Eq. (18-3)

FLEXURE

Fig. 3.9.6 Values of f_{ps} by stress-strain relationship — bonded wire

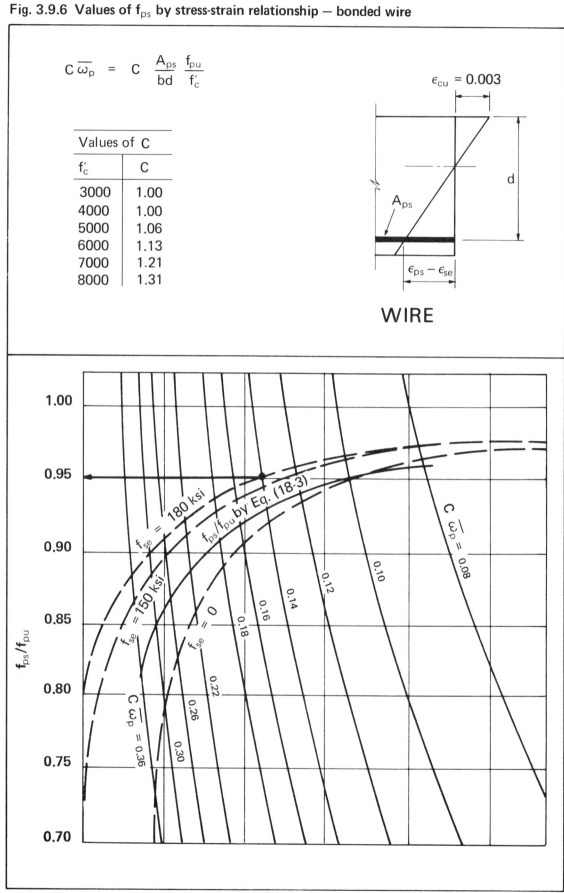

$$C\,\overline{\omega}_p = C\,\frac{A_{ps}}{bd}\,\frac{f_{pu}}{f'_c}$$

Values of C

f'_c	C
3000	1.00
4000	1.00
5000	1.06
6000	1.13
7000	1.21
8000	1.31

$\epsilon_{cu} = 0.003$

d

A_{ps}

$\epsilon_{ps} - \epsilon_{se}$

WIRE

f_{ps}/f_{pu}

$f_{se} = 180$ ksi

f_{ps}/f_{pu} by Eq. (18-3)

$f_{se} = 150$ ksi

$f_{se} = 0$

$C\,\overline{\omega}_p = 0.08$

0.10, 0.12, 0.14, 0.16, 0.18, 0.22, 0.26, 0.30

$C\,\overline{\omega}_p = 0.36$

PCI Design Handbook

FLEXURE

Fig. 3.9.7 Values of f_{ps} by stress-strain relationship — bonded bars

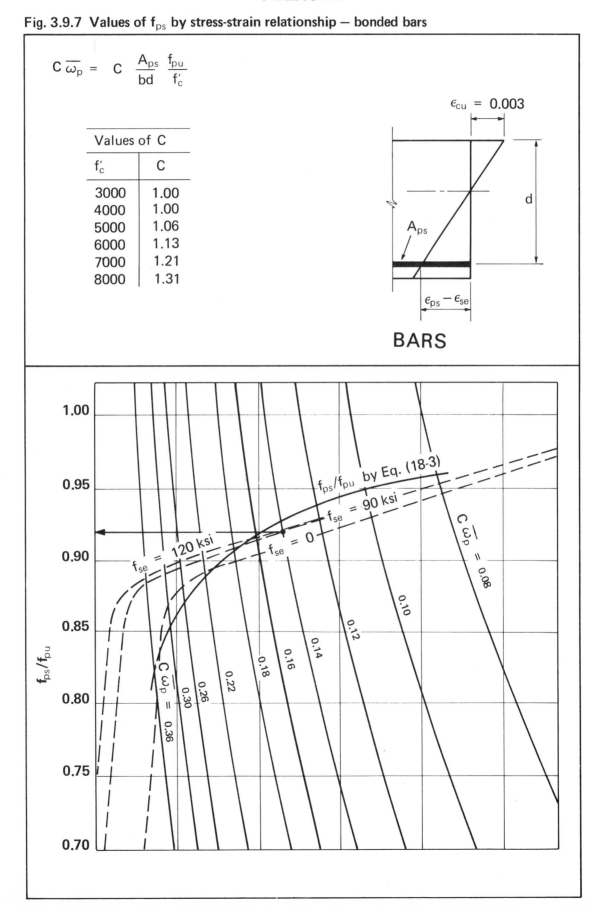

$$C\,\overline{\omega_p} = C\,\frac{A_{ps}}{bd}\,\frac{f_{pu}}{f_c'}$$

Values of C

f_c'	C
3000	1.00
4000	1.00
5000	1.06
6000	1.13
7000	1.21
8000	1.31

$\epsilon_{cu} = 0.003$

d

A_{ps}

$\epsilon_{ps} - \epsilon_{se}$

BARS

f_{ps}/f_{pu} by Eq. (18-3)

$f_{se} = 90$ ksi

$f_{se} = 120$ ksi

$f_{se} = 0$

$C\,\omega_p = 0.08$

0.10

0.12

0.14

0.16

0.18

0.22

0.26

0.30

$C\,\omega_p = 0.36$

f_{ps}/f_{pu}

FLEXURE

Fig. 3.9.8 Top tensile force at transfer of prestress

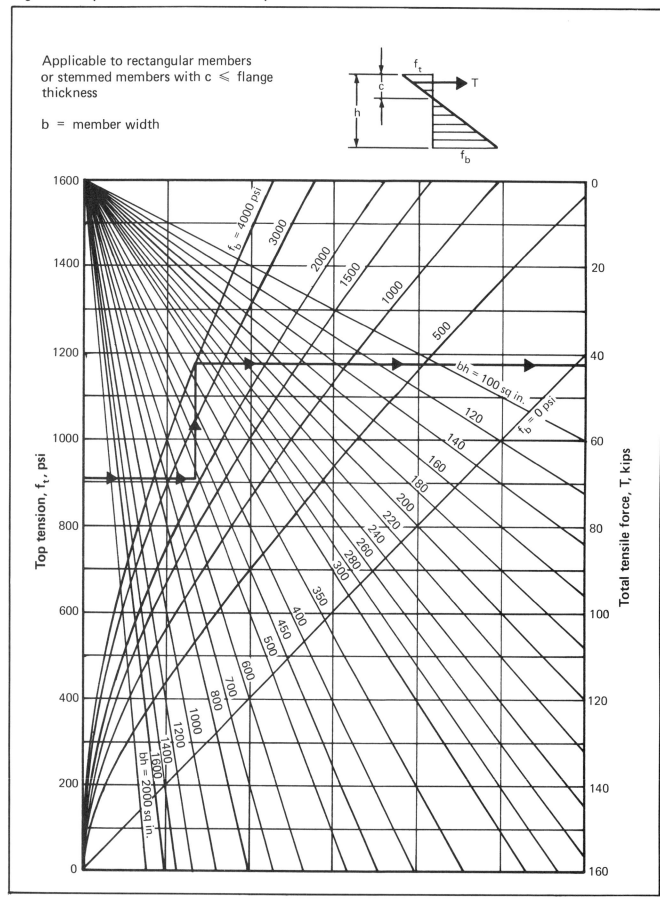

Applicable to rectangular members or stemmed members with c ≤ flange thickness

b = member width

PCI Design Handbook

FLEXURE

Fig. 3.9.9 Design stress for underdeveloped strand

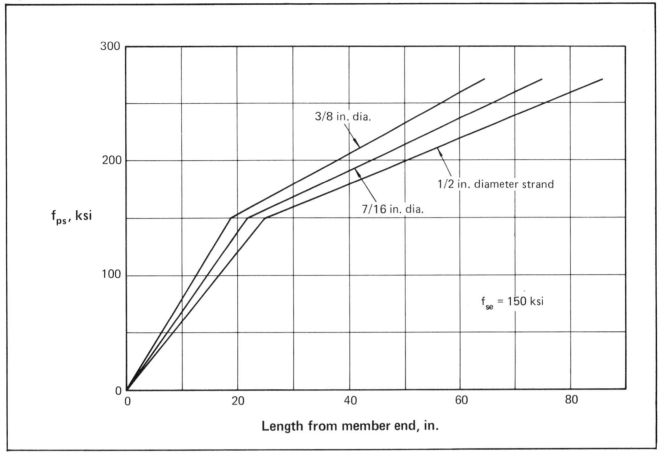

Curves based on Sect. 12.10, ACI 318-77

SHEAR

Fig. 3.9.10 Concrete shear strength by Eq. 11-10 (ACI 318-77)

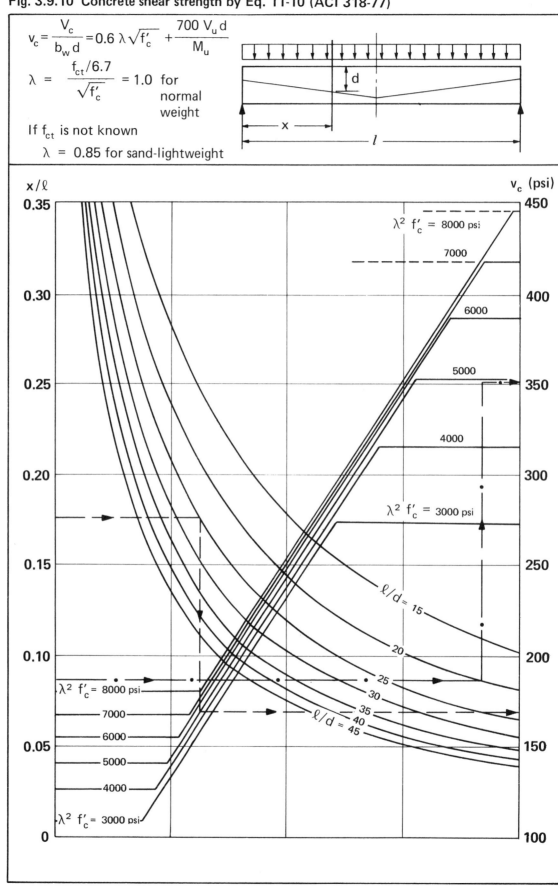

PCI Design Handbook

SHEAR

Fig. 3.9.11 Shear design by Eq. 11-10 (ACI 318-77)　　　　　　Straight strands

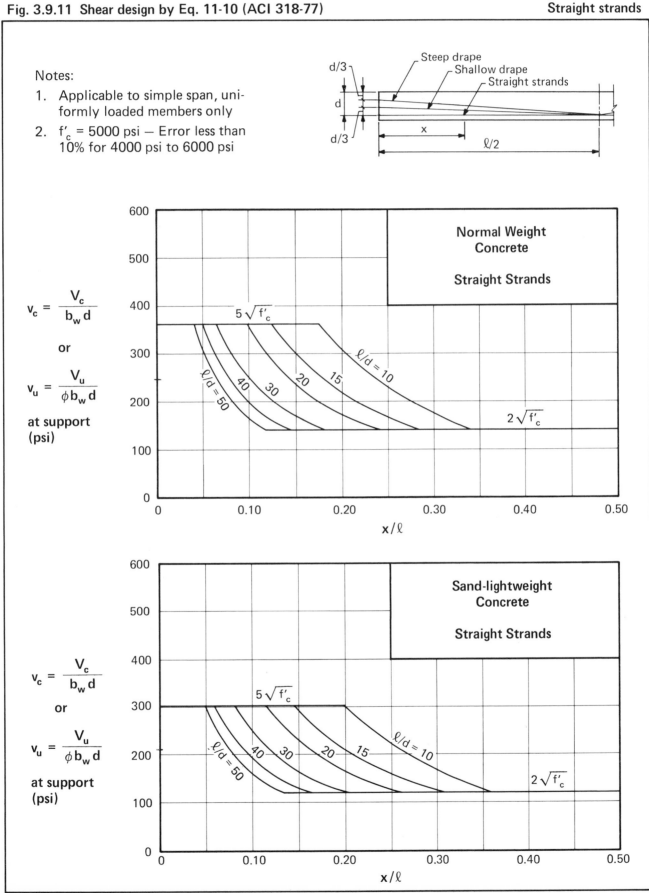

Notes:

1. Applicable to simple span, uniformly loaded members only

2. f'_c = 5000 psi — Error less than 10% for 4000 psi to 6000 psi

$$v_c = \frac{V_c}{b_w d}$$

or

$$v_u = \frac{V_u}{\phi b_w d}$$

at support (psi)

Normal Weight Concrete

Straight Strands

$$v_c = \frac{V_c}{b_w d}$$

or

$$v_u = \frac{V_u}{\phi b_w d}$$

at support (psi)

Sand-lightweight Concrete

Straight Strands

SHEAR

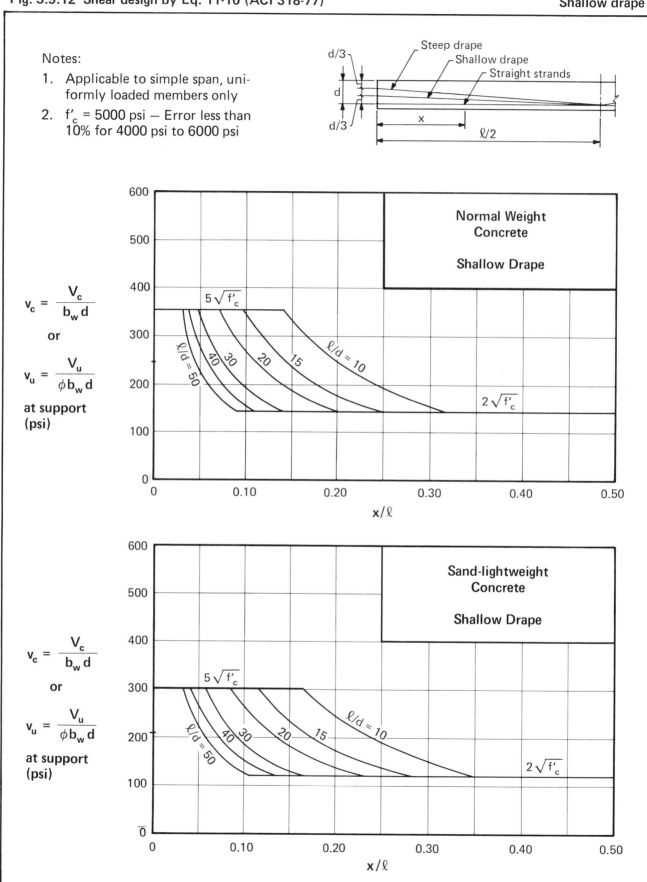

Notes:

1. Applicable to simple span, uniformly loaded members only
2. f'_c = 5000 psi — Error less than 10% for 4000 psi to 6000 psi

$$v_c = \frac{V_c}{b_w d}$$

or

$$v_u = \frac{V_u}{\phi b_w d}$$

at support (psi)

Normal Weight Concrete

Shallow Drape

Sand-lightweight Concrete

Shallow Drape

SHEAR

Fig. 3.9.13 Shear design by Eq. 11-10 (ACI 318-77)   Steep drape

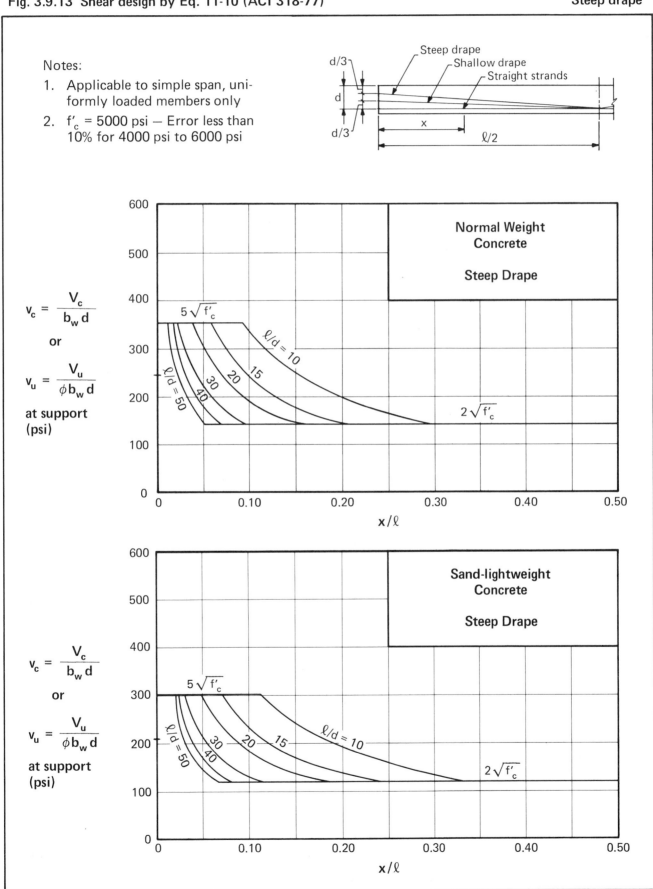

Notes:

1. Applicable to simple span, uniformly loaded members only
2. f'_c = 5000 psi — Error less than 10% for 4000 psi to 6000 psi

$$v_c = \frac{V_c}{b_w d}$$

or

$$v_u = \frac{V_u}{\phi b_w d}$$

at support (psi)

Normal Weight Concrete

Steep Drape

Sand-lightweight Concrete

Steep Drape

SHEAR

Fig. 3.9.14 Concrete shear strength at support by Eq. 11-13 (ACI 318-77)

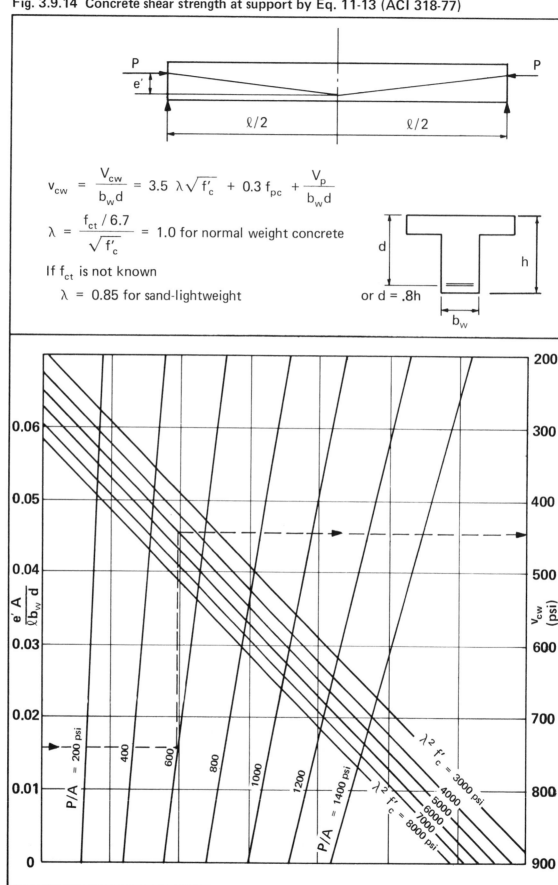

$$v_{cw} = \frac{V_{cw}}{b_w d} = 3.5 \; \lambda \sqrt{f'_c} \; + \; 0.3 \, f_{pc} \; + \; \frac{V_p}{b_w d}$$

$$\lambda = \frac{f_{ct} / 6.7}{\sqrt{f'_c}} = 1.0 \text{ for normal weight concrete}$$

If f_{ct} is not known

$\lambda = 0.85$ for sand-lightweight

or $d = .8h$

SHEAR

Fig. 3.9.15 v_{cw} corresponding to principal tensile stress $= 4\lambda\sqrt{f'_c}$

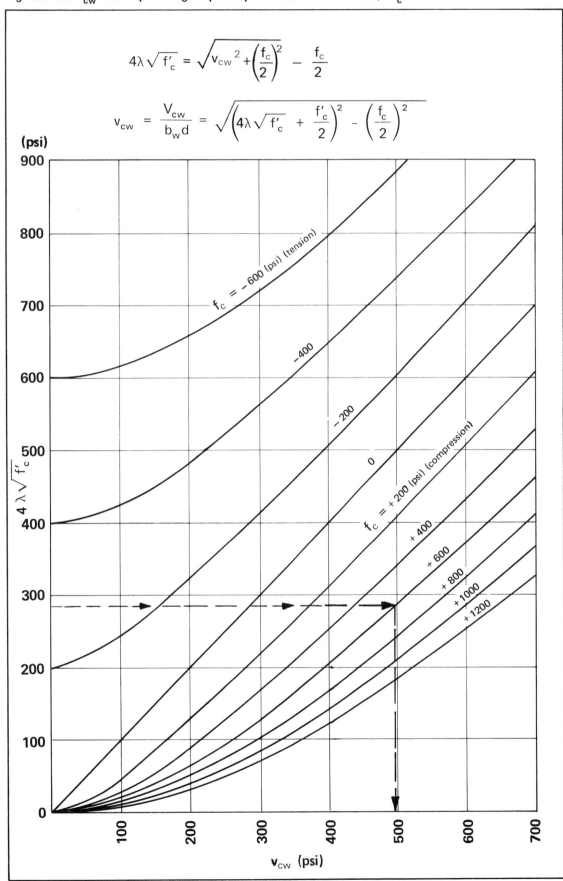

$$4\lambda\sqrt{f'_c} = \sqrt{v_{cw}^2 + \left(\frac{f_c}{2}\right)^2} - \frac{f_c}{2}$$

$$v_{cw} = \frac{V_{cw}}{b_w d} = \sqrt{\left(4\lambda\sqrt{f'_c} + \frac{f'_c}{2}\right)^2 - \left(\frac{f_c}{2}\right)^2}$$

SHEAR

Fig. 3.9.16 Minimum shear reinforcement by Eq. 11-15 (ACI 318-77)

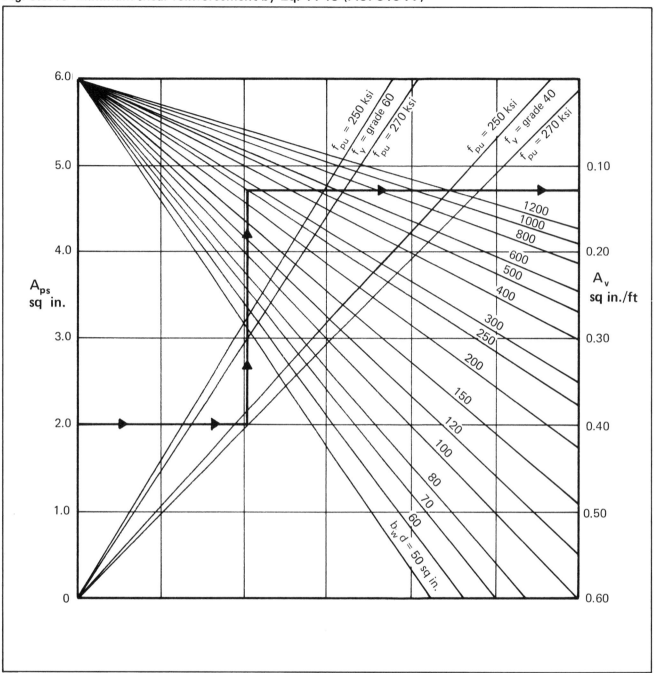

PCI Design Handbook

SHEAR

Table 3.9.17 Shear reinforcement

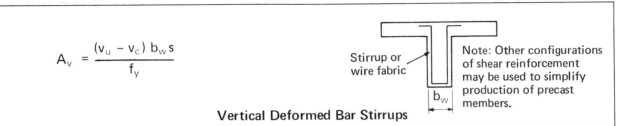

$$A_v = \frac{(v_u - v_c)\, b_w\, s}{f_y}$$

Stirrup or wire fabric

Note: Other configurations of shear reinforcement may be used to simplify production of precast members.

Vertical Deformed Bar Stirrups

Stirrup spacing (in.)	Maximum values of $(v_u - v_c)\, b_w$ (lb./in.)						Stirrup spacing (in.)
	f_y = 40,000 psi			f_y = 60,000 psi			
	No. 3 A_v = 0.22	No. 4 A_v = 0.40	No. 5 A_v = 0.62	No. 3 A_v = 0.22	No. 4 A_v = 0.40	No. 5 A_v = 0.62	
2.0	4400	8000	12400	6600	12000	18600	2.0
2.5	3520	6400	9920	5280	9600	14880	2.5
3.0	2933	5333	8267	4400	8000	12400	3.0
3.5	2514	4571	7086	3771	6857	10629	3.5
4.0	2200	4000	6200	3300	6000	9300	4.0
4.5	1956	3556	5511	2933	5333	8267	4.5
5.0	1760	3200	4960	2640	4800	7440	5.0
5.5	1600	2909	4509	2400	4364	6764	5.5
6.0	1467	2667	4133	2200	4000	6200	6.0
7.0	1257	2286	3543	1886	3429	5314	7.0
8.0	1100	2000	3100	1650	3000	4650	8.0
9.0	978	1778	2756	1467	2667	4133	9.0
10.0	880	1600	2480	1320	2400	3720	10.0
11.0	800	1455	2255	1200	2182	3382	11.0
12.0	733	1333	2067	1100	2000	3100	12.0
13.0	677	1231	1908	1015	1846	2862	13.0
14.0	629	1143	1771	943	1714	2657	14.0
15.0	587	1067	1653	880	1600	2480	15.0
16.0	550	1000	1550	825	1500	2325	16.0
17.0	518	941	1459	776	1412	2188	17.0
18.0	489	889	1378	733	1333	2067	18.0
20.0	440	800	1240	660	1200	1860	20.0
22.0	400	727	1127	600	1091	1691	22.0
24.0	367	667	1033	550	1000	1550	24.0

Welded Wire Fabric as Shear Reinforcement (f_y = 60,000 psi)

Spacing of vertical wire (in.)	Maximum values of $(v_u - v_c) b_w$ (lb./in.)								Spacing of vertical wire (in.)
	One row				Two rows				
	Vertical wire				Vertical wire				
	W7.5 A_v = 0.074	W5.5 A_v = 0.054	W4 A_v = 0.040	W2.9 A_v = 0.029	W7.5 A_v = 0.148	W5.5 A_v = 0.108	W4 A_v = 0.080	W2.9 A_v = 0.058	
2	2220	1620	1200	870	4440	3240	2400	1740	2
3	1480	1080	800	580	2960	2160	1600	1160	3
4	1110	810	600	435	2220	1620	1200	870	4
6	740	540	400	290	1480	1080	800	580	6

CAMBER AND DEFLECTION

Fig. 3.9.18 Camber equations for typical strand profiles

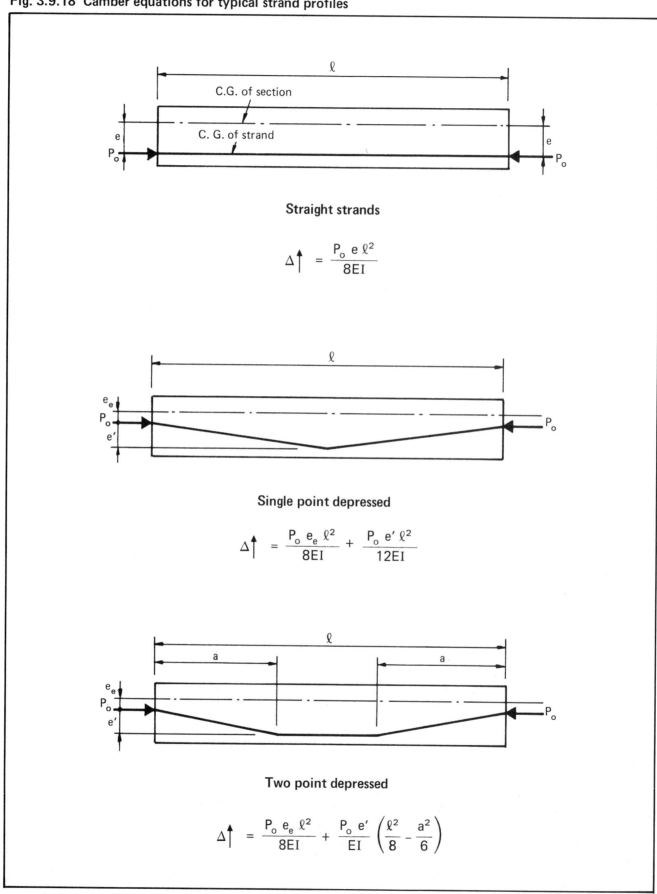

Straight strands

$$\Delta\uparrow = \frac{P_o\, e\, \ell^2}{8EI}$$

Single point depressed

$$\Delta\uparrow = \frac{P_o\, e_e\, \ell^2}{8EI} + \frac{P_o\, e'\, \ell^2}{12EI}$$

Two point depressed

$$\Delta\uparrow = \frac{P_o\, e_e\, \ell^2}{8EI} + \frac{P_o\, e'}{EI}\left(\frac{\ell^2}{8} - \frac{a^2}{6}\right)$$

PCI Design Handbook

CAMBER AND DEFLECTION

Table 3.9.19 Moment of inertia of transformed section

$$I_{cr} = nA_{ps}\,d^2\,(1 - \sqrt{\rho_p}\,)$$

$$= C \text{ (from table)} \times bd^3$$

$$\rho_p = \frac{A_{ps}}{bd}$$

Values of Coefficient, C

	ρ_p	f'_c, psi					
		3000	4000	5000	6000	7000	8000
Normal Weight Concrete	.0005	.0040	.0035	.0031	.0029	.0027	.0025
	.0010	.0080	.0069	.0062	.0057	.0053	.0049
	.0015	.0119	.0103	.0092	.0084	.0078	.0073
	.0020	.0158	.0137	.0123	.0112	.0104	.0097
	.0025	.0197	.0170	.0152	.0139	.0129	.0120
	.0030	.0235	.0203	.0182	.0166	.0154	.0144
	.0035	.0273	.0236	.0211	.0193	.0179	.0167
	.0040	.0310	.0269	.0240	.0219	.0203	.0190
	.0045	.0348	.0301	.0269	.0246	.0228	.0213
	.0050	.0385	.0333	.0298	.0272	.0252	.0236
	.0055	.0422	.0365	.0327	.0298	.0276	.0258
	.0060	.0458	.0397	.0355	.0324	.0300	.0281
	.0065	.0495	.0429	.0383	.0350	.0324	.0303
	.0070	.0531	.0460	.0411	.0376	.0348	.0325
	.0075	.0567	.0491	.0439	.0401	.0371	.0347
	.0080	.0603	.0522	.0467	.0427	.0395	.0369
	.0085	.0639	.0553	.0495	.0452	.0418	.0391
	.0090	.0675	.0584	.0523	.0477	.0442	.0413
	.0095	.0710	.0615	.0550	.0502	.0465	.0435
	.0100	.0745	.0645	.0577	.0527	.0488	.0456
Sand-Lightweight Concrete	.0005	.0060	.0052	.0047	.0043	.0039	.0037
	.0010	.0119	.0103	.0093	.0084	.0078	.0073
	.0015	.0178	.0154	.0138	.0126	.0116	.0109
	.0020	.0236	.0204	.0183	.0167	.0154	.0144
	.0025	.0293	.0254	.0227	.0207	.0192	.0179
	.0030	.0350	.0303	.0271	.0247	.0229	.0214
	.0035	.0406	.0352	.0315	.0287	.0266	.0249
	.0040	.0462	.0400	.0358	.0327	.0303	.0283
	.0045	.0518	.0449	.0401	.0366	.0339	.0317
	.0050	.0573	.0496	.0444	.0405	.0375	.0351
	.0055	.0628	.0544	.0487	.0444	.0411	.0385
	.0060	.0683	.0591	.0529	.0483	.0447	.0418
	.0065	.0737	.0638	.0571	.0521	.0483	.0451
	.0070	.0791	.0685	.0613	.0560	.0518	.0485
	.0075	.0845	.0732	.0655	.0598	.0553	.0518
	.0080	.0899	.0778	.0696	.0635	.0588	.0550
	.0085	.0952	.0824	.0737	.0673	.0623	.0583
	.0090	.1005	.0870	.0778	.0711	.0658	.0615
	.0095	.1058	.0916	.0819	.0748	.0692	.0648
	.0100	.1110	.0962	.0860	.0785	.0727	.0680

DEFLECTION AND CAMBER

Fig. 3.9.20 Effective moment of inertia by Eq. 9-7 (ACI 318-77)

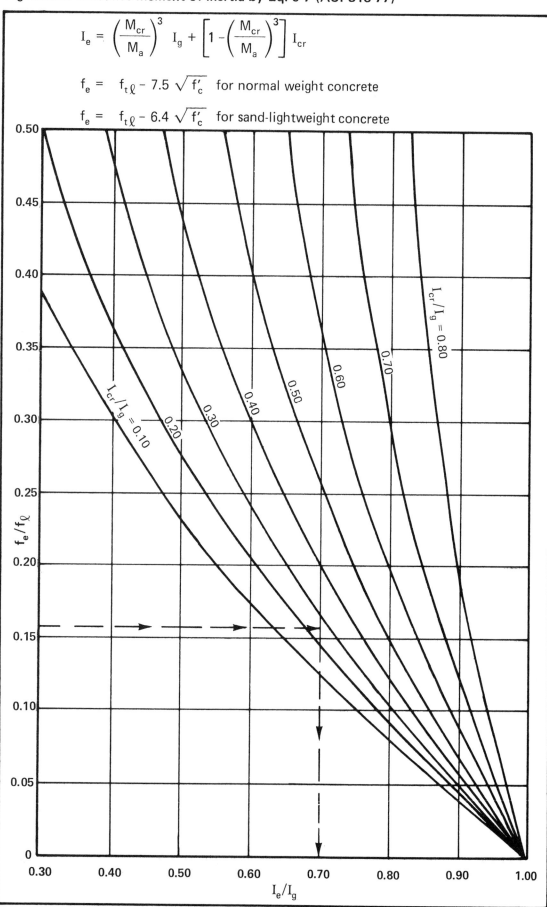

$$I_e = \left(\frac{M_{cr}}{M_a}\right)^3 I_g + \left[1 - \left(\frac{M_{cr}}{M_a}\right)^3\right] I_{cr}$$

$f_e = f_{t\ell} - 7.5 \sqrt{f'_c}$ for normal weight concrete

$f_e = f_{t\ell} - 6.4 \sqrt{f'_c}$ for sand-lightweight concrete

TORSION

Fig. 3.9.21 Values of v_{tu} below which torsion can be neglected

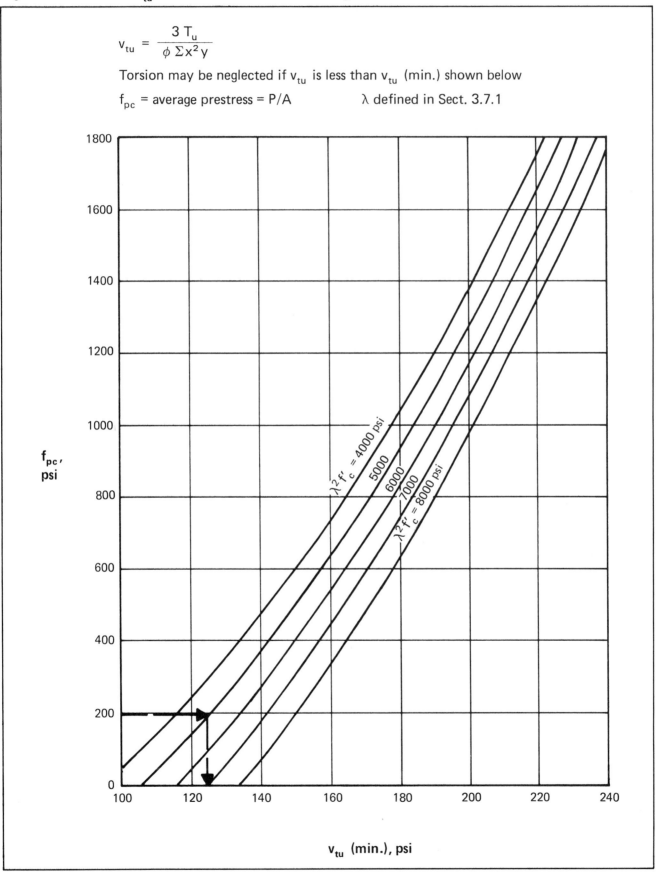

$$v_{tu} = \frac{3\,T_u}{\phi\,\Sigma x^2 y}$$

Torsion may be neglected if v_{tu} is less than v_{tu} (min.) shown below

f_{pc} = average prestress = P/A λ defined in Sect. 3.7.1

TORSION

Fig. 3.9.22 Maximum allowable torsion in combination with shear

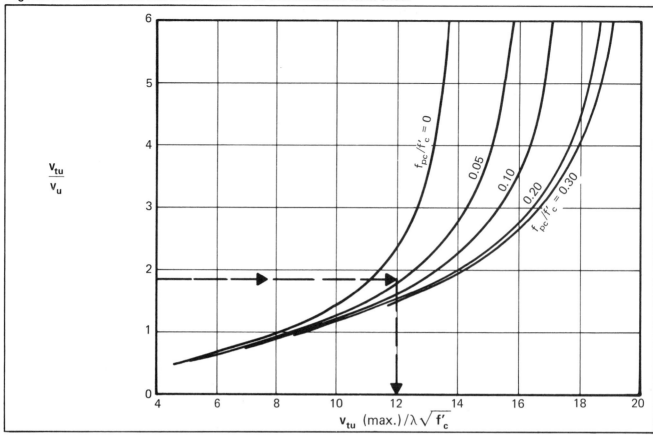

Fig. 3.9.23 Maximum allowable shear in combination with torsion

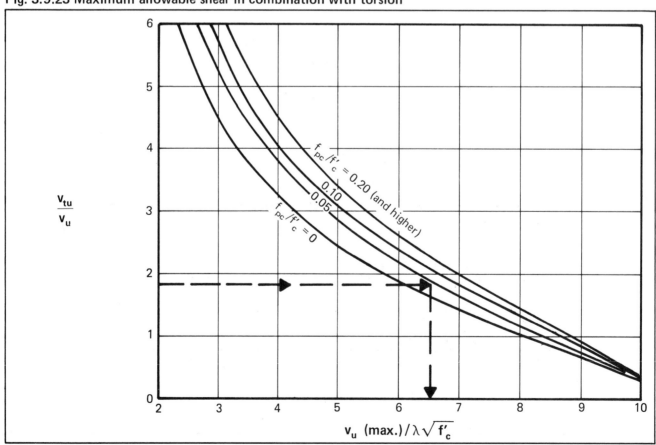

TORSION

Fig. 3.9.24 Coefficients, R_t, for determining v_{tc}

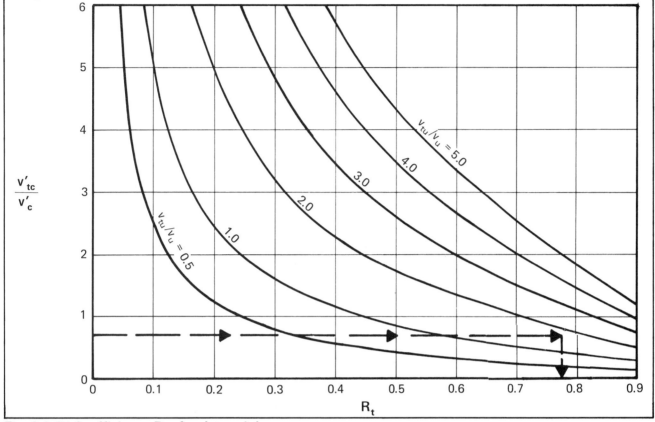

Fig. 3.9.25 Coefficients, R_v for determining v_c

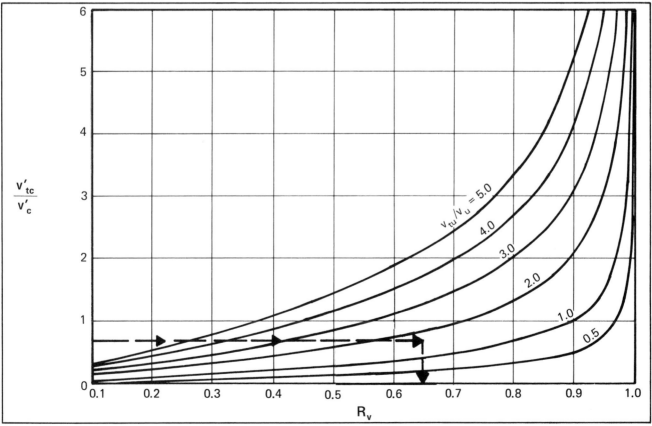

PART 4
ANALYSIS AND DESIGN OF PRECAST, PRESTRESSED CONCRETE STRUCTURES

ANALYSIS AND DESIGN OF
PRECAST, PRESTRESSED CONCRETE STRUCTURES

4.1 General

4.1.1 Notation

Note: Notation for earthquake analysis is in section 4.8.1.

A	=	area (with subscripts)
A_{vf}	=	area of shear-friction reinforcement
b	=	width of a section or structure
C	=	compressive force
C_m	=	a factor relating actual moment to equivalent uniform moment
C_u	=	factored compressive force
D	=	dead load
e	=	eccentricity of axial load
E	=	modulus of elasticity of a beam (with subscripts)
E_b	=	modulus of elasticity of beam concrete
E_c	=	modulus of elasticity of column concrete
f'_c	=	concrete compressive stress
f'_m	=	compressive strength of masonry
f_t	=	unfactored tensile stress
f_{ut}	=	factored tensile stress
f_y	=	yield strength of non-prestressed reinforcement
F_b	=	degree of base fixity (decimal)
F_i	=	lateral force at bay i or in shear wall i
g	=	assumed length over which elongation of the anchor bolt takes place
G	=	shear modulus of elasticity (modulus of rigidity)
h	=	column width in direction of bending
h_s	=	story height
H_u	=	total factored lateral force within a story
I	=	moment of inertia
I_b	=	moment of inertia of a beam
I_{bp}	=	moment of inertia of base plate (vertical cross-section dimensions)
I_c	=	moment of inertia of a column
I_f	=	moment of inertia of the footing (plan dimensions)
I_g	=	uncracked moment of inertia
k	=	effective length factor

k_b, k_f, k_m	=	coefficients used to determine forces and moments in beams and columns
k_s	=	coefficient of subgrade reaction
K	=	stiffnesses (with subscripts)
K_ℓ	=	constant used for the calculation of equivalent creep and shrinkage shortening
K_r	=	relative stiffness
K_t	=	constant used for the calculation of equivalent temperature shortening
ℓ	=	length of span or structure
ℓ_n	=	clear span
ℓ_s	=	distance from column to center of stiffness
ℓ_w	=	length of weld
M	=	unfactored moment
M_R	=	resisting moment
N	=	unfactored horizontal force
N_u	=	factored horizontal force
P	=	applied axial load
P_c	=	critical load
P_o	=	prestressing force after assumed initial loss
P_u	=	factored axial load
Q	=	stability index
Q	=	statical moment
r	=	radius of gyration
R_{du}	=	factored dead load reaction
t	=	thickness
T	=	tensile force
T_u	=	factored tensile force
v_r	=	unit shear on panel edge
v_u	=	factored unit shear
V_n	=	nominal shear strength
V_R	=	shear at right support
V_u	=	factored shear force
V_w	=	total wind shear
W	=	total lateral load
x_1	=	distance from face of column to center of anchor bolts
x_2	=	distance from face of column to base plate anchorage

γ = flexibility coefficient (with subscripts)

δ = moment magnifier

δ = volume change shortening (with subscripts)

Δ = total equivalent shortening or column deflection

Δ_u = deflection due to factored loads

μ_e = effective shear-friction coefficient

ϕ = strength reduction factor

ϕ = rotation (with subscripts)

ψ = ratio of column to beam stiffnesses

4.1.2 Introduction

Prestressed concrete is unlike other structural framing materials in that with prestressing, the designer has some control over not only the strength, but also the service load behavior of the structure or building frame.

The design of any building frame requires an understanding of the factors that influence the selection of the framing method. Proper understanding of these factors permits the engineer to design the frame components in a manner such that they will behave satisfactorily.

Selection of the framing method requires consideration of:

1. Design criteria.
2. Types of loads.
3. Response of structure to loads.
4. Selection of structural members.
5. Production of precast members.
6. Construction.
7. Service behavior.
8. Connections.

The items listed above are, of course, interrelated. Information on these items needed for an engineering analysis is included in this and other parts of the Handbook.

4.1.3 Design Criteria

Building Function

The function of the building plays a key role in the initial development of the framing concept. Framing methods can be influenced by:

1. Load requirements in addition to those specified by codes.
2. Number, location, and size of interior and exterior wall openings, as well as openings for mechanical trades in floors and roofs.
3. Location of structural members as dictated by the building space function.

4. Location and height of walls and partitions.
5. Thermal and acoustical considerations.

Building Codes

Building codes may impose certain requirements that affect selection of the framing method by:

1. Live load requirements depending on the occupancy.
2. Wind, seismic, and other lateral load requirements depending on geographic location.
3. Dimensional requirements such as the minimum thickness of walls, maximum span to depth ratio of beams and slabs, maximum distance between fire walls, and minimum cover thickness for reinforcement.
4. Fire resistance requirements.

Esthetic Requirements

The architectural requirements of a building often dictate the selection of the framing method. Esthetic considerations can influence:

1. Type of member if exposed.
2. Column sizes.
3. Beam sizes.
4. Slab thicknesses.
5. Location of structural framing members.
6. Connection type if connections are exposed.

4.1.4 Types of Loads

The types of loads imposed on a precast, prestressed concrete building can limit the choice of the framing method. A review of all the applied loadings, both external and internal, should be made during the conceptual development of the framing method.

Gravity Loads

It is important that the dead and live loads be categorized so that the designer understands the following:

1. Total superimposed dead and live loads for each part of the building.
2. The portion of the superimposed load that is dead load.
3. The portion of the live load that is sustained.
4. Proper load factors for each load.

Wind Loads

The design requirements for wind vary according to the geographic area. Factors that affect the degree to which wind influences the framing method are:

1. Building height.
2. Building shape.

3. Magnitude of wind pressure and wind suction.

4. Lateral wind deflection (drift).

Seismic Loads

Seismic considerations can restrict the selection of a framing method. The following factors should be considered:

1. Seismic zone of the geographic area.

2. Building height.

3. The fundamental period of vibration of the building.

4. Whether the lateral resisting elements are similar or are a combination of several types (i.e., combination space frame and shear walls).

5. Type of special loads the building must support such as elevated tanks, mechanical systems, etc.

6. What parts of the building resist the total seismic load, and what parts do not.

7. Seismic deformations.

Forces Due to Restrained Volume Changes of Concrete

Restraint of concrete volume change deformations may have a marked influence on the behavior of the framing system. Volume changes of concrete are those resulting from shrinkage, creep, and temperature changes.

Loads are induced in frames by volume change deformations, but only when such deformations are restrained. The effects of restrained volume change deformations are influenced mainly by the size of the building and stiffness of the framing members. Thus the designer can control the combinations of possible loadings from restrained volume changes by choosing the locations at which restraint can develop.

The following considerations concerning volume change deformations may influence the choice of the framing system:

1. Volume changes resulting from creep, shrinkage, and temperature change are time dependent. The volume changes that affect the building frame are those that occur after construction.

2. The potential unrestrained volume change deformation is a function of the concrete type, amount and type of reinforcement, ratio of stress to strength, and the ratio of volume to surface area of the member.

3. Two types of volume change deformations occur — axial shortening and rotational movements.

4. Restraint of volume change deformations creates loads that act on the building frame, in addition to the specified loads.

5. Axial forces created by restraint of volume changes are not uniform throughout the structural frame. They are generally greatest near the center of multi-bay buildings and diminish toward the exterior. The magnitude of such forces is affected by the location of rigid frames, wind bracing and shear walls, story height of columns, length of beams, column rigidity, and number of bays and stories.

Miscellaneous Lateral Forces

Selection of the framing method may be influenced by the location and magnitude of lateral loads peculiar to the building function. Examples of these loadings include overhead cranes, spectator seating in gymnasiums, and vibrating machinery.

4.1.5 Response of Structure to Loads

The combined effect of gravity loads, lateral loads, and restrained volume change forces are important to the selection of the framing system. Consideration of the loading combinations may indicate the use of a particular framing system.

The resistance or response of the building frame to loads affects the framing as follows:

1. Gravity loads can influence column spacing and might indicate that the flexural members should be continuous.

2. Lateral loads (other than those induced by volume changes) can be resisted by any of the framing methods listed below, or a combination thereof:

   a. Multi-story buildings with a limited number of bays in width — rigid or semi-rigid frames, diagonal bracing, or bracing by shear walls and cores.

   b. Low rise buildings without many large openings in exterior walls — exterior shear walls, interior shear walls, or cores.

   c. Low rise buildings with many large openings in exterior walls — interior shear walls, cores, rigid or semi-rigid frames, or columns cantilevered from the foundations.

   d. Low rise buildings with expansion joints parallel to the direction of lateral loads — shear walls, cores, rigid or semi-rigid frames, or columns cantilevered from the foundations.

Generally, simple span construction is more economical for pretensioned precast concrete members because of connection, production, and construction simplicity. Unless analyses indicate otherwise, the framing system should utilize only the number of rigid frames required to resist lateral loads, and the rest of the building should consist of simple spans.

3. Volume change lateral loads result from restraint of deformations, and as such can be controlled by design decisions.

   a. The magnitude of restrained volume change forces depends upon the location within the structure where restraint occurs.

   b. Stiffness or resistance of building frame components influence the magnitude of volume change loadings.

   c. Forces developed by restrained volume changes are minimized when the restraint occurs at the center of the building's lateral stiffness. This usually coincides with the center of the building.

   d. To minimize restrained volume change loadings, use only those shear walls, cores, or rigid frames needed to resist external lateral loads, i.e., loads other than those resulting from restrained volume changes.

4. The manner by which lateral loads are resisted by structural building components is affected by several factors:

   a. The moments at the base of a column resulting from lateral loads are reduced by rotational restraint at the top of the column and by the number of columns participating in resisting lateral forces.

   b. Column moments due to external lateral forces vary approximately as the square of the column height.

   c. The sum of the top and bottom column moments due to external lateral loads is a constant, irrespective of the end restraint.

   d. The sum of the top and bottom column moments due to restrained volume changes is markedly affected by the restraint at the ends of the columns.

   e. The moment at the base of a column created by restrained volume changes varies proportionally with both the rotational restraint of the column top and the length of the restrained beam framing into the column.

   f. Column moments due to expansion or contraction of a beam framed into the column top vary inversely as the square of the column height.

4.1.6 Selection of Structural Members

Precast, prestressed concrete framing members are readily available to satisfy the requirements of most framing systems. If the selection of the framing members is not governed by the factors discussed in Sections 4.1.1., 4.1.2., or 4.1.3. above, the following general criteria can be used:

1. Check availability of precast concrete structural and architectural members in the geographic area of construction. For maximum economy, select a bay size that is a multiple of the width of the member to be used.

2. Decide whether architectural wall panels should be load-bearing or curtain walls.

3. Typical span to depth ratios of flexural precast prestressed concrete members are:

Hollow-core floor slabs	30 to 40
Hollow-core roof slabs	40 to 50
Stemmed floor slabs	25 to 35
Stemmed roof slabs	35 to 40
Beams	10 to 20

   The required depth of a beam or slab is influenced by the ratio of live load to total load. Where this ratio is high, deeper sections may be needed.

4. Precast concrete columns are commonly available in sizes from 12 in. x 12 in. to 24 in. x 24 in., in 2-in. increments.

5. Framing members should be wide enough to provide for continuity connections and seating connections.

6. Check availability of materials to fabricate structural members such as lightweight aggregate, type of pretensioning strand, reinforcement grades and bar sizes, use of welded headed studs or deformed welded studs, types of inserts, plates, and handling equipment.

4.1.7 Production of Precast Members

Maximum economy can be achieved in the selection of precast concrete structural members by utilizing sections and details that permit efficient production:

1. Use available standard sections.

2. For slabs, use repetitive types of members, reinforcement details, strand patterns and profiles, embedded plate details, plate anchorages, inserts, blockouts, and concrete strengths.

3. Minimize the number of special slabs with openings or blockouts. In slabs, straight parallel strands are generally more economical, but if depressed strands are needed, use single point depression when possible since multiple point depression is more costly.

4. For beams with pretensioned reinforcement, straight parallel strands are most economical.

Consider the production process regarding shear reinforcement, strand profile, blockouts, and concrete strengths.

5. Columns are often cast in continuous steel forms, so base plates, cap plates, and blockout pockets should, wherever possible, be contained within the cross-sectional dimensions of the column.

6. Provide for tolerances necessary for economical production as given in the PCI *Manual for Quality Control for Plants and Production of Precast Prestressed Concrete Products* (MNL-116-77) or the local standards of practice.

4.1.8 Construction Techniques

Significant economies can be achieved when the designer takes into account the construction requirements. The following points should be considered:

1. Availability of equipment to haul and erect precast concrete structural members.

2. Size and weight limitations of precast members by regulations governing transportation.

3. Access, maneuverability, and positioning of erection equipment at the construction site.

4. Temporary bracing and shoring requirements. Economies might be achieved if bracing and shoring can be eliminated or minimized by designing all members to resist construction loads.

5. Tolerances and clearances necessary for economical construction.

4.1.9 Service Behavior

The desired performance of the building can be assured by appropriate consideration of the building component's deformation under service conditions.

1. Camber and deflection of flexural members are affected by concrete volume changes and consequently vary after time of construction. Maximum camber may occur at time of construction. Camber and deflection may affect:

    a. Level of floors and drainage of roofs.

    b. Details of shear wall connections if movement must be accommodated.

    c. Rotations at the ends of members at connections.

    d. Other building components such as windows, partitions, door openings, and mechanical items.

2. Axial deformations of members can be either isolated to individual members or accumulated throughout the length or height of the structure depending on the fixity of the details. Axial deformations, whether isolated or accumulated, can affect flashing details, roofing details, glass or window details, caulking, connection details, flexural deformation, floor toppings, continuity, and partitions.

4.1.10 Connections

In precast concrete building frames, connections between members are often the most significant factors that influence service behavior. Ideally, connections should control the service behavior without sacrificing economy.

The following items should be considered in the design of connections to assure satisfactory service behavior and maximum economy:

1. Connections are locations of stress concentration. All possible loadings and movements at the connection should be accounted for in the design.

2. Tolerances of fabricating connection parts and of members framing into a connection cause variations in the location of forces acting at the connection. Connections should be designed to withstand forces acting anywhere within the tolerance area.

3. Connection economies are achieved through repetitious use of details. For similar types of connections, the details should be identical wherever possible.

4. Connection details and types vary depending on local practice. It is advisable to use details that are standard within a geographic area.

4.2 Tolerances and Clearances

Because of precision limitations in the manufacture and erection of precast products, it is necessary to plan for variation from the planned dimensions of the products. Often, the dimensional precision of materials with which the precast product must interface is subject to greater variation in planned dimensions than is the precast piece. This is especially true when the interfacing material is cast-in-place concrete. Proper recognition of these facts and, hence, the realistic specification, design, detailing, and enforcement with regard to tolerances and clearances will have a significant effect on the performance, construction speed and economy of the structure.

4.2.1 Definitions

The following definitions apply to this section:

Tolerance: The allowable variation from a planned dimension or position.

Clearance: Specified interface space to allow for tolerance accumulations and structural movements caused by volume changes or elastic deformations.

4.2.2 Effect of Variations from Planned Dimensions

It is evident that variations from planned dimensions can affect the structural behavior (e.g., eccentricities greater than planned) and the appearance of the structure. It can also greatly affect the erection procedures and the time required for construction. The prudent designer will anticipate that dimensions may be somewhat "out of tolerance." He will also plan ample clearances with connections that permit easy and rapid field adjustment through the use of shims, grout, slotted bolt holes, field welding, etc.

Each project should be reviewed and tolerances and clearances specified that are consistent with the design requirements. The tolerances and clearances suggested in the following section and other publications should be considered as guidelines for an acceptability range and not limits for rejection.

If these suggested values are exceeded the product will usually be acceptable if:

1. Exceeding the suggested value does not affect the structural integrity or architectural performance of the structure.
2. The product can be brought within the suggested limits by structurally and architecturally satisfactory means.
3. The total erected assembly can be modified to meet all structural and architectural requirements.

4.2.3 Sources of Dimensional Variation

Following is a list of possible sources of dimensional variation which provides information for the design and selection of clearance dimensions.

1. *Product Manufacture.* Precast, prestressed concrete manufacturing tolerances are given in the *Manual for Quality Control for Plants and Production of Precast, Prestressed Concrete Products* MNL-116-77.* Combining the various component tolerances (end squareness, horizontal alignment, etc.) the following dimensional variations may be anticipated for design and the establishment of clearances:

Length ± 1 in.

---

*Available from Prestressed Concrete Institute.

Width ± 1/2 in.

Depth ± 1/4 in.

2. *Connections.* There are no standard tolerances for connections. The following values are suggested for purposes of interface design and clearance selection:

Location

Vertical	±1/2 in.
Skewed bearing	3/16 in.
Projection of corbels and other connecting elements	±1/4 in.

Internal reinforcement location

Minimum cover	−1/4 in.
Maximum cover	+3/4 in.
Bar spacing	±3/4 in.

When designing connections, it is also important that possible dimensional variations from all other sources be evaluated for the determination of connection load centroid for design and the effect of non-uniform bearings.

3. *Elastic Deformation.* The deformation under load of structural members, particularly during erection, can be the cause of dimensional variations. Thus, the analysis of structural deflections at each stage of the erection sequence is important. A typical example is a multi-story exterior column supporting spandrel beams loaded by long-span tees. Eccentric loads delivered to the columns by the beam reactions can cause column top deflections significant enough to cause erection problems at later stages of construction.

Another example of elastic deformations causing dimensional variations is the slight rolling of a spandrel beam under a one-side load application, causing non-uniform beam end bearing conditions. A similar condition of non-parallel bearing surfaces can result at the bearing interface between deck members and the spandrel beam. The slight beam roll combined with deck member end rotations due to camber may cause non-uniform bearing.

Specific values cannot be assigned to elastic deformation effects as they are a function of dimensions and loading sequence. Yet, by selecting ample clearances after considering realistic dimensional variations, the influence of elastic deformations can be minimized. Multi-story and long-span structures are especially sensitive to problems of structural deformations, therefore, it is suggested that their deformations be carefully reviewed to determine the influence on design of the members and connections.

4. *Volume changes.* Creep, shrinkage, and temperature change can also cause dimensional variations which should be considered in the design. This is discussed in detail in Section 4.4.

5. *Interfacing with other products and materials.* Also to be considered in light of the previous discussion is the problem of interfacing with other construction materials, such as cast-in-place concrete, steel, windows, doors, HVAC and plumbing openings, etc. Each of the various construction materials has a specific tolerance associated with its installation into the complete structure. Appropriate consideration must be given to each of these interface tolerances to allow for the cost effective installation of the different materials.

The best procedure available for accommodating interfacing is for the designer to rely on his experience, supplemented by discussions with suppliers and contractors and select ample minimum clearances between precast components and other construction materials. Experience with actual structures suggests the following guideline clearance dimensions:

Beam to column	1 in. to 1-1/2 in.
Slab to beam	1 in.
Column to footing	2 in. to 3 in.
Large exterior wall opening	2 in. clear all sides
Variation from planned elevation	±1 in.

4.3 Expansion Joints

4.3.1 General

Joints are placed in structures to limit the magnitude of forces which result from volume change deformations (temperature changes, shrinkage and creep), and to permit movements (volume change deformations) of structural elements. If the forces generated by temperature rise are significantly greater than shrinkage and creep forces, a true "expansion joint" is needed. However, in concrete structures, true expansion joints are seldom required. Instead, joints that permit contraction of the structure are needed to relieve the strains caused by temperature drop and restrained creep and shrinkage, which are additive. Such joints are properly called contraction or control joints but are commonly referred to as expansion joints.

It is desirable to have as few expansion joints as possible. Expansion joints are often located by "rules of thumb" without considering the structural framing method. The purpose of this section is to present guidelines for determining the spacing and width of expansion joints.

4.3.2 Spacing of Expansion Joints

There is a wide divergence of opinion concerning the spacing of expansion joints. Typical practice in concrete structures, prestressed or non-prestressed, is to locate expansion joints at distances between 150 and 200 ft. However, reinforced concrete buildings exceeding these limits have performed well without expansion joints. Recommended joint spacings for precast concrete buildings are generally based on experience and may not consider several important items. Among these items are the types of connections used, the column stiffnesses in simple span structures, the relative stiffness between beams and columns in framed structures, and the weather exposure conditions. Non-heated structures such as parking garages, are subjected to greater temperature changes than occupied structures, so shorter distances between expansion joints are warranted.

Sections 4.4 and 4.6 present methods for analyzing the potential movement of framed structures, and the effect of restraint of movement on the connections and structural frame. This information along with the connection design methods in Part 5 can aid in determining spacing of expansion joints.

Fig. 4.3.1 shows joint spacing as recommended by the Federal Construction Council, and is adapted from *Expansion Joints in Buildings*, Technical Report No. 65, prepared by the Standing Committee on Structural Engineering of the Federal Construction Council, Building Research Advisory Board, Division of Engineering, National Research Council, National Academy of Sciences, 1974. Note that the spacings obtained from the graph in Fig. 4.3.1 should be modified for various conditions as shown in the notes beneath the graph. Values for the design temperature change can be obtained from Section 4.4.

When expansion joints are required in non-rectangular structures, they should always be located at places where the plan or elevation dimensions change radically.

4.3.3 Width of Expansion Joints

The width of the joint can be calculated theoretically using a coefficient of expansion of 6 x

Fig. 4.3.1 Maximum building length without use of expansion joints

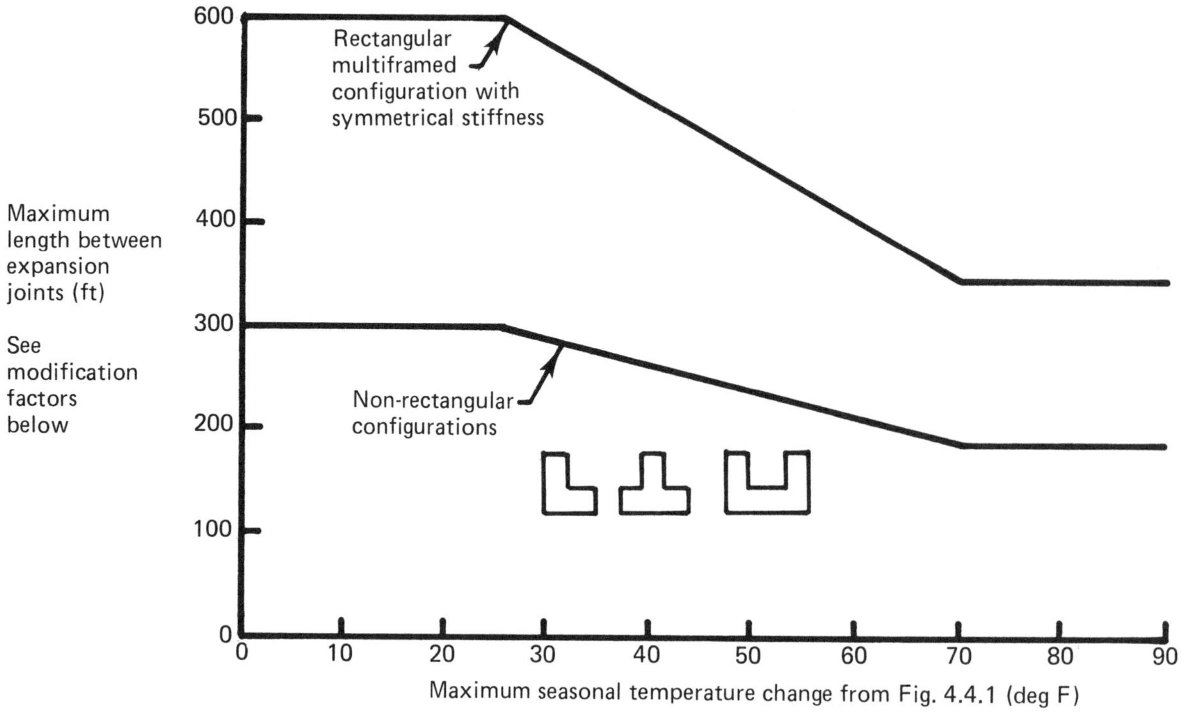

These curves are directly applicable to buildings of beam-and-column construction, hinged at the base, and with heated interiors. When other conditions prevail, the following rules are applicable:

(a) If the building will be heated only and will have hinged-column bases, use the allowable length as specified;

(b) If the building will be air conditioned as well as heated, increase the allowable length by 15 percent (provided the environmental control system will run continuously);

(c) If the building will be unheated, decrease the allowable length by 33 percent;

(d) If the building will have fixed-column bases, decrease the allowable length by 15 percent;

(e) If the building will have substantially greater stiffness against lateral displacement at one end of the plan dimension, decrease the allowable length by 25 percent.

When more than one of these design conditions prevail in a building, the percentile factor to be applied should be the algebraic sum of the adjustment factors of all the various applicable conditions.

<u>Source</u>: "Expansion Joints in Buildings," Technical Report No. 65, National Research Council, National Academy of Sciences, 1974.

10^{-6} in./in./deg F for normal weight and 5×10^{-6} in./in./deg F for lightweight concrete. The Federal Construction Council report referenced above recommends a minimum width of 1 in. However, since the primary problem in concrete buildings is contraction rather than expansion, joints that are too wide may result in problems with reduced bearing or loss of filler material.

4.4 Volume Changes

The strains resulting from creep, shrinkage and temperature change, and the forces caused by restraining these strains have important effects on connections, service load behavior and ultimate capacity of precast prestressed structures. Consequently, these strains and forces must be considered in the design.

4.4.1 Volume Change Data

Tables 4.4.1 through 4.4.5 and Figs. 4.4.1 and 4.4.2 provide the data needed to determine volume change strains.*

Example 4.4.1 — Calculation of volume change shortening

Given:

Heated structure in Denver, Colorado
Normal weight concrete beam — 12RB28
12-1/2-in. diameter, 270K strands
Assume initial prestress loss = 10 percent
Release strength = 4500 psi (accelerated cure)
Length = 24 ft

Problem:

Determine the actual shortening that can be anticipated from:

a. Casting to erection at 50 days.

b. Erection to the end of service life.

Solution:

From Figs. 4.4.1 and 4.4.2:

Design temperature = 70 F
Average ambient relative humidity = 55 percent
Prestress force:

$$A_{ps} = 12 \times 0.153 = 1.836 \text{ sq in.}$$
$$P_o = 1.836 \times 270 \times 0.70 \times 0.90$$
$$= 312.3 \text{ kips}$$
$$P_o /A = 312.3 (1000) / (12 \times 28)$$
$$= 929 \text{ psi}$$

Volume/surface ratio = $[12 \times 28] / [(2 \times 12) + (2 \times 28)] = 4.2$ in.

a. At 50 days

From Table 4.4.1:

Creep strain = 268×10^{-6} in./in.

Shrinkage strain = 267×10^{-6} in./in.

From Table 4.4.2:

Creep correction factor = 0.71 + (129/200) (0.88 – 0.71) = 0.82

From Table 4.4.3:

Creep correction = 1.17 – 0.5 (1.17 – 1.08)
= 1.13

---

*Tables 4.4.1 through 4.4.3 are based on standard creep and shrinkage equations, discussed in detail in *Deformation of Concrete Structures* by Dan E. Branson (McGraw-Hill International Book Co., New York, 1977.) For development of the tables, figures, and equations in this section, see "Volume Changes in Precast Prestressed Concrete Structures," *PCI Journal,* September-October, 1977.

Shrinkage correction = 1.29 – 0.5 (1.29 – 1.14)
= 1.22

From Table 4.4.4:

Creep correction = 0.46 – 0.2 (0.46 – 0.34)
= 0.44

Shrinkage correction = 0.44 – 0.2 (0.44 – 0.29)
= 0.41

(Note: Temperature shortening is not significant for this calculation.)

Total strain:

Creep = $268 \times 10^{-6} \times 0.82 \times 1.13 \times 0.44$
= 109×10^{-6} in./in.

Shrinkage = $267 \times 10^{-6} \times 1.22 \times 0.41$
= 134×10^{-6} in./in.

Total: 243×10^{-6} in./in.

Total shortening:
= $243 \times 10^{-6} \times 24 \times 12$
= 0.07 in.

b. At final

Factors from Tables 4.4.2 and 4.4.3 same as for 50 days

From Table 4.4.1:

Creep strain = 524×10^{-6} in./in.

Shrinkage strain = 560×10^{-6} in./in.

From Table 4.4.4:

Creep correction = 0.77 – 0.2 (0.77 – 0.74)
= 0.76

Shrinkage correction = 0.75 – 0.2 (0.75 – 0.64)
= 0.73

From Table 4.4.5:

Temperature strain = 210×10^{-6} in./in.

Total creep and shrinkage strain:
Creep = $524 \times 10^{-6} \times 0.82 \times 1.13 \times 0.76$
= 369×10^{-6} in./in.

Shrinkage = $560 \times 10^{-6} \times 1.22 \times 0.73$
= 499×10^{-6} in./in.

Total: 868×10^{-6} in./in.

Difference from 50 days to final
= 868 – 243
= 625×10^{-6} in./in.

Total strain
= 625 + 210
= 835×10^{-6} in./in.

Total shortening
= $835 \times 10^{-6} \times 24 \times 12$
= 0.24 in.

Fig. 4.4.1 Maximum seasonal climatic temperature change, deg F

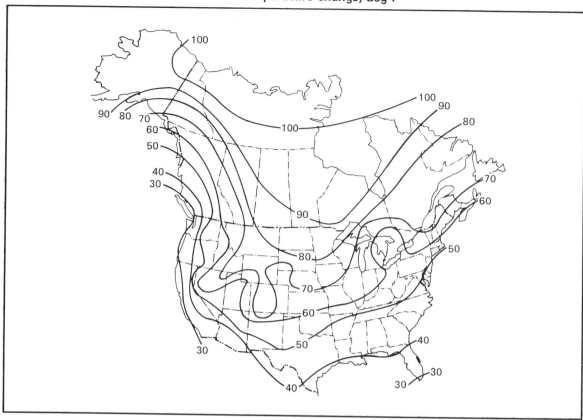

Fig. 4.4.2 Annual average ambient relative humidity, percent

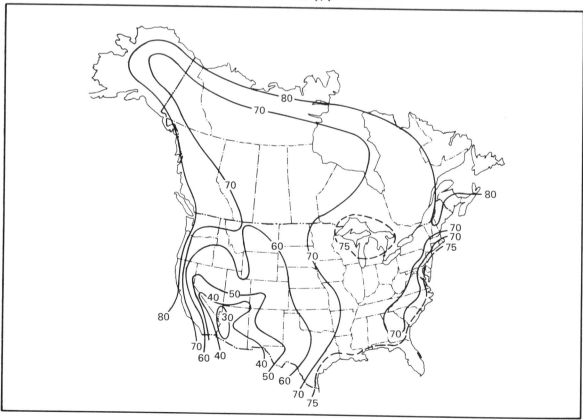

PCI Design Handbook

Table 4.4.1 Creep and shrinkage strains (millionths)[1]

Time Days[2]	Creep		Shrinkage	
	Normal weight	Lightweight	Accelerated cure	Moist cure
1	48	71	10	14
3	85	126	29	40
5	109	162	47	64
7	127	190	63	85
9	143	212	79	104
10	149	222	86	113
20	197	294	149	185
30	228	340	198	235
40	250	373	236	272
50	268	399	267	300
60	282	420	292	322
70	294	438	314	340
80	305	454	332	355
90	313	467	348	367
100	321	479	361	378
200	370	551	439	434
1 Yr	406	605	487	465
3 Yr	456	679	533	494
5 Yr	472	703	544	500
Final	524	781	560	510

Concrete Release Strength = 3500 psi
Average Prestress = 1000 psi
Relative Humidity = 70 percent
Volume / Surface Ratio = 1.5 in.

1. These factors are based on average values of a large amount of data with significant scatter (see "Volume Changes in Precast Prestressed Concrete Structures", *PCI Journal,* September-October, 1977). Thus, they may not apply in particular situations. The use of reliable creep and shrinkage properties, based on local concretes and conditions, may be used when available. Table values may be modified by simple proportion when using other creep and shrinkage factors.
2. Refers to days after release of prestress for creep and shrinkage-accelerated cure; and to days after completion of moist curing (of 5 to 7 days) for shrinkage-moist cure.

Table 4.4.2 Correction factors for prestress and concrete strength (creep only)

Ave. P/A (psi)	Release strength, f_{ci} (psi)						
	2500	3000	3500	4000	4500	5000	6000
0	0.00	0.00	0.00	0.00	0.00	0.00	0.00
200	0.24	0.22	0.20	0.19	0.18	0.17	0.15
400	0.47	0.43	0.40	0.37	0.35	0.33	0.31
600	0.71	0.65	0.60	0.56	0.53	0.50	0.46
800	0.95	0.86	0.80	0.75	0.71	0.67	0.61
1000	1.18	1.08	1.00	0.94	0.88	0.84	0.76
1200	1.42	1.30	1.20	1.12	1.06	1.00	0.92
1400	1.66	1.51	1.40	1.31	1.23	1.17	1.07
1600		1.73	1.60	1.50	1.41	1.34	1.22
1800		1.94	1.80	1.68	1.59	1.51	1.37
2000			2.00	1.87	1.76	1.67	1.53
2200				2.06	1.94	1.84	1.68
2400				2.24	2.12	2.01	1.83
2600					2.29	2.18	1.99
2800						2.34	2.14
3000						2.51	2.29

Table 4.4.3 Correction factors for relative humidity

Ave. Ambient R.H. (from Fig. 4.4.2)	Creep	Shrinkage
40	1.25	1.43
50	1.17	1.29
60	1.08	1.14
70	1.00	1.00
80	0.92	0.86
90	0.83	0.43
100	0.75	0.00

Table 4.4.4 Correction factors for volume/surface ratio

Time, days	Creep V/S 1	2	3	4	5	6	Shrinkage V/S 1	2	3	4	5	6
1	1.30	0.78	0.49	0.32	0.21	0.15	1.25	0.80	0.50	0.31	0.19	0.11
3	1.29	0.78	0.50	0.33	0.22	0.15	1.24	0.80	0.51	0.31	0.19	0.11
5	1.28	0.79	0.51	0.33	0.23	0.16	1.23	0.81	0.52	0.32	0.20	0.12
7	1.28	0.79	0.51	0.34	0.23	0.16	1.23	0.81	0.52	0.33	0.20	0.12
9	1.27	0.80	0.52	0.35	0.24	0.17	1.22	0.82	0.53	0.34	0.21	0.12
10	1.26	0.80	0.52	0.35	0.24	0.17	1.21	0.82	0.53	0.34	0.21	0.13
20	1.23	0.82	0.56	0.39	0.27	0.19	1.19	0.84	0.57	0.37	0.23	0.14
30	1.21	0.83	0.58	0.41	0.30	0.21	1.17	0.85	0.59	0.40	0.26	0.16
40	1.20	0.84	0.60	0.44	0.32	0.23	1.15	0.86	0.62	0.42	0.28	0.17
50	1.19	0.85	0.62	0.46	0.34	0.25	1.14	0.87	0.63	0.44	0.29	0.19
60	1.18	0.86	0.64	0.48	0.36	0.26	1.13	0.88	0.65	0.46	0.31	0.20
70	1.17	0.86	0.65	0.49	0.37	0.28	1.12	0.88	0.66	0.48	0.32	0.21
80	1.16	0.87	0.66	0.51	0.39	0.29	1.12	0.89	0.67	0.49	0.34	0.22
90	1.16	0.87	0.67	0.52	0.40	0.31	1.11	0.89	0.68	0.50	0.35	0.23
100	1.15	0.87	0.68	0.53	0.42	0.32	1.11	0.89	0.69	0.51	0.36	0.24
200	1.13	0.90	0.74	0.61	0.51	0.42	1.08	0.92	0.75	0.59	0.44	0.31
1 Yr	1.11	0.91	0.77	0.67	0.58	0.50	1.07	0.93	0.79	0.64	0.50	0.38
3 Yr	1.10	0.92	0.81	0.73	0.67	0.62	1.06	0.94	0.82	0.71	0.59	0.47
5 Yr	1.10	0.92	0.82	0.75	0.70	0.66	1.06	0.94	0.83	0.72	0.61	0.49
Final	1.09	0.93	0.83	0.77	0.74	0.72	1.05	0.95	0.85	0.75	0.64	0.54

Table 4.4.5 Design temperature strains* (millionths)

Temperature zone from Fig 4.4.1	Normal weight Heated	Unheated	Lightweight Heated	Unheated
10	30	45	25	38
20	60	90	50	75
30	90	135	75	113
40	120	180	100	150
50	150	225	125	188
60	180	270	150	225
70	210	315	175	263
80	240	360	200	300
90	270	405	225	338
100	300	450	250	375

* Based on accepted coefficients of thermal expansion, reduced to account for thermal lag.
(See referenced committee report, *PCI Journal,* September-October, 1977)

Table 4.4.6 Volume change strains for typical buildings (millionths)

Prestressed members (P/A = 1000 psi), normal weight concrete

Temp. zone (Fig. 4.4.1)	Actual strain — Ave. relative humidity (Fig. 4.4.2)					Equivalent strain — Ave. relative humidity (Fig. 4.4.2)				
	40	50	60	70	80	40	50	60	70	80
Heated buildings										
0	703	643	583	524	464	176	161	146	131	116
10	733	673	613	554	494	196	181	166	151	136
20	763	703	643	584	524	216	201	186	171	156
30	793	733	673	614	554	236	221	206	191	176
40	823	763	703	644	584	256	241	226	211	196
50	853	793	733	674	614	276	261	246	231	217
60	883	823	763	704	644	296	281	266	252	237
70	913	853	793	734	674	316	301	287	272	257
80	943	883	823	764	704	336	322	307	292	277
90	973	913	853	794	734	357	342	327	312	297
100	1003	943	883	824	764	377	362	347	332	317
Unheated structures										
0	703	643	583	524	464	176	161	146	131	116
10	748	688	628	569	509	206	191	176	161	146
20	793	733	673	614	554	236	221	206	191	176
30	838	778	718	659	599	266	251	236	221	206
40	883	823	763	704	644	296	281	266	252	237
50	928	868	808	749	689	326	312	297	282	267
60	973	913	853	794	734	357	342	327	312	297
70	1018	958	898	839	779	387	372	357	342	327
80	1063	1003	943	884	824	417	402	387	372	357
90	1108	1048	988	929	869	447	432	417	402	387
100	1153	1093	1033	974	914	477	462	447	432	418

Table 4.4.7 Volume change strains for typical buildings (millionths)

Prestressed members (P/A = 1000 psi), lightweight concrete

Temp. zone (Fig. 4.4.1)	Actual strain — Ave. relative humidity (Fig. 4.4.2)					Equivalent strain — Ave. relative humidity (Fig. 4.4.2)				
	40	50	60	70	80	40	50	60	70	80
Heated buildings										
0	861	791	721	650	580	215	198	180	163	145
10	886	816	746	675	605	232	215	197	179	162
20	911	841	771	700	630	249	231	214	196	179
30	936	866	796	725	655	266	248	230	213	195
40	961	891	821	750	680	282	265	247	230	212
50	986	916	846	775	705	299	281	264	246	229
60	1011	941	871	800	730	316	298	281	263	246
70	1036	966	896	825	755	333	315	297	280	262
80	1061	991	921	850	780	349	332	314	297	279
90	1086	1016	946	875	805	366	349	331	313	296
100	1111	1041	971	900	830	383	365	348	330	313
Unheated structures										
0	861	791	721	650	580	215	198	180	163	145
10	899	829	758	688	618	240	223	205	188	170
20	936	866	796	725	655	266	248	230	213	195
30	974	904	833	763	693	291	273	256	238	220
40	1011	941	871	800	730	316	298	281	263	246
50	1049	979	908	838	768	341	323	306	288	271
60	1086	1016	946	875	805	366	349	331	313	296
70	1124	1053	983	913	843	391	374	356	338	321
80	1161	1091	1021	950	880	416	399	381	364	346
90	1199	1128	1058	988	918	441	424	406	389	371
100	1236	1166	1096	1025	955	467	449	431	414	396

Table 4.4.8 Volume change strains for typical buildings (millionths)

Non-prestressed members, normal weight concrete

	Heated buildings									
	Actual strain					Equivalent strain				
Temp. zone (Fig. 4.4.1)	Ave. relative humidity (Fig. 4.4.2)					Ave. relative humidity (Fig. 4.4.2)				
	40	50	60	70	80	40	50	60	70	80
0	329	296	263	230	197	82	74	66	58	49
10	359	326	293	260	227	102	94	86	78	69
20	389	356	323	290	257	122	114	106	98	90
30	419	386	353	320	287	143	134	126	118	110
40	449	416	383	350	317	163	154	146	138	130
50	479	446	413	380	347	183	175	166	158	150
60	509	476	443	410	377	203	195	186	178	170
70	539	506	473	440	407	223	215	206	198	190
80	569	536	503	470	437	243	235	227	218	210
90	599	566	533	500	467	263	255	247	238	230
100	629	596	563	530	497	283	275	267	259	250
	Unheated structures									
0	329	296	263	230	197	82	74	66	58	49
10	374	341	308	275	242	112	104	96	88	79
20	419	386	353	320	287	143	134	126	118	110
30	464	431	398	365	332	173	164	156	148	140
40	509	476	443	410	377	203	195	186	178	170
50	554	521	488	455	422	233	225	217	208	200
60	599	566	533	500	467	263	255	247	238	230
70	644	611	578	545	512	293	285	277	269	260
80	689	656	623	590	557	323	315	307	299	291
90	734	701	668	635	602	354	345	337	329	321
100	779	746	713	680	647	384	375	367	359	351

Table 4.4.9 Volume change strains for typical buildings (millionths)

Non-prestressed members, lightweight concrete

	Heated buildings									
	Actual strain					Equivalent strain				
Temp. zone (Fig. 4.4.1)	Ave. relative humidity (Fig. 4.4.2)					Ave. relative humidity (Fig. 4.4.2)				
	40	50	60	70	80	40	50	60	70	80
0	329	296	263	230	197	82	74	66	58	49
10	354	321	288	255	222	99	91	83	74	66
20	379	346	313	280	247	116	108	99	91	83
30	404	371	338	305	272	132	124	116	108	100
40	429	396	363	330	297	149	141	133	125	116
50	454	421	388	355	322	166	158	150	141	133
60	479	446	413	380	347	183	175	166	158	150
70	504	471	438	405	372	199	191	183	175	167
80	529	496	463	430	397	216	208	200	192	183
90	554	521	488	455	422	233	225	217	208	200
100	579	546	513	480	447	250	242	233	225	217
	Unheated structures									
0	329	296	263	230	197	82	74	66	58	49
10	366	334	301	268	235	107	99	91	83	74
20	404	371	338	305	272	132	124	116	108	100
30	441	409	376	343	310	158	149	141	133	125
40	479	446	413	380	347	183	175	166	158	150
50	516	484	451	418	385	208	200	191	183	175
60	554	521	488	455	422	233	225	217	208	200
70	591	559	526	493	460	258	250	242	233	225
80	629	596	563	530	497	283	275	267	259	250
90	666	634	601	568	535	308	300	292	284	275
100	704	671	638	605	572	333	325	317	309	301

4.4.2 Equivalent Volume Change

If a horizontal framing member is connected at the ends, such that the volume change shortening is restrained, a tensile force is built up in the member and transmitted to the supporting elements. However, since the shortening takes place gradually over a period of time, the effect of the shortening on the shears and moments of the support is lessened because of creep and micro-cracking of the member and its support.

For ease of design, the volume change shortenings can be treated in the same manner as short term elastic deformations by using a concept of "equivalent" shortening.

Thus, the following relations can be assumed:

$$\delta_{ec} = \delta_c / K_\ell$$
$$\delta_{es} = \delta_s / K_\ell$$

where

δ_{ec}, δ_{es} = equivalent creep and shrinkage shortenings, respectively

δ_c, δ_s = calculated creep and shrinkage shortenings, respectively

K_ℓ = a constant for design purposes which varies from 3 to 5

The value of K_ℓ will be near the lower end of the range when the members are heavily reinforced, and near the upper end when they are lightly reinforced. For most common structures, a value of $K_\ell = 4$ is sufficiently conservative.

Shortening due to temperature change* will be similarly modified. However, the maximum temperature change will usually occur over a much shorter time, probably within 60 to 90 days.

Thus

$$\delta_{et} = \delta_t / K_t$$

where

δ_{et} and δ_t = the equivalent and calculated temperature shortening, respectively

K_t = a constant; recommended value = 1.5

The total equivalent shortening to be used for design is:

$$\Delta = \delta_{ec} + \delta_{es} + \delta_{et}$$
$$= \frac{\delta_c + \delta_s}{K_\ell} + \frac{\delta_t}{K_t}$$

---

* Temperature change is, of course, a reversible effect; increases cause expansion and are important in design and location of expansion joints (see Sect. 4.3). Temperature differentials in roof and wall elements should also be considered.

When the equivalent shortening is used in the frame analysis for determining shears and moments in the supporting elements, the actual modulus of elasticity of the members is used, rather than a reduced modulus as used in other methods.

Example 4.4.2 — Calculation of column moment caused by volume change shortening of a beam

Given: The beam of Example 4.4.1 is supported and attached to two 16 x 16-in. columns as shown in the sketch.

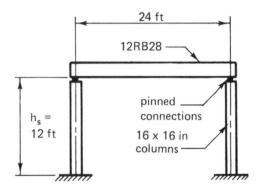

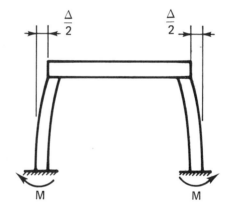

f'_c (col.) = 5000 psi

E_c = 4.3 x 10⁶ psi

Problem: Determine the horizontal force at the top of the column and the moment at the base of the column caused by volume change shortening of the beam.

Solution:

$I_c = bh^3/12 = 16^4/12 = 5461$ in.⁴

From Example 4.4.1:

Total volume change shortening from erection to final is 0.24 in. or 0.12 in. each end.

Calculate the equivalent shortening:

$$\Delta = \frac{\delta_c + \delta_s}{K_\ell} + \frac{\delta_t}{K_t}$$

$$= \left[\frac{(369 - 109 + 499 - 134)}{4} + \frac{210}{1.5} \right] (10^{-6})(24)(12)$$

$$= 296 \times 10^{-6} \times 288 = 0.085 \text{ in.}$$

$$\Delta/2 = 0.085/2 = 0.043 \text{ in. each end}$$

$$\Delta/2 = N h_s^3 / 3 E_c I_c$$

$$N = 3 E_c I_c (\Delta/2)/h_s^3$$

$$= 3 (4.3 \times 10^6)(5461)(0.043)/ (12 \times 12)^3$$

$$= 1014 \text{ lb}$$

$$M = N h_s$$

$$= 1014 \times 144$$

$$= 146,016 \text{ in.-lb (or 12.2 ft-kips)}$$

4.4.3 Usual Design Criteria

The behavior of actual structures indicates that reasonable estimates of volume change characteristics are satisfactory for the design of most structures even though test data relating volume changes to the variables shown in Tables 4.4.1 through 4.4.5 exhibit a considerable scatter. Therefore, it is possible to reduce the variables and use approximate values as shown in Tables 4.4.6 through 4.4.9.

Example 4.4.3 — Determination of actual and equivalent volume change shortenings by Tables 4.4.6 through 4.4.9

Given: Same as Examples 4.4.1 and 4.4.2.

Solution: For prestressed, normal weight concrete, in a heated building, use Table 4.4.6.

For 55 percent relative humidity and 70 F temperature, interpolating from Table 4.4.6:

Actual strain = 823×10^{-6} in./in.

Equivalent strain = 294×10^{-6} in./in.

These results compare with values of 835×10^{-6}, and 296×10^{-6} respectively, calculated from Tables 4.4.1 through 4.4.5.

4.4.4 Influence of Connections on Volume Change Restraint Forces

Design of connections should consider the forces that can result from the restraint of volume changes. For buildings with normal spans, floor heights and section dimensions, connections which use properly designed bearing pads, as outlined in Part 5 of this Handbook, will usually permit sufficient movement that the effect on the frame can be neglected. However, it is recommended that the connection and the connecting elements of all structures be designed for a minimum factored horizontal force, N_u, of:

$$\text{min } N_u = 0.2 R_{du}$$

where R_{du} is the factored dead load reaction

Unusual framing conditions, such as long beam spans on very stiff columns, should be analyzed for the effect of volume changes.

When determining the amount of potential movement to allow for in the detailing of connections and control joints, actual rather than the equivalent strain values from Tables 4.4.1 through 4.4.9 should be used.

Connections and framing members in full moment-resisting frames should be designed for the full accumulation of forces that results from the restraint of the equivalent volume change deformations.

4.5 Diaphragm Design

Horizontal loads from wind or earthquake are usually transmitted to shear walls or moment-resisting frames through the roof and floors acting as horizontal diaphragms.

4.5.1 Method of Analysis

The diaphragm is analyzed by considering the roof or floor as a deep horizontal beam, analogous to a plate girder or I-beam. The shear walls or structural frames are the supports for this analogous beam. Thus the lateral loads are transmitted to these supports as reactions. As in a beam, tension and compression are induced in the chords or "flanges" of the analogous I-beam as shown in Fig. 4.5.1.

When precast concrete members which span parallel to the supporting shear walls or frames are used for the diaphragm, it is apparent that the shear in the analogous beam must be transferred between adjacent members and also to the supporting elements. The "web" shear must also be transferred to the chord elements. Thus the design of a diaphragm is essentially a connection design problem.

4.5.2 Shear Transfer Between Members

In floors or roofs without composite topping, the shear transfer between members is usually ac-

Fig. 4.5.1 Analogous beam design of a diaphragm

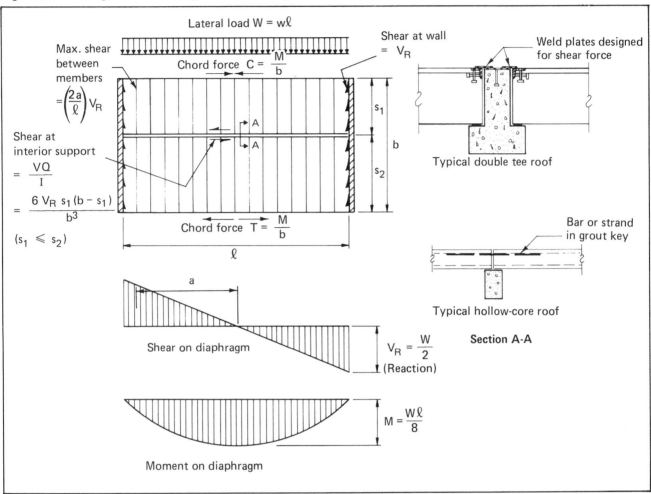

complished by weld plates or grout keys, depending on the member.

Weld plates may be analyzed as illustrated in Fig. 4.5.2. In addition to the hardware details shown, many others are used by precast concrete manufacturers.

For members connected by grout keys, a conservative value of 40 psi can be used for the average design strength of the grouted key. If necessary, reinforcement placed as shown in Fig. 4.5.3 can be used to transfer the shear. This steel is designed by the shear-friction principles discussed in Part 5.

In floor or roofs with composite topping, the topping itself can act as the diaphragm, if it is adequately reinforced. Reinforcement requirements can be determined by shear-friction analysis.

It should be noted that the connections between members often serve functions in addition to the transfer of shear for lateral loads. For example, weld plates in flanged members are often used to adjust differential camber. Grout keys may be called upon to distribute concentrated loads.

Connections which transfer shear from the diaphragm to the shear walls or moment-resisting frame are analyzed in the same manner as the connections between members.

4.5.3 Chord Forces

Chord forces are calculated as shown in Fig. 4.5.1. For roofs with intermediate supports as shown, the shear stress is carried across the beam with weld plates or bars in grout keys as shown in Section A-A. Bars are designed by shear-friction. Stresses are usually quite low, and only as many bars or weld plates as required should be used.

In flanged deck members the chord tension at the perimeter of the building is usually transferred between members by the same type of connection used for shear transfer (Fig. 4.5.2). Between connections, within the member flange, the tension is usually assumed to be taken by the tensile strength of the concrete. When forces are high, such as in design for earthquake, transverse reinforcing bars may be placed in the flange and attached to the connection device by welding or lapping with the

Fig. 4.5.2 Typical flange weld plate details

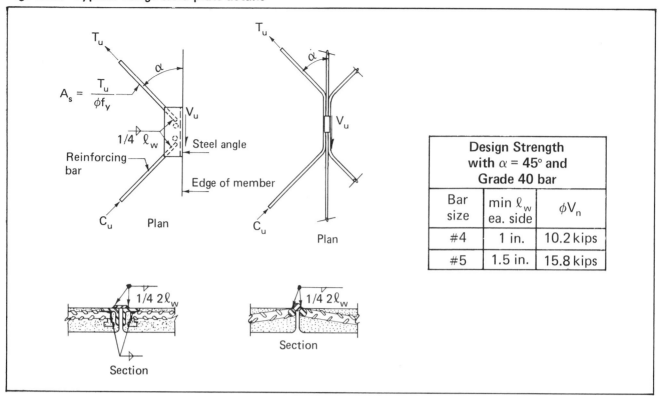

Design Strength with $\alpha = 45°$ and Grade 40 bar		
Bar size	min ℓ_w ea. side	ϕV_n
#4	1 in.	10.2 kips
#5	1.5 in.	15.8 kips

connection bars. Examples of each of these are given in Sections 4.7 and 4.8.

Static friction as discussed in Section 5.5 can be used to transfer wind loads to walls. Note that the static coefficients of friction (Table 5.5.1) should be divided by 5 when used for this purpose. Static friction is not reliable for transferring earthquake forces.

In some bearing wall buildings, a minimum amount of perimeter reinforcement is recommended for resistance to "abnormal loads."* When abnormal load design is required by the building code or owner, these minimum requirements may be more than enough to resist the chord tension.

4.6 Buildings With Moment-Resisting Frames

4.6.1 General

Precast, prestressed concrete beams and deck members âre usually most economical when they can be designed and connected into a structure as simple-span members. This is because:

1. Positive moment-resisting capacity is much easier and less expensive to attain with pretensioned members than negative moment capacity at supports.

2. Connections which achieve continuity at the

supports are usually complex and costly.

3. The restraint to volume changes that occurs in rigid connections may cause serious cracking and unsatisfactory performance or, in extreme cases, even structural failure.

Therefore, it is most desirable when designing precast, prestressed concrete structures to have connections which allow lateral movement and rotation, i.e., pinned ends, and achieve lateral stability through the use of floor and roof dia-

Fig. 4.5.3 Use of perimeter reinforcement as shear-friction steel

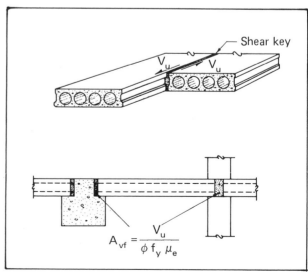

$$A_{vf} = \frac{V_u}{\phi f_y \mu_e}$$

*Speyer, Irwin J., "Considerations for the Design of Precast Concrete Bearing Wall Buildings to Withstand Abnormal Loads," *PCI Journal*, March-April, 1976.

phrams and shear walls.

However, in some structures, adequate shear walls interfere with the function of the building, or are more expensive than alternate solutions. In these cases, the lateral stability of the structure depends on the moment-resisting capacity of either the column bases, a beam-column frame, or both.

When moment connections between beams and columns are required to resist lateral loads, it is desirable, when possible, to make the moment connection after most of the dead loads have been applied. This requires careful detailing, specification of the construction process, and inspection. If such details are possible, the moment connections need only resist the negative moments from live load, lateral loads and volume changes, and will then be less costly.

4.6.2 Moment Resistance of Column Bases

Single-story and some low-rise buildings without shear walls may depend on the fixity of the column base to resist lateral loads. The ability of a spread footing to resist moments caused by lateral loads is dependent on the rotational characteristics of the base. The total rotation of the column base is a function of rotation between the footing and soil, bending in the base plate, and

Fig. 4.6.1 Assumptions used in derivation of rotational coefficients for column bases

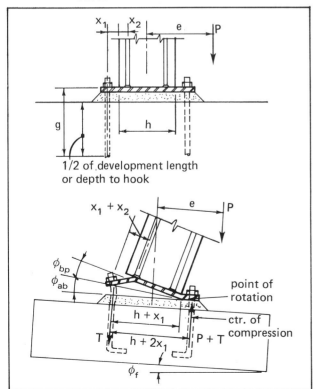

Fig. 4.6.2 Approximate relationship between allowable soil bearing value and coefficient of subgrade reaction, k_s

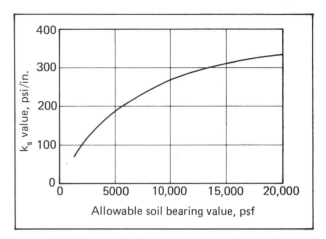

elongation of the anchor bolts, as shown in Fig. 4.6.1.

The total rotation of the base is

$$\phi_b = \phi_f + \phi_{bp} + \phi_{ab}$$

If the axial load is large enough so that there is no tension in the anchor bolts, ϕ_{bp} and ϕ_{ab} are zero, and:

$$\phi_b = \phi_f$$

Rotational characteristics can be expressed in terms of flexibility or stiffness coefficients:

$$\phi = \gamma M = M/K$$

where M = applied moment = Pe

e = eccentricity of the applied load, P

γ = flexibility coefficient

K = stiffness coefficient = $1/\gamma$

If bending of the base plate and strain in the anchor bolts are assumed as shown in Fig. 4.6.1, the flexibility coefficients for the base can be derived, and the total rotation of the base becomes:

$$\phi_b = M(\gamma_f + \gamma_{bp} + \gamma_{ab}) = Pe(\gamma_f + \gamma_{bp} + \gamma_{ab})$$

$$\gamma_f = 1/k_s I_f \qquad \text{(Eq. 4.6-1)}$$

$$\gamma_{bp} = \frac{(x_1 + x_2)^3 \; [2e/(h + 2x_1) - 1]}{6e \, E_s \, I_{bp} \, (h + x_1)} \geqslant 0$$
$$\text{(Eq. 4.6-2)}$$

$$\gamma_{ab} = \frac{g \; [2e/(h + 2x_1) - 1]}{2e \, A_b \, E_s \, (h + x_1)} \geqslant 0 \qquad \text{(Eq. 4.6-3)}$$

where

$\gamma_f, \gamma_{bp}, \gamma_{ab}$ = flexibility coefficients of the footing/soil interaction, the base plate

Table 4.6.1 Flexibility coefficients for footing soil interaction

Flexibility of base = $\gamma_b = \gamma_f + \gamma_{ab} + \gamma_{bp}$

Rotation of base = $\gamma_b \times Pe$ in inch-lbs

Stiffness of base = $K_b = 1/\gamma_b$

Fixity of base = $K_b/(K_c + K_b)$

K_c = Column stiffness = $4E_c I_c/h_s$

E_c = Modulus of elasticity of column concrete, psi

I_c = Moment of inertia of column, in.^{4}

h_s = Story height, in.

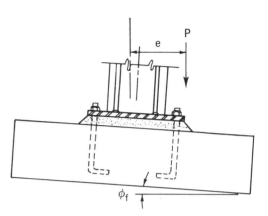

γ_f, 1/in-lb X 10^{-10} for square footings

Footing Size (ft)	k_s				
	100	150	200	250	300
2.0 X 2.0	3616.9	2411.3	1808.4	1446.8	1205.6
2.5 X 2.5	1481.5	987.7	740.7	592.6	493.8
3.0 X 3.0	714.4	476.3	357.2	285.8	238.1
3.5 X 3.5	385.6	257.1	192.8	154.3	128.5
4.0 X 4.0	226.1	150.7	113.0	90.4	75.4
4.5 X 4.5	141.1	94.1	70.6	56.5	47.0
5.0 X 5.0	92.6	61.7	46.3	37.0	30.9
5.5 X 5.5	63.2	42.2	31.6	25.3	21.1
6.0 X 6.0	44.7	29.8	22.3	17.9	14.9
6.5 X 6.5	32.4	21.6	16.2	13.0	10.8
7.0 X 7.0	24.1	16.1	12.1	9.6	8.0
7.5 X 7.5	18.3	12.2	9.1	7.3	6.1
8.0 X 8.0	14.1	9.4	7.1	5.7	4.7
9.0 X 9.0	8.8	5.9	4.4	3.5	2.9
10.0 X 10.0	5.8	3.9	2.9	2.3	1.9
11.0 X 11.0	4.0	2.6	2.0	1.6	1.3
12.0 X 12.0	2.8	1.9	1.4	1.1	0.9

and the anchor bolts, respectively

k_s = coefficient of subgrade reaction from Fig. 4.6.2

I_f = moment of inertia of the footing (plan dimensions)

E_s = modulus of elasticity of steel

I_{bp} = moment of inertia of the base plate (vertical cross-section dimensions)

A_b = total area of anchor bolts which are in tension

h = width of the column in the direction of bending

x_1 = distance from face of column to the center of the anchor bolts, positive when anchor bolts are outside the column, and negative when an-chor bolts are inside the column

x_2 = distance from the face of the column to base plate anchorage

g = assumed length over which elongation of the anchor bolt takes place = 1/2 of development length + projection for deformed anchor bolts or the length to the hook + projection for smooth anchor bolts. (See Fig. 4.6.1)

Rotation of the base may cause an additional eccentricity of the loads on the columns, causing moments which must be added to the moments induced by the lateral loads.

Note that in Eqs. 4.6-2 and 4.6-3, if the eccentricity, e, is less than $h/2 + x_1$ (inside the center of compression), γ_{bp} and γ_{ab} are less than zero, meaning that there is no rotation between the col-

PCI Design Handbook

Table 4.6.2 Flexibility coefficients for anchor bolts and base plates

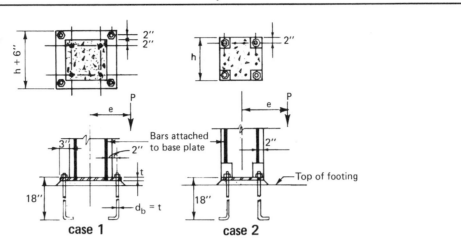

case 1   case 2

Column size, h (in)	e (in)	Case 1: Exterior anchor bolts Base plate thickness & anc. bolt diameter				Case 2: Interior anchor bolts Base plate thickness & anc. bolt diameter			
		.75	1.00	1.25	1.50	.75	1.00	1.25	1.50
12 × 12	4	.0	.0	.0	.0	.0	.0	.0	.0
	6	.0	.0	.0	.0	29.3	16.5	10.5	7.3
	8	.0	.0	.0	.0	43.9	24.7	15.8	11.0
	10	16.6	7.9	4.5	2.9	52.7	29.6	19.0	13.2
	12	27.7	13.2	7.5	4.8	58.5	32.9	21.1	14.6
	14	35.7	16.9	9.6	6.1	62.7	35.3	22.6	15.7
	16	41.6	19.8	11.2	7.2	65.8	37.0	23.7	16.5
	18	46.2	22.0	12.5	8.0	68.3	38.4	24.6	17.1
16 × 16	6	.0	.0	.0	.0	.0	.0	.0	.0
	8	.0	.0	.0	.0	10.5	5.9	3.8	2.6
	10	.0	.0	.0	.0	16.7	9.4	6.0	4.2
	12	7.7	3.7	2.1	1.4	20.9	11.8	7.5	5.2
	14	13.1	6.3	3.6	2.3	23.9	13.4	8.6	6.0
	16	17.2	8.3	4.8	3.1	26.1	14.7	9.4	6.5
	18	20.4	9.8	5.7	3.6	27.9	15.7	10.0	7.0
	20	23.0	11.1	6.4	4.1	29.3	16.5	10.5	7.3
20 × 20	8	.0	.0	.0	.0	.0	.0	.0	.0
	10	.0	.0	.0	.0	4.9	2.7	1.8	1.2
	12	.0	.0	.0	.0	8.1	4.6	2.9	2.0
	14	4.1	2.0	1.2	.7	10.5	5.9	3.8	2.6
	16	7.1	3.5	2.0	1.3	12.2	6.9	4.4	3.0
	18	9.5	4.6	2.7	1.7	13.5	7.6	4.9	3.4
	20	11.4	5.6	3.2	2.1	14.6	8.2	5.3	3.7
	22	13.0	6.3	3.7	2.4	15.5	8.7	5.6	3.9
24 × 24	10	.0	.0	.0	.0	.0	.0	.0	.0
	12	.0	.0	.0	.0	2.7	1.5	1.0	.7
	14	.0	.0	.0	.0	4.6	2.6	1.6	1.1
	16	2.4	1.2	.7	.5	6.0	3.4	2.2	1.5
	18	4.3	2.1	1.2	.8	7.1	4.0	2.6	1.8
	20	5.8	2.8	1.7	1.1	8.0	4.5	2.9	2.0
	22	7.0	3.4	2.0	1.3	8.7	4.9	3.1	2.2
	24	8.0	3.9	2.3	1.5	9.3	5.2	3.4	2.3

$\gamma_{ab} + \gamma_{bp}$, 1/in.-lb $\times$ 10^{-10} for typical details

umn and the footing, and only the rotation from soil deformation (Eq. 4.6-1) need be considered.

Values of Eqs. 4.6-1 through 4.6-3 are tabulated for typical cases in Tables 4.6.1 and 4.6.2.

Example 4.6.1 — Stability analysis of an unbraced frame

Given:

The column shown in Fig. 4.6.3

Soil bearing capacity = 5000 psf

P = 80 kips dead load, 30 kips live load

W = 2 kips wind load

Determine the column design loads and moments for stability as an unbraced frame.

Solution:

ACI 318-77 requires that the column be designed for the following conditions:

(1) 1.4 D + 1.7 L

(2) 0.75 (1.4 D + 1.7 L + 1.7 W)

(3) 0.9 D + 1.3 W

The maximum eccentricity would occur when (3) is applied. Moment at base of column = 2 x 16 = 32 ft-kips = 384 in.-kips

0.9 D = 0.9 (80) = 72 kips

1.3 W = 1.3 (384) = 499.2 in.-kips

Eccentricity due to wind load

$$= \frac{M_u}{P_u} = \frac{499.2}{72} = 6.93 \text{ in.}$$

To determine the moments caused by base rotation, an iterative procedure is required.

Estimate eccentricity due to rotation = 0.25 in.
e = 6.93 + 0.25 = 7.18 in.

Check rotation between column and footing
$h/2 + x_1 = 20/2 + (-2) = 8$ in. > 7.18, thus there is no tension in the anchor bolts and no rotation between the column and footing.

$$I_f = (6 \times 12)^4/12 = 2.24 \times 10^6 \text{ in.}^4$$

From Fig. 4.6.2: $k_s \approx 200$ psi/in.²

$$\gamma_b = 1/k_s I_f = 1/(200 \times 2.24 \times 10^6)$$
$$= 2.23 \times 10^{-9}$$

(Note: this could also be read from Table 4.6.1)

$$M_u = 72 (7.18) = 517 \text{ in.-kips}$$
$$\phi_b = \gamma_f M_u = 2.23 \times 10^{-9} \times 517 \times 10^3$$
$$= 0.00115 \text{ radians}$$

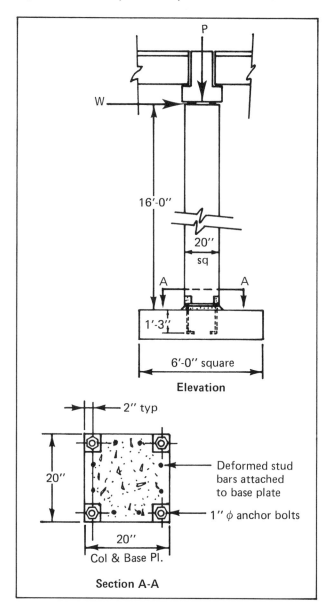

Fig. 4.6.3 Examples 4.6.1, 4.6.2 and 4.6.5

16'-0"

20"
sq

1'-3"

6'-0" square

Elevation

2" typ

20"

20"

Col & Base Pl.

Deformed stud bars attached to base plate

1" φ anchor bolts

Section A-A

Eccentricity caused by rotation

$$= \phi_b h_s = 0.00115 (16 \times 12) = 0.22 \text{ in.} \approx 0.25$$
no further trial is required

Design requirements for 0.9 D + 1.3 W:

P_u = 72 kips

M_u = 517 in.-kips = 43.1 ft-kips

Check for 0.75 (1.4 D + 1.7 L + 1.7 W)

P_u = 0.75 (1.4 D + 1.7 L) = 0.75 [1.4 (80) + 1.7 (30)] = 122.3 kips

M_u = 0.75 (1.7 W) = 0.75 [1.7 (384)] = 489.6 in.-kips

$$e = \frac{489.6}{122.3} = 4.0 \text{ in.}$$

Estimate eccentricity due to rotation = 0.22 in.

M_u = 122.3 (4.22) = 516.1 in.-kips

ϕ_b = $\gamma_f M_u$ = 2.23 × 10^{-9} × 516.1 × 10^3

= 0.00115 radians

$\phi_b h_s$ = 0.00115 (16 × 12) = 0.22 in. OK

Design requirements for 0.75 [1.4 D + 1.7 L + 1.7 W]:

P_u = 122.3 kips

M_u = 516.1 in.-kips = 43.0 ft-kips

Section 10.11.5 (ACI 318-77) also requires that the moment caused by a minimum eccentricity of 0.6 + 0.03h be considered when designing for 1.4 D + 1.7 L

P_u = 1.4 D + 1.7 L = 1.4 (80) + 1.7 (30)

= 163 kips

e = 0.6 + 0.03h = 0.6 + 0.03 (20) = 1.2 in.

Estimate eccentricity due to rotation = 0.1 in.

M_u = P_u e = 163 (1.2 + 0.1) = 211.9 in.-kips

ϕ_b = 2.23 × 10^{-9} × 211.9 × 10^3 = 0.000473 radians

$\phi_b h_s$ = 0.000473 (16 × 12) = 0.09 in. ≈ 0.1 in. OK

Design requirements for 1.4 D + 1.7 L:

P_u = 163 kips

M_u = 212 in.-kips = 17.7 ft-kips

4.6.3 Fixity of Column Bases

The degree of fixity of a column base is the ratio of the rotational stiffness of the base to the sum of the rotational stiffnesses of the column plus the base:

$$F_b = \frac{K_b}{K_b + K_c}$$

where

F_b = degree of base fixity, expressed as a decimal

K_b = 1/γ_b

K_c = $\dfrac{4 E_c I_c}{h_s}$

E_c = modulus of elasticity of the column concrete

I_c = moment of inertia of the column

h_s = column height

Example 4.6.2 — Calculation of degree of fixity

Determine the degree of fixity of the column

base in Example 4.6.1, E_c = 4300 ksi:

K_b = 1/γ_b = 1/(2.23 × 10^{-9}) = 4.48 × 10^8

I_c = 20^4/12 = 13,333

K_c = $\dfrac{4 (4.3 \times 10^6) (13,333)}{16 \times 12}$ = 11.94 × 10^8

F_b = $\dfrac{4.48}{4.48 + 11.94}$ = 0.27

4.6.4 Volume Change Effects in Moment-Resisting Frames

The restraint of volume changes in moment-resisting frames causes tension in the girders and deflections and moments in the columns. The magnitude of these tensions, moments and deflections is dependent on the distance from the *center of stiffness* of the frame.

The center of stiffness is that point of a frame, which is subject to a uniform unit shortening, at which no lateral movement will occur. For frames which are symmetrical with respect to bay sizes, story heights and member stiffnesses, the center of stiffness is located at the midpoint of the frame, as shown in Fig. 4.6.4.

Tensions in girders are maximum in the bay nearest the center of stiffness. Deflections and moments in columns are maximum furthest from the center of stiffness. Thus in Fig. 4.6.4:

$F_1 < F_2 < F_3$

$\Delta_1 > \Delta_2 > \Delta_3$

$M_1 > M_2 > M_3$

The degree of fixity of the column base as described in Section 4.6.3 has a great effect on the magnitude of the forces and moments caused by volume change restraint. An assumption of a fully fixed base in the analysis of the structure may result in significant overestimation of the restraint forces, whereas assuming a pinned base may have the opposite effect. The degree of fixity used in the volume change analysis should be consistent with that used in the analysis of the column for other loadings, and determination of slenderness effects.

4.6.4.1 Calculation of Volume Change Restraint Forces in Moment-Resisting Frames

Several computer analysis programs such as STRESS or STRUDL are available that will allow the input of not only gravity and lateral loads, but also the shortening strains of members from volume changes. The equivalent strains as described in Section 4.4 can be input directly into such programs. While these programs have methods of inputing partially fixed bases, they are somewhat complex. The designer may prefer to make two computer runs, one with the base pinned and one with the base fixed and then interpolate between the results.

For frames that are approximately symmetrical, the coefficients from Tables 4.6.3 and 4.6.4 can be used with small error. The use of these tables is described in Fig. 4.6.5.

Example 4.6.3 — Volume Change Restraint Forces

Given:

The 4-bay, 2 story frame shown
Beam modulus of elasticity = E_b = 4300 ksi
Column modulus of elasticity = E_c = 4700 ksi
Column bases 20% fixed (see Sect. 4.6.3)
Design R.H. = 70%
Design temperature change = 70 deg. F

Problem:

Determine the maximum tension in the beams and the maximum moment in the columns caused by volume change restraint.

Solution:

1. Determine relative stiffness between columns and beams:

$$I_b = 12 (24)^3 /12 = 13,824 \text{ in.}^4$$
$$E_b I_b /\ell = 4300 (13,824) / (24 \times 12)$$
$$= 206,400$$
$$I_c = 16 (16)^3 /12 = 5461 \text{ in.}^4$$
$$E_c I_c /h_s = 4700 (5461) / (16 \times 12)$$
$$= 133,681$$

$$K_r = \frac{E_b I_b /\ell}{E_c I_c /h_s} = \frac{206,400}{133,681} = 1.5$$

2. Determine deflections:
From Table 4.4.6: $\delta_e = 272 \times 10^{-6}$
$$\Delta_B = \delta_e \ell = 0.000272 (24) (12) = 0.078 \text{ in.}$$

$$\Delta_A = \delta_e (2\ell) = 0.157 \text{ in.}$$

3. Determine maximum beam tension:
Maximum tension is nearest the center of stiffness, i.e., beams BC and CD, 2nd floor.
From Table 4.6.3:
For n = 4 and i = 2, k_b = 3.00
From Table 4.6.4:
For K_r = 1.0, fixed base, k_f = 11.2
For K_r = 2.0, fixed base, k_f = 11.6
Therefore for K_r = 1.5, k_f = 11.4
For pinned based, k_f = 3.4
(for K_r = 1.0 and 2.0)
For 20% fixed:
$$k_f = 3.4 + 0.20 (11.4 - 3.4) = 5.0$$
$$F_2 = k_f k_b \Delta_i E_c I_c / h_s^3$$
$$= 5.0 \times 3.0 \times 0.078 \times 4700 \times 5461 / (16 \times 12)^3$$
$$= 4.24 \text{ kips}$$

4. Determine maximum column moments:
For base moment, M_1:
From Table 4.6.4, by interpolation similar to above:
$$k_m \text{ (fixed)} = (4.9 + 5.2) / 2 = 5.05$$
$$k_m \text{ (pinned)} = 0$$
$$k_m \text{ (20\% fixed)} = 0 + 0.20 (5.05) = 1.0$$
$$M_1 = k_m \Delta_i E_c I_c / h_s^2$$
$$= 1.0 \times 0.157 \times 4700 \times 5461 / (16 \times 12)^2$$
$$= 109 \text{ in.-kips}$$

For second floor moment, M_{2L}:
$$k_m \text{ (fixed)} = (3.9 + 4.5) / 2 = 4.2$$
$$k_m \text{ (pinned)} = (2.1 + 2.4) / 2 = 2.25$$
$$k_m \text{ (20\% fixed)} = 2.25 + 0.20 (4.20 - 2.25) = 2.64$$
$$M_{2L} = 2.64 \times 0.157 \times 4700 \times 5461 / (16 \times 12)^2$$
$$= 289 \text{ in.-kips}$$

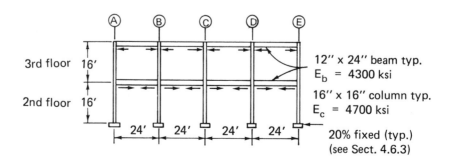

PCI Design Handbook

Fig. 4.6.4   Effect of volume change restraints in   building frames

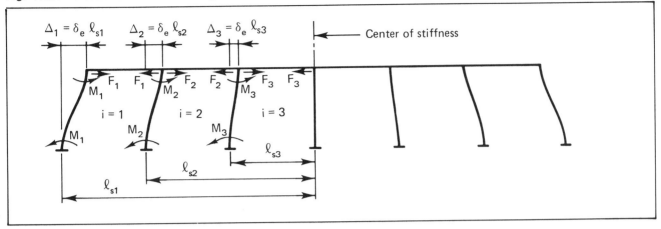

Table 4.6.3 Build-up of restraint forces in beams (k_b)

Total number of bays (n)	Number of bays from end (i)							
	1	2	3	4	5	6	7	8
2	1.00							
3	1.00	4.00						
4	1.00	3.00						
5	1.00	2.67	9.00					
6	1.00	2.50	6.00					
7	1.00	2.40	5.00	16.00				
8	1.00	2.33	4.50	10.00				
9	1.00	2.29	4.20	8.00	25.00			
10	1.00	2.25	4.00	7.00	15.00			
11	1.00	2.22	3.86	6.40	11.67	36.00		
12	1.00	2.20	3.75	6.00	10.00	21.00		
13	1.00	2.18	3.67	5.71	9.00	16.00	49.00	
14	1.00	2.17	3.60	5.50	8.33	13.50	28.00	
15	1.00	2.15	3.55	5.33	7.86	12.00	21.00	64.00
16	1.00	2.14	3.50	5.20	7.50	11.00	17.50	36.00

4.6.5   Eccentrically Loaded Columns

Many precast concrete structures utilize multi-story columns with simple-span beams resting on haunches. Fig. 4.6.6 and Table 4.6.5 are provided as aids for determining the various combinations of load and moment that can occur with such columns.

The following conditions and limitations apply to Fig. 4.6.6 and Table 4.6.5:

1. The coefficients are only valid for braced columns. Lateral stability must be achieved by other shear walls or moment-resisting frames.

2. For partially fixed column bases (see Section 4.6.3), a straight line interpolation between the coefficients for pinned and fixed bases can be used with small error.

3. For higher columns, the coefficients for the 4-story columns can be used with small error.

4. The coefficients in the "Σ Max" line will yield the maximum required restraining force, F_i, and column moments caused by loads (equal at each level) which can occur on either side of the column, for example, live loads on interior columns. The maximum force will not necessarily occur with the same loading pattern that causes the maximum moment.

5. The coefficients in "Σ One Side" line will yield the maximum moments which can occur if the column is loaded on only one side, such as the end column in a bay.

Fig. 4.6.5 Use of Table 4.6.4

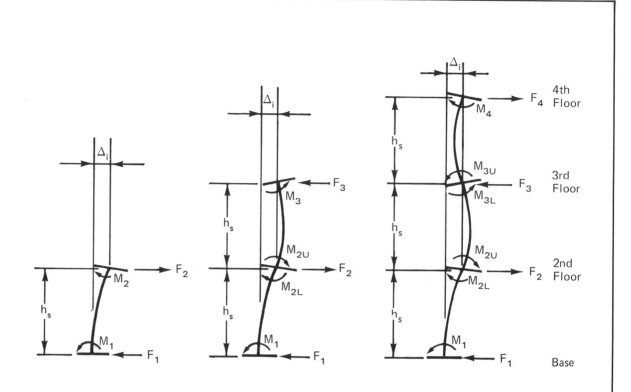

$$\Delta_i \;=\; \delta_e \ell_s$$

$$F_i \;=\; k_f k_b \Delta_i E_c I_c / h_s^3$$

$$M_i \;=\; k_m \Delta_i E_c I_c / h_s^2$$

where: δ_e = equivalent unit strain (see Section 4.4)

ℓ_s = distance from column to center of stiffness

F_i = F_1, F_2, etc., as shown above

k_f, k_m = coefficients from Table 4.6.4

k_b = $i\left(\dfrac{n+1-i}{n+2-2i}\right)$ (or from Table 4.6.3)

n = no. of bays

i = as shown in Fig. 4.6.4

E_c = Modulus of elasticity of the column concrete

I_c = Moment of inertia of the column

Table 4.6.4 Coefficients k_f and k_m for forces and moments caused by volume change restraint forces (see Fig. 4.6.5 for notation)

No. of Stories	$K_r = \dfrac{\Sigma E_b I_b / \ell}{\Sigma E_c I_c / h_s}$	Base Fixity	Values of k_f				Values of k_m					
			F_1	F_2	F_3	F_4	Base M_1	2nd floor		3rd floor		4th
								M_{2L}	M_{2U}	M_{3L}	M_{3U}	M_4
1	0	Fixed	3.0	3.0			3.0	0				
		Pinned	0	0			0	0				
	0.5	Fixed	6.0	6.0			4.0	2.0				
		Pinned	1.2	1.2			0	1.2				
	1.0	Fixed	7.5	7.5			4.5	3.0				
		Pinned	1.7	1.7			0	1.7				
	2.0	Fixed	9.0	9.0			5.0	4.0				
		Pinned	2.2	2.2			0	2.2				
	4.0 or more	Fixed	10.1	10.1			5.4	4.7				
		Pinned	2.5	2.5			0	2.5				
2	0	Fixed	6.8	9.4	2.6		4.3	2.6	2.6	0		
		Pinned	0	3.0	1.5		0	1.5	1.5	0		
	0.5	Fixed	8.1	10.7	2.6		4.7	3.4	2.1	0.4		
		Pinned	1.9	3.4	1.4		0	1.9	1.2	0.2		
	1.0	Fixed	8.9	11.2	2.3		4.9	3.9	1.8	0.5		
		Pinned	2.1	3.4	1.3		0	2.1	1.0	0.3		
	2.0	Fixed	9.7	11.6	1.9		5.2	4.5	1.4	0.5		
		Pinned	2.4	3.4	1.0		0	2.4	0.8	0.3		
	4.0 or more	Fixed	10.4	11.9	1.4		5.5	5.0	1.0	0.4		
		Pinned	2.6	3.4	0.8		0	2.6	0.5	0.2		
3 or more	0	Fixed	7.1	10.6	4.1	0.7	4.4	2.8	2.8	0.7	0.7	0
		Pinned	1.6	3.6	2.4	0.4	0	1.6	1.6	0.4	0.4	0
	0.5	Fixed	8.2	11.1	3.5	0.5	4.7	3.5	2.2	0.7	0.4	0.09
		Pinned	1.9	3.6	1.9	0.3	0	1.9	1.2	0.4	0.2	0.05
	1.0	Fixed	8.9	11.4	2.9	0.4	5.0	3.9	1.9	0.7	0.3	0.09
		Pinned	2.2	3.5	1.6	0.2	0	2.2	1.0	0.4	0.2	0.05
	2.0	Fixed	9.7	11.7	2.2	0.2	5.2	4.7	1.4	0.6	0.2	0.06
		Pinned	2.4	3.5	1.2	0.1	0	2.4	0.8	0.3	0.1	0.03
	4.0 or more	Fixed	10.4	11.9	1.5	0.04	5.5	5.0	1.0	0.5	0.04	0.01
		Pinned	2.6	3.4	0.8	0.02	0	2.6	0.5	0.2	0.02	0.00

Fig. 4.6.6 Use of Table 4.6.5

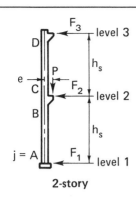

2-story

$F_i = k_f\, Pe/h_s$

$M_j = k_m\, Pe$

where

F_i = Restraining force at level i

M_j = Moment at point j

k_f, k_m = Coefficients from Table 4.6.5

P = Vertical load acting at eccentricity e

See text for limitations and design example

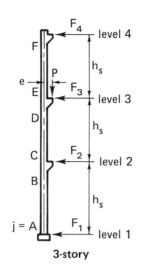

3-story

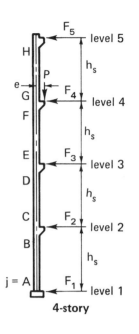

4-story

Example 4.6.4 Use of Fig. 4.6.6 and Table 4.6.5

Using Table 4.6.5, determine the maximum restraining force and moment in the lowest story of a 3-story frame for:

a. An interior column in a multi-bay frame

b. An exterior column

Beam reactions to column haunch at each level:

D.L. = 50 kips
L.L. = 20 kips

Eccentricity, e, = 14 in.
Story height, h_s, = 16 ft
Column base is 65% fixed

Solution:

Factored loads: D.L. = 1.4 x 50 = 70.0 kips
   L.L. = 1.7 x 20 = $\underline{34.0}$ kips
   104.0 kips

a. For the interior column, the dead load reaction would be the same on either side, thus no moment results. The live load could occur on any one side at any floor, hence use the coefficients in the "Σ Max" line:

$P_u e$ = 34.0 x 14 = 476 in.-kips = 39.7 ft-kips

To determine the maximum moment at point B:

For a pinned base, k_m = 0.67
For a fixed base, k_m = 0.77
For 65% fixed, k_m = 0.67 + 0.65
   (0.77 – 0.67)
  = 0.74

$M_u = k_m\, P_u e$ = 0.74 (39.7) = 29.4 ft-kips

Maximum restraining force at level 2
  = $k_f P_u e/h_s$

k_f = 1.40 + 0.65 (1.62 – 1.40) = 1.54

F_u = 1.54 (39.7) / 16 = 3.82 kips (tension or compression)

b. For the exterior column, the total load is eccentric on the same side of the column, hence use the coefficients in the "Σ One Side" line:

$P_u e$ = 104.0 x 14 = 1456 in.-kips
  = 121.3 ft-kips

To determine the maximum moment at point B:

For a pinned base, k_m = 0.40
For a fixed base, k_m = 0.46

Table 4.6.5 Coefficients k_f and k_m for determining moments and restraining forces on eccentrically loaded columns braced against sidesway

+ Indicates clockwise moments on the columns and compression in the restraining beam															
No. of stories	Base Fixity	P acting at level	k_f at level					k_m at point							
			1	2	3	4	5	A	B	C	D	E	F	G	H
2	PINNED	3	+ 0.25	− 1.50	+ 1.25			0	− 0.25	+ 0.25	+ 1.0				
		2	− 0.50	0	+ 0.50			0	+ 0.50	+ 0.50	0				
		Σ Max	± 0.75	± 1.50	± 1.75			0	± 0.75	± 0.75	± 1.0				
		Σ One Side	− 0.25	− 1.50	+ 1.75			0	+ 0.25	+ 0.75	+ 1.0				
	FIXED	3	+ 0.43	− 1.72	+ 1.29			− 0.14	− 0.29	+ 0.29	+ 1.0				
		2	− 0.86	+ 0.43	+ 0.43			+ 0.29	+ 0.57	+ 0.43	0				
		Σ Max	± 1.29	± 2.15	± 1.72			± 0.43	± 0.86	± 0.72	± 1.0				
		Σ One Side	− 0.43	− 1.29	+ 1.72			+ 0.15	+ 0.28	+ 0.72	+ 1.0				
3	PINNED	4	− 0.07	+ 0.40	− 1.60	+ 1.27		0	+ 0.07	− 0.07	− 0.27	+ 0.27	+ 1.0		
		3	+ 0.13	− 0.80	+ 0.20	+ 0.47		0	− 0.13	+ 0.13	+ 0.53	+ 0.47	0		
		2	− 0.47	− 0.20	+ 0.80	− 0.13		0	+ 0.47	+ 0.53	+ 0.13	− 0.13	0		
		Σ Max	± 0.67	± 1.40	± 2.60	± 1.87		0	± 0.67	± 0.73	± 0.93	± 0.87	± 1.0		
		Σ One Side	− 0.41	− 0.60	− 0.60	+ 1.61		0	+ 0.40	+ 0.60	+ 0.40	+ 0.60	+ 1.0		
	FIXED	4	− 0.12	+ 0.47	− 1.62	+ 1.27		+ 0.04	+ 0.08	− 0.08	− 0.27	+ 0.27	+ 1.0		
		3	+ 0.23	− 0.92	+ 0.23	+ 0.46		− 0.08	− 0.15	+ 0.15	+ 0.54	+ 0.46	0		
		2	− 0.81	+ 0.23	+ 0.70	− 0.12		+ 0.27	+ 0.54	+ 0.46	+ 0.12	− 0.12	0		
		Σ Max	± 1.16	± 1.62	± 2.55	± 1.85		± 0.38	± 0.77	± 0.69	± 0.92	± 0.85	± 1.0		
		Σ One Side	− 0.70	− 0.22	− 0.69	+ 1.61		+ 0.23	+ 0.46	+ 0.54	+ 0.38	+ 0.62	+ 1.0		
4	PINNED	5	+ 0.02	− 0.11	+ 0.43	− 1.61	+ 1.27	0	− 0.02	+ 0.02	+ 0.07	− 0.07	− 0.27	+ 0.27	+ 1.0
		4	− 0.04	+ 0.22	− 0.86	+ 0.22	+ 0.46	0	+ 0.04	− 0.04	− 0.14	+ 0.14	+ 0.54	+ 0.46	0
		3	+ 0.13	− 0.75	0	+ 0.75	− 0.12	0	− 0.13	+ 0.13	+ 0.50	+ 0.50	+ 0.12	− 0.12	0
		2	− 0.46	− 0.22	+ 0.86	− 0.22	+ 0.04	0	+ 0.46	+ 0.54	+ 0.14	− 0.14	− 0.04	+ 0.04	0
		Σ Max	± 0.65	± 1.30	± 2.15	± 2.80	± 1.89	0	± 0.64	± 0.72	± 0.86	± 0.86	± 0.97	± 0.89	± 1.0
		Σ One Side	− 0.35	− 0.86	+ 0.43	− 0.86	+ 1.65	0	+ 0.35	+ 0.65	+ 0.57	+ 0.43	+ 0.35	+ 0.65	+ 1.0
	FIXED	5	+ 0.03	− 0.12	+ 0.43	− 1.61	+ 1.27	− 0.01	− 0.02	+ 0.02	+ 0.07	− 0.07	− 0.27	+ 0.27	+ 1.0
		4	− 0.06	+ 0.25	− 0.87	+ 0.22	+ 0.46	+ 0.02	+ 0.04	− 0.04	− 0.14	+ 0.14	+ 0.54	+ 0.46	0
		3	+ 0.22	− 0.87	+ 0.03	+ 0.74	− 0.12	− 0.07	− 0.14	+ 0.14	+ 0.51	+ 0.50	+ 0.12	− 0.12	0
		2	− 0.80	+ 0.21	+ 0.74	− 0.18	+ 0.03	+ 0.27	+ 0.54	+ 0.46	+ 0.12	− 0.12	− 0.03	+ 0.03	0
		Σ Max	± 1.11	± 1.45	± 2.07	± 2.75	± 1.88	± 0.37	± 0.74	± 0.67	± 0.84	± 0.83	± 0.96	± 0.88	± 1.0
		Σ One Side	− 0.61	− 0.53	+ 0.33	− 0.83	+ 1.64	+ 0.21	+ 0.41	+ 0.59	+ 0.56	+ 0.44	+ 0.36	+ 0.64	+ 1.0

For 65% fixed, k_m = 0.40 + 0.65
(0.46 − 0.40)
= 0.44

$M_u = k_m P_u e = 0.44 (121.3) = 53.4$ ft-kips

Maximum restraining force at level 2
= $k_f P_u e / h_s$

$k_f = -0.60 + 0.65 (-0.60 + 0.22) = -0.35$

$F_u = -0.35 (121.3) / 16 = -2.65$ kips
(tension)

4.6.6 Slenderness Effects in Compression Members

4.6.6.1 Approximate Evaluation of Slenderness Effects

ACI 318-77 permits an approximate evaluation of slenderness in Section 10.11. Application of this section of the Code for members braced against sidesway is shown in Example 4.6.5, and for unbraced frames in Section 4.6.7. A more rigorous approach, which meets Section 10.10.1 of the Code is discussed briefly in Section 4.6.6.2.

The effective length factor, k, can be determined from the alignment charts, Fig. 4.6.8. For column bases, the value of ψ for use in these charts can be calculated from the rotational stiffness coefficients described in Section 4.6.2, with ψ base = K_c/K_b. For most structures, ψ base should not be taken less than 1.0. For column bases which are assumed pinned in the frame analysis, ψ base can be assumed equal to 10 when using Fig. 4.6.8.

Example 4.6.5   Moment magnifier for a column in a braced frame

Given:

The interior column of Example 4.6.4

Column size = 14 x 14 in.
E_c = 4700 ksi
P_u = 624 kips
h_s = 16 ft
Braced frame

Base stiffness coefficient, K_b = 10.0 x 10^8

Problem:

Determine moment magnifier

Solution:

$$K_c = \frac{4 E_c I_c}{h_s} = \frac{4 (4.7 \times 10^6) (3201)}{(16 \times 12)}$$

$$= 3.13 \times 10^8$$

From Fig. 4.6.8:

$$\psi_A = \frac{3.13}{10} = 0.31 \text{ use min. of } 1.0$$

$\psi_B = \infty$ (pinned connection)

k = 0.87

Slenderness may be neglected when $k\ell_u/r$ is less than $34 - 12 M_1/M_2$

r = 0.3 x 14 = 4.2

$k\ell_u/r = 0.87 (16 \times 12) / 4.2 = 39.8$

From Table 4.6.5:

M_1 = Moment at the base
= (% fixity) $k_m P_u e$
= 0.65 (0.38) (39.7) = 9.8 ft-kips

M_2 = 29.4 ft-kips (see Example 4.6.4)

Note: M_1 and M_2 are opposite direction, therefore:

$M_1/M_2 = -(9.8/29.4) = -0.33$

$34 - 12 (-0.33) = 38.0 < 39.8$

Therefore slenderness must be considered

$$\beta_d = \frac{\text{Factored dead load moment}}{\text{Factored total load moment}}$$

$$\approx \frac{70 (14)}{104 (14)} = 0.67$$

Note: β_d is a factor that takes into account creep due to sustained loads. When the moment to be magnified is caused by short-term loads, such as wind or earthquake, β_d may be considered equal to zero.

Using Eq. 10-10 of ACI 318-77:

$$EI = (E_c I_g/2.5)/(1 + \beta_d)$$
$$= 4700 (3201)/(2.5 \times 1.67)$$
$$= 3.60 \times 10^6 \text{ k} - \text{in}^2$$

$$P_c = \frac{\pi^2 E I}{(k\ell_u)^2} = \frac{\pi^2 (3.60 \times 10^6)}{(0.87 \times 16 \times 12)^2}$$

$$= 1273 \text{ kips}$$

$$C_m = 0.6 + 0.4 M_1/M_2 = 0.6 + 0.4 (-0.33)$$
$$= 0.47$$

$$\delta = \frac{C_m}{1 - \dfrac{P_u}{\phi P_c}} = \frac{0.47}{1 - \dfrac{624}{0.7 (1273)}} = 1.57$$

Fig. 4.6.7 could also be used for this example.

Fig. 4.6.7 Slenderness effects by Sec. 10-11 (ACI 318-77)

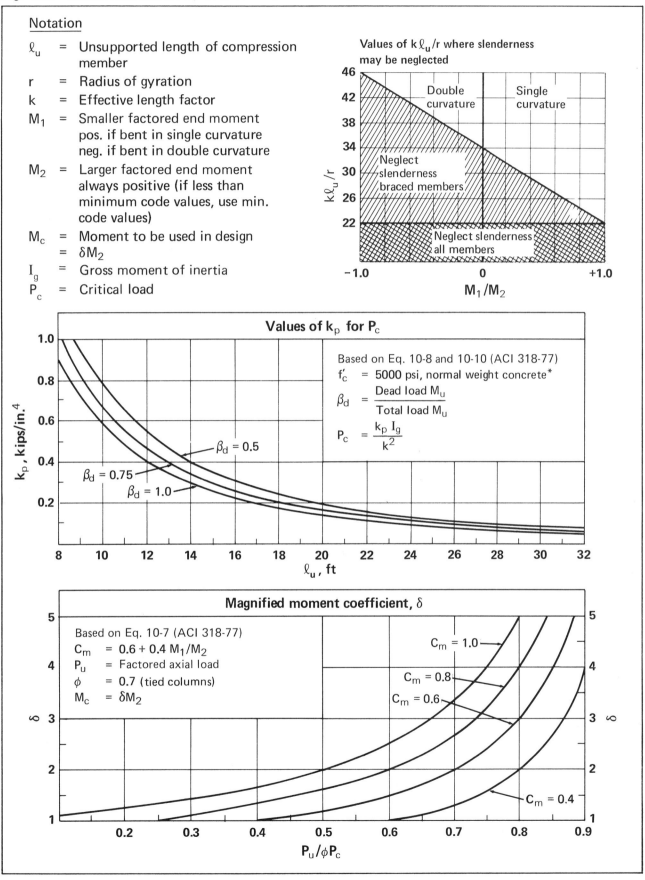

Notation

ℓ_u = Unsupported length of compression member

r = Radius of gyration

k = Effective length factor

M_1 = Smaller factored end moment
pos. if bent in single curvature
neg. if bent in double curvature

M_2 = Larger factored end moment
always positive (if less than minimum code values, use min. code values)

M_c = Moment to be used in design
= δM_2

I_g = Gross moment of inertia

P_c = Critical load

Values of $k\ell_u/r$ where slenderness may be neglected

Double curvature   Single curvature

Neglect slenderness braced members

Neglect slenderness all members

Values of k_p for P_c

k_p, kips/in.^{4}

$\beta_d = 0.5$

$\beta_d = 0.75$

$\beta_d = 1.0$

ℓ_u, ft

Based on Eq. 10-8 and 10-10 (ACI 318-77)

f'_c = 5000 psi, normal weight concrete*

$\beta_d = \dfrac{\text{Dead load } M_u}{\text{Total load } M_u}$

$P_c = \dfrac{k_p I_g}{k^2}$

Magnified moment coefficient, δ

Based on Eq. 10-7 (ACI 318-77)

C_m = 0.6 + 0.4 M_1/M_2

P_u = Factored axial load

ϕ = 0.7 (tied columns)

M_c = δM_2

$C_m = 1.0$

$C_m = 0.8$

$C_m = 0.6$

$C_m = 0.4$

$P_u/\phi P_c$

*For other concretes, $P_c = k_p I_g E_c/4300k^2$, E_c = modulus of elasticity, ksi

Fig. 4.6.8 Alignment charts for determining effective length factors

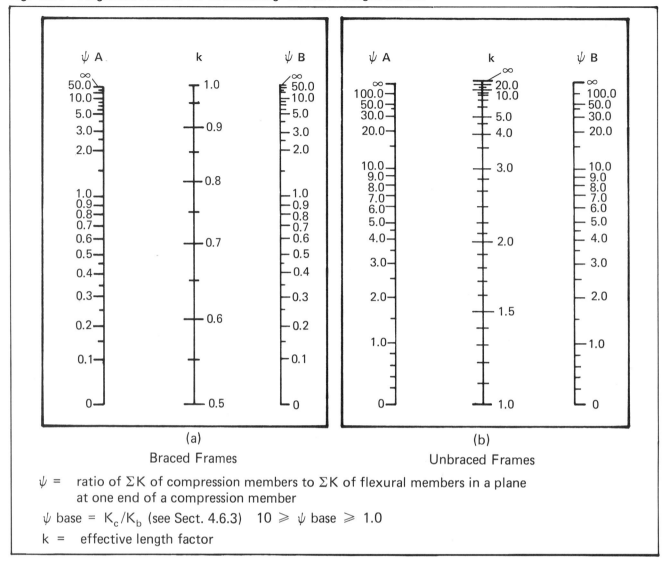

(a)
Braced Frames

(b)
Unbraced Frames

ψ = ratio of ΣK of compression members to ΣK of flexural members in a plane at one end of a compression member

ψ base = K_c / K_b (see Sect. 4.6.3) $\quad 10 \geqslant \psi$ base $\geqslant 1.0$

k = effective length factor

When a story of a structure fails in a lateral instability mode, one floor translates relative to another as a unit. Therefore, in frames not braced against sidesway, ACI 318-77 specifies that P_u and P_c in the above equation for δ be replaced by ΣP_u and ΣP_c for all columns in a story.

When considering the unbraced frame, only those column moments which are cumulative need be magnified. For example, moments caused by volume change restraints are in opposite directions either side of the stiffness center, and need not be magnified. However, such moments should be considered when analyzing individual columns. Note that ACI 318-77 requires that the moment magnifier be the larger of the values computed for the entire story, or as computed for the individual column assumed braced.

4.6.6.2 Slenderness Effects by Structural Analysis

In lieu of the approximate evaluation of slenderness effects, the requirements of Section 10.10.1 of ACI 318-77 can be met by using an iterative PΔ analysis that is suitable for computer solution,* or by direct PΔ analysis.†

An iterative P Δ analysis accounts for the effects of frame drift due to slenderness by designing for frame moments which correspond to the calculated shears, which are required for equilibrium for an initial sway.

*Gouwens, A.J., "Lateral Load Analysis of Multistory Frames with Shear Walls," Bulletin AEC2, Portland Cement Association Computer Program.

†MacGregor, J.C., and Hage, S.E., "Stability Analysis and Design of Concrete," *Proceedings, Journal of the Structural Division,* ASCE, October, 1977.

Fig. 4.6.9 P△ effects in a frame

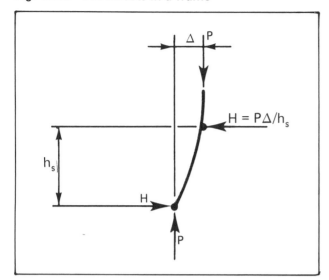

A short explanation of this procedure follows:

1. Select an initial frame deflection to account for out-of-plumbness. (1/4 of 1% has been suggested in the past.)

2. Calculate approximate story shears resulting from this initial deflection. See Fig. 4.6.9.

3. Determine the frame deflection corresponding to the above shears. Note that this deflection must be based on a reduced EI to account for creep due to sustained loads and cracking. See Section 10.10.1 of ACI 318-77 Commentary.

4. Calculate the shears resulting from the increased deflection determined in Step 3. Once again determine the corresponding deflection. (This may be quickly done using a ratio of shears.)

5. The addition to the initial shears reduces after each successive iteration (if it does not then the structure has a stability problem and member dimensions should be reconsidered) and generally an upper bound value for the design story shear can be selected following several iterations.

6. The computed forces corresponding to these story shears account for the effects of column slenderness on the frame. Now each column may be designed without moment magnifiers for the moments resulting from gravity loads acting at an eccentricity of △.

7. When designing for the case of wind, an additional story shear results from the wind deflection and must be accounted for. For this case, use the actual EI, as wind effects are short term. Due to the nature of volume change effects, generally no additional story shear will result from these forces as they tend to cancel.

Section 10.11.2 of the Code Commentary refers to the direct P△ solution by MacGregor and Hage and allows an unbraced frame to be considered braced when the stability index, Q, is equal to or less than 0.04, where:

$$Q = \frac{\Sigma P_u \Delta_u}{H_u h_s}$$

ΣP_u = summation of factored loads in a given story

Δ_u = elastically-computed first order lateral deflection due to H_u (neglecting P△ effects) at the top of the story relative to the bottom of the story

H_u = total factored lateral force within the story

h_s = height of story, center to center of floors or roof

The direct P△ analysis by MacGregor and Hage also provides the following simple method of computing the moment magnifier, δ, when $0.04 < Q \leqslant 0.22$:

$$\delta = \frac{1}{1 - Q} = \frac{1}{1 - \dfrac{\Sigma P_u \Delta_u}{H_u h_s}}$$

4.6.7 Example — Seven-Story Parking Structure

The wind load analysis and design of a multi-level parking structure located in Chicago, Illinois, are illustrated by the schematic drawings shown in Fig. 4.6.10. Deck members are lightweight single tees with 3 in. (min.) normal weight concrete topping. Columns and beams are also precast, prestressed members. Unfactored loads are given as follows:

Dead loads:

Tees and topping	=	94 psf
Beams, columns, misc.	=	75
		169 psf

Live loads:

Roof	75 psf
Floor	50 psf
Wind	20 psf

(Gross projected area)

Wind load analysis, east-west direction

Lateral stability in the east-west direction was

Fig. 4.6.10   7-story structure of Section 4.6.7

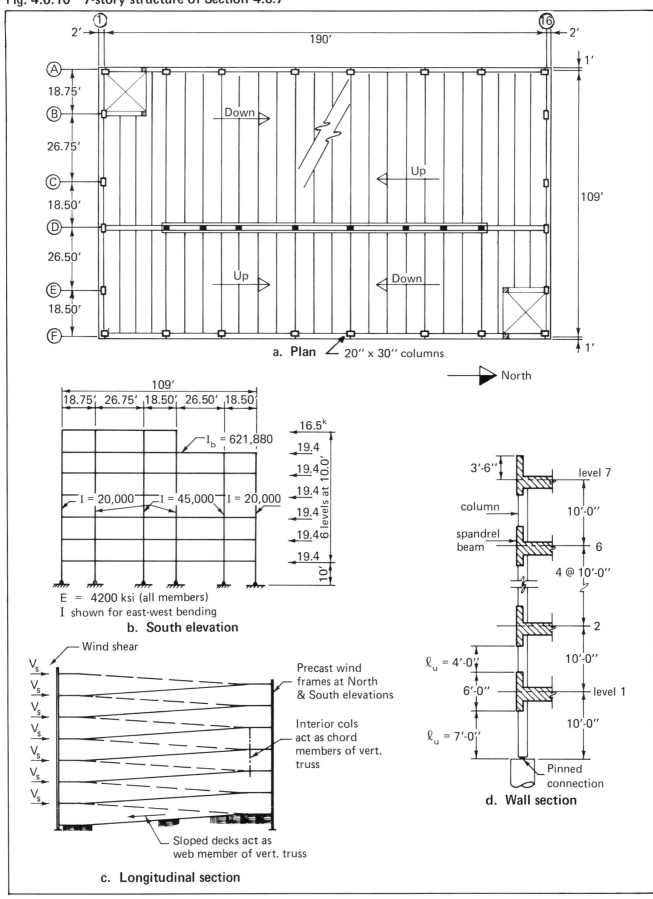

a.   Plan

20″ x 30″ columns

North

b.   South elevation

$I_b = 621,880$

$I = 20,000$ $I = 45,000$ $I = 20,000$

16.5^k

19.4

19.4

19.4

19.4

19.4

19.4

6 levels at 10.0′

10′

E = 4200 ksi (all members)
I shown for east-west bending

c.   Longitudinal section

Wind shear

V_s

Precast wind frames at North & South elevations

Interior cols act as chord members of vert. truss

Sloped decks act as web member of vert. truss

d.   Wall section

3′-6″

column

spandrel beam

$\ell_u = 4'-0''$

6′-0″

$\ell_u = 7'-0''$

level 7

10′-0″

6

4 @ 10′-0″

2

10′-0″

level 1

10′-0″

Pinned connection

achieved by providing moment connections between the spandrel beams and columns on lines 1 and 16.

Story wind shears per frame:

Roof: P = 0.02 ksf x 8.5 ft x 194/2
= 16.5 kips

Floor: P = 0.02 x 10.0 x 194/2
= 19.4 kips/floor

The wind analysis was done by computer. Output moments are shown in Fig. 4.6.11. (Note: because of the size of the beams on this structure, uncracked section properties were used in the analysis.)

Wind drift:

From the computer analysis, deflection at the first floor due to wind is 0.10 in. The "drift index" is then:

$$\frac{0.10}{10 \times 12} = 0.00083 \text{ in./in.} < 0.0025^*$$

Wind load analysis, north-south direction:

The full analysis for wind in the north-south direction is not shown here, but the assumptions used are shown in Fig. 4.6.10c.

Volume change analysis, east-west direction:

From Figs. 4.4.1 and 4.4.2, the structure should be designed for a temperature change of 70 deg. F, and an average ambient R.H. of 70 percent.

Beams and columns are prestressed, normal weight concrete, so use Table 4.4.6 to find volume change strains. For an unheated structure:

Actual strain = 839×10^{-6} in./in.

Equivalent strain = 342×10^{-6} in./in.

Because of the stiffening effect of the scissor ramps on the interior columns at line D, and the deck diaphragm, the center of stiffness of the structure can be assumed to be approximately at line D.

From Table 4.6.4, for a structure with 3 or more floors, pinned column base and $K_r > 4$,

k_m at second floor = 2.6

Column moment = $k_m \Delta_i E_c I_c / h_s^2$

For column A, 64 ft from stiffness center:

$\Delta_A = 342 \times 10^{-6} \times 64 \times 12 = 0.263$ in.

---

*Several values for permissible wind drift have been published, varying from 0.0015 in./in. to 0.0025 in./in.

M_A = 2.6 (0.263) (4200) (20,000) / $(10 \times 12)^2$ = 3989 in.-kips
= 332 ft-kips

Δ_B = 0.186 in., M_B = 526 ft-kips

Δ_C = 0.076 in., M_C = 216 ft-kips

Δ_E = 0.109 in., M_E = 310 ft-kips

Δ_F = 0.185 in., M_F = 234 ft-kips

Volume change analysis, north-south direction:

Δ_{max} = $342 \times 10^{-6} \times 12 \times 190/2 = 0.390$ in.

(at lines 1 and 16)

From Table 4.6.4, with $K_r = 0$:

k_m at second floor = 1.6

For columns at lines A and F:

I = 45,000 in.^{4} (bending in 30 in. direction)
M = 1.6 (0.390) (4200) (45,000) / $(10 \times 12)^2$ = 8190 in.-kips
= 682 ft-kips

For columns at lines B, C, D, and E:

I = 20,000 in.^{4} (bending in 20 in. direction)
M = 303 ft-kips

Moments at upper levels can be calculated in a similar manner.

Column slenderness effects — east-west bending:

Using the approximate evaluation, it is necessary to calculate two moment magnifiers, one for moments caused by gravity loads (in this case the required minimum eccentricity), another for moments caused by wind.

Sum of all columns loads:

P_{story} (kips)

Story	D.L.	L.L.	T.L.
7	2633	1169	3802
6	3507	1037	4544
5	3507	1037	4544
4	3507	1037	4544
3	3507	1037	4544
2	3507	1037	4544
1	3507	1037	4544
Total	23,675	7391	31,066

P_u = 1.4 (23,675) + 1.7 (7391)
= 33,145 + 12,565
= 45,710 kips

First level moment magnifier:

ℓ_u = 7'0" = 84 in. (see Fig. 4.6.10d)

For moments caused by gravity:

β_d = 33,145/45,710 = 0.73

For moments caused by wind:

β_d = 0

For columns at lines A and F:

$E_c = E_b$ = 4200 ksi
I_c = 20,000 in.[4]
I_b = 621,880 in.[4]

From Fig. 4.6.8:

ψ_A (base) = 10.0 (see Sect. 4.6.6.1)

In using Fig. 4.6.8, I_b can be considered equal to 0.5 I_g (see Commentary to ACI 318-77).

$$\psi_B = \frac{\Sigma E_c I_c / h_s}{\Sigma E_b I_b / \ell}$$

$$= \frac{2 \, (20,000) / (10 \times 12)}{0.5 \, (621,880) / (18.75 \times 12)}$$

= 0.24

k = 1.75 (unbraced frame)

EI = $(E_c I_c / 2.5) / (1 + \beta_d)$

Gravity loads:

EI = 4200 (20,000) / (2.5 x 1.73)
  = 19.42 x 10^6

P_c = $\pi^2 EI / (k\ell_u)^2 = \pi^2 (19.42 \times 10^6) /$ (1.75 x 84)²
  = 8870 kips

Wind loads:

EI = 4200 (20,000) / (2.5 x 1.0)
  = 33.60 x 10^6

P_c = π^2 (33.60 x 10^6) / (1.75 x 84)²
  = 15,346 kips

For columns at lines B, C, D, and E:

Assume column B as typical

I_c = 45,000 in.[4]
ψ_A = 10.0

$$\psi_B = \frac{2 \, (45,000) / (10 \times 12)}{0.5 \, (621,880 \, / \, 225 + 621,880 \, / \, 321)}$$

  = 0.32

k = 1.80

Gravity loads:

EI = 4200 (45,000) / (2.5 x 1.73)
  = 43.70 x 10^6

P_c = π^2 (43.70 x 10^6) / (1.80 x 84)²
  = 18,866 kips

Wind loads:

EI = 4200 (45,000) / (2.5 x 1.0)
  = 75.60 x 10^6

P_c = π^2 (75.60 x 10^6) / (1.80 x 84)²
  = 32,638 kips

Calculate the sum of P_c of all columns which resist lateral loads:

Gravity loads:

ΣP_c = 4 (8870) + 8 (18,866) = 186,408 kips

Note: This neglects any contribution toward stability of the interior columns, which is theoretically correct, since they are assumed pinned at each end. Actual connections seldom behave as full pins, so the above assumption is conservative.

$$\delta = \frac{C_m}{1 - \dfrac{\Sigma P_u}{\phi \, (\Sigma P_c)}} = \frac{1.0}{1 - \dfrac{45,710}{0.7 \, (186,408)}} = 1.54$$

Wind loads:

ΣP_c = 4 (15,346) + 8 (32,638) = 322,488 kips

$$\delta = \frac{1.0}{1 - \dfrac{0.75 \, (45,710)}{0.7 \, (322,488)}} = 1.18$$

The magnifier for moments caused by wind loads can also be computed by the direct PΔ analysis of MacGregor and Hage as follows:

Δ = first level wind drift from the computer analysis
  = 0.10 in.

Δ_u = 0.75 [1.7 (0.10)] = 0.13 in.

ΣP_u = 0.75 (45,710) = 34,282 kips

H = 6 (19.4) + 16.5 = 132.9 kips

H_u = 0.75 [1.7 (132.9)] = 169.4 kips

h_s = story height = 120 in.

$$Q = \frac{\Sigma P_u \Delta_u}{H_u h_s} = \frac{34,282 \, (0.13)}{169.4 \, (120)}$$

  = 0.219

$$\delta = \frac{1}{1 - Q} = \frac{1}{1 - 0.219}$$

  = 1.28

To complete the analysis for slenderness effects, the moment magnifier for the columns assumed braced should also be calculated. In this case, it can be shown that for braced conditions, $k\ell_u/r < 22$, so slenderness can be neglected.

Lines B, C, D and E upper level effective length factors:

$$\psi_A = \psi_B = 0.32$$

From Fig. 4.6.8, k = 1.10

$$\ell_u = 4'0'' = 48'' \text{ (see Fig. 4.6.10d)}$$

$$k\ell_u/r = 1.10 (48) / (0.3 \times 30) = 5.9 < 22$$

Lines A and F upper level effective length factors:

$$\psi_A = \psi_B = 0.24$$

$$k = 1.08$$

$$k\ell_u/r = 1.08(48) / (0.3 \times 20) = 8.6 < 22$$

Therefore slenderness may be neglected at upper levels.

Column slenderness effects — north-south bending

Braced frame, $k \leqslant 1.0$ (see Fig. 4.6.10d)

Max. $k\ell_u/r = 1.0 (84) / (0.3 \times 30) = 9.3$

Slenderness may be neglected in north-south direction.

Gravity loads and moments on columns:

A detailed analysis of the gravity loads and moments on the columns is not shown here, but is summarized in Tables 4.6.6 and 4.6.7 for typical columns. The moments in the east-west direction are calculated using moment distribution or other suitable indeterminate structural analysis, and are caused by the dead load of the spandrel beam. The gravity moments in the north-south direction are caused by the eccentric reaction of the interior beams on column corbels, and may be calculated using the coefficients from Table 4.6.5.

ACI 318-77 requires that the moment magnifiers be applied to a moment caused by a minimum eccentricity of:

$$e_{min} = 0.6 + 0.03h$$

where h is the column dimension in the direction of bending. Thus for the columns on lines A and F:

$$e_{min} \text{ (east-west)} = 0.6 + 0.03 (20) = 1.2 \text{ in.}$$

$$e_{min} \text{ (north-south)} = 0.6 + 0.03 (30) = 1.5 \text{ in.}$$

Table 4.6.6 Summary of column A-16 design forces — east-west bending

Strength Equations ACI 318-77 Sect. 9.2		Floor						
		1	2	3	4	5	6	7
Unfactored forces	Axial dead load, P, kips	140	120	100	80	60	40	20
	Axial live load, P, kips	42.0	36.4	30.8	25.2	19.6	14.0	8.4
	Wind moment, W, ft-kips	135	56	45	36	26	17	11
	Volume change moment, T, ft-kips	332	64	3	—	—	—	—
1.4D + 1.7L	Calculated M_u, ft-kips	33	22	22	22	22	22	18
	Minimum M_u, ft-kips	27	23	19	16	12	8	4
	Moment magnifier, δ	1.54	1.0	1.0	1.0	1.0	1.0	1.0
	Design M_u, ft-kips	51	23	22	22	22	22	18
	Design P_u, kips	267	230	192	155	117	80	42
0.75 (1.4D + 1.7L + 1.7W) (Unbraced)	Moment magnifier, δ	1.18	1.0	1.0	1.0	1.0	1.0	1.0
	Design M_u, ft-kips	232	88	74	62	50	38	28
	Design P_u, kips	200	173	144	116	88	60	32
0.9D + 1.3W (Unbraced)	Moment magnifier, δ	1.10	1.0	1.0	1.0	1.0	1.0	1.0
	Design M_u, ft-kips	209	85	69	56	43	31	22
	Design P_u, kips	126	108	90	72	54	36	18
0.75 (1.4D + 1.4T + 1.7L) (Braced)	Design M_u, ft-kips	373	89	21	17	17	17	14
	Design P_u, kips	200	173	144	116	88	60	32
1.4 (D + T) (Braced)	Design M_u, ft-kips	487	109	20	15	15	15	12
	Design P_u, kips	196	168	140	112	84	56	28

Table 4.6.7 Summary of column D-1 design forces — north-south bending

Strength Equations ACI 318-77 Sect. 9.2		Floor						
		1	2	3	4	5	6	7
Unfactored forces	Axial dead load, P, kips	1106	963	794	625	456	287	118
	Axial live load, P, kips	329	286	236	186	135	85	35
	Volume change moment, T, ft-kips	303	303	76	—	—	—	—
1.4D + 1.7L	Calculated M_u, ft-kips	128	237	157	183	208	157	297
	Minimum M_u, ft-kips	211	183	151	119	87	55	23
	Design M_u, ft-kips	211	237	157	183	208	157	297
	Design P_u, kips	2108	1834	1513	1191	868	546	225
0.75 (1.4D + 1.4T + 1.7L) (Braced)	Design M_u, ft-kips	414	496	198	137	156	118	222
	Design P_u, kips	1580	1376	1134	893	651	410	169
1.4 (D + T) (Braced)	Design M_u, ft-kips	518	598	222	134	153	115	218
	Design P_u, kips	1548	1348	1112	875	638	402	165

Moment connections at lines 1 and 16:

Negative beam moments can be determined by a computer analysis or by approximate methods shown here.

Volume change moments:

From Table 4.6.4, k_m for the column above the second floor is 0.5. Thus the net moment to be distributed to the beams is (2.6 + 0.5) / 2.6 times the calculated column moment below the second floor. This is distributed to the beam on each side of the column in proportion to the stiffness. For example at column E:

$$M_E = 310 \text{ ft-kips (see p. 4—37)}$$

$$\Sigma M_{Beams} = \frac{2.6 + 0.5}{2.6} \ (310)$$

$$= 370 \text{ ft-kips}$$

$$M_{EF} = \frac{26.50}{18.50 + 26.50} \ (370)$$

$$= 218 \text{ ft-kips}$$

$$M_{ED} = 370 - 218 = 152 \text{ ft-kips}$$

Gravity load moments:

In lieu of a more precise analysis, the approximate moment coefficients given in Section 8.3 of ACI 318-77 can be used to calculate the gravity load moment. For example, at column E, with the spandrel weight equal to 1.43 kips/ft:

Ave. clear span = ℓ_n = 20.2 ft.

$$M_{EF} = \frac{w\ell_n{}^2}{10} = \frac{1.43 \ (20.2)^2}{10}$$

$$= 58 \text{ ft-kips}$$

$$M_{ED} = \frac{w\ell_n{}^2}{11} = \frac{1.43 \ (20.2)^2}{11}$$

$$= 53 \text{ ft-kips}$$

Fig. 4.6.11 Wind moments in members — from computer output (level 1)

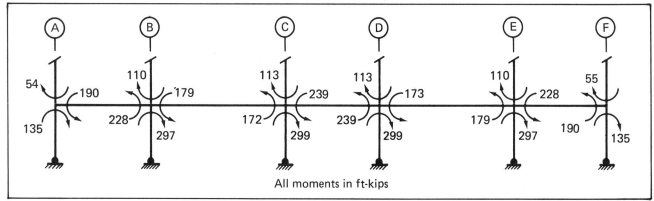

All moments in ft-kips

Fig. 4.6.12 Moment connection details for example of Section 4.6.7

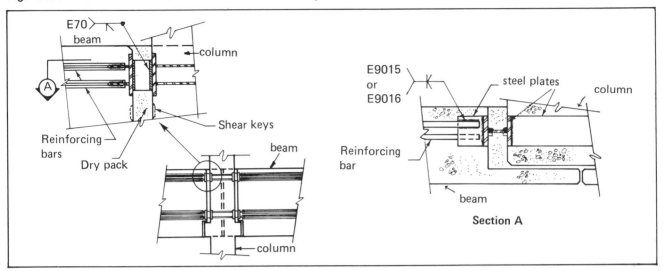

Other moments:

In this example, the wind load moments in the beams can be taken from the computer output shown in Fig. 4.6.11. In some cases, other moments, such as those caused by differential foundation settlement should also be considered. In this example, an analysis of the effects of differential settlement, which included the moderating effect of creep, resulted in an additional negative moment of 300 ft-kips. The wind moment in beam EF (see Fig. 4.6.11) is 228 ft-kips. Thus the total factored negative moment at beam EF is the greater of:

$$M_u = 0.75(1.4D + 1.7L)$$
$$= 0.75[1.4(58) + 1.7(228)] = 352 \text{ ft-kips}$$
$$M_u = 1.4(D + T)$$
$$= 1.4(58 + 147 + 300)$$
$$= 707 \text{ ft-kips (controls)}$$

The moment connection used in this example is shown conceptually in Fig. 4.6.12.

4.7 Shear Wall Buildings

4.7.1 General

In most precast, prestressed concrete buildings, it is desirable to resist lateral loads with shear walls. The shear walls can be of precast concrete, cast-in-place concrete or unit masonry. This section illustrates the analysis and design of two of the most common applications of precast concrete shear walls to resist lateral wind loads. More severe loading caused by earthquake is shown in Section 4.8.

Shear walls can be the exterior wall system, interior walls, or walls of elevator, stairway, mechanical shafts, or cores. The transfer of load from horizontal diaphragm to the shear wall can be achieved either through connections or by direct bearing.

4.7.2 Lateral Load Distribution to Shear Walls

Lateral loads are distributed to each shear wall in proportion to its shear and flexural stiffness:

$$F_i = V_W \frac{K_i}{\Sigma K}$$

= load induced into shear wall i

K_i = stiffness of wall i where $\dfrac{1}{K_i} = \dfrac{1}{K_{si}} + \dfrac{1}{K_{fi}}$

ΣK = sum of the stiffnesses of all shear walls acting to resist the wind load
$= \Sigma(K_s + K_f)$

V_W = total wind shear acting at the floor under consideration

K_{si} = shear stiffness of wall i $= \dfrac{G_i A_i}{1.2 h_s}$

K_{fi} = flexural stiffness of wall i $= \dfrac{12 E_i I_i}{h_s^3}$

(this assumes double curvature in the wall)

G_i = shear modulus of elasticity (modulus of rigidity) of wall i
= 0.4 E_c for concrete and 400 f'_m but not to exceed 1.2×10^6 psi for concrete masonry, where f'_m is the assumed compressive strength of the masonry.

A_i = the plan area of the web of shear wall i parallel to the load (length times thickness)

Fig. 4.7.1 Unsymmetrical shear walls

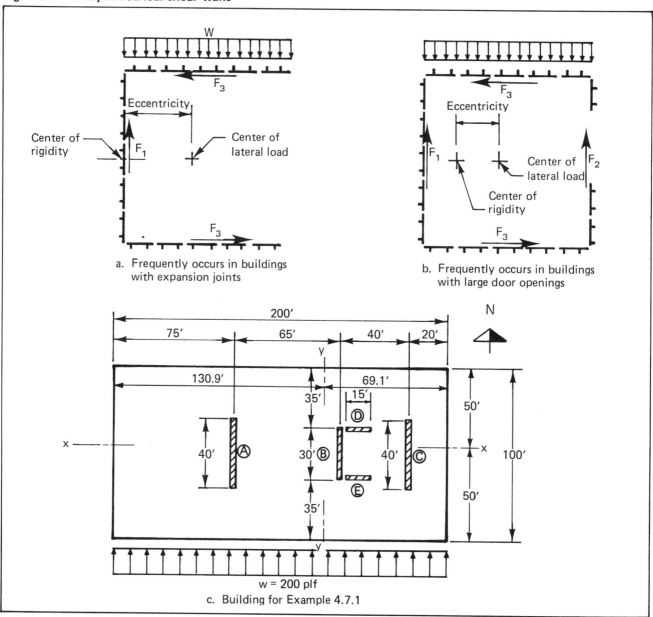

a. Frequently occurs in buildings with expansion joints

b. Frequently occurs in buildings with large door openings

c. Building for Example 4.7.1

E_i = modulus of elasticity of the wall material

= E_c for concrete and 1000 f'_m, but not to exceed 3.0×10^6 psi for concrete masonry

I_i = moment of inertia of the plan area of wall i

h_s = story height

For a structure with rectangular shear walls of the same material, with a wall height to length ratio of less than about 0.3, the flexural stiffness can be neglected, and the distribution made in accordance with the wall web areas (A_i). If the height to length ratio is greater than about 3.0, the shear stiffness can be neglected, and the distri-

bution made in accordance with the moments of inertia.

A similar analysis can be made when shear walls and columns interact to resist the lateral loads. In most such cases, the shear wall stiffness is much greater than the column stiffness, and the contribution of the columns can be neglected with small error.

4.7.3 Unsymmetrical Shear Walls

Structures which have shear walls placed unsymmetrically with respect to the center of the lateral load should take the torsional effect into account. Typical examples are shown in Fig. 4.7.1. For wind loading on most structures, a simplified method of determining the torsional resistance may be used in lieu of more exact design. The method is similar to the design of rivet groups in steel connections, and is illustrated in the following example.

Example 4.7.1 — Design of unsymmetrical shear walls

Given the structure of Fig. 4.7.1c. All walls are 8 ft high and 8 in. thick. Determine the shear in each wall, assuming the floors and roof are rigid diaphragms. Walls D and E are not connected to Wall B.

Solution:

Maximum height to length ratio of north-south walls = 8/30 < 0.3. Thus for distribution of the direct wind shear, neglect flexural stiffness. Since walls are the same thickness and material, distribute in proportion to length.

Total lateral load, W = 0.20 x 200 = 40 kips

Determine center of rigidity:

$$\bar{x} = \frac{40(75) + 30(140) + 40(180)}{40 + 30 + 40}$$

$$= 130.9 \text{ ft from left}$$

$\bar{y}$ = center of building, since walls D and E are placed symmetrically about the center of the building in the north-south direction

Torsional moment, T = 40(130.9 − 200/2) = 1236 ft-kips.

Determine the polar moment of inertia of the shear wall group about the center of rigidity:

$$I_p = I_{xx} + I_{yy}$$

$$I_{xx} = \Sigma \ell y^2 \text{ of east-west walls}$$
$$= 2(15)(15)^2 = 6750 \text{ ft}^3$$

$$I_{yy} = \Sigma \ell y^2 \text{ of north-south walls}$$
$$= 40(130.9 - 75)^2 + 30(140 - 130.9)^2$$
$$\quad + 40(180 - 130.9)^2$$
$$= 223{,}909 \text{ ft}^3$$

$$I_p = 6750 + 223{,}909 = 230{,}659 \text{ ft}^3$$

Shear in north-south walls = $\dfrac{W\ell}{\Sigma \ell} + \dfrac{Tx\ell}{I_p}$

$$\text{Wall A} = \frac{40(40)}{110} + \frac{1236(130.9 - 75)(40)}{230{,}659}$$

$$= 14.5 + 12.0 = 26.5 \text{ kips}$$

Fig. 4.7.2 Effective width of walls perpendicular to shear walls

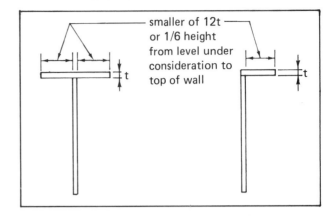

smaller of 12t or 1/6 height from level under consideration to top of wall

$$\text{Wall B} = \frac{40(30)}{110} + \frac{1236(-9.1)(30)}{230{,}659}$$

$$= 10.9 - 1.5 = 9.4 \text{ kips}$$

$$\text{Wall C} = \frac{40(40)}{110} + \frac{1236(-49.1)(40)}{230{,}659}$$

$$= 14.5 - 10.5 = 4.0 \text{ kips}$$

Shear in east-west walls = $\dfrac{Ty\ell}{I_p} = \dfrac{1236(15)(15)}{230{,}659}$

$$= 1.2 \text{ kips}$$

4.7.4 Precast Concrete Shear Walls

A shear wall need not consist of a single element. It can be composed of independent units such as double tee or hollow-core wall panels. If such units have adequate shear ties between them, they can be designed to act as a single unit, greatly increasing their shear resistance. Connecting such units can, however, result in a build-up of volume change forces, so it is usually desirable to connect only as many units as necessary to resist the overturning moment. Connecting as few units as necessary near mid-length of the wall will minimize the volume change restraint forces.

Connection of rectangular wall units to form "T" or "L" shaped walls increases their flexural rigidity, but does little to increase shear rigidity. The effective flange width that can be assumed for such walls is illustrated in Fig. 4.7.2. The designer should evaluate whether such connections are worth the extra cost. In some structures, such as the example in Section 4.7.6, it may be desirable to provide shear connections between the non load-bearing and load-bearing shear walls in order to increase the dead load resistance to the moments caused by lateral loads.

Fig. 4.7.3 Example of Section 4.7.5

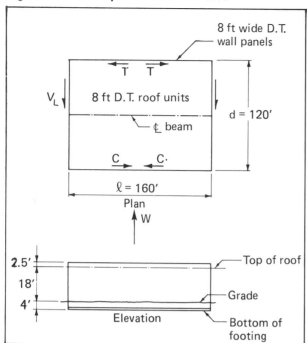

Plan

Elevation

Small openings in walls can usually be neglected. Walls with large openings are usually designed as independent walls separated by the opening, or as coupled walls, considering the stiffness of the connecting beam.

4.7.5 Example — One-Story Building

The wind load analysis and design of a typical one-story industrial type building are illustrated by the structure shown in Fig. 4.7.3. 8-ft wide double tees are used for both the roof and walls. The local building code specifies that a wind load of 25 psf be used for buildings of this height.

1. Calculate forces, reactions, shears and moments:

    Total wind force to roof:

    $W = 25$ psf x 160 ft x (18/2 + 2.5)

    $\quad = 46,000$ lb = 46 kips

    $V_L = V_R = 23$ kips

    Diaphragm moment $= \dfrac{W\ell}{8} = \dfrac{46 \times 160}{8}$

    $\quad\quad = 920$ ft-kips

2. Check sliding resistance of the shear wall:

    Determine dead load on the footing:

8DT12 wall $= 37$ psf x 23.5 x 120

$\quad\quad\quad = 104,340$ lb

12'' x 18'' footing $= 1 \times 1.5 \times 150$ pcf x 120

$\quad\quad\quad = 27,000$ lb

Assume 2 ft backfill x 100 pcf x 1.5 x 120

$\quad\quad\quad = \underline{36,000}$ lb

$\quad\quad$ Total $= 167,340$ lb

$\quad\quad = 167.34$ kips

Assume coefficient of friction against granular soil, $\mu_s = 0.5$

Sliding resistance $= \mu_s N = 0.5\,(167.34)$

$\quad\quad\quad = 83.67$ kips

Factor of safety $= 83.67/23 = 3.64$ OK
(Note: A factor of 1.5 is specified by many building codes.)

3. Check overturning resistance:

    Applied overturning moment $= 23\,(4 + 18)$

    $\quad\quad\quad = 506.0$ ft-kips

    Resistance to overturning:
    Assume axis of rotation at leeward edge of the building.

    (Note: This is as recommended by the National Building Code, 1976. Other model building codes do not specify the axis of rotation. Some engineers prefer to use the more conservative assumption of an axis at d/5, d/4 or d/3 from the leeward edge, depending on the foundation conditions.)

    Resisting moment $= 167.34\,(120/2)$
    $\quad\quad\quad = 10,040$ ft-kips
    Factor of safety $= 10,040/506$
    $\quad\quad\quad = 19.8 > 1.5$ OK

4. Analyze connections:

    a. Shear ties in double tee roof joint:
       (Maximum load at next to last joint)
       Applied shear $= (80 - 8)/80\,(23)$

       $\quad\quad\quad = 20.7$ kips

       Load factor by ACI 318-77 $= 1.3$

       Connection load factor (see Sect. 5.3)
       $\quad\quad = 1.3$

V_u = 20.7 x 1.3 x 1.3 = 35.0 kips

v_u = 35.0/120 = 0.292 kips/ft

Use # 4 ties as shown in Fig. 4.5.2

ϕV_n = 10.2 kips

Required spacing = 10.2/0.292 = 34.9 ft

(Note: Most engineers and precasters prefer a maximum connection spacing of about 8 to 10 ft.)

b. Shear ties at the shear walls:

V_u = 23 x 1.3 x 1.3 = 38.9 kips

v_u = 38.9/120 = 0.324 kips/ft

A connection as shown in Fig. 4.5.2 is designed similar to the shear tie between tees. This would require a spacing of 10.2 / 0.324 = 31.5 ft. In order to distribute the load to the wall panels, at least one connection per panel is required. From Fig. 4.7.4a it is apparent that these connections should occur at the tee stems. Thus a spacing of 4 ft or 8 ft would be used in this case.

Other types of connections using short welded headed studs are commonly used for this application. Design of studs is shown in Part 5 of this Handbook.

In some cases, the designer may find it necessary to provide a connection that permits vertical movement of the roof member. Such a connection is illustrated in Fig. 4.7.4b.

c. Chord force (see Fig. 4.7.5)

T = $C = M / d = 920 / 120$

  = 7.67 kips

T_u = 1.3 x 7.67 = 9.97 kips

This force can be transmitted between members by ties at the roof tees, wall panels or a combination, as illustrated in Fig. 4.7.5. The force through the member flanges can be transmitted by concrete tension, using reasonable assumptions of effective areas and tensile strength of the concrete. A conservative value of $3 \phi\sqrt{f'_c}$ is suggested for the tensile strength of the concrete.

For higher forces, or where more ductility is required, reinforcing bars can be placed in the flanges. Design procedures are shown in Section 4.8.

Fig. 4.7.4 Connection of roof tee to wall

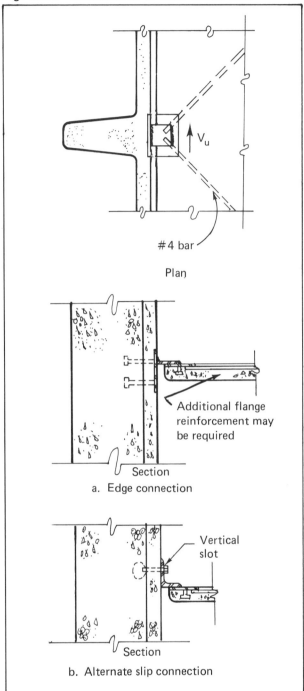

Plan

a. Edge connection

Additional flange reinforcement may be required

b. Alternate slip connection

Vertical slot

4 bar

d. Wall panel connections:

This shear wall may be designed to act as a series of independent units, without ties between the panels. The shear force is assumed to be distributed equally among the wall panels. (See Fig. 4.7.6.)

n = 120/8 = 15 panels

V = V_R /n = 23/15 = 1.53 kips

Fig. 4.7.5 Chord forces

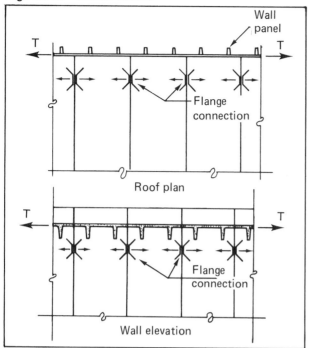

Roof plan

Wall elevation

Fig. 4.7.6 Panels acting as individual units in a shear wall

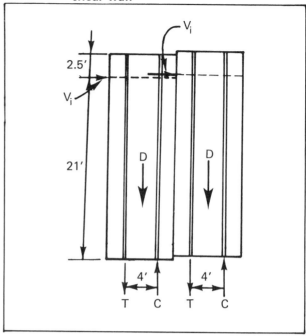

D = 37 psf x 8 x 23.5 = 6956 lb = 6.96 kips

Design base connection for 1.3W – 0.9D

T_u = (1.3 x 1.53 x 21 – 0.9 x 6.96 x 2) / 4
= 7.31 kips tension

As an alternative, the shear wall may be designed with 2 or more panels connected together, as described in Section 4.7.4. The following illustrates the connection design if all panels are connected (see Fig. 4.7.7).

Shear ties between panels:

v_r = V_R /d = 23/120 = 0.192 klf

The unit shear stress, v_r, is equal on all sides of the panel (See Section 4.8.6).

v_u = 0.192 x 1.3 x 1.3 = 0.324 klf

Using the same shear ties as on the roof:
maximum spacing = 10.2/0.324 = 31.5 ft
(Note: Most designers prefer a minimum of 2 or 3 ties per panel, at a maximum spacing of 8 to 10 ft.)

Check for tension using factored loads:

By ACI 318-77, the required load factor equation to use under this condition is 1.3W – 0.9D. The tensile stress would be:

f_t = $\dfrac{M}{Z}$ – $\dfrac{P}{A}$

Z = ℓ^2 /6 = 120^2 /6 = 2400 ft²

Fig. 4.7.7 Panels connected together as a monolithic shear wall

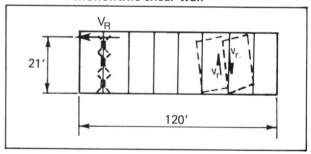

A = ℓ = 120 ft

M = $V_R h_s$ = 23 x 21 = 483 ft-kips

P = D.L. of wall = 104.34 kips

f_{ut} = $\dfrac{1.3M}{Z}$ – $\dfrac{0.9P}{A}$

= $\dfrac{1.3\ (483)}{2400}$ – $\dfrac{0.9\ (104.34)}{120}$

= –0.521 klf (compression)

Thus no tension connections are required.

4.7.6 Example — Four-Story Building

The wind load analysis and design of a typical four-story residential building are illustrated by the structure shown in Fig. 4.7.8. 8-in. deep hollow-core units are used for the floors and roof, and 8-in. thick precast concrete walls are used for

Fig. 4.7.8 Four-story design example - Section 4.7.6

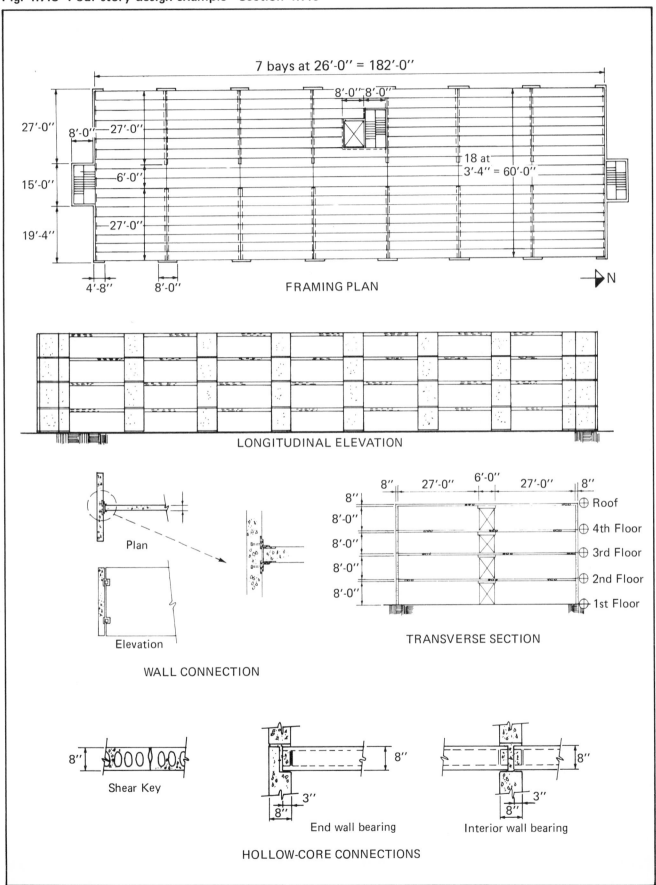

7 bays at 26'-0'' = 182'-0''

27'-0''

8'-0'' 27'-0''

15'-0''

6'-0''

8'-0'' 8'-0''

18 at
3'-4'' = 60'-0''

19'-4''

27'-0''

4'-8'' 8'-0''

FRAMING PLAN

N

LONGITUDINAL ELEVATION

Plan

Elevation

WALL CONNECTION

8'' 27'-0'' 6'-0'' 27'-0'' 8''

8''

8'' ⊕ Roof

8'-0'' ⊕ 4th Floor

8'-0'' ⊕ 3rd Floor

8'-0'' ⊕ 2nd Floor

8'-0'' ⊕ 1st Floor

TRANSVERSE SECTION

8''
Shear Key

8''
End wall bearing
3''
8''

8''
Interior wall bearing
3''
8''

HOLLOW-CORE CONNECTIONS

Fig. 4.7.9 Loads to transverse walls — 4-story design example

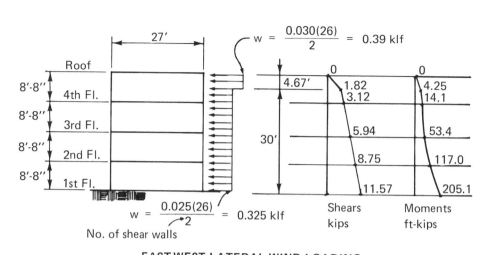

EAST-WEST LATERAL WIND LOADING

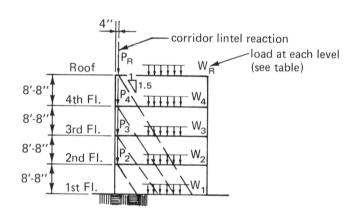

GRAVITY LOADS ON BEARING WALL

Summary of gravity loads

Load Mark	Tributary Area	Unit loads, psf		Wall weight, klf	Total unfactored loads		
		L.L.	D.L.		L.L.	D.L.	T.L.
P_R	78 sq. ft.	30	74	—	2.3 kips	5.8 kips	8.1 kips
P_4	78 sq. ft.	100	64	—	7.8 kips	5.0 kips	12.8 kips
P_3	78 sq. ft.	100	64	—	7.8 kips	5.0 kips	12.8 kips
P_2	78 sq. ft.	100	64	—	7.8 kips	5.0 kips	12.8 kips
W_R	26 lin. ft.	30	74	—	0.78 klf	1.92 klf	2.70 klf
W_4	26 lin. ft.	16*	74	0.8	0.42 klf	2.72 klf	3.14 klf
W_3	26 lin. ft.	16*	74	0.8	0.42 klf	2.72 klf	3.14 klf
W_2	26 lin. ft.	16*	74	0.8	0.42 klf	2.72 klf	3.14 klf
W_1	N / A	—	—	0.8	0	0.80 klf	0.80 klf

* Includes live load reduction allowed by codes

PCI Design Handbook

all walls shown. Unfactored loads are given as follows:

1. Gravity loads:

	L.L.	D.L.
Roof	30	
Roofing, mechanical, etc.		10
Hollow-core slabs		64
	30 psf	74 psf
Typical floor		
Living areas	40	
Corridors & stairs	100	
Partitions		10
Hollow-core slabs		64
		74 psf
Walls		100 psf
Stairs	100	130 psf

2. Wind loads:

0 to 30 ft above grade = 25 psf
30 to 34'8" above grade = 30 psf

For wind in the transverse (east-west) direction, normal practice for this structure would be to conservatively neglect the resistance provided by the stair, elevator and longitudinal walls. Thus, two 27 ft long interior bearing walls can be assumed to resist the wind on one 26 ft bay. The wind and gravity loads on the wall are shown in Fig. 4.7.9.

Concentrated loads from the corridor lintels can be assumed to be distributed as shown in Fig. 4.7.9. In this example, these loads are conservatively neglected to simplify the calculations.

Check overturning of shear wall:

D.L. resisting moment about toe of wall

$= 27 (27/2) [1.92 + 3(2.72) + 0.8]$

$= 3966$ ft-kips

Factor of safety $= 3966/205.1$

$= 19.3 > 1.5$ OK

Check for tension using factored loads:
Dead weight on wall

$= P = [1.92 + 3(2.72) + 0.8] 27$

$= 293.8$ kips

Maximum moment at foundation

$= 205.1$ ft-kips

$$f_{ut} = \frac{1.3M}{(\ell^2/6)} - \frac{0.9 (293.8)}{27}$$

$$= \frac{1.3 (205.1)}{(27^2/6)} - \frac{0.9 (326.2)}{27}$$

$$= -7.60 \text{ klf (compression)}$$

No tension connections required between panels and the foundation.* Thus the building is stable under wind loads in east-west direction.

For wind in the longitudinal (north-south) direction, the shear walls will be connected to the load bearing walls. The assumed resisting elements are shown in Fig. 4.7.10 and a summary of the properties is shown in Table 4.7.1. Sample calculations of these properties are given below for element A.

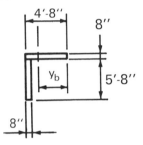

Effective width of perpendicular wall (see Fig. 4.7.2) is the smaller of $12t = 12(8) = 96$ in. or $1/6(34.67 \times 12) = 69.3$ in. — use 5'-8".

Area of web $= 4.67 \times 0.67 = 3.11$ ft^2
Area of flange $= 5.67 \times 0.67 = \underline{3.78}$
                                    6.89 ft^2

$$y_b = \frac{3.11 (4.67/2) + 3.78 (4.67 - 0.33)}{6.89}$$

$= 3.43$ ft

$y_t = 4.67 - 3.43 = 1.24$ ft

$$I = \frac{0.67 (4.67)^3}{12} + 3.11 (3.43 - 2.33)^2$$
$$+ 3.78 (1.24 - 0.33)^2$$
$= 12.58$ ft^4

$$K_s = \frac{GA}{1.2h_s} = \frac{0.4 \, EA_w}{1.2h_s} = \frac{EA_w}{3h_s}$$

$$K_f = \frac{12EI}{h_s^3}$$

$$\frac{1}{K_s} + \frac{1}{K_f} = \frac{3h_s}{EA_w} + \frac{h_s^3}{12 \, EI}$$

*Note: Other design considerations may dictate the use of minimum vertical ties. See "Considerations for the Design of Precast Concrete Bearing Wall Buildings to Withstand Abnormal Loadings," *PCI Journal*, March-April, 1976.

Fig. 4.7.10 Wind resisting elements for north-south wind

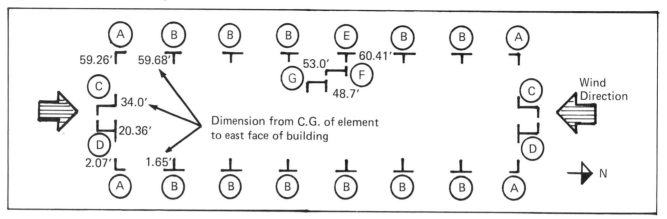

$$= \frac{h_s}{E} \left[\frac{3}{A_w} + \frac{h_s^2}{12 \, I} \right] \qquad h_s = 8.0 \text{ ft}^*$$

The relative stiffness coefficient for this problem is:

$$\frac{1}{K_r} = \frac{3}{A_w} + \frac{5.33}{I}$$

For element A:

$$\frac{1}{K_r} = \frac{3}{3.11} + \frac{5.33}{12.58} = 1.39$$

$$K_r = \frac{1}{1.39} = 0.72$$

% Distribution to element A (see Table 4.7.1)

$$= \frac{0.72 \, (100)}{29.90} = 2.41\%$$

The shears and moments in the north-south direction are shown in Fig. 4.7.11, and the distributions are shown in Table 4.7.2.

To check overturning, consider element B at the first floor. From Fig. 4.7.9 the dead load on the 6'-4'' portion of element B is 1.92 + 3 (2.72) + 0.8 = 10.88 kips/ft. The dead load on the 8'-0'' portion of element B is the weight of the wall = 34.67 x 0.1 = 3.47 kips/ft. The resisting moment is then:

$$M_R = 10.88 \, (5.67) \, (4) + 3.47 \, (8) \, (4)$$
$$= 358 \text{ ft-kips} \times 11 \text{ elements}$$
$$= 3938 \text{ ft-kips}$$

Factor of safety = 3938/966.9 = 4.1 > 1.5 OK

(Note: This conservatively neglects the contribution of the other elements.)

To check for tension, also consider element B:

Total dead weight on the wall
$$= 10.88 \, (5.67) + 3.47 \, (8)$$
$$= 89.45 \text{ kips}$$

Total wall area
$$= (8.0 + 5.67) \, 0.67$$
$$= 9.16 \text{ ft}^2$$

$$M = 43.3 \text{ ft-kips (see Table 4.7.2)}$$

$$f_{ut} = \frac{1.3M \, (d/2)}{I} - \frac{0.9P}{A}$$

$$= \frac{1.3 \, (43.3) \, (4.0)}{28.7} - \frac{0.9 \, (89.45)}{9.16}$$

$$= -0.94 \text{ ksf (compression)}$$

No tension connections are required between panels and the foundation. Thus the building is stable under wind loads in the north-south direction.

The connections required to assure that the elements will act in a composite manner as assumed can be designed by considering element A. The unit stress at the interface is determined using the classic equation for horizontal shear:

$$v_h = \frac{VQ}{I}$$

Fig. 4.7.11 Wind load in north-south direction

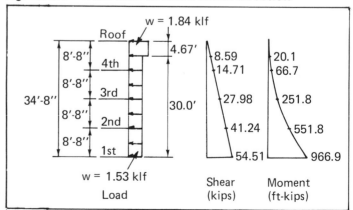

---

*Either the clear height or the story height can be used to calculate the relative stiffness.

PCI Design Handbook

Table 4.7.1 Properties of resisting elements for wind in longitudinal direction

Element	A_w	I	y_b	K_r	No. of elements	nK_r	$\frac{K_r}{\Sigma nK_r}(100)$	$\Sigma\bar{y}$	$K_r(\Sigma\bar{y})$
A	3.11	12.6	3.43	0.72	4	2.88	2.41	123	89
B	5.36	28.7	4.0	1.34	11	14.74	4.48	308	413
C	5.81	158.1	4.34	1.82	2	3.64	6.09	68	124
D	5.81	205.6	3.45	1.84	2	3.68	6.15	41	75
E	5.36	29.0	4.0	1.35	1	1.35	4.52	60	81
F	5.81	114.1	2.72	1.78	1	1.78	5.95	53	94
G	5.81	171.6	4.09	1.83	1	1.83	6.12	49	90
						$\Sigma nK_r = 29.90$			$\Sigma = 966$

Center of rigidity = 966 / 29.90 = 32.31 ft from east
Note: The north-south wind load is slightly eccentric by 32.31 − 61.33/2 = 1.65 ft.
Torsion due to this eccentricity is neglected in calculating shears and moments in Table 4.7.2.

$Q = 5.67 (0.67) (1.24 - 0.33) = 3.46$ ft^3

$v_h = \dfrac{VQ}{I} = \dfrac{1.31 \times 3.46}{12.6} = 0.36$ kips/ft

Total shear = 0.36 x 8.0 = 2.88 kips

Connections similar to those shown in Fig. 4.7.8 can be designed using the principles outlined in Part 5 of this Handbook.

Design of floor diaphragm:

Analysis procedures for the floor diaphragm are described in Section 4.5. For this example refer to Fig. 4.7.12:

The factored wind load for a typical floor is:

$W_u = 1.3 \times 25$ psf x 8.67 ft = 282 plf

For wind from the east or west:

$V_{Ru} = \dfrac{0.282 \times 26}{2} = 3.67$ kips

Table 4.7.2   Distribution of wind shears and moments (north-south direction)

Element	% Dist.	4th floor		3rd floor		2nd floor		1st floor	
		Shear	Moment	Shear	Moment	Shear	Moment	Shear	Moment
		14.71 kips	66.7 ft-kips	27.98 kips	251.7 ft-kips	41.24 kips	551.8 ft-kips	54.51 kips	966.9 ft-kips
A	2.41	0.35	1.6	0.67	6.1	0.99	13.3	1.31	23.3
B	4.48	0.66	3.0	1.25	11.3	1.85	24.7	2.44	43.3
C	6.09	0.90	4.1	1.70	15.3	2.51	33.6	3.32	58.9
D	6.15	0.90	4.1	1.72	15.5	2.54	33.9	3.35	59.5
E	4.52	0.66	3.0	1.26	11.4	1.86	24.9	2.46	43.7
F	5.95	0.88	4.0	1.66	15.0	2.45	32.8	3.24	57.5
G	6.12	0.90	4.1	1.71	15.4	2.52	33.8	3.34	59.2

The relative stiffness and percent distribution for the elements in this table are assumed the same for all stories. The exact values may be slightly different for each story because the values change due to the reduced flange width (see Fig. 4.7.2).

$$C_u = T_u = \frac{M_u}{\ell} = \frac{0.282\,(26)^2}{8\,(56.67)}$$

$$= 0.42 \text{ kips}$$

The reaction V_{Ru} is transferred to the shear wall by static friction:

Dead load of floor $= 26/2\,(64 + 10) \times 60$
$$= 57,720$$
Dead load of wall $= 800/2(54)$ $= \underline{21,600}$
$$79,320$$

Static coefficient of friction from Table 5.5.1 (hardboard to concrete) = 0.5. Reduce by factor of 5 as recommended in Sect. 5.5.

$\mu = 0.5/5 = 0.10$

Resisting force = 0.10 (79.3)
$\qquad = 7.93 > 3.67$ OK

The chord tension, T_u, is resisted by the tensile strength of the floor slab. The grout key between slabs must also resist approximately the same force.

Assume area of exterior slab = 218 in.^{2}, and the grout key is 3 in. deep. Concrete $f'_c = 5000$ psi. Use a resisting tensile strength of $3\phi \sqrt{f'_c} = 191$ psi. Grout key resisting strength is 40 psi (see Sect. 4.5.2).

Resisting tensile strength of slab
$\qquad = 218\,(0.191) = 41.6 \text{ kips} > 0.67$ OK

Resisting strength of grout key
$\qquad = 26\,(12)\,(3)\,(0.040)$
$\qquad = 37.4 \text{ kips} > 0.42$ OK

For wind from the north or south:

$$V_{Ru} = \frac{0.282\,(61.33)}{2} = 8.65 \text{ kips}$$

Resisting force in the first joint
$\qquad = 181.33\,(12)\,(3)\,(0.04) = 261 \text{ kips}$ OK

$$C_u = T_u = \frac{0.282\,(61.33)^2}{8\,(181.33)}$$

$$= 0.73 \text{ kips} < 7.93 \text{ OK}$$

In this example, only the resistance to wind loading was analyzed. Any other required loading (including "abnormal" loads) must be reviewed for a complete analysis.

Fig. 4.7.12   Diaphragm analysis

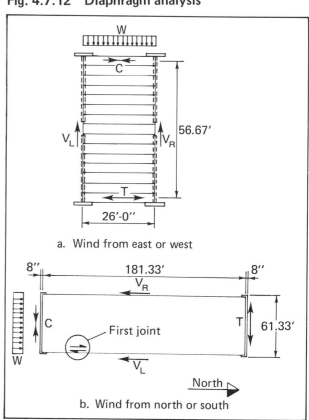

a. Wind from east or west

b. Wind from north or south

4.8 Earthquake Analysis

4.8.1 Notation

A_ℓ = Cross sectional area in linear measure

A_s = Area of non-prestressed reinforcement

A_{vf} = Area of shear-friction reinforcement

b = Width of panel (Fig. 4.8.6)

C = Coefficient for base shear (UBC-76); total compressive force

C_p = Coefficient for horizontal force (UBC-76)

C_o = Compressive chord force (Fig. 4.8.2)

C_1 = Overturning couple force

C_u = Factored compressive force

D = Dimension of building in direction parallel to applied lateral force (UBC-76)

D_s = Plan dimension of the vertical lateral force resisting system (UBC-76)

d = Dimension of building; distance from extreme compression fiber to centroid of tension reinforcement

F_p = Lateral force on the part of the structure and in the direction under consideration (UBC-76)

F_t = That portion of V considered concentrated at the top of the structure, level n (UBC-76)

F_x = Forces in x direction; force at level x

F_y = Forces in y direction

f'_c = Concrete compressive strength

f_s = Stress in steel

f_y = Yield strength of non-prestressed reinforcement

H = Horizontal force needed to overcome friction

h = Height of member

h_i, h_n, h_x = Height above base level to level "i", "n", or "x", respectively

I = Occupancy importance factor (UBC-76)

K = Coefficient relating to type of construction (UBC-76)

ℓ = Length of building or member

M = Moment

M_1 = Overturning moment

M_R = Overturning moment resistance

m_1, m_2 = Dimensions (Fig. 4.8.6)

N = Force normal to friction plane

n = Uppermost level in the structure

P_1, P_2, P_3, P_4 = Forces (Fig. 4.8.6)

R_o, R_1 = Reactions (Fig. 4.8.2)

R_S = Resistance to sliding

S = Coefficient for site-structure resonance (UBC-76)

s = Spacing of weld clips

T = Fundamental period of vibration of the building in the direction under consideration (UBC-76); total tensile capacity or force

T_o = Tensile chord force (Fig. 4.8.2)

T_1 = Overturning couple force

T_S = Characteristic site period (UBC-76)

T_u = Factored tensile force

V = Total lateral load or shear at the base (UBC-76); shear force

V_{Ru} = Design shear strength

v = Unit shear stress

v_o, v_1, v_2, v_3 = Unit shear stresses (Fig. 4.8.6)

v_{ru} = Design unit shear strength

v_u = Factored shear stress

W = Total dead load of building (UBC-76)

W_p = Total weight of a part or portion of a structure (UBC-76)

w_i, w_x = That portion of "W" which is located at or is assigned to level "i" or "x" respectively

Z = Coefficient dependent upon the seismic zone (UBC-76)

Z_ℓ = Section modulus in linear measure

ϕ = Capacity reduction factor (ACI-318-77)

μ = Shear-friction coefficient

μ_s = Static coefficient of friction

4.8.2 General

Earthquake design is unlike gravity and wind load design, in that the seismic loads which may reasonably be expected during the life of the structure are much larger than the design loads given in the building codes. It is expected that a structure will suffer damage under these overloads during a severe earthquake. To protect life and property, it is important that this damage not cause collapse. This is accomplished by designing members and connections which can deform inelastically without fracture.

Load tests of prestressed concrete members

have consistently shown that large deflections occur as the design strength is approached. Because of prestress, the transition from linear to nonlinear response is gradual and smooth. Cyclic load tests have shown that prestressed concrete beams can undergo several cycles of intense load reversals and still maintain their design strengths.

Prestressed concrete structures can be designed to withstand the effects of earthquakes in accordance with the requirements of building codes. For some types of buildings, box-type structures offer an economical solution. In other cases, ductile moment-resisting frames can be used.

This section deals with design of precast concrete buildings located in Seismic Zones 2, 3 and 4 as affected by the requirements for resistance to seismic forces. Mainly, this discussion deals with the connections necessary to establish that the structure has adequate resistance to lateral loads, and that the concrete dimensions are sufficient to receive and transmit the forces funneled through the connections.

4.8.3 Building Code Requirements

In the example problems that follow, the seismic requirements of the 1976 Uniform Building Code (referred to herein as UBC-76) are used, except as noted.

The response of a structure to the ground motion of an earthquake depends on the structural system with its damping characteristics, and on the distribution of its mass. With mathematical idealization a designer can determine the probable response of the structure to an imposed earthquake. UBC-76 requires a dynamic analysis for structures which have highly irregular shapes or framing systems and allows it for other structures. However, most buildings have structural systems and shapes which are more or less regular, and many designers use the equivalent static load method for these structures. Calculations must be supplemented with engineering judgment.

In its simplest form, UBC-76 requires that a total base shear, V, be applied to the building in any horizontal direction, where

$$V = ZIKCSW$$

in which

Z is based on the expected earthquake intensity

I is based on the type of occupancy anticipated

K is based on the type of framing

C is based on the flexibility of the structure

S is based on the relative natural vibration periods of the site and the structure

W is the dead load of the structure

This total shear is divided among the story levels, with the upper stories being assigned more of the horizontal load than the lower stories. This is the method of *equivalent static loads.*

Also, UBC-76 requires that parts of the structure such as roofs, floors, walls, and their connections be designed locally for either the distributed base shear or the lateral forces determined by the expression:

$$F_p = ZIC_pSW_p$$

in which the subscript p denotes the effects of the building part. For certain parts of buildings, this latter requirement is sometimes more severe than the base shear.

The occupancy importance factor, I, has the effect of making structures that would be essential during an earthquake disaster (i.e., hospitals, fire stations, etc.), or those that could house large numbers of people less likely to be severely damaged.

The coefficient for site-structure resonance, S, is a function of the ratio of the fundamental period of vibration of the building, T, and the characteristic site period, T_s. It can be calculated by equations given in UBC-76 or, if T_s is not known, the value of S is taken as 1.5.

4.8.4 Concept of Box-Type Buildings

A box-type building consists of a roof, floor diaphragms, and shear walls. These are connected wherever they meet, and if the arrangement is suitable, the structure is resistant to lateral loads.

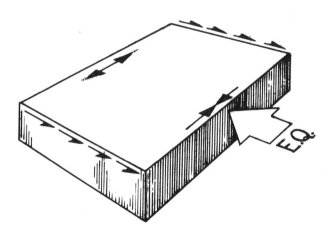

Since an earthquake is a ground motion reacting with the inertia of the building and its parts, the equivalent static loads are applied at the centroids of the parts. The internal forces that link the applied loads and the ground reactions follow the stiffest paths consistent with equilibrium and compatibility of deflection. With box-type buildings of moderate height-to-width ratio, the stiffnesses of diaphragms and walls in their own planes greatly exceed other resistances. The load paths are then along diaphragms and walls rather than through moment-resisting frames.

In multi-story box-type buildings, the equivalent static loads find resistances in the several diaphragms and walls. If a load originates on a floor, the load path is through that floor to adjacent connected walls, and downward to the footings. After the load path has entered a wall, it does not leave that wall unless the wall is discontinued or its stiffness is reduced substantially. The designer should try to arrange the path of resistance to be direct.

A diaphragm made up of precast elements requires a special effort to be assured that it is strong enough in shear and moment. Walls may be full of windows. The designer must judge if the wall should be considered a shear wall or a moment resisting frame depending on the number and size of windows and doors. The UBC-76 requirements for a box-type structure are severe, that is, they call for a short period of vibration and high values for the coefficients "C" (≈ 0.1) and "K" ($=1.33$).

4.8.5 Structural Layout and Connections

A box-type structure may have a large number of precast concrete elements that are assembled into walls, floors, roof, and maybe frames. Proper connections between the many pieces create the diaphragms and shear walls, and the connections between these, in turn, create the box-type structure. In the seismic design of a box-type structure, there are two fundamental and different requirements of the two groups of connections:

a) First, one group of connections that transmit forces between elements within a horizontal diaphragm or a shear wall, and

b) Second, another group of connections that transmit forces between a horizontal diaphragm and a shear wall.

In seismic design, several demands are placed on the connections. Tension must be transmitted directly from steel to steel, and not from steel through concrete to steel. Compression must be transmitted either from steel to steel or from concrete to concrete — not through elastomeric pads. Concrete dimensions must be ample, so that the hardware of the connection is confined, and the connection thus can transmit accidental forces that are normal to the usual plane of the load path. Finally, a connection should be such that if

it were to yield, it will do so in a ductile manner, i.e., without loss of load-carrying capacity when the concrete cracks.

4.8.6 Example — One-Story Building

4.8.6.1 General

By taking advantage of walls already present, one-story buildings usually can be designed to resist lateral loads (wind or earthquake) by shear-wall and diaphragm action. If a shear-wall and diaphragm concept is feasible, it is generally the most economical concept. This section of the Handbook is intended to assist a designer with the shear-wall/diaphragm concept for precast prestressed concrete buildings.

To show a fairly complete design, the simple one-story example building in Fig. 4.8.1 will be illustrated. It is 128 ft x 160 ft in plan, and has 16-ft clear height inside. It is entirely precast above the floor, using 16-in. double tees for the walls and 24-in. double tees for the roof. The double tees are 8-ft wide, with stems 4 ft on centers. The flanges on the roof tees are 2-in. thick, and on the wall tees, 4-in. thick.

Because all loads must funnel through the connections, gravity and lateral loads must be considered together. Thus, this example shows both gravity and lateral load connections. The example emphasizes both free bodies and the concept of load path. Most serious design errors are ones of concept caused by not drawing enough free-body diagrams, and failure to provide a continuous load path.

4.8.6.2 Load Analysis

Earthquakes impose lateral and vertical ground motions upon a structure. The structure responds to these motions with its own deflections. These deflections are accompanied by corresponding strains and the resulting stresses. However, most designers are used to thinking of stress as caused by load rather than as caused by deflection. Consequently, all the common methods of design use a set of static lateral loads intended to be equivalent to (i.e., produce the same stresses as) the real, dynamic loads caused by deflection. One such method is used in UBC-76. The example building is analyzed here for N-S earthquake only. In a real design situation, it would also have to be analyzed for E-W earthquake.

The example building resists lateral load by diaphragm and shear-wall action. Inertia loads (mass x acceleration) are delivered to the roof diaphragm. The diaphragm acts like a plate girder laid flat, spanning between the shear walls. In de-

Fig. 10.1 One-story example building

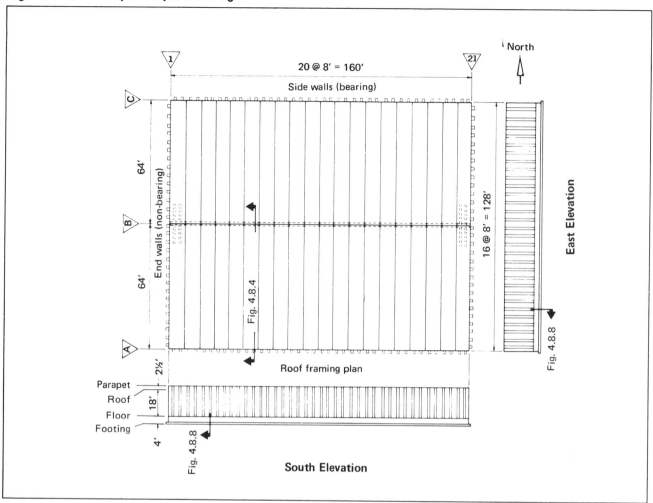

termining the equivalent static loads on this diaphragm, the mass tributary to the diaphragm must be determined. This is done in terms of dead weight, W, as follows:

N and S Walls:

half ht + parapet	= 11.5 ft
total length = 2 x 160	= 320 ft
weight = 11.5 x 320 x 0.080	= 295 kips

Roof:

weight = 128 x 160 x 0.075	= 1535 kips
Total, W	= 1830 kips

The equivalent lateral load, V, is computed by multiplying W by an acceleration. The acceleration is determined by multiplying together the five constants, ZIKCS. The "Z" factor, denoting geographical zones of equal probability of serious earthquake, is taken here as 0.75, representing the second-most earthquake prone areas, Zone 3. The occupancy importance factor, I, is assumed as 1.0 and the "K" factor, used to indicate the performance certain types of framing have shown in

actual earthquakes is taken as 1.33. Since the characteristic site period, T_s, is assumed unknown in this example, the value of "S" is taken as 1.5. The "C" factor is used to allow for the fact that flexible long-period (low natural frequency) buildings tend to absorb ground motion with less damage than do stiff buildings. The example building is very stiff and rates the highest "C" value of 0.12, however, the product of CS need not exceed 0.14. The coefficient used in design is thus:

ZIKCS = 0.75 x 1.0 x 1.33 x 0.14 = 0.14

The shear on the diaphragm due to earthquake is:

V = ZIKCS x W = 0.14 x 1830 = 256 kips

In comparison, the shear caused by a 20 psf wind load is:

V = 11.5 x 160 x 0.020 = 37 kips

This building is definitely governed by earthquake rather than by wind. A word of caution is in order. The base shear due to earthquake as calculated above is a "service" load. Accelerations in real earthquakes may be many times the "ZIKCS" shown above. This will cause elements of the structure to be stressed to the yield point and beyond in a severe earthquake. The structure must have deformation capability to absorb overloads of short duration without failure.

To guard against accidental horizontal torsion effects, UBC-76 requires that a minimum 5 percent eccentricity be assumed between centroid of resistance and center of mass as shown in Fig. 4.8.2. With this in mind, the forces internal to the roof diaphragm are:

Max shear reaction, $R_o = 0.55V = 0.55 \times 256$ = 141 kips

Max. shear intensity, $v_o = 141/128 = 1.10$ klf

Max. bending moment, $M_o = V\ell/8$ = 256(160/8) = 5120 ft-kips

Max. chord forces, $C_o = T_o = M_o/d = 5120/128$ = 40.0 kips

The shear forces are analogous to those in the web of a plate girder. The chord forces are analogous to those in the flanges of a plate girder.

Considering the roof as a free body, Fig. 4.8.2 shows that equilibrium is maintained by the reactions R_o from the tops of the shear walls. One must also consider the shear wall as a free body. Fig. 4.8.2 shows that the wall can be in equilibrium only if sufficient sliding resistance and overturning capacity are provided. The forces acting which must be resisted by the shear wall are:

Sliding force, $R_o = 141$ kips $= R_1$

Overturning moment $M_1 = R_1 h = 141 \times 22$ = 3102 ft-kips

Overturning couple $C_1 = T_1 = M_1/\ell$ = 3102/128 = 24.2 kips

The weight of the end wall, tributary roof, floor, backfill, etc., is about $N = 410$ kips. The available resistance to overturning, then, is:

$M_R = N(d/2) = 410 (128/2) = 26,240$ kip-ft

The sliding force is resisted by friction in the bottom of the wall footing. Assuming a granular soil, the coefficient of sliding friction is about $\mu_s N = 0.5 \times 410 = 205$ kips.

The factors of safety (load factors) for overturning and sliding are seen to be sufficient.

$M_R/M_1 = 26240/3102 = 8.46$

$R_s/R_o = 205/141 = 1.45$

The designer must be careful to follow the loads all the way down into the ground. This is the "load-path" concept. The designer *must* provide for a complete and continuous load path.

Fig. 4.8.2 Forces acting on roof and walls.

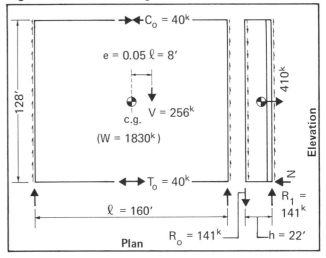

4.8.6.3 Strength Analysis

In this example, strengths of concrete components are analyzed using the load factors of ACI 318-77, and strength of structural steel parts by working-stress methods. Where welded, rebar is A615, Grade 40; otherwise Grade 60. (See Part 5 for guides in welding reinforcing bars.)

Following the load path, the diaphragm is first analyzed for shear. The applied design shear in the double-tee flanges is:

$v_u = 1.4 v_o = 1.4 \times 1.10 = 1.54$ klf

(The load factor, 1.4, is derived from Sects. 9.2.2 and 9.2.3 of ACI 318-77. From Eq. 9-2, $U = 0.75 (1.7 \times 1.1E) = 1.4E$, and from Eq. 9-3, $U = 1.3 \times 1.1E \approx 1.4E$. Also, UBC-76 stipulates a load factor of 1.4.)

The design shear strength of the reinforced concrete in the double-tee flanges is:

$v_{ru} = (2 \phi \sqrt{f'_c}) t = (2 \times 0.85 \sqrt{6000}) \times 2$ = 263 lb/in. = 3.16 klf

This is greater than required, therefore satisfactory.

The double tee flanges must be connected at their edges to each other and to the end shear walls. This is analogous to providing shear strength along vertical joints in the web of a plate girder, and is done by weld clips as shown in Figs. 4.8.3 (a), (b), and (c). This clip is analyzed by truss analogy, illustrated in Fig. 4.8.3 (d). The design forces in the rebars, and their resultant along the double tee edge, are:

$C_u = T_u = \phi f_y A_s = 0.9 \times 40 \times 0.31 = 11.2$ kips, and

$V_{Ru} = (C_u + T_u) \cos 45° = (11.2 + 11.2) \times 0.707$ = 15.8 kips

Hence at the junction with the end walls, the

spacing between clips (as limited by horizontal or diaphragm shear) must be no more than:

s = V_{Ru}/v_u = 15.8/1.54 = 10.3 ft, say 10 ft

Ratioing by distance from the center of the roof, the shear between the first and second double tee and the corresponding spacing between clips, is:

v_u = (72/80) x 1.54 = 1.4 klf and

s = 15.8/1.4 = 11.3 ft; say 10 ft on centers

Ten feet is a reasonable maximum spacing. However, in earthquake, the double-tee next to the end wall is subject to vertical bouncing. This could destroy the essential connection of diaphragm to shear wall, unless the connection is strengthened sufficiently to yield the double-tee flange as a cantilever in bending where it joins the first interior stem. The flanges of the double tees used here are reinforced with 12 x 6-W1.4 x W2.5 welded wire fabric (A_s = 0.050 in.^{2}/ft), 1/2 in. clear top. If standard weld clips are placed at 4 ft on centers (2 per wall panel), they will easily force uniform yield in the flange. Hence, the clips are spaced:

s = 10 ft on centers typically, and

s = 4 ft on centers at end walls

The double tee flanges must also be connected to the north and south walls, in order to transfer the "VQ/Ib" type web shears to the chords. This is analogous to the connection between web and flange of a plate girder. This same connection must also function as a tension tie, by holding the wall panels onto the roof against wind and earthquake, and by holding the building together against shrinkage and related forces.

In holding the wall panel onto the roof, the connection must be designed for an earthquake force, F_p, which is obtained by multiplying the tributary weight of this part of the building, W_p, by an acceleration, ZIC_pS. Z is the same zone factor as before. C_p is a factor used to correlate for past good and bad experience in earthquakes. The UBC-76 requires a C_p of 2.00 for connections of wall panels, but also states that if C_p is 1.00 or more, the values of I and S need not exceed 1.0. Hence, the design acceleration is:

ZIC_pS = 1.0 x 1.0 x 2.00 x 1.0 = 2.00

The tributary weight of an 8-ft wide wall panel is:

W_p = 11.5 (half ht + parapet) x 8 x 0.080
   = 7.4 kips

The corresponding lateral earthquake force is:

F_p = $ZIC_p SW_p$ = 2.00 x 7.4 = 14.8 kips, or 7.4 kips per double tee stem

Fig. 4.8.3 Connections between flanges of roof double tees.

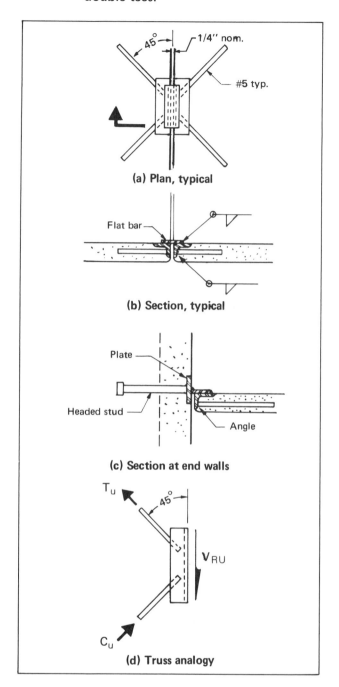

(a) Plan, typical

(b) Section, typical

(c) Section at end walls

(d) Truss analogy

As this is equivalent to a wind-force of 160 psf, wind is no problem. The connection must transmit diaphragm shear and at the same time allow for the live load end rotation of the simple span roof tees. This is done by welding the roof and wall tees together at the top of the roof tee, but *not* at the bearing, Fig. 4.8.4. The bearing will then slide under live load end rotation, and friction at the bearing produces tension in the top connection.

Fig. 4.8.4 Connections between roof double tees at (a) end walls and (b) ledger beam

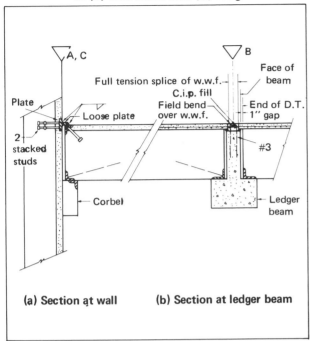

(a) Section at wall   **(b) Section at ledger beam**

The horizontal friction at each bearing, and the corresponding tension tie strength required, assuming a coefficient of friction of 0.5 is:

$$H = \mu_s N = 0.5 \times 9.6 = 4.8 \text{ kips}$$

This H is smaller than the $F_p = 7.4$ kips required for earthquakes. The wall panel to roof connection is thus designed for a horizontal pull-out force of 7.4 kips. The rebar anchorage is sized thus:

$$T_u = 1.4H = 1.4 \times 7.4 = 10.4 \text{ kips}$$
$$A_s = T_u/\phi f_y = 10.4/(0.9 \times 40) = 0.29 \text{ in.}^2$$

Two #4's ($A_s = 0.40$ in.^{2}) would be sufficient. Checking this connection for diaphragm shear, by shear-friction principles (see Part 5):

$$V_{Ru} = A_{vf}\, \phi\, f_y \mu = 0.40 \times 0.85 \times 40 \times 1.0$$
$$= 13.6 \text{ kips}$$
$$v_{ru} = V_{Ru}/s = 13.6/4 = 3.4 \text{ klf}$$

This is greater than the $v_u = 1.54$ klf required. The flange, or "chord" rebar is sized for tension as follows:

$$T_u = C_u = 1.4T_o = 1.4 \times 40.0 = 56 \text{ kips}$$
$$A_s = T_u/\phi\, f_y = 56/(0.9 \times 40) = 1.55 \text{ in.}^2, \text{ or}$$
two #8's ($A_s = 1.58$ in.^{2}).

These bars are located at the top of the wall panels, and must be spliced at the vertical joints as shown in Fig. 4.8.5. The cross-sectional area of the angle should be sufficient to assure prior tensile yield in the rebar, and of size to provide reasonable concentricity between rebar and angle.

The weld is field-made and is designed as shown in Part 5. The #3 hairpins shown provide enough shear-friction capacity to transmit the maximum diaphragm shear into the two #8 chord rebars. The bars should extend an anchorage length below the double tee-to-wall tension ties. This provides a load-path for the diaphragm shears between the chord rebars and the roof weld clips.

The diaphragm must be continuous in shear over the centerline ledger beam. See Fig. 4.8.4 (b). This bearing connection is designed very much like that of Fig. 4.8.4 (a). By shear-friction, the required diaphragm continuity steel area is:

$$A_{vf} = v_u/\phi\, f_y \mu$$
$$= 1.54/(0.85 \times 60 \times 1.4) = 0.022 \text{ in.}^2/\text{ft}$$

The connection should also have a horizontal tensile capacity, as was required at the exterior wall. Select a 6 x 6-W2.9 x W2.9, $A_s = 0.058$ in.^{2}/ft. This will provide about the same tensile capacity as provided at the exterior wall.

This completes the strength analysis of the diaphragm for N-S earthquake. For E-W earthquake, chords would also be needed in the east and west walls.

Following the load path, the shear walls are analyzed next. For simplicity, it is assumed the walls have no openings. Thus, there are interior and exterior (corner) wall panels, as shown in Fig. 4.8.6. The distinction between "interior" and

Fig. 4.8.5 Flange or "chord" reinforcement at top of wall panels

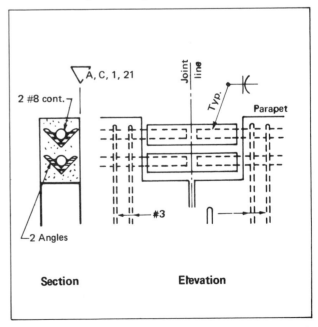

Section   **Elevation**

Note: This example assumes a mild climate. In colder climates, it would be preferable to locate the tensile chord in a location less exposed to the elements — say, under the roof insulation at the roof-to-wall junction.

Fig. 4.8.6 Forces acting on wall panels

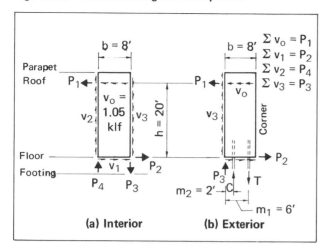

(a) Interior　　**(b) Exterior**

"exterior" is that an interior panel has weld clips on both vertical edges for transfer of shear. An exterior panel lacks the capability for shear transfer on one vertical edge. As will be seen, an interior panel requires no provision for overturning; an exterior panel does.

An interior panel is analyzed first. See Fig. 4.8.6 (a). The diaphragm shear $v_o = 1.10$ klf is applied from the roof. The distributed shears v_1, v_2, and v_3 resist v_o. These distributed shears have resultants P_1, P_2, P_3, and P_4. Treating the panel as a free body of zero weight:

$$\Sigma F_x = 0: \quad P_2 = P_1; \; v_1 b = v_o b; \; v_1 = v_o, \text{ and}$$

$$\Sigma F_y = 0: \quad P_3 = P_4; \; v_2 h = v_3 h; \; v_2 = v_3, \text{ and}$$

$$\Sigma M = 0: \quad P_1 h = P_3 b; \; P_3 = P_1 h/b, \text{ or } v_3 h$$
$$= v_o bh/b, \text{ so}$$

$v_3 = v_o$, and $v_o = v_1 = v_2 = v_3 = 1.10$ klf.

Weld-clips similar to Fig. 4.8.3 (b) are used at the vertical joints. The number required at each joint is determined by dividing the factored shear force by the design strength of the clip:

$$P_3 = v_o h = 1.10 \times 20 = 22.0 \text{ kips}$$
$$P_{3u} = 1.4 \times P_3 = 1.4 \times 22 = 30.8 \text{ kips}$$

required number = 30.8/15.8 = 1.9, use 3 min.

The shear at the panel bottom is taken by projecting dowels. These dowels are anchored into grouted sleeves, sufficient to develop full tensile yield, and act in shear-friction. Per double tee stem, the shear-friction steel required is:

$$v_u = 1.4 \; v_o = 1.4 \times 1.10 = 1.54 \text{ klf}$$
$$V_u = v_u \ell = 1.54 \times 4.0 = 6.16 \text{ kips}$$
$$A_{vf} = V_u / \phi \; f_y \mu = 6.16/(0.85 \times 60 \times 0.7)$$
$$= 0.17 \text{ in.}^2, \text{ or one \#4 bar}$$

By judgment, this size dowel is too delicate. Use one #6 at each double tee stem, typically.

An exterior, or corner, panel, is analyzed next.

See Fig. 4.8.6 (b). To obtain reliable strength, the anchor rebars are located at the double tee stems, rather than at the flange edge. Again taking the panel as a free body of zero weight,

$$\Sigma F_x = 0: \quad P_2 = P_1 = v_o b = 1.10 \times 8 = 8.8 \text{ kips}$$
$$\Sigma F_y = 0: \quad C = T - P_3 = T - 22.0$$
$$\Sigma M = 0: \quad T m_1 = P_1 h + C m_2 = P_1 h +$$
$$(T - 22.0) m_2 \text{ so}$$

$$T (m_1 - m_2) = P_1 h - 22.0 \; m_2$$

$$T = \frac{P_1 h - 22.0 \; m_2}{m_1 - m_2}$$

$$= \frac{(8.8 \times 20) - (22.0 \times 2)}{6 - 2} = 33.0 \text{ kips}$$

$$C = 33.0 - 22.0 = 11.0 \text{ kips}$$

Fig. 4.8.7 Wall double-tee connections

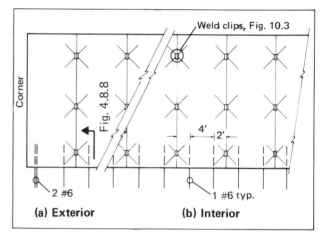

(a) Exterior　　**(b) Interior**

Fig. 4.8.8 Wall-footing connection

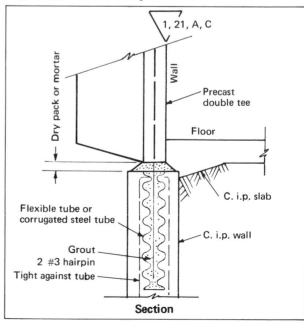

Section

The rebar required to take T is determined as follows:

$$T_u = 1.4T = 1.4 \times 33.0 = 46.2 \text{ kips}$$
$$A_s = T_u / \phi f_y = 46.2/(0.9 \times 60) = 0.86 \text{ in.}^2,$$

or two #6 ($A_s = 0.88$ in.^{2})

Hence, at each double tee stem, use one #6 typically and two #6's at corners. See Figs. 4.8.7 and 4.8.8.

In this example, the wall panels act together as a unit because of the weld clips. In locations subject to severe volume change deformations, the weld clips might cause local cracking. To avoid this problem, the designer may choose to utilize only those panels located near the middle of the wall to resist shear forces. These panels could be tied together with weld clips, but the other panels would be fastened only at the top and bottom.

4.8.7 Example: 4-Story Building

General

The following is an example analysis of a four-story building in which the structural system resisting lateral forces is of the box type, that is, floors and roof are horizontal diaphragms, and exterior walls are shear walls. The building is rectangular in plan, and dimensions and layout are shown in Figs. 4.8.9 and 4.8.10. Floors and roof are made up of 8-ft wide precast, prestressed concrete double tees, which are supported by exterior bearing walls and a central interior frame of precast, reinforced concrete beams and columns. There are smaller frames at the two elevator-stairway shafts, one at each end of the building. The interior frames are made up of full height columns and short beams. The exterior walls are 8-ft wide precast, prestressed concrete double tees placed on end, and are continuous from top of footing to top of parapet.

Approximations and Partial Designs

In the example, several approximations are made, but they are all on the conservative side. The designs are not complete in all respects, but in the case of several similar parts the more critical part is analyzed to illustrate the procedure. Some items, such as number or spacing of connections, are in some cases determined by judgment.

Weights

The unit dead loads are:

Roof: 100 psf
Floors: 110 psf
Walls: 90 psf max., 67 psf average

The dead load assigned to each level of the building is the weight of the roof or floor plus the walls from midstory to midstory (only walls for first level).

This gives values of w_x as listed in Table 4.8.1 Column 3.

Table 4.8.1

(1) Level	(2) h_x, ft	(3) w_x, kips	(4) F_x, kips	(5) F_p, kips
Roof	50	831	92.4*	56.1
4th	39	926	80.4*	62.5
3rd	28	926	57.7	62.5*
2nd	17	962	36.4	64.9*
1st	2	178	6.8	0
Totals		3823	267.7	

*These are used for design of diaphragms.

Equivalent Static Loads

The building is located in a Seismic Zone 2; Z = 3/8.

The structural system is of the box type; K = 1.33.

The fundamental period of vibration in the longitudinal direction of the building is $T = 0.05 h_n / \sqrt{D} = 0.05 \times 50 / \sqrt{92} = 0.26$ sec, and the corresponding coefficient for base shear is $C = 1/15 \sqrt{T} = 1/15 \sqrt{0.26} = 0.131$. In the transverse direction, C = 0.121. However, C need not be greater than 0.12 nor CS greater than 0.14. CS = 0.14 governs. The total dead load of the building is W = 3823 kips, as shown in Table 4.8.1. The base shear is then V = ZIKCSW = 0.375 × 1.0 × 1.33 × 0.14 × 3823 = 267 kips.

The base shear is distributed over the height of the building. In this example, the extra lateral force at the top level is $F_t = 0$, since T < 0.7 sec. The distribution to the several levels is then

$$F_x = \frac{(V - F_t) w_x h_x}{\sum\limits_{i=1}^{n} w_i h_i} = \frac{267 w_x h_x}{120,000}$$

which gives the values of F_x listed in Table 4.8.1, Column 4. These lateral loads are used for design of the building as a whole, like shear in and overturning of walls. They are also used for design of building parts unless they are exceeded by the loads developed in the following:

$$F_p = ZIC_p SW_p$$

When applied to the design of the roof and floor diaphragms, use $C_p = 0.12$ which gives $F_p = 0.375 \times 1.0 \times 0.12 \times 1.4 \times W_p = 0.0675 W_p$. The result-

Fig. 4.8.9 Four-story example building — elevations

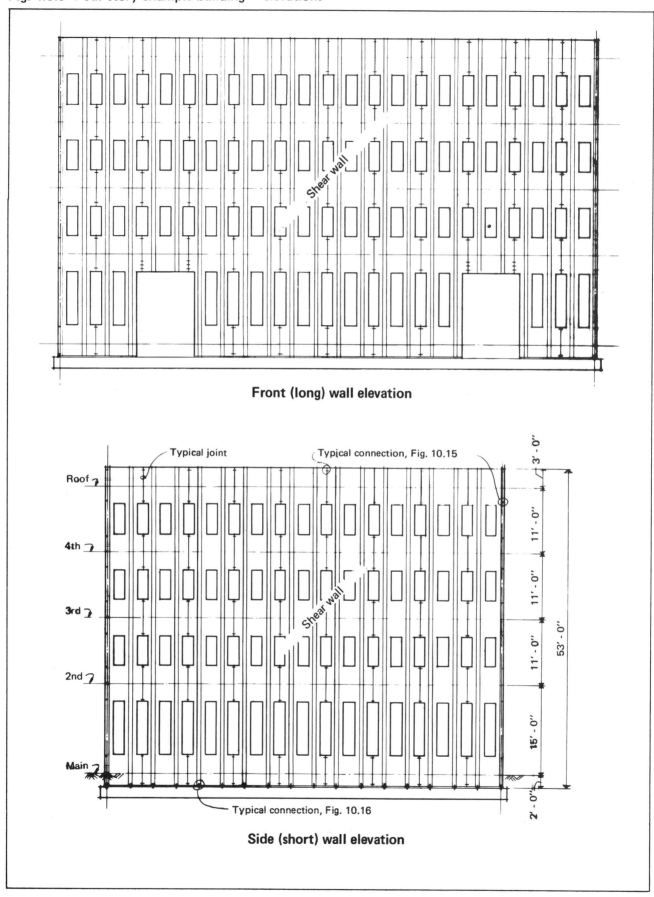

Front (long) wall elevation

Side (short) wall elevation

Fig. 4.8.10 Four-story example building — typical floor plan

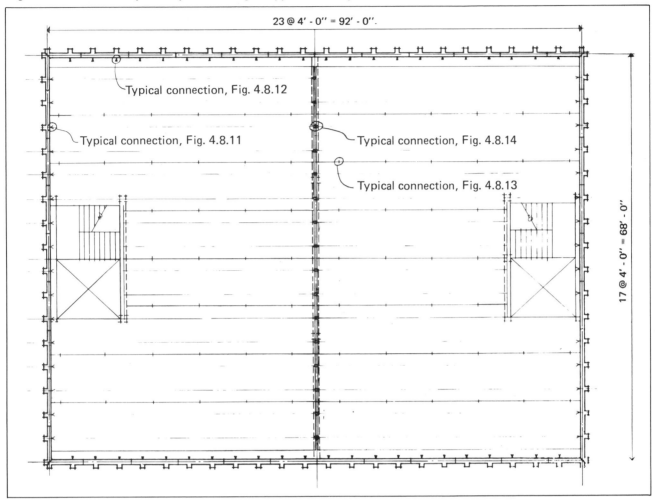

23 @ 4' - 0'' = 92' - 0''.

17 @ 4' - 0'' = 68' - 0''

Typical connection, Fig. 4.8.12

Typical connection, Fig. 4.8.11

Typical connection, Fig. 4.8.14

Typical connection, Fig. 4.8.13

ing values of F_p are listed in Table 4.8.1, Column 5. However, at each level the larger of the loads F_x or F_p must be used; this is indicated by asterisks in Table 4.8.1. Note that these loads include the effects of the weights of walls that are parallel to the direction of motion, and although this is not really required, the design is somewhat conservative.

Chord Forces in Diaphragm

The roof is the most heavily loaded diaphragm, with a total lateral load of 92.4 kips. With the diaphragm spanning in the long direction, the moment T d = 1/8 Wℓ and the chord force

$$T = \frac{W\ell}{8d} = \frac{92.4 \times 92}{8 \times (68 - 2)} = 16.1 \text{ kips}$$

Design of the chord reinforcement: $A_s = T_u/\phi f_y = 1.4 \times 16.1/(0.9 \times 60) = 0.42$ in.^{2} Use two #5 bars placed as shown in Figs. 4.8.11 and 4.8.12.

Requirements at lower levels and for the diaphragm spanning in the short direction are even less, but, for simplicity, the same diaphragm chord reinforcement is used throughout.

Shear Stresses in Diaphragm

This example building is assumed to be symmetrical, and the centers of mass and rigidity coincide. It is then necessary to include the effect of an arbitrary torsion, that is, the lateral load is offset from the center of rigidity by 5 percent of the maximum dimension of the diaphragm.

The torsional moment is resisted by all four exterior walls, depending on rigidities and distances from the center or rigidity. However, it is conservative to assume that the torsional moment is resisted by the two walls that are parallel to the direction of motion. For earthquake in the transverse direction, the maximum total shear to one wall is $R_o = 0.55$ (F_x or F_p) = $0.55 \times 92.4 = 50.8$ kips. Deducting the length of the service shafts, the maximum unit shear is $v_o = R_o/\ell = 50.8/ (68 - 20) = 1.06$ klf.

For earthquake in the longitudinal direction, the maximum unit shear to the long walls is less.

Fig. 4.8.11 Typical connection between diaphragm and wall at end of double tee

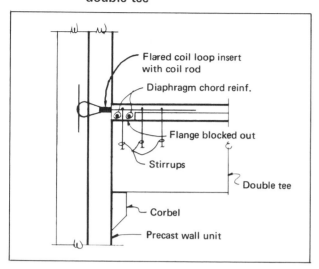

Fig. 4.8.12 Typical connection between diaphragm and wall at side of double tee

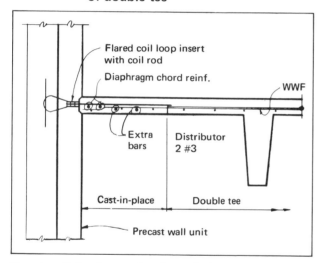

For simplicity in production of elements and erection at the site, the 1.06 klf is used throughout all diaphragms and for their connections to the walls.

Typical Connections for Interior of Diaphragm

A typical connection transmitting shear between double tees is illustrated in Fig. 4.8.13. In each edge there is a small angle anchored with two rebars ar 45°. The shear strength is calculated from force components in the anchor bars, as in the previous example.

$$V_{Ru} = 15.8 \text{ kips}$$

To equalize cambers and deflections in neighboring elements, a maximum spacing should not exceed about 8 ft. This provides a resisting unit shear of $v_u = 15.8/8 = 1.98$ klf which compares to the requirement of $1.4 \times 1.06 = 1.48$ klf.

The cast-in-place strip along the longitudinal wall is reinforced with 8 x 4 – W2.9 x W2.9 projecting out of the double tee flange. By shear-friction, this is good for

$$V_{Ru} = \phi A_s f_y \mu = 0.85 \times 0.087 \times 60 \times 1.0 = 4.43 \text{ klf which is several times that required.}$$

Across the center of the building, the double tee elements are supported by a beam, and they meet end to end. It is desirable to make the shear path stay at the level of the flange, that is, making the path as direct as feasible. The connections to take care of this may be angle splices on rebars projecting out of the tops of the ribs. The gap is filled with concrete, and shear-friction is applicable. Such a connection is shown in Fig. 4.8.14.

It should be noted that structural topping containing a rather light welded wire fabric could be used in place of the direct connections between neighboring elements. Even so, it may be desirable to use some of the hardware type connections to make the building resistant to lateral loads during construction.

Fig. 4.8.13 Typical diaphragm connections between double tees

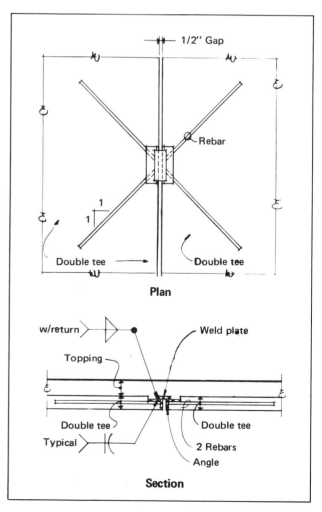

Fig. 4.8.14 Typical end connection between double tees

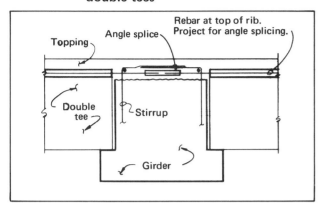

4.8.7.6 Connection of Diaphragm to Wall

First consider the connections between the ends of the double-tees and the short wall. The ribs of the double tees and of the wall elements are aligned, and the vertical loads are carried by corbels projecting from the interior wall surface. Two requirements must be satisfied: a) horizontal shear between diaphragm and wall, and b) direct horizontal tension caused by the tendency of the wall to fall away from the building. The two requirements will be superimposed because earthquake motion may occur in any horizontal direction, and is not limited to the principal axes of the building.

It is convenient to space the connections 4 ft apart so that they will occur at wall ribs. At each connection the forces are:

a) Horizontal shear, $V = 1.06 \times 4 = 4.24$ kips

This can be handled by shear-friction. Then the tensile capacity must be $V_u/\mu = 1.4 \times 4.24/1.0 = 5.9$ kips.

b) Direct horizontal tension is given by
$$F_p = ZIC_p SW_p$$

UBC-76 is not explicit on what the value of C_p should be in this situation, but $C_p = 2.0$ is certainly a conservative interpretation. For the weight of wall contributing to one connection, a unit weight of 90 psf over an area of $4 \times 14 = 56$ ft^2 will be used, so that $W_p = 0.090 \times 56 = 5.04$ kips. Thus $F_p = 0.375 \times 2.0 \times 5.04 = 3.78$ kips and the design tensile strength is $1.4 \times 3.78 = 5.29$ kips.

A choice must be made of the type and size of connection. The coil-loop insert with continuous threaded coil rod is a simple and useful type, and there are several proprietary makes. The installation is illustrated in Fig. 4.8.11. A note of caution is in order: the coil-loop insert and the coil rod

should each be amply anchored in their respective parts so that failure by yielding in the coil rod is assured.

The connections between the edge of the double tee and the long wall are of the same kind, and are calculated in the same way. The primary difference is caused by the cast-in-place strip of diaphragm adjacent to the wall. The threaded coil rods extend into this strip and overlap the welded wire fabric that projects out of the edge of the double tee flange. See Fig. 4.8.12.

4.8.7.7 Special Situation at Service Shafts

From the typical floor plan, Fig. 4.8.10, it is seen that several wall ribs are not tied to the floor, and the wall, because of vertical joints, is not suited to span horizontally past the shaft openings. This situation is readily alleviated with a strong-back type beam placed horizontally at each floor level, and spanning the shaft opening. The adjacent connections at each side of the shaft will have to be stronger than the standard connection designed in the preceding section.

4.8.7.8 Shear Wall — General

The exterior walls of the building provide the horizontal reactions for the diaphragms. These forces are in the plane of the wall and become horizontal shears, therefore the term shear wall. The end walls in the example building do double duty in that they are also bearing walls.

The design of the individual wall elements can be important, but will not be covered here. For the example, next examine the shear in the joints, and overall stability. The equivalent static loads, which are assumed to act horizontally at each floor level, cause a tendency for the wall to overturn. Gravity loads and footing will stabilize the wall and prevent overturning, but internal shear and direct stresses will develop. In this case, no advantage will be taken of the slightly reduced overturning permitted by the UBC-76, but the full value of the cantilever moment will be used.

Table 4.8.2

Level	V_x, kips	M_x, ft-kips
Roof	46.2	0
4th	86.4	508
3rd	115.3	1458
2nd	133.5	2726
1st	133.9	4728
Top of footing		4996

Fig. 4.8.15 Typical connection, wall to wall

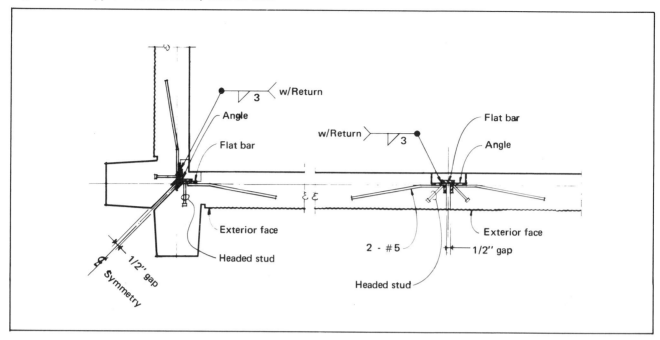

Results of the calculations are shown in Table 4.8.2 for one wall.

4.8.7.9 Shear in Shear Wall

The maximum shear in the short end wall due to the equivalent static loads and the arbitrary 5 percent eccentricity is V = 0.55 x 267 = 147 kips, or v = 147/68 = 2.16 klf. UBC-76 requires a load factor of 2.0 for shear in shear walls, so v_u = 2.0 x 2.16 = 4.32 klf. This unit shear occurs near the bottom of the wall, and is both horizontal and vertical.

A typical connection for the vertical joints between wall elements is shown in Fig. 4.8.15. In each edge there is a small angle anchored with two #5 bars. With 6-in. thick walls, it is possible to drypack the joints effectively. Apply the shear-friction concept, with the anchor bars placed at 90° to the joint. The shear capacity of one connection is:

$$V_{Ru} = \phi \, A_s \, f_y \, \mu = 0.85 \times 2 \times 0.31 \times 40 \times 1.0 = 21.1 \text{ kips}$$

Near the bottom of the wall, the average vertical spacing should not be more than 21.1/4.32 = 4.9 ft. Use three connections in each story. Since windows cross the joints, be sure to allow for this when locating the connections.

4.8.7.10 Moment in Shear Wall

The maximum moment in the short end wall, with allowance for the arbitrary eccentricity, is M = 4996 x 1.10 = 5496 ft-kips. The maximum grav-ity load is

W of one end wall, 68 x 53 x 0.067 = 241

W of ¼ of roof, 68 x $\frac{92}{4}$ x 0.100 = 156

W of ¼ of floors, 68 x $\frac{92}{4}$ x 0.110 x 3 = 516

$\qquad\qquad$ Total W = 913 kips

With wall ribs neglected, section properties in linear measure are, area A_ℓ = 68 ft, and section modulus Z_ℓ = 1/6 x 68^2 = 771 ft^2. Fiber stresses at bottom of wall are

$$f = \frac{W}{A} \pm \frac{M}{Z} = \frac{913}{68} \pm \frac{5496}{771} = \begin{array}{l} 20.6 \text{ klf} \\ 6.3 \text{ klf} \end{array}$$

These values are a good measure of the stresses in the gross wall section, and can also be used to calculate the necessary width of wall footing. Additional refinement of stress calculations are necessary at window openings.

4.8.7.11 Other Shear Walls

The long shear walls, which provide the resistance to earthquake in the longitudinal direction of the building, are less critical than the short walls, and they will not be discussed. Details of design will be the same as for the short walls.

There is a special situation at the building's entrance doors. The wall panel above the doorway is supported by connections to neighboring units. The necessary strength can be provided by

Fig. 4.8.16 Typical connection of wall to footing

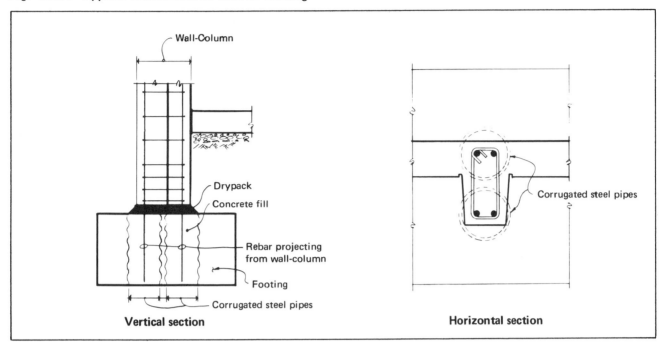

Vertical section

Horizontal section

two additional connections in each joint above the doorway.

4.8.7.12 Connections of Walls to Footings

The footings can be proportioned on the basis of reactions from the walls, as indicated above. Each wall element can be connected to the footing by projecting the main rebars of the ribs downward into blockouts in the footings, and filling the blockouts with concrete. The space between wall element and footing must be dry-packed as shown in Fig. 4.8.16.

4.8.8 Example — 12-Story Building

Box-type buildings can be built to a height of 160 ft, according to UBC-76. For buildings approaching that height, the structural layout and connection details may differ from those of the preceding example, but the design procedure is the same. The base shear, its distribution as equivalent static loads over the height of the building, and the load paths down to the foundations, are calculated as illustrated in Sect. 4.8.7.

Fig. 4.8.17 shows a 12-story building which is square in plan. Wall units are 12 ft high and 30 ft or 15 ft in width. The "running bond" pattern illustrated is not an essential feature of seismic design, but it does help tie the building together, thus reducing the likelihood of progressive collapse in case of accidental overloads.

Fig. 4.8.17 Twelve-story building — elevation

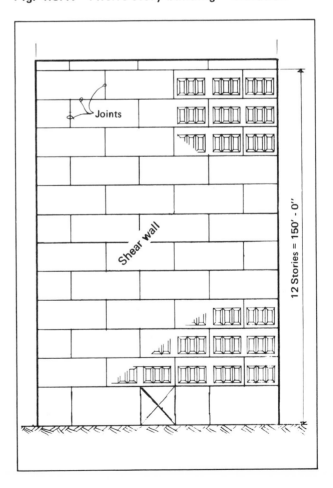

Fig. 4.8.18 Typical connections for 12-story example building — seismic loads

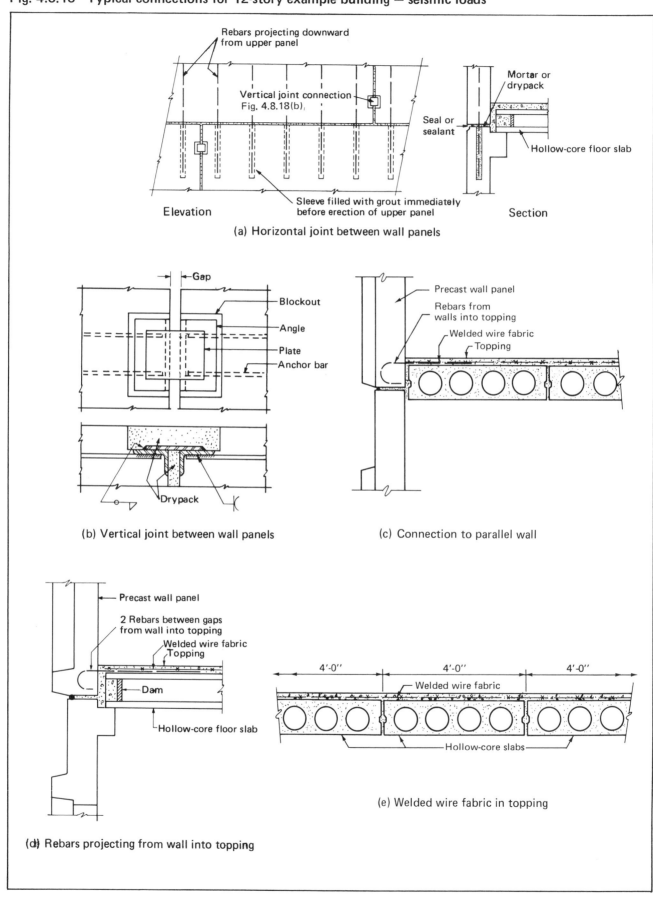

(a) Horizontal joint between wall panels

(b) Vertical joint between wall panels

(c) Connection to parallel wall

(d) Rebars projecting from wall into topping

(e) Welded wire fabric in topping

PCI Design Handbook

Fig. 4.8.19 Twelve-story building — isometric view of assembly

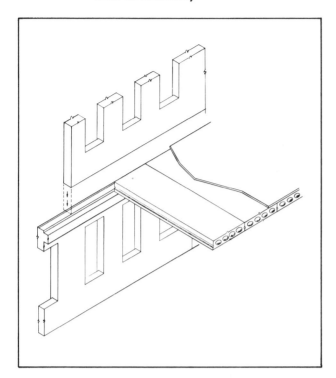

Floors consist of precast, prestressed hollow-core slabs, 12 in. deep and 4 ft wide. They bear on continuous corbels on the walls and on the interior beams.

Buildings of this type are generally erected one story at a time, and locations of joints, temporary bracing, and connection details must be correlated with the erection sequence. Fig. 4.8.19 indicates the erection cycle. The walls of the story below have been erected, braced, and fully connected. The hollow-core slabs are then erected, bearing on wall corbels and interior beams. The wall panels of the next story are then erected; these have rebars projecting downward into sleeves filled with grout in the wall panels below (Fig. 4.8.18a), hardware in the vertical joints for welded connections (Fig. 4.8.18 c, d, and e). The next step is to place continuous welded wire fabric for the topping, and finally place the topping. The erection cycle can be repeated as soon as the grout and concrete have adequate strength, and the welding of the connections between the wall panels has been completed.

The connections shown in Fig. 4.8.18 are illustrations of methods of providing the diaphragms and continuous load paths discussed in the previous examples. Many other details have been used successfully on similar structures in high-intensity earthquake areas.

4.8.9 Example — 23-Story Building

Figs. 4.8.20 and 4.8.21 show a 23-story building 300 ft high. The building was designed for earthquake resistance with a ductile moment-resisting space frame.

All floors consist of 10-ft wide single-tee units that were precast and prestressed. Adjacent tees were connected by means of hardware in the flanges, but these connections were not considered as diaphragm connections. The tees have a structural concrete topping reinforced with welded wire fabric. The diaphragm is provided by the

Fig. 4.8.20 23-story example building — elevation

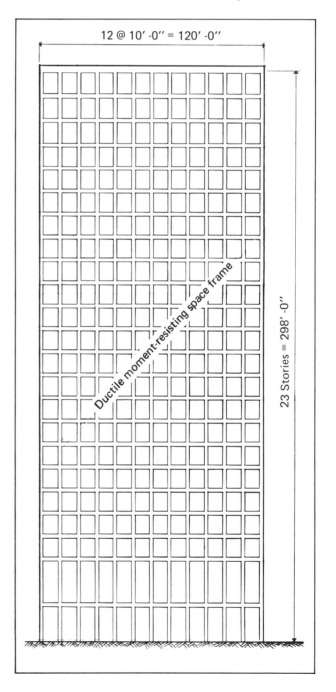

12 @ 10' -0'' = 120' -0''

Ductile moment-resisting space frame

23 Stories = 298' -0''

Fig. 4.8.21 23-story example building — typical floor plan

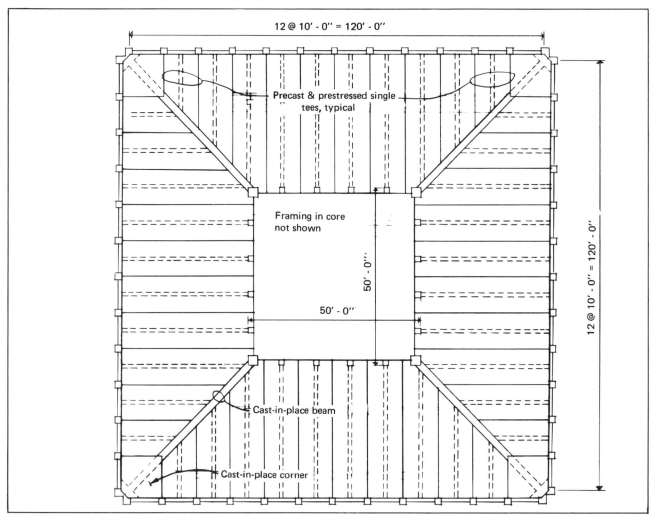

topping rather than by the flanges of the tees.

The core of the building has no structural walls or frames that contribute measurably to lateral stability of the building. Each single tee is supported at the core by a cast-in-place column, but the connection is immaterial for earthquake resistance. The core has additional columns and partial slabs, but in effect, the core creates a large square hole in the floor diaphragm. Reinforcement must be provided in the diaphragm around the core in the same manner as around any large hole in a slab.

The exterior walls of the building are the moment-resisting frames. They consist of reinforced concrete columns and spandrel beams. For these frames to be ductile, they must have closely spaced spirals in the columns and ties in the spandrels. At the intersections, the congestion of reinforcement is so great that it is not feasible to have the tees align with the columns. Instead they are located midway between columns, but again the connection of the tee stem to the spandrel is immaterial to the earthquake resistance. The vital connection here is between the flange and the spandrel. This was accomplished by rebars projecting out of the end of the tee flange, which are later embedded in the cast-in-place concrete of the spandrel.

PCI Design Handbook

PART 5
DESIGN OF CONNECTIONS

DESIGN OF CONNECTIONS

5.1 Notation

a = shear span

A_b = area of bar or stud

A_{cr} = area of crack face

A_n = area of reinforcement required to resist axial tension

A_o = lateral surface area of failure surface

A_s = area of reinforcement

A'_s = area of vertical reinforcement near end of steel haunch

A_{sh} = area of reinforcement for horizontal or diagonal cracks

A_v = diagonal tension reinforcement in dapped end

A_{vf} = area of horizontal ties

A_{vh} = area of reinforcement nominally perpendicular to crack plane

A_w = area of weld

b = width

C.E. = carbon equivalent

C_r = reduction coefficient used in Eq. 5.7-1

C_{es} = reduction coefficient for edge distance

d = depth to centroid of reinforcement

d_b = bar or stud diameter

d_e = edge distance in direction of load

d_h = head diameter of stud

e_i = center of bolt to horizontal reaction

e_v = eccentricity of vertical load

f = unit stress

f_{bu} = factored bearing stress

f_{ct} = splitting tensile strength of concrete

f'_c = compressive strength of concrete

f_v = yield strength of A_v

f_y = yield strength of reinforcement or structural steel

f_{ys} = yield strength of A_{sh}

f_{yv} = yield strength of A_{vh}

F_s = factored friction force

ΣF = greatest sum of factored anchor bolt forces on one side of the column

g = gage of angle

G = shear modulus

G_t = long term shear modulus

h = total depth

j_u = for resisting lever arm, used in $j_u d$

ℓ_d = development length

ℓ_e = embedment length

ℓ_p = projection of corbel, beam ledge or dapped end

ℓ_w = length of weld

n = number of studs in a group

N = unfactored horizontal or axial force

N_u = factored horizontal or axial force

P_c = nominal tensile strength of concrete element

P_s = nominal tensile strength of steel element

P_u = factored tension load

s = distance from free edge to center of bearing

t = thickness

t_c = column dimension

t_t = total thickness of pad or pad assembly

t_w = effective throat thickness of weld

V = unfactored vertical or shear force

V_c = nominal shear strength of concrete element

V_d = unfactored vertical dead load

V_n = nominal bearing or shear strength of an element

V_r = nominal strength provided by reinforcement

V_s = nominal shear strength of steel element

V_u = factored vertical force

V_{up} = factored force parallel to assumed crack plane

w = dimension (see specific application)

x, y = surface dimensions of assumed failure plane

x_c = distance from centerline of bolt to face of column

x_o = base plate projection

x_t = distance from centerline of bolt to centerline of reinforcement

Z_s = plastic section modulus of structural steel section

Δ = horizontal deformation of bearing pad

α = angle of reinforcement placement

θ = angle of assumed crack plane

λ = coefficient for use with lightweight concrete (see Sect. 5.6)

μ = shear-friction coefficient

μ_e = effective shear-friction coefficient

μ_s = static coefficient of friction

ϕ = strength reduction factor

5.2 General

This part of the Handbook presents concepts of analysis and equations for design of connections for precast concrete. The material was prepared by the PCI Committee on Connection Details and represents the collective experience of the members of that committee. Design equations were developed through field experience, laboratory tests, and structural analysis. The recommendations take into consideration current design office and plant practice and are intended as reasonable guidelines for the analysis and design of connections. Other types of connections are in use and some have been extensively tested. Continuing research will lead to improved details and analyses. Therefore, engineers should not necessarily be limited to the design equations or connection details included in this part.

Practical and economical connection design must consider production of the units and construction situations pertinent to precast concrete buildings, as well as the performance of the connections in service. Detailed discussions and recommendations of these aspects may be found in the PCI publications, "Manual on Design of Connections for Precast Prestressed Concrete," "Manual for Quality Control for Plants and Production of Precast, Prestressed Concrete Products," "Architectural Precast Concrete," and "Manual for Structural Design of Architectural Precast Concrete."

5.3 Loads and Load Factors

With certain noted exceptions, such as bearing pad design, the equations in this part are based on strength design relationships, incorporating the load factors and strength reduction factors (ϕ - factors) specified in ACI 318-77.

In addition to gravity loads, wind, earthquake, and forces from restraint of volume changes should be considered. Calculation of volume change restraint forces is described in Part 4 of this Handbook. It is recommended that connections of flexural members be designed for a minimum tensile force of 0.20 times the vertical dead load, unless properly designed bearing pads are used (See Section 5.4).

It is undesirable for the connection to be the weak link in a precast framing system. Therefore, it is recommended that most connections be designed with an additional load factor of at least 1.3. Insensitive connections, such as column bases, do not need the additional factor.

5.4 Bearing Pads

Bearing pads are used to distribute vertical loads over the bearing area. Some types of pads also reduce force build-up at the connection by permitting small displacements and rotations. Their use is encouraged wherever applicable.

The performance of most bearing pads is a function of their deformation characteristics under service loads. Hence, these pads are designed using unfactored (service) loads.

There are a number of suitable materials and combinations of materials that can be used for bearing pads. A few are described below with some design recommendations. In some cases, various grades of bearing pads can satisfy these descriptions, but exhibit widely different properties and behavior. In case of doubt, the pad supplier or precast product manufacturer should be consulted for proper selection of the pad. Most pad manufacturers have technical brochures available to aid the designer.

1. Commercial grade elastomeric (neoprene) pads are readily available and inexpensive. However, these pads exhibit wide variations in shear deformation characteristics and bearing strength, so they should not be used unless performance test data is available.

2. Structural grade neoprene pads are those which meet the requirements of Section 25, Division 2 of the AASHTO Standard Specifications for Highway Bridges (1977). For optimum economy, their use should be limited to places where uniform bearing is important, or when it is desired to reduce volume change restraints. Design recommendations are shown in Fig. 5.4.1.

   For high compressive stresses and/or large horizontal displacements, laminated pads consisting of layers of elastomer bonded between

Fig. 5.4.1 Design of structural grade elastomeric bearing pads

Design Recommendations
1. Use unfactored loads for design
2. Maximum compressive stress = 1000 psi
3. Maximum shear stress = 100 psi
4. Maximum shear deformation = t/2
5. Maximum compressive strain = 15%
6. $w \geqslant 5t$ or 4 in.
7. $t_t \geqslant 1/4$ in. for stems, 3/8 in. for beams

Notation
b = dimension perpendicular to beam span, in.
w = dimension parallel to beam span, in.
t = thickness of pad, or of each lamination in pads laminated with bonded steel plates, in.
t_t = total thickness of pad or pad assembly, in.
V = unfactored vertical reaction, lb
N = unfactored axial tension, lb
f = unfactored compressive stress, psi
G = shear modulus, psi
G_t = long term shear modulus = 0.5G, psi
Δ = shear deformation, in.

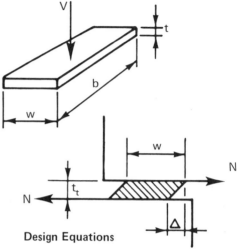

Design Equations

$$\text{Shape factor} = \frac{w\,b}{2(w + b)t}$$

$$f = \frac{V}{wb}$$

$$N = \frac{\Delta wbG_t}{t_t}$$

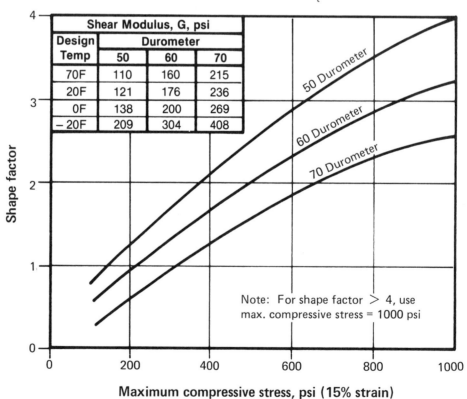

Shear Modulus, G, psi			
Design Temp	Durometer		
	50	60	70
70F	110	160	215
20F	121	176	236
0F	138	200	269
−20F	209	304	408

Note: For shape factor > 4, use max. compressive stress = 1000 psi

Shape factor

Maximum compressive stress, psi (15% strain)

steel plates can be used. Each layer behaves in compression like an individual pad, but the shear deformation is a function of the thickness of the total assembly.

It should be cautioned that in unheated buildings, such as parking structures, the unfactored bearing stress should be a *minimum* of 400 psi to keep the pad from "walking" due to cyclical temperature conditions.

3. Laminated fabric bearing pads composed of multiple layers of 8-oz cotton duck with a high quality natural rubber binder can sustain unfactored compressive stresses up to 2000 psi. They are designed in a manner similar to elastomeric pads except the shape factor need not be considered. These pads do not deform as readily as elastomeric pads, so do not provide the same stress reducing characteristics. The shear modulus, G, may be assumed to be 550 psi, unless more specific data is available.

4. Preformed pads composed of synthetic fibers and a rubber body are designed as in (3) above, except that unfactored compression should be limited to 1500 psi. The shear modulus can be assumed equal to $525 - 4v/3$ unless more specific data is available (v = unit unfactored shear stress). This type pad should meet the requirement of Section 2.10.3 (L) of the AASHTO Standard Specifications for Highway Bridges (1977).

5. Tetrafluorethylene (TFE) bearing pads reduce horizontal stresses because of their low coefficients of friction. The unfactored bearing stress should not exceed 1000 psi when unreinforced or 2000 psi when reinforced with glass fibers or similar material. TFE pads are sometimes used in combination with elastomeric or fabric pads, or can be bonded to steel plates.

6. A multipolymer plastic bearing strip is manufactured expressly for bearing purposes. It is a commonly used material for the bearing support of hollow-core slabs, and is highly suitable for this application. The material has a compressive strength higher than the typical design range of concrete used in precast construction.

Tempered hardboard strips are also used with hollow-core slabs to prevent concrete to concrete bearing.

Example 5.4.1: Design of elastomeric bearing pad

Given:

A 16 inch wide prestressed beam located in a heated building in Pocatello, Idaho, with the following end reactions:

Dead load = 25 kips

Live load = $\dfrac{30}{55}$ kips

Beam span = 30 ft

Problem:

Design a 50 durometer elastomeric bearing pad to support the beam and determine horizontal force the pad develops.

Solution: (See Fig. 5.4.1):

For 16 in. beam width use pad 15 in. wide.

$$f = \frac{V}{w(b)} = 1000 \text{ psi max}$$

$$w = \frac{V}{f(b)} = \frac{55000}{1000(15)} = 3.7 \text{ in.}$$

For beams, min. thickness recommended is 3/8 in.

Try pad 4 in. x 15 in. x 3/8 in.

$$\text{Shape factor} = \frac{wb}{2(w+b)t} = \frac{4(15)}{2(4+15)(3/8)}$$
$$= 4.2$$

From Fig. 5.4.1, since shape factor > 4, use of f = 1000 is OK

From Figs. 4.4.1, 4.4.2 and Table 4.4.6, the unit strain is 793×10^{-6} in./in.

Δ (total) = $793 \times 10^{-6} (30 \times 12) = 0.285$ in.

Δ (each end) = $0.285/2 = 0.143 < t/2$ OK

From Fig. 5.4.1

$G = 110$ psi, $G_t = 55$ psi

$$N = \frac{\Delta \, wb \, G_t}{t_t} = \frac{0.143(4)(15)(55)}{3/8} = 1258 \text{ lb}$$

$N_u = 1.4 \times 1.3 \times 1258 = 2290$ lb

5.5 Friction

The static coefficients of friction shown in Table 5.5.1 are conservative values for use in determining the *upper limit* of volume change forces for members without "hard" connections. Thus, the maximum force resulting from the frictional restraint of axial movements can be determined by:

$$F_s = \mu_s V_{ud} \qquad \qquad \text{(Eq. 5.5-1)}$$

where

F_s = factored friction force

μ_s = static coefficient of friction as given in

Table 5.5.1

V_{ud} = factored dead load force normal to the friction face.

The coefficients in Table 5.5.1 should be divided by 5 if friction is to be depended upon for support of temporary loads.

Table 5.5.1 Static coefficients of friction of dry materials

Material	μ_s
Elastomeric to steel or concrete	0.7
Laminated cotton duck fabric to concrete	0.6
Concrete to concrete	0.8
Concrete to steel	0.4
Steel to steel (not rusted)	0.25
TFE to TFE	0.05[1]
Hardboard to concrete	0.5
Multipolymer plastic (non-skid) to concrete	1.2[2]
Multipolymer plastic (smooth) to concrete	0.4[2]

(1) Repeated movement may increase μ_s by exposing the reinforcing fibers in reinforced TFE.

(2) Courtesy Koro Corp.

5.6 Shear-Friction

Use of the shear-friction theory is recognized by Sect. 11.7 of ACI 318-77, which states that shear friction "may be applied where it is appropriate to consider shear transfer across a given plane such as an existing or potential crack, an interface between dissimilar materials, or an interface between two concrete surfaces cast at different times."

Shear-friction is an extremely useful tool in connection design, and other applications in precast, prestressed concrete structures (see Sect. 3.3.4 – "Horizontal shear transfer in composite members.").

A basic assumption used in applying the shear-friction concept is that concrete within the direct shear area of the connection will crack in the most undesirable manner. Ductility is achieved by placing reinforcement across this anticipated crack so that the tension developed by the reinforcing bars will provide a force normal to the crack. This normal force in combination with "friction" at the crack interface provides the shear resistance. The shear-friction analogy can be adapted to designs for reinforced concrete bearing, corbels, daps, composite sections, and other connection devices.

As a result of the analysis of recent test data*, the PCI Committee on Connection Details recommends the use of an "effective shear-friction coefficient," μ_e, when the concept is applied to precast concrete connections. The shear-friction reinforcement nominally perpendicular to the assumed crack plane can be determined by:

$$A_{vf} = \frac{V_{up}}{\phi\, f_y\, \mu_e} \qquad \text{(Eq. 5.6-1)}$$

where

ϕ = 0.85

A_{vf} = area of reinforcement nominally perpendicular to the assumed crack plane, in.^{2}

f_y = yield strength of A_{vf}, psi (equal to or less than 60,000 psi)

V_{up} = Applied factored shear force, parallel to the assumed crack plane, lb (limited by the values given in Table 5.6.1)

$$\mu_e = \frac{1000\, \lambda^2\, A_{cr}\, \mu}{V_{up}} \qquad \text{(Eq. 5.6-2)}$$

λ = 1.0 for normal weight concrete

= $(f_{ct}/6.7)\sqrt{f'_c}$ for sand-lightweight or all-lightweight concrete. If f_{ct} is unknown:

0.85 for sand-lightweight concrete

0.75 for all-lightweight concrete

f_{ct} = splitting tensile strength of concrete, psi

μ = value from Table 5.6.1

A_{cr} = area of the crack interface (varies depending on the type of connection) sq. in.

The shear-friction reinforcement should not be less than:

$$A_{vf} = \frac{120\, A_{cr}}{f_y} \qquad \text{(Eq. 5.6-3)}$$

unless A_{vf} calculated by Eq. 5.6-1 is increased by one-third.

When axial tension is present, additional reinforcement area should be provided:

$$A_n = \frac{N_u}{\phi\, f_y} \qquad \text{(Eq. 5.6-4)}$$

---

* Shaikh, A. F., "Proposed Revisions to Shear-Friction Provisions," *PCI Journal,* March-April, 1978.

Table 5.6.1 Shear-friction coefficients

	Crack interface condition	Recommended μ	Maximum V_{up}, lb
1	Concrete to concrete, cast monolithically	1.4	$0.30 \, \lambda^2 \, f'_c \, A_{cr} \leqslant 1000 \, \lambda^2 \, A_{cr}$
2	Concrete to hardened concrete with roughened surface	1.0	$0.25 \, \lambda^2 \, f'_c \, A_{cr} \leqslant 1000 \, \lambda^2 \, A_{cr}$
3	Concrete to concrete, smooth interface	0.4	$0.15 \, \lambda^2 \, f'_c \, A_{cr} \leqslant 600 \, \lambda^2 \, A_{cr}$
4	Concrete to steel	0.6	$0.20 \, \lambda^2 \, f'_c \, A_{cr} \leqslant 800 \, \lambda^2 \, A_{cr}$

where

A_n = area of reinforcement required to resist axial tension, sq in.

N_u = applied factored horizontal tensile force nominally perpendicular to the assumed crack plane, lb

ϕ = 0.85

All reinforcement, either side of the assumed crack plane, should be properly anchored by development length, welding to angles or plates, or hooks.

5.7 Bearing on Plain Concrete

For uniform bearing on plain concrete the design bearing strength (Fig. 5.7.1) is:

$$\phi \, V_n = \phi \, C_r \, 70 \, \lambda \sqrt{f'_c} \, (s/w)^{1/3} \, bw \quad \text{(Eq. 5.7-1)}$$

where

$\phi \, V_n$ = design bearing strength*, lb

ϕ = 0.70

---

*ACI 318-77 has introduced a new system of notation to be used in strength design of reinforced concrete. Previously, the subscript "u" denoted either the applied factored forces (M_u, V_u, P_u, etc.) or the design strength (termed "ultimate strength" prior to 1971). In the 1977 edition of the code, the subscript "u" denotes only the applied factored forces. The subscript "n" denotes the "nominal strength." The "design strength" (or "ultimate strength") is the nominal strength multiplied by the strength reduction factor, ϕ, for example ϕM_n, ϕV_n, ϕP_n. Thus, the design of a member or component requires that $M_u \leqslant \phi M_n$, $V_u \leqslant \phi V_n$, $P_u \leqslant \phi P_n$, etc.

In presenting equations for strength design, this leads to the apparent algebraic redundancy of having the term ϕ on both sides of the equation. The Committee on Design Handbook decided, however, that to avoid the inadvertent neglect of the ϕ-factor, the equations should be in terms of "design strength" rather than "nominal strength."

Fig. 5.7.1 Bearing on plain concrete

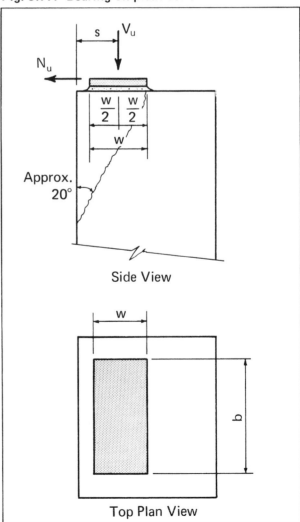

Side View

Top Plan View

b = dimension of bearing area parallel to free edge, in.

s = distance from free edge to center of bearing, in.

w = dimension of bearing area perpendicular to the free edge, in.

$C_r = \left(\dfrac{sw}{200}\right)^{N_u/V_u}$ = 1.0 when reinforcement is provided in direction of N_u, in accordance with Sect. 5.8, or when N_u is zero. sw should not be taken greater than 9.0 sq in.

λ = as defined in Sect. 5.6

Table 5.20.1 is a design aid for calculation of uniform bearing stress.

If uniform bearing is not ensured by a grout bed or a properly selected pad, the term (s/w) in Eq. 5.7-1 should not exceed 0.5.

The bearing strength should meet the limits provided in Sect. 10.16 of ACI 318-77:

$$\phi V_n = 0.85\, \phi f'_c\, bw \qquad \text{(Eq. 5.7-2)}$$

$$\phi = 0.70$$

To avoid accidental spalling or cracking at the ends of thin stemmed members, minimum reinforcement equal to $N_u/\phi f_y$, but not less than one #3 bar, is recommended when the bearing area is less than 20 sq in.

5.8 Reinforced Concrete Bearing

If the applied load, V_u, exceeds the design bearing strength, ϕV_n, as calculated by Eq. 5.7-1, reinforcement is required in the bearing area. This reinforcement can be designed by shear-friction as discussed in Sect. 5.6. Referring to Fig. 5.8.1, the reinforcement $A_{vf} + A_n$ nominally parallel to the direction of the axial load, N_u, is determined by Eqs. 5.6-1 through 5.6-4, with $V_{up} = V_u \cos\theta$ and A_{cr} the lesser of $bw/\sin\theta$ or $bh/\cos\theta$.

Tables 5.20.2 and 5.20.3 are design aids which may be used to determine A_{vf}.

With reinforcement in the direction of the axial tension, the design bearing strength can be calculated by Eq. 5.7-1 with $C_r = 1.0$.

Vertical reinforcement across potential horizontal cracks can be calculated by:

$$A_{sh} = \dfrac{(A_{vf} + A_n)\, f_y}{\mu_e\, f_{ys}} \qquad \text{(Eq. 5.8-1)}$$

where

$$\mu_e = \dfrac{1000\lambda^2\, A_{cr}\, \mu}{(A_{vf} + A_n) f_y} \qquad \text{(Eq. 5.8-2)}$$

f_{ys} = yield strength of A_{sh}, psi

A_{cr} = $1.7\ell_d\, b$, sq in.

b = average member width, in.

ℓ_d = development length of A_{vf} bars, in.

Fig. 5.8.1 Reinforced concrete bearing

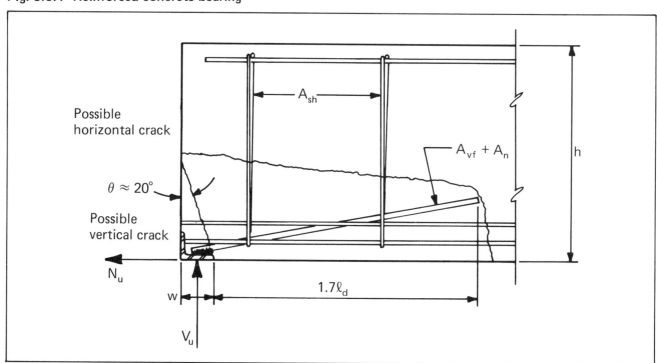

with $\mu = 1.4$, f_y and f_{ys} in ksi, Eqs. 5.8-1 and 5.8-2 can be combined so that:

$$A_{sh} = \frac{[(A_{vf} + A_n)f_y]^2}{\lambda^2(1.4\,f_{ys})(1.7\,\ell_d b)} \qquad \text{(Eq. 5.8-3)}$$

The minimum requirements are:

$$A_{sh}(min) = \frac{120\,(1.7\,\ell_d)b}{f_{ys}} \qquad \text{(Eq. 5.8-4)}$$

unless 1/3 more reinforcement than required by Eq. 5.8-3 is provided.

Stirrups or mesh used for diagonal tension reinforcement can be considered to act as A_{sh} reinforcement.

Table 5.20.4 may be used to determine A_{sh} requirements.

When members are subjected to bearing stresses in excess of the limits indicated in Sect. 10.16 of ACI 318-77, confinement reinforcement in all directions may be required. Design of this steel is beyond the scope of this manual.

Example 5.8.1 Reinforced bearing for a rectangular beam

Given:

PCI standard rectangular beam 16RB28

V_u = 100 kips (includes all load factors)

N_u = 20 kips

bearing pad = 4 in. x 14 in.

f_y for all reinforcement = 40,000 psi

f'_c = 5000 psi (normal weight)

Find: Reinforcement requirements for the end of the member.

Solution:

Check plain concrete bearing by Eq. 5.7-1

Assume s = w/2 = 2 in.

$$C_r = \left(\frac{sw}{200}\right)^{N_u/V_u} = \left(\frac{2 \times 4}{200}\right)^{0.20}$$
$$= 0.525$$

$$\phi V_n = \phi C_r\, 70\,\lambda\sqrt{f'_c}\,(s/w)^{1/3}bw$$
$$= 0.70\,(0.525)(70)(1)(\sqrt{5000})(2/4)^{1/3}(4 \times 14)/1000$$
$$= 80.8 \text{ kips} < 100, \text{ therefore reinforcement is required.}$$

(This could also be determined from Table 5.20.1)

Assume $\theta = 20°$

$$A_{cr} = \text{lesser of } \frac{bw}{\sin\theta} \text{ or } \frac{bh}{\cos\theta}$$

$$\frac{bw}{\sin\theta} = 16(4)/0.342 = 187 \text{ sq in.}$$

$$\frac{bh}{\cos\theta} = 16(28)/0.940 = 477 \text{ sq in.}$$

Use A_{cr} = 187 sq in.

$$V_{up} = V_u \cos\theta = 100,000\,(0.940) = 94,000 \text{ lb}$$

By Eq. 5.6-2

$$\mu_e = \frac{1000\lambda^2 A_{cr}\,\mu}{V_{up}} = \frac{1000(1)(187)(1.4)}{94,000}$$
$$= 2.79$$

By Eq. 5.6-1

$$A_{vf} = \frac{V_{up}}{\phi f_y\,\mu_e} = \frac{94,000}{0.85(40,000)(2.79)}$$
$$= 0.99 \text{ sq in.}$$

Alternately, interpolating from Table 5.20.2

$$A_{vf} = 0.93 + \frac{200-187}{200-175}(1.06-0.93)$$
$$= 1.00 \text{ sq in.}$$

Check minimum by Eq. 5.6-3:

$$A_{vf}(min) = \frac{120\,A_{cr}}{f_y} = \frac{120\,(187)}{40,000} = 0.56$$
$$< 0.99 \text{ sq in.}$$

By Eq. 5.6-4

$$A_n = \frac{N_u}{\phi f_y} = \frac{20,000}{0.85(40,000)}$$
$$= 0.59 \text{ sq in.}$$

$$A_{vf} + A_n = 0.99 + 0.59 = 1.58 \text{ sq in.}$$

Use 4 – #6 = 1.76 sq in.

By Eq. 5.8-3:

$$A_{sh} = \frac{[(A_{vf} + A_n)\,f_y]^2}{\lambda^2(1.4\,f_{ys})(1.7\,\ell_d b)}$$

$1.7\,\ell_d$ from Table 8.2.7 = 20.4 in.

$$A_{sh} = \frac{[(1.58)(40)]^2}{(1)(1.4)(40)(20.4)(16)} = 0.22 \text{ sq in.}$$

By Eq. 5.8-4

$$A_{sh}(min) = \frac{120\,(1.7\,\ell_d)b}{f_{ys}} = \frac{120\,(20.4)(16)}{40,000}$$
$$= 0.98 \text{ sq in.}$$

Thus provide 1/3 more than required by Eq. 5.8-3:

$$= 4/3\,(0.22) = 0.29 \text{ sq in.}$$

Alternately using Table 5.20.4, for b = 16 in. and $A_{vf} + A_n = 4 - \#6$, read:

$$A_{sh} = 0.37 \text{ sq in.}$$

Note: Table 5.20.4 will usually be more conservative because it assumes A_{vf} to be the steel provided rather than the steel required.

5.9 Dapped-End Connections

Design of connections which are recessed, or dapped into the end of the member, requires the investigation of several potential failure modes. These are illustrated in Fig. 5.9.1 and listed below with the reinforcement required for each consideration.

1) Flexure (cantilever bending) and axial tension in the extended end. Provide flexural reinforcement, A_s, plus axial tension reinforcement, A_n.

2) Direct shear at the junction of the dap and the main body of the member. Provide shear-friction reinforcement composed of A_s and A_{vh}, plus axial tension reinforcement, A_n.

3) Diagonal tension emanating from the reentrant corner. Provide shear reinforcement, A_{sh}.

4) Diagonal tension in the extended end. Provide shear reinforcement composed of A_{vh} and A_v.

5) Bearing on the extended end. If plain concrete bearing strength is exceeded, use A_s as shear-friction reinforcement.

Each of these potential failure modes should be investigated separately. The reinforcement requirements are not cumulative, that is, A_s is the greater of that required by 1, 2 or 5, not the sum. A_{vh} is the greater of that required by 2 or 4.

5.9.1 Flexure and Axial Tension in the Extended End

The flexural and axial tension horizontal reinforcement can be determined by:

$$A_s + A_n = \frac{1}{\phi f_y} \left[\frac{V_u a + N_u (h - d)}{d} + N_u \right]$$
(Eq. 5.9-1)

where:

$\phi = 0.85*$

a = shear span, in. (can be assumed = $3/4\, \ell_p$)

ℓ_p = dap projection, in.

h = depth of the member above the dap, in.

d = distance from top to center of the reinforcement, A_s, in.

f_y = yield strength of the flexural reinforcement, psi

For design convenience, Eq. 5.9-1 can be rearranged as follows:

$$A_s + A_n = \frac{1}{\phi f_y} \left[V_u \left(\frac{a}{d} \right) + N_u \left(\frac{h}{d} \right) \right]$$
(Eq. 5.9-1a)

Table 5.20.5 may be used to determine the steel requirements.

5.9.2 Direct Shear

The potential vertical crack shown in Fig. 5.9.1 is resisted by a combination of $(A_s + A_n)$ and A_{vh}. This reinforcement can be calculated by Eqs. 5.9.2 through 5.9.4:

$$A_s = \frac{2V_u}{3\, \phi f_y \mu_e}$$
(Eq. 5.9-2)

$$A_n = \frac{N_u}{\phi f_y}$$
(Eq. 5.9-3)

$$A_{vh} = \frac{V_u}{3\, \phi f_{yv} \mu_e}$$
(Eq. 5.9-4)

where

$\phi = 0.85$

f_y = yield strength of A_s and A_n, psi

f_{yv} = yield strength of A_{vh}, psi

$$\mu_e = \frac{1000 \lambda^2\, bh\, \mu}{V_u}$$
(Eq. 5.9-5)

(See Sect. 5.6 for definition of λ)

The recommended minimum reinforcement requirements are:

$$A_s \text{ (min)} = \frac{80\, bh}{f_y}$$
(Eq. 5.9-6)

$$A_{vh} \text{ (min)} = \frac{40\, bh}{f_{yv}}$$
(Eq. 5.9-7)

unless one-third more than that required by either Eq. 5.9-2 or 5.9-4 is provided.

Reinforcement A_{vh} should be uniformly distributed within 2/3 d of reinforcement $A_s + A_n$.

---

* To be theoretically correct, Eq. 5.9-1 should have $j_u d$ in the denominator. The use of $\phi = 0.85$ instead of 0.90 (flexure) compensates for this approximation.

Fig. 5.9.1 Required reinforcement in dapped-end connections

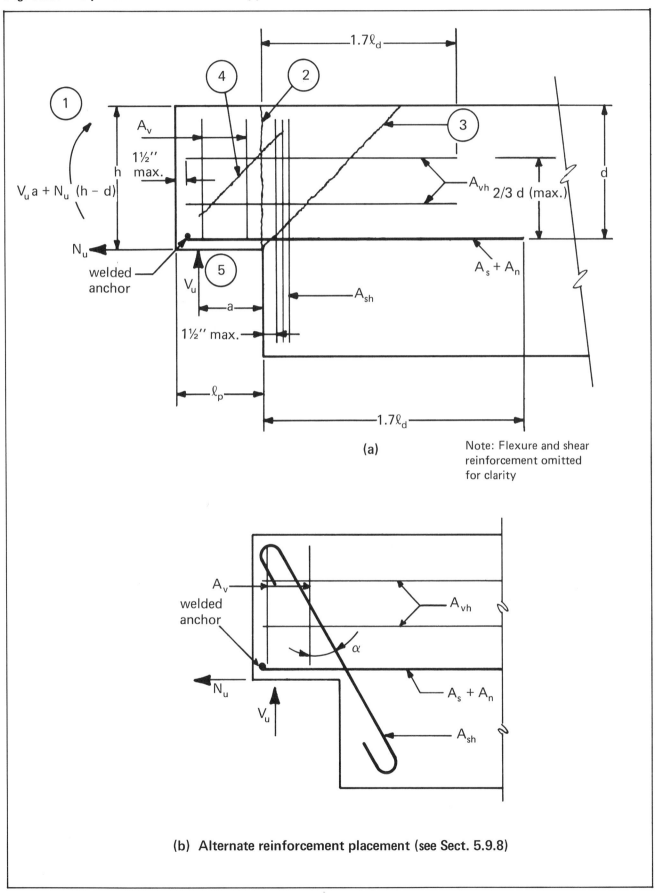

(a)

Note: Flexure and shear
reinforcement omitted
for clarity

(b) **Alternate reinforcement placement (see Sect. 5.9.8)**

For design convenience, Eq. 5.9-5 can be combined with Eqs. 5.9-2 and 5.9-4 to yield:

$$A_s = \frac{V_u^2}{1.78 f_y \lambda^2 bh} \qquad \text{(Eq. 5.9-2a)}$$

$$A_{vh} = \frac{V_u^2}{3.57 f_{yv} \lambda^2 bh} \qquad \text{(Eq. 5.9-4a)}$$

where

$\mu = 1.4$

V_u is in kips

f_y and f_{yv} are in ksi

5.9.3 Diagonal Tension at Reentrant Corner

The reinforcement required to resist diagonal tension cracking starting from the reentrant corner can be calculated from:

$$A_{sh} = \frac{V_u}{\phi f_{ys}} \qquad \text{(Eq. 5.9-8)}$$

where

$\phi = 0.85$

V_u = applied factored load, lb

A_{sh} = vertical or diagonal bars across potential diagonal tension crack, sq in.

f_{ys} = yield strength of A_{sh}, psi

$V_u/\phi bd$ should not exceed $8\lambda \sqrt{f'_c}$.

5.9.4 Diagonal Tension in the Extended End

Additional reinforcement is required in the extended end, as shown in Fig. 5.9.1, such that:

$$\phi V_n = \phi (A_v f_v + A_{vh} f_{yv} + 2 \lambda bd \sqrt{f'_c}) \qquad \text{(Eq. 5.9-9)}$$

where

f_v = yield strength of A_v

Tests on dapped-end beams* indicate that at least one half of the reinforcement required in this area should be placed vertically. Thus:

$$\min A_v = \frac{1}{2 f_v} \left(\frac{V_u}{\phi} - 2 \lambda bd \sqrt{f'_c} \right) \qquad \text{(Eq. 5.9-10)}$$

5.9.5 Bearing on the Extended End

The bearing on the extended end should be checked against the plain concrete bearing limita-

---

* Test performed by Raths, Raths and Johnson, Hinsdale, IL (results unpublished).

tion of Eq. 5.7.1. If the limits are exceeded, then the capacity should be checked for reinforced concrete bearing as described in Sect. 5.8:

$$A_s + A_n = \frac{1}{\phi f_y} \left(\frac{V_u \cos\theta}{\mu_e} + N_u \right) \qquad \text{(Eq. 5.9-11)}$$

$$A_{sh} f_{ys} + A_v f_v = \frac{A_s f_y}{\mu_e} \qquad \text{(Eq. 5.9-12)}$$

5.9.6 Anchorage of Reinforcement

Horizontal bars $A_s + A_n$ should be extended a minimum of $1.7\ell_d$ past the end of the dap, and anchored at the end of the beam by welding to cross bars, angles or plates. Horizontal bars A_{vh} should be extended a minimum of $1.7\ell_d$ past the end of the dap, and anchored at the end of the beam by hooks or other suitable means. Vertical or diagonal bars A_{sh} and A_v should be properly anchored by hooks as required by ACI 318-77. Welded wire fabric may be used for reinforcement, and should be anchored in accordance with ACI 318-77.

5.9.7 Detailing Considerations

Experience has shown that the depth of the extended end should not be less than one-half the depth of the beam, unless the beam is significantly deeper than necessary for architectural reasons.

Diagonal tension reinforcement, A_{sh}, should be placed as closely as practical to the reentrant corner. This reinforcement requirement is not additive to other shear reinforcement requirements.

Reinforcement requirements may be met with welded headed studs, deformed bar anchors or welded wire fabric.

If the flexural stress, calculated for the full depth of section using factored loads and gross section properties, exceeds $6\sqrt{f'_c}$ immediately beyond the dap, longitudinal reinforcement should be placed in the beam to develop the required flexural strength.

5.9.8 Alternate Placement of Reinforcement

As an alternate to placing reinforcement as shown in Fig. 5.9.1a, diagonal bars can be placed as shown in Fig. 5.9.1b. The requirements for reinforcement placed in this manner can be determined by:

$$A_{sh} = \frac{V_u}{\phi f_{ys} \cos\alpha} \approx \frac{V_u \sqrt{a^2 + d^2}}{\phi f_{ys} d} \qquad \text{(Eq. 5.9-13)}$$

$$A_s + A_n = \frac{N_u h}{\phi f_y d} \qquad \text{(Eq. 5.9-14)}$$

Fig. 5.9.2 Dapped-end beam of Example 5.9.1

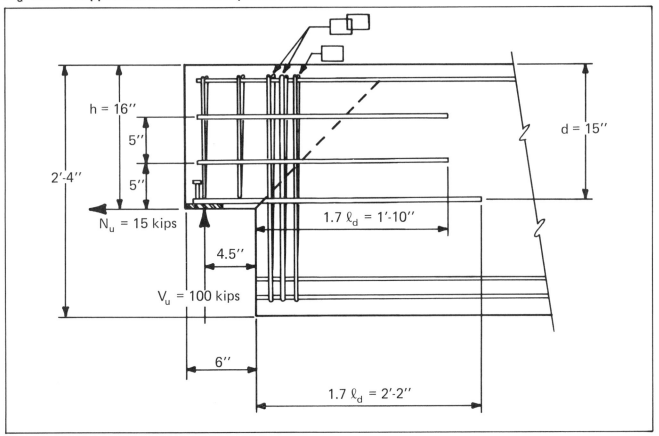

but not less than that determined in sections 5.9.2 or 5.9.5.

If the diagonal bars can be adequately anchored into the extended end, they may also be used as at least partial replacement for A_v and A_{vh} requirements shown in Sect. 5.9.4.

Example 5.9.1 Reinforcement for dapped-end beam

Given: The 16RB28 beam with a dapped end as shown in Fig. 5.9.2.

V_u = 100 kips (includes all load factors)

N_u = 15 kips

f'_c = 5000 psi (normal weight)

f_y for all reinforcement = 60 ksi (weldable)

Problem: Determine the requirements for reinforcement $A_s + A_n$, A_{vh}, A_{sh}, and A_v shown in Fig. 5.9.1.

Solution:

Assume: Shear span, a = 3/4 (6) = 4.5 in.

d = 15 in.

1) Flexure in extended end:

By Eq. 5.9-1a:

$$A_s + A_n = \frac{1}{\phi f_y}\left[V_u\left(\frac{a}{d}\right) + N_u\left(\frac{h}{d}\right)\right]$$

$$= \frac{1}{0.85 \times 60}\left[100\left(\frac{4.5}{15}\right) + 15\left(\frac{16}{15}\right)\right] = 0.90 \text{ sq in.}$$

Use 3 – #5, A_s = 0.93 sq in.

This could also be obtained from Table 5.20.5

2) Direct shear:

By Eq. 5.9-2a:

$$A_s = \frac{V_u{}^2}{1.78 f_y \lambda^2 bh} = \frac{(100)^2}{1.78(60)(1)(16)(16)}$$

$$= 0.37 \text{ sq in.}$$

By Eq. 5.9-6:

$$A_s \text{ (min)} = \frac{80 \, bh}{f_y} = \frac{80(16)(16)}{60,000}$$

$$= 0.34 \text{ sq in.} < 0.37$$

By Eq. 5.9-3:

$$A_n = \frac{N_u}{\phi f_y} = \frac{15,000}{(0.85)(60,000)} = 0.29 \text{ sq in.}$$

$A_s + A_n = 0.37 + 0.29 = 0.66 \text{ sq in.} < 0.93$

Therefore, flexure governs.

By Eq. 5.9-4a:

$$A_{vh} = \frac{V_u^2}{3.57 f_{yv} \lambda^2 bh}$$

$$= \frac{(100)^2}{3.57 (60)(1)(16)(16)} = 0.18 \text{ sq in.}$$

From Eq. 5.9-7:

$$A_{vh} \text{ (min)} = \frac{40 \, bh}{f_y} = \frac{40 \, (16)(16)}{60,000}$$

$$= 0.17 \text{ sq in.} < 0.18 \text{ sq in.}$$

Try 2 – #3 U-bars, $A_{vh} = 0.44 \text{ sq in.}$

3) Diagonal tension at reentrant corner:

By Eq. 5.9-8:

$$A_{sh} = \frac{V_u}{\phi f_y} = \frac{100}{0.85(60)} = 1.96 \text{ sq in.}$$

Use 5 – #4 closed ties = 2.00 sq in.

Check $V_u/\phi bd = 100/(0.85 \times 16 \times 15)$

$$= 0.490 \text{ ksi} < 8\sqrt{f'_c} = 0.566 \text{ ksi} \quad OK$$

4) Diagonal tension in the extended end:

Concrete capacity = $2\lambda \sqrt{f'_c} \, bd$

$= 2 (1)\sqrt{5000} \, (16)(15)/1000 = 33.9 \text{ kips}$

By Eq. 5.9-10:

$$A_v = \frac{1}{2 f_v}\left(\frac{V_u}{\phi} - 2\lambda\sqrt{f'_c} \, bd\right)$$

$$= \frac{1}{2 (60)}\left(\frac{100}{0.85} - 33.9\right) = 0.70 \text{ sq in.}$$

Try 2 – #4 = 0.80 sq in.

Check Eq. 5.9-9

$\phi V_n = \phi (A_v f_v + A_{vh} f_{yv} + 2\lambda\sqrt{f'_c} \, bd)$

$= 0.85 [0.80 (60) + 0.44 (60) + 33.9]$

$= 92.1 \text{ kips} < 100$

Change A_{vh} to 2 – #4

$\phi V_n = 110.4 \text{ kips} > 100 \quad OK$

5) Check concrete bearing by Eq. 5.7-1:

$w = 4.5 \text{ in.}$

$s = 4.5/2$ (worst case)

$sw = 4.5 (4.5/2) = 10.1$, use 9 max

$s/w = 0.5$

$$C_r = \left(\frac{sw}{200}\right)^{N_u/V_u} = (9/200)^{0.15} = 0.63$$

$\phi V_n = \phi C_r \, 70 \, \lambda\sqrt{f'_c} \, (s/w)^{1/3} \, bw$

$\phi V_n = 0.7 \, (0.63)(70)(1) \sqrt{5000} \, (0.5)^{1/3}$
$(16)(4.5)/1000 = 124.7 \text{ kips} > 100 \quad OK$

Reinforcement is not required for bearing (Could also be determined from Table 5.20.1)

6) Check anchorage requirements:

$A_s + A_n$ bars:

From Table 8.2.7:

$f_y = 60,000 \text{ psi}, f'_c = 5000 \text{ psi}, #5 \text{ bars}$

$1.7\ell_d = 26 \text{ in. beyond dap}$

A_{vh} bars:

From Table 8.2.7, for #4 bars,

$1.7\ell_d = 20 \text{ in. beyond dap}$

5.10 Beam Ledges

The design shear strength of continuous beam ledges supporting concentrated loads, as illustrated in Fig. 5.10.1, can be determined by the lesser of Eq. 5.10-1 and 5.10-2:

for $s > b + h$

$$\phi V_n = 3 \phi h \, \lambda\sqrt{f'_c} \, (2\ell_p + b + h) \quad \text{(Eq. 5.10-1)}$$

$$\phi V_n = \phi h \, \lambda\sqrt{f'_c} \, (2\ell_p + b + h + 2d_e)$$
$$\text{(Eq. 5.10-2)}$$

for $s < b + h$, and equal concentrated loads, use the lesser of Eqs. 5.10-1a, 5.10-2a or 5.10-3

$$\phi V_n = 1.5 \phi h \, \lambda\sqrt{f'_c} \, (2\ell_p + b + h + s)$$
$$\text{(Eq. 5.10-1a)}$$

$$\phi V_n = \phi h \, \lambda\sqrt{f'_c} \, \left(\ell_p + \frac{b + h}{2} + d_e + s\right)$$
$$\text{(Eq. 5.10-2a)}$$

where:

h = depth of the beam ledge, in.

ℓ_p = ledge projection, in.

b = width of bearing area, in.

s = spacing of concentrated loads, in.

d_e = distance from center of load to the end of the beam, in.

Fig. 5.10.1 Design of beam ledges

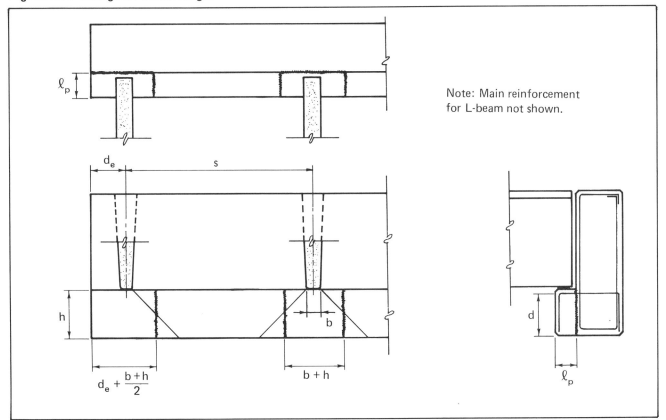

Note: Main reinforcement for L-beam not shown.

Note: Eq. 5.10-2 or 2a will be critical when $d_e < (2\ell_p + b + h)$

If the applied factored load exceeds the strength as determined by Eq. 5.10-1 or 5.10-2, the ledge should be designed in accordance with Sect. 5.9.

If the ledge supports a continuous load or closely spaced concentrated loads, the design shear strength is:

$$\phi V_n = 24 \, \phi h \, \lambda \sqrt{f'_c} \qquad \text{(Eq. 5.10-3)}$$

where ϕV_n is the design shear strength in pounds per foot.

Flexural reinforcement, A_s, computed by Eq. 5.9-1 or 5.9-1a, should be provided in the beam ledge. Such reinforcement may be uniformly spaced over a width of 6h on either side of the bearing, but not to exceed 1/2 the distance to the next load. Bar spacing should not exceed the ledge depth, h, or 18 in.

A minimum reinforcement equal to 200 (b + h) d/f_y should be placed in this area unless the reinforcement provided is 1/3 greater than that calculated by Eq. 5.9-1 or 5.9-1a.

The provisions for plain or reinforced concrete bearing (Sect. 5.7 or 5.8) should also be met.

Example 5.10.1 – Design of a beam ledge

Given: 8 ft wide double tees resting on a standard L-beam similar to that shown in Fig. 5.10.1. Lay-out of tees is irregular so that a stem can be placed at any point on the ledge.

V_u per stem = 18 kips

N_u per stem = 3 kips

bearing pad, b = 3 in., w = 4 in.

h = 12 in.

d = 11 in.

ℓ_p = 6 in.

s = 48 in.

f'_c = 5000 psi (normal weight)

f_y = 40 ksi

Problem: Investigate shear strength and determine reinforcement for the ledge.

Solution:

Min d_e = b/2 = 1.5 in.

Since s > b + h and $d_e < 2\ell_p + b + h$, use Eq. 5.10-2

$$\phi V_n = \phi h \lambda \sqrt{f'_c} \ (2\ell_p + b + h + 2d_e)$$
$$= 0.85 \,(12)(1) \sqrt{5000} \ [2\,(6) + 3 + 12 + 2\,(1.5)]\,/1000 = 21.6 \text{ kips} > 18$$

Shear span, a = $3\ell_p/4$ = 4.5 in.

By Eq. 5.9-1a:

$$A_s = \frac{1}{\phi f_y}\left[V_u\left(\frac{a}{d}\right) + N_u\left(\frac{h}{d}\right)\right]$$

$$= \frac{1}{0.85(40)}\left[18(4.5/11) + 3(12/11)\right]$$

$$= 0.31 \text{ in.}$$

(Could also be obtained from Table 5.20.5)

$$A_s \text{ (min)} = 200(b+h)d/f_y$$

$$= 200(3+12)(11)/40,000$$

$$= 0.83 \text{ sq in.}$$

or $4/3 \times 0.31 = 0.41$ sq in.

$6h = 6 \text{ ft} > s/2$

Therefore distribute reinforcement over s/2 each side of the load.

$s/2 \times 2 = 4$ ft

Maximum bar spacing = h = 12 in.

#3 @ 12 in. = 0.44 sq in. in each 4 ft

Place 2 additional bars at the beam end to provide equivalent reinforcement for stem placed near the end.

Check plain concrete bearing:

By Eq. 5.7-1:

min s = w/2 = 2 in.

$$C_r = \left(\frac{sw}{200}\right)^{N_u/V_u} = \left(\frac{2\times4}{200}\right)^{3/18} = 0.58$$

$$\phi V_n = \phi\, C_r\, 70\lambda\sqrt{f'_c}\,(s/w)^{1/3}\,bw$$

$$= 0.70(0.58)(70)(1)\sqrt{5000}\,(2/4)^{1/3}$$

$$(3)(4)/1000 = 19.1 \text{ kips} > 18 \text{ kips}$$

Reinforcement is not required for bearing

5.11 Concrete Brackets or Corbels*

Design of reinforced concrete brackets or corbels should include the investigation of flexure, axial tension, direct shear and bearing in a manner similar to that shown for dapped-end members in Sect. 5.9. Thus, referring to Fig. 5.11.1, the main tensile reinforcement, $A_s + A_n$, is the greater of that determined from Eqs. 5.11-1 and 5.11-2:

$$A_s + A_n = \frac{1}{\phi f_y}\left[V_u\left(\frac{a}{d}\right) + N_u\left(\frac{h}{d}\right)\right] \quad \text{(Eq. 5.11-1)}$$

$$A_s + A_n = \frac{1}{\phi f_y}\left[\frac{2V_u}{3\mu_e} + N_u\right] \quad \text{(Eq. 5.11-2)}$$

---

* As a result of recent test data, the PCI Committee on Connection Details suggests the design method shown here as a substitute for that shown in Sect. 11.9 of ACI 318-77.

$$A_{vh} = \frac{V_u}{3\phi f_{yv}\mu_e} \quad \text{(Eq. 5.11-3)}$$

The minimum reinforcement requirements of Sect. 5.9.2 should also be met.

The horizontal reinforcement, A_{vh}, should be distributed over a depth of d/2 from the flexural reinforcement, A_s.

All reinforcement should be adequately anchored by welding, hooks, or development length.

The section should also be investigated for plain or reinforced concrete bearing, as shown in Sect. 5.7 or 5.8.

Table 5.20.6 gives corbel capacities for various widths, projections, and reinforcement.

Example 5.11.1 Reinforced concrete corbel

Given: A concrete corbel similar to that shown in Fig. 5.11.1

V_u = 80 kips (includes all load factors)

N_u = 15 kips

f_y = Grade 40 (weldable)

f'_c = 5000 psi (normal weight)

Bearing pad – 14 in. x 6 in.

b = 14 in.

ℓ_p = 8 in.

Find: Corbel depth and reinforcement

Solution:

Try h = 14 in., d = 13 in.

$a = 3/4\,\ell_p = 6$ in.

By Eq. 5.11-1:

$$A_s + A_n = \frac{1}{\phi f_y}\left[V_u\left(\frac{a}{d}\right) + N_u\left(\frac{h}{d}\right)\right]$$

$$= \frac{1}{0.85(40)}\left[80\left(\frac{6}{13}\right) + 15\left(\frac{14}{13}\right)\right]$$

$$= 1.56 \text{ sq in.}$$

(Could also be obtained from Table 5.20.5)

By Eq. 5.11-2:

$$\mu_e = \frac{1000\,bh\,\mu}{V_u} = \frac{1000(14)(14)(1.4)}{80,000}$$

$$= 3.43$$

$$A_s + A_n = \frac{1}{\phi f_y}\left[\frac{2V_u}{3\mu_e} + N_u\right]$$

$$= \frac{1}{0.85(40)}\left[\frac{2(80)}{3(3.43)} + 15\right]$$

$$= 0.90 < 1.56$$

Provide 3 – #7 = 1.80 sq in.

Fig. 5.11.1 Design of concrete corbels

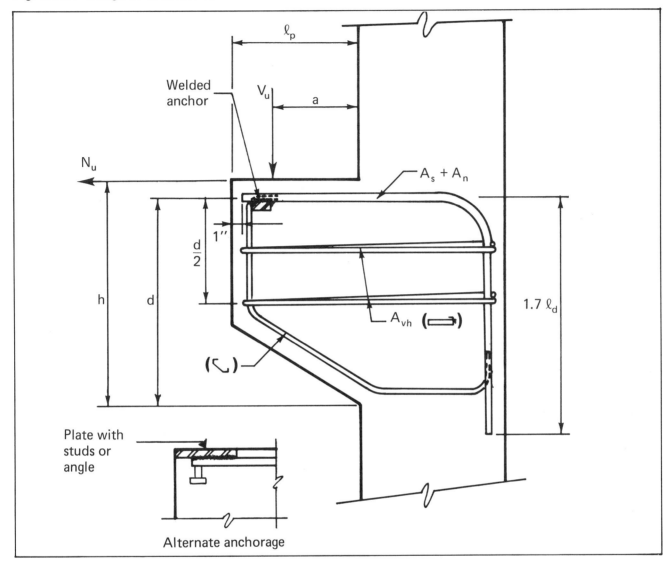

Alternately, the solution could be obtained from Table 5.20.6:

For b = 14 in. and ℓ_p = 8 in., interpolation shows that for h = 14 in., the corbel would have a strength of about 89 kips with A_s = 3 – #7.

By Eqs. 5.9-6 and 5.9-3:

$$A_s \text{ (min)} + A_n = \frac{80\,bh}{f_y} + \frac{N_u}{\phi\,f_y}$$

$$= \frac{80\,(14)(14)}{40,000} + \frac{15,000}{0.85\,(40,000)}$$

$$= 0.39 + 0.44 = 0.83 \text{ sq in.} < 1.56$$

By Eq. 5.11-3:

$$A_{vh} = \frac{V_u}{3\,\phi\,f_{yv}\,\mu_e}$$

$$= \frac{80,000}{3\,(0.85)(40,000)(3.43)} = 0.23 \text{ sq in.}$$

$$A_{vh} \text{ (min)} = \frac{40\,bh}{f_{yv}} = \frac{40\,(14)(14)}{40,000}$$

$$= 0.20 \text{ sq in.} < 0.23$$

Provide 2 – #3 closed ties = 0.44 sq in.
Check plain concrete bearing:

Assume s = w/2 = 3 in.

N_u / V_u = 15/80 = 0.19 say 0.20
From Table 5.20.1:

$$\phi V_n = 1480\,(14)(6)/1000 = 124 \text{ kips} > 80$$

No bearing reinforcement is required.

For anchorage, weld to a #7 cross bar. (Note: The requirements for weldability in Sect. 5.18 must be met.)

Tension in each bar $\dfrac{A_s + A_n}{\text{No. of bars}}$ (f_y)

$$= \dfrac{1.56}{3}\ (40) = 20.8 \text{ kips per bar}$$

From Table 5.18.1:

$$A_w = \dfrac{20.8}{42} = 0.49 \text{ sq in.}$$

From Fig. 5.18.1

Effective throat $= 1/3\ d_b = 1/3\ (7/8)$
$= 0.29$ in.

required $\ell_w = 0.49/0.29 = 1.69$ in.

For a #7 cross bar, $\ell_w = 0.875$ in.

Weld on two sides $= 2(0.875) = 1.75$ in.
> 1.69 OK

5.12 Structural Steel Haunches

Structural steel shapes such as wide flange beams, double channels, tubes or vertical plates often serve as haunches or brackets as illustrated in Fig. 5.12.1. In lieu of a more precise analysis, the capacity of these members can be calculated by statics, using the conservative assumptions shown in Fig. 5.12.1.

The design strength of the section is then:

$$V_c = \dfrac{0.85\ f'_c\ b\ell_e}{3.67 + 4a/\ell_e} \qquad \text{(Eq. 5.12-1)}$$

V_c = nominal strength of the section controlled by concrete, lb

a = shear span, in.

ℓ_e = embedment depth, in.

b = effective width of the compression block (See Fig. 5.12.2)

The effective width of the compression block for double flanged members can be as shown in Fig. 5.12.2(a) provided steps are taken to assure good compaction of the concrete and/or confinement of the concrete around the section. Holes (more than 1 in. diameter) aid compaction, the steel studs provide confinement.

It should be noted that there must be adequate concrete and/or superimposed axial dead load above and below the haunch to develop the compressive forces indicated in Fig. 5.12.1 in order for Eq. 5.12-1 to be valid. If not, it may be possible to develop the force couple by using reinforcing bars in tension, in a manner similar to that shown below.

Additional capacity can be obtained by weld-

ing vertical reinforcing bars to the steel section, as shown in Fig. 5.12.2(b). In lieu of a more precise analysis[1] the additional capacity can be calculated by conservatively assuming that the reinforcement acts at the centers of compression, and that reinforcement nearest to the applied load is balanced by reinforcement near the end of the steel member. Thus:

$$V_r = \dfrac{3\ A_s\ f_y}{3.67 + 4a/\ell_e} \qquad \text{(Eq. 5.12-2)}$$

$$A'_s = \dfrac{A_s\ (12a + 2\ell_e)}{12a + 11\ \ell_e} \qquad \text{(Eq. 5.12-3)}$$

where:

V_r = additional nominal strength of the section provided by reinforcement

A_s = area of vertical reinforcement nearest to the applied load (theoretically located at $\ell_e/6$ from face)

A'_s = area of vertical reinforcement near the end of the steel section (theoretically located at $11\ \ell_e/12$ from face)

f_y = yield strength of the reinforcement

The total design strength of the section is thus:

$$\phi V_n = \phi\ (V_c + V_r) \qquad \text{(Eq. 5.12-4)}$$

$$\phi = 0.85$$

The design strength of the steel section can be determined by:

Flexural design strength:

$$\phi V_n = \dfrac{\phi Z_s\ f_y}{a} \qquad \text{(Eq. 5.12-5)}$$

Shear design capacity:

$$\phi V_n = \phi\ (0.55\ f_y\ h\ t) \qquad \text{(Eq. 5.12-6)}$$

where:

Z_s = plastic section modulus of the steel section (see table 5.12.1)

f_y = yield strength of the steel

h, t = depth and thickness of steel web, respectively.

$\phi = 0.90$

Note: Plastic design criteria for structural steel does not require the use of a ϕ- factor. However, the load factors used are 1.7 (D + L). Therefore, when using steel plastic design with concrete load

---

[1] Raths, Charles H., "Embedded Structural Steel Connections," *PCI Journal,* May-June, 1974.

PCI Design Handbook

Fig. 5.12.1 Embedded structural steel shape

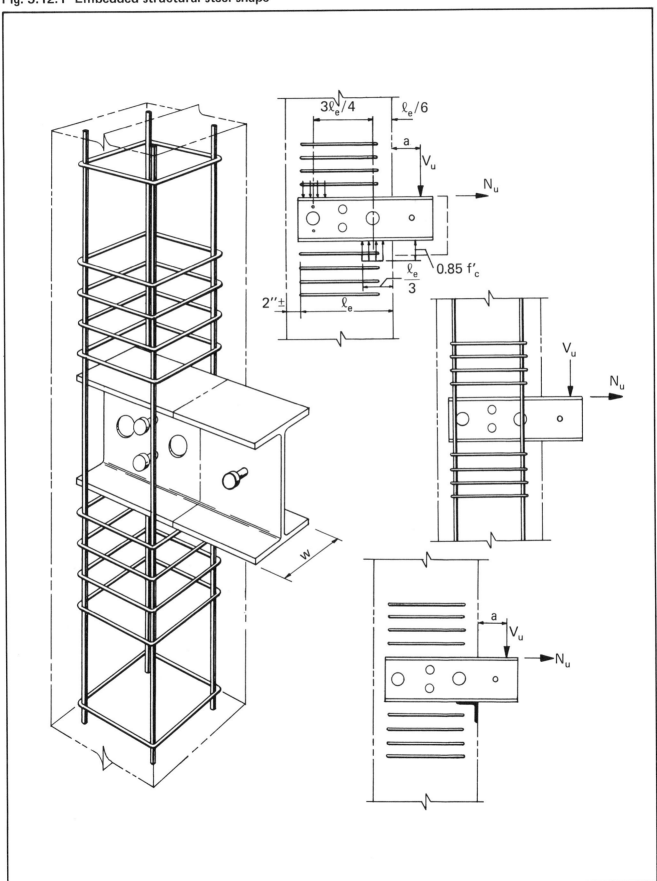

Table 5.12.1 Plastic section moduli and shape factors

Section	Plastic section modulus, Z_s, in.^{3}	Shape factor
	$$\dfrac{bh^2}{4}$$	1.5
	x-x axis: $$bt\,(h-t)+\dfrac{w}{4}\,(h-2t)^2$$	1.12 (approx)
	y-y axis: $$\dfrac{b^2 t}{2}+\dfrac{(h-2t)w^2}{4}$$	1.55 (approx)
	$$bt\,(h-t)+\dfrac{w(h-2t)^2}{4}$$	1.12 (approx)
	$$\dfrac{h^3}{6}$$	1.70
	$$\dfrac{h^3}{6}\left[1-\left(1-\dfrac{2t}{h}\right)^3\right]$$ th^2 for $t\ll h$	$$\dfrac{16}{3\pi}\left[\dfrac{1-\left(1-\dfrac{2t}{h}\right)^3}{1-\left(1-\dfrac{2t}{h}\right)^4}\right]$$ 1.27 for $t\ll h$
	$$\dfrac{bh^2}{4}\left[1-\left(1-\dfrac{2w}{b}\right)\left(1-\dfrac{2t}{h}\right)^2\right]$$	1.12 (approx) for thin walls
	$$\dfrac{bh^2}{12}$$	2

Fig. 5.12.2 Effective width of embedded shapes

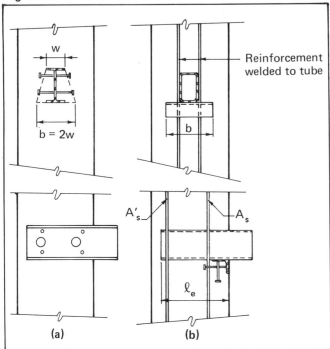

| (a) | (b) |

factors (1.4 D + 1.7 L), the use of $\phi = 0.90$ is recommended by the PCI Connections Committee in order to provide approximately the same overall factor of safety.

For steel shapes projecting equally from each side of the member, with approximately symmetrical loading, the design strength on each side as governed by the capacity of the concrete can be calculated by:

$$\phi V_n = \phi \frac{0.85 \, f'_c \, bt_c}{3} \qquad \text{(Eq. 5.12-7)}$$

Horizontal forces, N_u, are resisted by bond on the perimeter of the section. If the bond stress resulting from factored loads exceeds 250 psi, headed studs or reinforcing bars can be welded to the section.

Example 5.12.1 Design of structural steel haunch

Given: The structural steel haunch shown

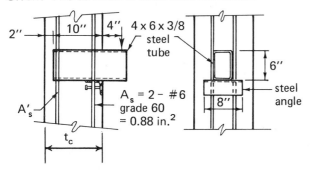

$f'_c = 5000$ psi

f_y (reinforcement) = 60,000 psi (weldable)

f_y (structural steel) = 36,000 psi

Find: The design strength

Solution:

$$V_c = \frac{0.85 \, f'_c \, b\ell_e}{3.67 + 4a/\ell_e} = \frac{0.85 \, (5)(8)(10)}{3.67 + 4 \, (4)/10}$$
$$= 64.5 \text{ kips}$$

$$V_r = \frac{3A_s \, f_y}{3.67 + 4a/\ell_e} = \frac{3 \, (0.88)(60)}{3.67 + 4(4)/10}$$
$$= 30.1 \text{ kips}$$

$$\phi V_n = 0.85 \, (64.5 + 30.1) = 80.4 \text{ kips}$$

$$A'_s = \frac{A_s \, (12a + 2\ell_e)}{12a + 11\ell_e} = \frac{0.88 \, (48 + 20)}{12 \, (4) + 11 \, (10)}$$
$$= 0.38 \text{ sq in.}$$

use 2 – #4

Alternate solution using Tables 5.20.7 and 5.20.8:

For b = 8 in., a = 4 in., t_c = 12 in.

Read ϕV_c = 55 kips (Table 5.20.7)

For A_s = 2 – #6, a = 4 in., t_c = 12 in.
f_y = 60,000 psi

Read ϕV_r = 26 kips (Table 5.20.8)

A'_s = 0.38 sq in.

ϕV_n = 55 + 26 = 81 kips

Plastic section modulus of tube (see Table 5.12.1)

$$Z_s = \frac{bh^2}{4} \left[1 - \left(1 - \frac{2w}{b}\right)\left(1 - \frac{2t}{h}\right)^2 \right]$$
$$= \frac{4 \, (6)^2}{4} \left[1 - \left(1 - \frac{0.75}{4}\right)\left(1 - \frac{0.75}{6}\right)^2 \right]$$

$$= 13.61 \text{ in.}^3$$

Flexural design strength:

$$\phi V_n = \frac{\phi Z_s \, f_y}{a} = \frac{0.9 \, (13.61)(36)}{4}$$
$$= 110.2 \text{ kips}$$

Shear design capacity

$$\phi V_n = \phi(0.55 \, f_y \, h \, t)$$
$$= 0.90(0.55)(36)(2)(6)(3/8)$$
$$= 80.2 \text{ kips}$$

Therefore, steel shear controls.

5.13 Welded Headed Studs

Welded headed studs are designed to resist direct shear, tension or a combination of the two. Either the strength of the concrete or of the steel may be critical, and both must be checked. The design equations and recommendations presented are derived from limited tests and material analysis. The equations are intended to apply to studs which are previously welded to steel plates or structural steel members, and embedded in unconfined concrete. Confinement of the concrete, either from applied compressive loads, or from reinforcement has been shown to substantially increase the capacity.

5.13.1 Shear

The concrete shear strength for studs not located near a free edge can be analyzed by shear-friction principles. Because of the lack of test data, a conservative shear-friction coefficient, μ, of 1.0 is recommended:

$$\phi V_c = \phi \mu A_b f_y \qquad \text{(Eq. 5.13-1)}$$

where:

ϕ = 0.85

μ = 1.0

A_b = cross-sectional area of the stud, sq in.

f_y = yield strength of the stud = $0.9 f_s$, psi

f_s = ultimate tensile strength of the stud material, psi

For most stud material, $f_s = 60,000$ psi and Eq. 5.13-1 becomes:

$$\phi V_c = 45,900 \, A_b, \text{ lb} \qquad \text{(Eq. 5.13-1a)}$$

A shear cone failure has also been observed in shear tests, so a further limitation is:

$$\phi V_c \leqslant \phi P_c \qquad \text{(Eq. 5.13-2)}$$

where ϕP_c is defined in Sect. 5.13.2.

If a stud is located near a free edge, the shear strength should be limited by:

$$\phi V_c = 3250 \, \phi (d_e - 1) \, \lambda \sqrt{\frac{f'_c}{5000}}$$
$$\text{(Eq. 5.13-3)}$$

where

ϕ = 0.85

d_e = distance from the stud to the free edge (see Fig. 5.13.1)

λ = as defined in Sect. 5.6

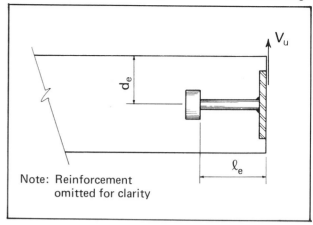

Fig. 5.13.1 Shear loading on stud near a free edge

Note: Reinforcement omitted for clarity

In no case should the edge distance, d_e, be less than 2 in.

The maximum design shear strength as governed by the steel strength is:

$$\phi V_s = 0.75 \, A_b f_s \qquad \text{(Eq. 5.13-4)}$$

Table 5.20.9 tabulates the maximum capacities from the above equations.

5.13.2 Tension

The design tensile strength of the concrete surrounding a headed stud is governed by the general equation:

$$\phi P_c = \phi A_o \left(4\lambda \sqrt{f'_c} \right) \qquad \text{(Eq. 5.13-5)}$$

where:

ϕ = 0.85

A_o = area of an assumed failure surface

For a single stud not located near a free edge, the failure surface can be conservatively assumed to be a 45° truncated cone (Fig. 5.13.2). Eq. 5.13-5 then becomes:

Fig. 5.13.2 Shear cone development for welded headed studs

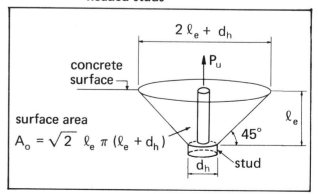

concrete surface

surface area
$A_o = \sqrt{2} \, \ell_e \, \pi \, (\ell_e + d_h)$

$2\ell_e + d_h$

P_u

ℓ_e

45°

d_h

stud

Fig. 5.13.3 Pullout surface areas for stud groups

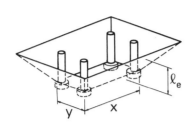

Case 1: Not near a free edge

$$A_o = xy + 2\ell_e(x+y)\sqrt{2} + 4\sqrt{2}\,\ell_e^2$$

$$\phi P_c = \phi\,4\lambda\sqrt{f'_c}\,A_o$$

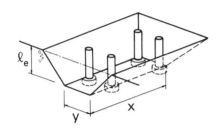

Case 2: Near a free edge on one side

$$A_o = xy + \ell_e(2x+y)\sqrt{2} + 2\sqrt{2}\,\ell_e^2$$

$$\phi P_c = \phi\,4\lambda\sqrt{f'_c}\,A_o$$

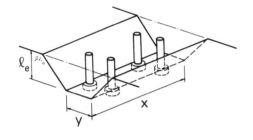

Case 3: Near a free edge on 2 opposite sides

$$A_o = xy + 2\ell_e\,x\sqrt{2}$$

$$\phi P_c = \phi\,4\lambda\sqrt{f'_c}\,A_o$$

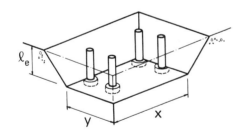

Case 4: Near a free edge on 2 adjacent sides

$$A_o = xy + \ell_e(x+y)\sqrt{2} + \sqrt{2}\,\ell_e^2$$

$$\phi P_c = \phi\,4\lambda\sqrt{f'_c}\,A_o$$

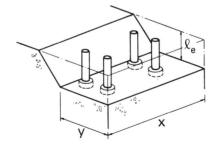

Case 5: Near a free edge on 3 sides

$$A_o = xy + \ell_e\,x\sqrt{2}$$

$$\phi P_c = \phi\,4\lambda\sqrt{f'_c}\,A_o$$

$$\phi P_c = 15.1 \, \ell_e \lambda \, (\ell_e + d_h) \sqrt{f'_c} \quad \text{(Eq. 5.13-5a)}$$

When studs are located near a free edge, a spalling type failure has been observed in tests, rather than a shear cone. For such cases it is recommended that Eq. 5.13-5 or Eq. 5.13-5a be multiplied by the reduction coefficient, C_{es}:

$$C_{es} = \frac{d_e}{\ell_e} \leqslant 1 \quad \text{(Eq. 5.13-6)}$$

where ℓ_e is the embedment length of the stud, and d_e is the distance to the edge.

> Note: If the stud is located in the corner of a concrete member, it may be necessary to apply Eq. 5.13-6 twice, once for each edge distance, d_e.

For groups of studs, if the spacing of the studs is less than $2\sqrt{2}\,\ell_e$, or if the distance to a free edge is less than $2\sqrt{2}\,\ell_e$, the 45° shear cone cannot be developed. In this case, a pull-out surface similar in shape to a truncated pyramid, as shown in Fig. 5.13.3 is recommended.

The concrete tensile strength should be the lesser of that calculated by Eq. 5.13-5a, times $C_{es}n$, where n is the number of studs in a group, and the capacity calculated by Eq. 5.13-5, using the surface of the truncated pyramid as A_o. Design aids, Tables 5.20.11 through 5.20.15 are provided to calculate these values.

The maximum design tensile strength as governed by the steel strength is:

$$\phi P_s = \phi A_b f_y = 54,000 \, A_b \quad \text{(Eq. 5.13-7)}$$
where
$$\phi = 1.0$$

Table 5.20.10 tabulates the maximum design strengths from the above equations.

5.13.3 Combined Shear and Tension

For combined shear and tension loading on headed studs, the concrete and steel strengths should be investigated separately by the following interaction equations:

Concrete:

$$\left(\frac{P_u}{\phi P_c}\right)^{4/3} + \left(\frac{V_u}{\phi V_c}\right)^{4/3} \leqslant 1 \quad \text{(Eq. 5.13-8)}$$

Steel:

$$\left(\frac{P_u}{\phi P_s}\right)^{2} + \left(\frac{V_u}{\phi V_s}\right)^{2} \leqslant 1 \quad \text{(Eq. 5.13-9)}$$

These equations are plotted in Fig. 5.20.16.

5.13.4 Plate Thickness

Plates to which studs are attached should be selected such that:

$$t \geqslant \frac{2}{3} \, d_b \quad \text{(Eq. 5.13-10)}$$

where:

t = plate thickness

d_b = nominal diameter of the stud

Example 5.13.1 Capacity of welded headed studs

Given: Bracket on column as shown

f'_c = 5000 psi (normal weight)

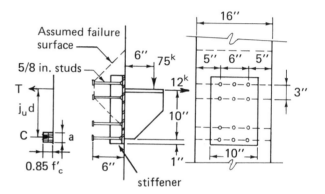

Problem: Determine if studs are adequate to resist the loads shown.

Solution:

Check concrete strength:

Tension (top group of studs)

from Table 5.20.10

d_e = 5 in., ℓ_e = 6 in., 5/8 in. studs

ϕP_c = 6 (38.7) = 232.2 kips

This is the cumulative capacity of six individual cones, reduced for edge distance. It can also be determined from Eqs. 5.13-5a and 5.13-6.

Or ϕP_c from Table 5.20.13 (case 3)

y = 3 in., x = 16 in. ℓ_e = 6 in.

ϕP_c = 77 kips

This is the capacity of a truncated pyramid accounting for the stud spacing and controls the design.

A moment-resisting couple is formed:

C = T = 0.85 f'_c ba = 77 kips

comp. block, $a = \dfrac{77}{0.85\,(5)(10)} = 1.81$ in.

$j_u d = 11 - 1.81/2 = 10.1$ in.

$P_u = T + N_u = M_u/j_u d + N_u = 75\,(6)/10.1$
$\qquad + 12 = 56.6$ kips

$P_u/\phi P_c = 56.6/77 = 0.74$

Check shear (all studs):

From Table 5.20.9

$f'_c = 5000$ psi, $d_e > 11$ in., 5/8 in. studs

$\phi V_c = 12\,(14.1) = 169.2$ kips

$V_u/\phi V_c = 75/169.2 = 0.44$

Combined capacity:

From Eq. 5.13-8

$(P_u/\phi P_c)^{4/3} + (V_u/\phi V_c)^{4/3} = 0.67 + 0.33$
$\qquad = 1.00$ OK

(Could also be obtained from Fig. 5.20.16)

Check steel strength:

Tension in top group of studs:

From Table 5.20.10 for 5/8 in. studs:

$\phi P_s = 6\,(16.6) = 99.6$ kips

(Could also be determined from Eq. 5.13-7)

$C = T = 0.85\,f'_c\,ba = 99.6$ kips

comp. block $a = \dfrac{99.6}{0.85\,(5)(10)} = 2.34$ in.

$j_u d = 11 - 2.34/2 = 9.83$ in.

$P_u = M_u/j_u d + N_u = 75\,(6)/9.83 + 12$
$\qquad = 57.8$ kips

Shear in studs:

From Table 5.20.9 for 5/8 in. studs:

$\phi V_s = 12\,(13.8) = 165.6$ kips

(Could also be determined from Eq. 5.13-4)

Combined capacity:

From Eq. 5.13-9

$(P_u/\phi P_s)^2 + (V_u/\phi V_s)^2 = (57.8/99.6)^2$
$\qquad + (75/165.6)^2 = 0.34 + 0.21$
$\qquad = 0.55 < 1.0$ OK

(Could also be obtained from Fig. 5.20.16)

5.14 Deformed Bar Anchors

Deformed bar anchors which are automatically welded (similar to headed studs) to steel plates should have their development length calculated by:

$$\ell_d = \frac{0.03\,d_b\,(f_y)}{\lambda\sqrt{f'_c}} \geqslant 12 \text{ in.} \qquad \text{(Eq. 5.14-1)}$$

where

$d_b = $ diameter of bar, in.

$f_y \leqslant 60{,}000$ psi

For $f_y > 60{,}000$ psi, the above value should be multiplied by the quantity:

$$\left(2 - \frac{60{,}000}{f_y}\right)$$

Horizontal bars placed so that more than 12 in. of concrete is below the bar (top bars) should have the above value of ℓ_d multiplied by 1.4.

Table 5.20.17 tabulates development length requirements.

5.15 Inserts Cast in Concrete

Loop inserts of the type shown in Fig. 5.15.1 can be investigated in a manner similar to that for welded studs, using Eqs. 5.15.1 and 5.15.2 for the concrete tensile and shear strengths.

$$\phi P_c = \phi\,4\,A_o\,\lambda\sqrt{f'_c} \qquad \text{(Eq. 5.15-1)}$$

$$\phi V_c = \phi\,(2500\,d_e - 3500) \qquad \text{(Eq. 5.15-2)}$$

Fig. 5.15.1 Shear cone development for loop inserts

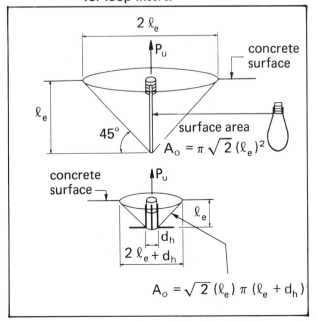

Fig. 5.16.1 Design relationships for connection angles

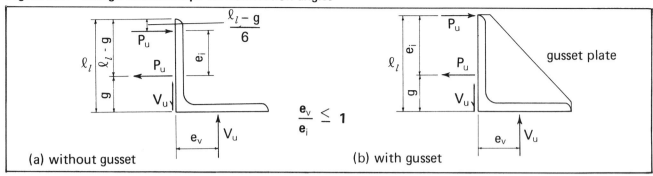

(a) without gusset

(b) with gusset

$$\frac{e_v}{e_i} \le 1$$

Actual strength of the insert will often be controlled by the mechanical strength of the insert, or the strength of the bolt or threaded rod used. Yield strengths of various sizes of wires commonly used in inserts, and typical strengths of bolts and rods are shown in Table 5.20.18. Information from the insert manufacturer and AISC recommendations for standard threaded members should be checked. If the insert is located $4\ell_e$ or greater from a free edge, the shear strength should be assumed equal to the pull-out strength; within $4\ell_e$ of a free edge, shear strength should be calculated by Eq. 5.15-2. Combined shear and tension strengths can be evaluated in a manner similar to that shown for headed studs.

5.16 Connection Angles

Angles used to support precast members can be designed by statics as shown in Fig. 5.16.1. In addition to the applied vertical and horizontal loads, the design should include all loads induced by restraint of relative movement between the precast member and the supporting member. The minimum thickness of non-gusseted angles loaded in shear as shown in Fig. 5.16.2 can be determined by:

$$t = \sqrt{\frac{4 V_u e_v}{\phi f_y b}} \qquad \text{(Eq. 5.16-1)}$$

where

$\phi = 0.90$

b = width of the angle

design e_v = specified e_v + 1/2 in.

The tension on the bolt can be calculated by:

$$P_u = V_u \frac{e_v}{e_i} \qquad \text{(Eq. 5.16-2)}$$

For angles loaded axially, Fig. 5.16.3, either in tension or compression the minimum thickness of non-gusseted angles can be calculated by:

$$t = \sqrt{\frac{4 N_u g}{\phi f_y b}} \qquad \text{(Eq. 5.16-3)}$$

where

$\phi = 0.90$

g = gage of the angle (see Fig. 5.16.3)

b = width of the angle

Fig. 5.16.2 Vertical loads on connection angle

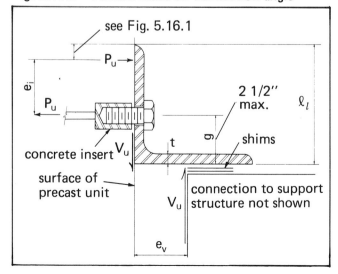

Fig. 5.16.3 Horizontal loads on connection angle

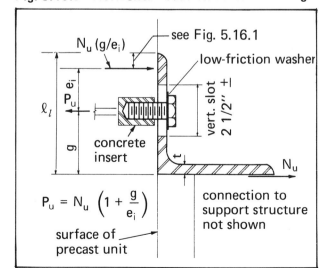

Tables 5.20.19 and 5.20.20 may be used for the design of connection angles.

Connections may be made by welding instead of bolting and the welds designed in accordance with AISC specifications.

5.17 Column Base Plates

Column bases must be designed for both erection loads and loads which occur in service, the former often being more critical. Two commonly used base plate details are shown in Fig. 5.17.1, although other details are also frequently used.

If in the analysis for erection loads or temporary construction loads before grout is placed under the plate, all the anchor bolts are in compression, the base plate thickness required to satisfy bending is determined from:

$$t = \sqrt{\frac{(\Sigma F) 4x_c}{\phi b f_y}} \qquad \text{(Eq. 5.17-1)}$$

where

$\phi = 0.90$

x_c, b from Fig. 5.17.1, in.

f_y = yield strength of the base plate, psi

ΣF = greatest sum of anchor bolt factored forces on one side of the column, lb

If the analysis indicates the anchor bolts on one or both sides of the column are in tension, the base plate thickness is determined by:

$$t = \sqrt{\frac{(\Sigma F) 4x_t}{\phi b f_y}} \qquad \text{(Eq. 5.17-2)}$$

x_t from Fig. 5.17.1

Under loads which occur at service, the base plate thickness may be controlled by bearing on the concrete or grout. In this case, the base plate thickness is determined by:

$$t = x_o \sqrt{\frac{2 f_{bu}}{\phi f_y}} \qquad \text{(Eq. 5.17-3)}$$

where:

x_o from Fig. 5.17.1

$f_{bu} = 0.70 (0.85) f'_c = 0.595 f'_c$

Table 5.20.21 may be used for base plate design.

Nominal base plate shearing stresses should not exceed $0.55 f_y$.

The anchor bolt diameter is determined by the tension or compression on the stress area of the

threaded portion of the bolt. Anchor bolts may be ASTM A307 bolts or, more frequently, threaded rods of ASTM A36 steel.

When the bolts are near a free edge, as in a pier or wall, the buckling of the bolts before grouting may be a consideration. Confinement reinforcement, as shown in Fig. 5.17.1 should be provided in such cases. A minimum of 4 – #3 ties at about 3 in. centers is recommended for confinement.

The strength of the concrete when the bolt is in tension may be critical and can be determined by assuming a shear cone pull-out failure as described for headed studs.

The length of the anchor bolt should be such that the concrete will develop the desired strength of the bolt in bond and bearing on the hook projection or bolt head. Bearing area of bolt heads can be increased by welding a washer or steel plate to the bolt head. Nominal bond stress on smooth anchor bolts should not exceed 250 psi. The nominal confined bearing stress on the hook or bolt head should not exceed $\phi f'_c$. The bottom of the bolt should be a minimum of 4 in. above the bottom of a footing, and above the footing reinforcement.

Compression on anchor bolts during erection can be substantially reduced by the use of steel shims. The required area of the shims can be determined by the bearing stress of the concrete.

5.18 Welding of Reinforcing Bars

Welding of reinforcement may be a practical method of developing the force transfer in many connections, provided the weldability of the bars can be assured.

Welding of reinforcing bars is covered by AWS D12.1-75, "Reinforcing Steel Welding Code" by the American Welding Society. Weldability is defined in that publication as a function of the chemical composition, as shown in the mill report, by the following formula:

$$\text{C.E.} = \%C + \frac{\%Mn}{6} + \frac{\%Cu}{40} + \frac{\%Ni}{20} + \frac{\%Cr}{10}$$

$$- \frac{\%Mo}{50} - \frac{\%V}{10} \qquad \text{(Eq. 5.18-1)}$$

where C.E. = carbon equivalent

AWS D12.1-75 indicates that most reinforcing bars can be welded. However, the preheat and other quality control measures that are required for bars with high carbon equivalents are very difficult to achieve. Except for welding shops with proven quality control procedures that meet AWS D12.1-75, it is recommended that carbon equivalents be limited to 0.45% for #7 and larger

Fig. 5.17.1 Column base connections

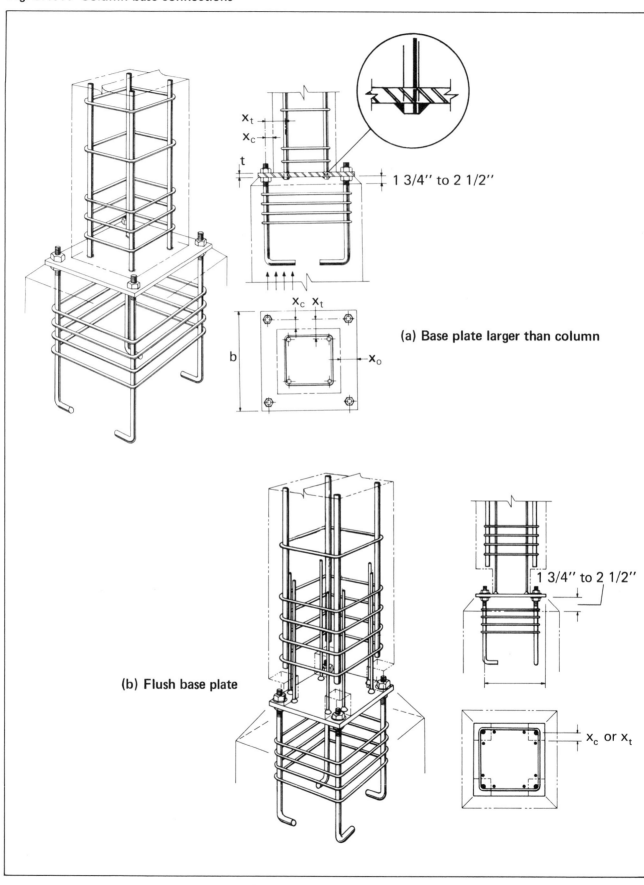

(a) Base plate larger than column

(b) Flush base plate

bars, and 0.55% for #6 and smaller bars.

Unfortunately, at the present time (1978) most reinforcing bars will not meet the above chemistry specifications, and special purchases may be required by the precast concrete manufacturer. There is an ASTM Specification (A706) for weldable Grade 60 bars, but availability of such bars is limited. The precast manufacturer should be consulted as to the availability of weldable bars, and mill reports should be obtained for all bars to be welded.

The nominal strength of full penetration groove welds (Fig. 5.18.1) can be considered the same as the nominal strength of the bar. Fillet, flare-V and bevel groove weld strengths are shown in Table 5.18.1. The weld area, A_w, shown in Table 5.18.1 is the product of the effective throat, t_w, and the length of the weld, ℓ_w.

5.19 Moment Connections

Moment connections are sometimes required

Table 5.18.1 Nominal strength of welds[1], kips

Grade of Bar	Electrode[2]	Nominal strength
40	E70	42 A_w
50	E80	48 A_w
60	E90	54 A_w
75	E100	60 A_w

(1) Bars must meet carbon equivalent requirements (see Sect. 5.18)

(2) Use low hydrogen electrodes, see AWS D12.1-75

in building frames, as discussed in Section 4.6. Moment-resisting capacity is attained by designing the connection so that a force couple can be developed. Tensile capacity through the connection can be attained by studs, deformed bar anchors, inserts, welding, post-tensioning or combinations of these.

Fig. 5.18.1 Typical reinforcing bar welds

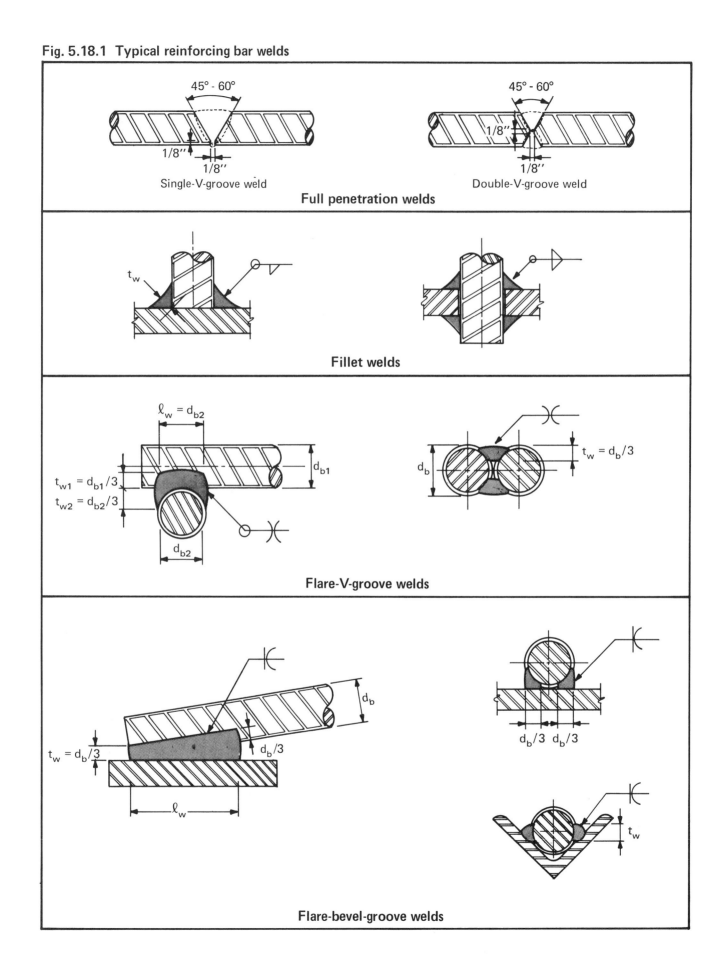

45° - 60°

1/8"

1/8"

Single-V-groove weld

45° - 60°

1/8"

1/8"

Double-V-groove weld

Full penetration welds

t_w

Fillet welds

$\ell_w = d_{b2}$

$t_{w1} = d_{b1}/3$

$t_{w2} = d_{b2}/3$

d_{b1}

d_{b2}

d_b

$t_w = d_b/3$

Flare-V-groove welds

d_b

$d_b/3$

$t_w = d_b/3$

ℓ_w

$d_b/3$ $d_b/3$

t_w

Flare-bevel-groove welds

CONNECTIONS

Table 5.20.1 Bearing on plain concrete

$$\phi V_n/bw = \phi\, C_r\, 70\lambda\sqrt{f'_c}\left(\frac{s}{w}\right)^{1/3}$$

$\phi\quad = 0.70$

$C_r\quad = \left(\dfrac{sw}{200}\right)^{N_u/V_u} = 1.0$ for $N_u = 0$

$\qquad\qquad\qquad sw \leqslant 9.0$ in.^{2}

$f'_c\quad = 5000$ psi*

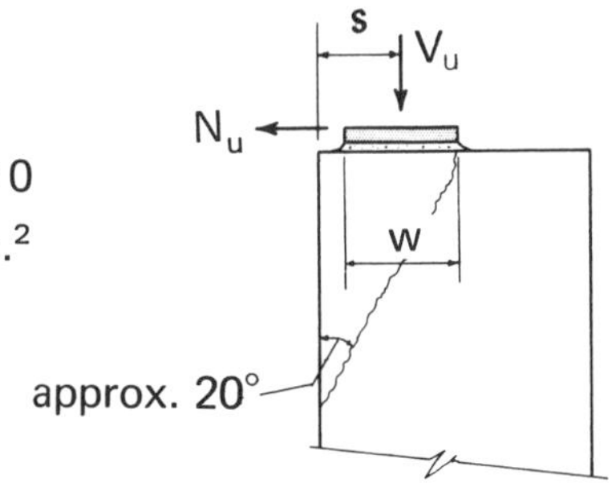

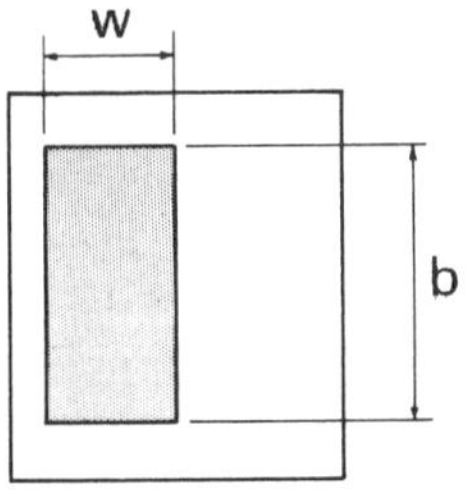

* Table values are for 5000 psi normal weight concrete. For other concretes multiply table values by $\lambda\sqrt{f'_c/5000}$.

Design Bearing Strength = $\phi\, V_n/bw$, psi

w (in.)	$N_u/V_u = 0$	$N_u/V_u = 0.2$	$N_u/V_u = 0.4$	$N_u/V_u = 0.6$	$N_u/V_u = 0.8$
			s = w/2		
2	2750	1100	440	180	70
3	2750	1290	600	280	130
4	2750	1450	760	400	210
5	2750	1480	800	430	230
6	2750	1480	800	430	230
7	2750	1480	800	430	230
8	2750	1480	800	430	230
9	2750	1480	800	430	230
10	2750	1480	800	430	230

w (in.)	$N_u/V_u = 0$	$N_u/V_u = 0.2$	$N_u/V_u = 0.4$	$N_u/V_u = 0.6$	$N_u/V_u = 0.8$
			s = w/2 + 1"		
2	3470	1590	730	330	150
3	3260	1690	880	450	240
4	3150	1690	910	490	260
5	3080	1650	890	480	260
6	3030	1630	880	470	250
7	2990	1610	870	470	250
8	2960	1590	860	460	250
9	2940	1580	850	460	250
10	2920	1570	850	460	250

Table 5.20.2 Reinforced concrete bearing — horizontal reinforcement $\qquad$ $f_y = 40,000$ psi

$$A_{vf} = \frac{V_u \cos \theta}{\phi \mu_e f_y}$$

$$\mu_e = \frac{A_{cr} \mu}{V_u \cos \theta}$$

$$A_{vf} \text{ (min)} = 120 A_{cr}/f_y$$

$$A_{cr} = \text{lesser of } \frac{bw}{\sin \theta} \text{ or } \frac{bh}{\cos \theta}$$

(Can be assumed 3 bw or bh)

b = beam width

ϕ = 0.85

Note: To table values add axial tension reinforcement

$$A_n = \frac{N_u}{\phi f_y}$$

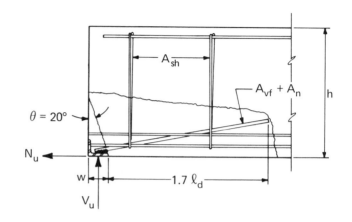

For sand-lightweight concrete, divide table values by $\lambda^2 = 0.72$

Values of A_{vf}, sq in.

V_u Kips	Area of crack interface, A_{cr} (sq in.)											
	10	15	20	25	30	35	40	45	50	55	60	65
10	.19	.12	.09	.08	.09	.11	.12	.13	.15	.16	.18	.19
20			.37	.30	.25	.21	.19	.16	.15	.16	.18	.19
30					.56	.48	.42	.37	.33	.30	.28	.26
40							.74	.66	.59	.54	.49	.46
50									.93	.84	.77	.71

V_u Kips	Area of crack interface, A_{cr} (sq in.)											
	70	80	90	100	125	150	175	200	225	250	275	300
10	.21	.24	.27	.30	.38	.45	.53	.60	.68	.75	.83	.90
20	.21	.24	.27	.30	.38	.45	.53	.60	.68	.75	.83	.90
30	.24	.24	.27	.30	.38	.45	.53	.60	.68	.75	.83	.90
40	.42	.37	.33	.30	.38	.45	.53	.60	.68	.75	.83	.90
50	.66	.58	.52	.46	.38	.45	.53	.60	.68	.75	.83	.90
60	.95	.83	.74	.67	.53	.45	.53	.60	.68	.75	.83	.90
70	1.30	1.14	1.01	.91	.73	.61	.53	.60	.68	.75	.83	.90
80		1.48	1.32	1.19	.95	.79	.68	.60	.68	.75	.83	.90
90			1.67	1.50	1.20	1.00	.86	.75	.68	.75	.83	.90
100				1.86	1.48	1.24	1.06	.93	.82	.75	.83	.90
125					2.32	1.93	1.66	1.45	1.29	1.16	1.05	.97
150						2.78	2.39	2.09	1.86	1.67	1.52	1.39
175							3.25	2.84	2.53	2.27	2.07	1.89
200								3.71	3.30	2.97	2.70	2.47
225									4.17	3.76	3.42	3.13
250										4.64	4.22	3.86
275											5.10	4.68
300												5.57

Table 5.20.3 Reinforced concrete bearing — horizontal reinforcement $f_y = 60,000$ psi

$$A_{vf} = \frac{V_u \cos \theta}{\phi \, \mu_e \, f_y}$$

$$\mu_e = \frac{A_{cr} \, \mu}{V_u \cos \theta}$$

$$A_{vf} \text{ (min.)} = 120 \, A_{cr}/f_y$$

$$A_{cr} = \text{lesser of } \frac{bw}{\sin \theta} \text{ or } \frac{bh}{\cos \theta}$$

(can be assumed 3 bw or bh)

b = beam width

ϕ = 0.85

Note: To table values add axial

tension reinforcement, $A_n = \dfrac{N_u}{\phi \, f_y}$

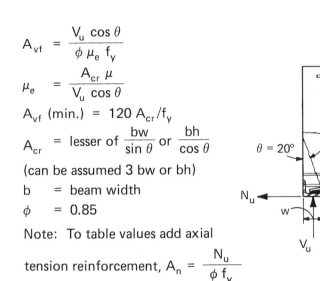

For sand-lightweight concrete divide table values by $\lambda^2 = 0.72$

Values of A_{vf}, sq in.

V_u Kips	Area of crack interface, A_{cr} (sq in.)											
	10	15	20	25	30	35	40	45	50	55	60	65
10	.12	.08	.06	.05	.06	.07	.08	.09	.10	.11	.12	.13
20			.25	.20	.16	.14	.12	.11	.10	.11	.12	.13
30				.37	.32	.28	.25	.22	.20	.19	.17	
40						.49	.44	.40	.36	.33	.30	
50								.62	.56	.52	.48	

V_u Kips	Area of crack interface, A_{cr} (sq in.)											
	70	80	90	100	125	150	175	200	225	250	275	300
10	.14	.16	.18	.20	.25	.30	.35	.40	.45	.50	.55	.60
20	.14	.16	.18	.20	.25	.30	.35	.40	.45	.50	.55	.60
30	.16	.16	.18	.20	.25	.30	.35	.40	.45	.50	.55	.60
40	.28	.25	.22	.20	.25	.30	.35	.40	.45	.50	.55	.60
50	.44	.39	.34	.31	.25	.30	.35	.40	.45	.50	.55	.60
60	.64	.56	.49	.45	.36	.30	.35	.40	.45	.50	.55	.60
70	.87	.76	.67	.61	.48	.40	.35	.40	.45	.50	.55	.60
80		.99	.88	.79	.63	.53	.45	.40	.45	.50	.55	.60
90			1.11	1.00	.80	.67	.57	.50	.45	.50	.55	.60
100				1.24	.99	.82	.71	.62	.55	.50	.55	.60
125					1.55	1.29	1.10	.97	.86	.77	.70	.64
150						1.86	1.59	1.39	1.24	1.11	1.01	.93
175							2.16	1.89	1.68	1.52	1.38	1.26
200								2.47	2.20	1.98	1.80	1.65
225									2.78	2.50	2.28	2.09
250										3.09	2.81	2.58
275											3.40	3.12
300												3.71

CONNECTIONS

Table 5.20.4 Reinforced concrete bearing — vertical reinforcement

$$A_{sh} = \frac{(A_{vf} + A_n)f_y}{\mu_e\, f_{ys}}$$

or

$$\text{min } A_{sh} = \frac{120(1.7\ell_d)b}{f_{ys}} \geq \frac{4(A_{vf} + A_n)f_y}{3\mu_e\, f_{ys}}$$

$f_{ys} = 40,000$ psi   For other values of f_{ys}, multiply table values by

$$\frac{40,000}{f_{ys}}$$

f_y = yield strength of $A_{vf} + A_n$
f_{ys} = yield strength of A_{sh}
b = width of beam

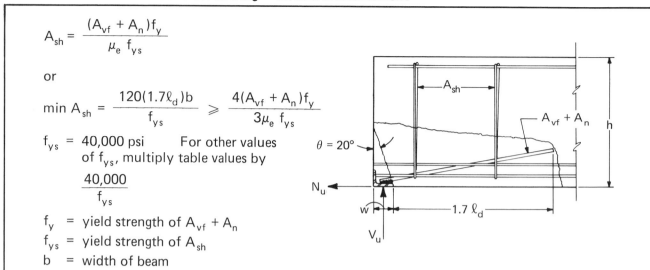

b (in)	No. of $A_{vf}+A_n$ bars	f_y = 40,000 psi						f_y = 60,000 psi					
		# 4	# 5	# 6	# 7	# 8	# 9	# 4	# 5	# 6	# 7	# 8	# 9
4	1	.03	.05	.09	.15	.19	.25	.04	.08	.14	.22	.29	.37
6	1	.02	.04	.06	.10	.13	.16	.03	.05	.09	.15	.19	.25
	2	.07	.14	.25	.39	.52	.66	.11	.21	.37	.59	.78	.98
8	1	.01	.03	.05	.07	.10	.12	.02	.04	.07	.11	.15	.18
	2	.05	.11	.18	.29	.39	.49	.08	.16	.28	.44	.58	.74
10	1	.01	.02	.04	.06	.08	.10	.02	.03	.06	.09	.12	.15
	2	.04	.09	.15	.23	.31	.39	.07	.13	.22	.35	.47	.59
	3	.10	.19	.33	.53	.70	.88	.15	.29	.50	.79	1.05	1.33
12	1	.01	.02	.03	.05	.06	.08	.01	.03	.05	.07	.10	.12
	2	.04	.07	.12	.20	.26	.33	.05	.11	.18	.29	.39	.49
	3	.08	.16	.28	.44	.58	.74	.12	.24	.41	.66	.87	1.11
	4	.15	.28	.49	.78	1.04	1.31	.22	.43	.74	1.17	1.55	1.97
16	1	.01	.01	.02	.04	.05	.06	.01	.02	.03	.05	.07	.09
	2	.03	.05	.09	.15	.19	.25	.04	.08	.14	.22	.29	.37
	3	.06	.12	.21	.33	.44	.55	.09	.18	.31	.49	.66	.83
	4	.11	.21	.37	.59	.78	.98	.16	.32	.55	.88	1.16	1.47
	5	.17	.33	.58	.91	1.21	1.54	.26	.50	.86	1.37	1.82	2.30
20	1	.01	.01	.02	.03	.04	.05	.01	.02	.03	.04	.06	.07
	2	.02	.04	.07	.12	.16	.20	.03	.06	.11	.18	.23	.29
	3	.05	.10	.17	.26	.35	.44	.07	.14	.25	.40	.52	.66
	4	.09	.17	.30	.47	.62	.79	.13	.26	.44	.70	.93	1.18
	5	.14	.27	.46	.73	.97	1.23	.20	.40	.69	1.10	1.46	1.84
	6	.20	.38	.66	1.05	1.40	1.77	.30	.58	1.00	1.58	2.10	2.65
24	1	.00	.01	.02	.02	.03	.04	.01	.01	.02	.04	.05	.06
	2	.02	.04	.06	.10	.13	.16	.03	.05	.09	.15	.19	.25
	3	.04	.08	.14	.22	.29	.37	.06	.12	.21	.33	.44	.55
	4	.07	.14	.25	.39	.52	.66	.11	.21	.37	.59	.78	.98
	5	.11	.22	.38	.61	.81	1.02	.17	.33	.58	.91	1.21	1.54
	6	.16	.32	.55	.88	1.16	1.47	.25	.48	.83	1.32	1.75	2.21

Required area of vertical reinforcement, A_{sh}, sq in. — Size of $A_{vf}+A_n$ bars

CONNECTIONS

Table 5.20.5 Tensile reinforcement for dapped beams, column corbels and beam ledges

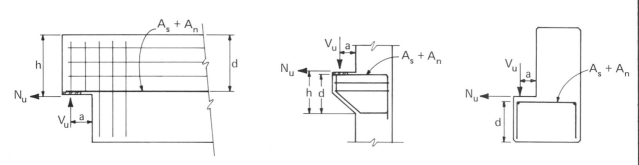

Note:
Maximum a/d = 1.0
Minimum h/d = 1.0

Procedure:

1. Enter table with V_u and a/d, read A_{s1}
2. Enter table with N_u and h/d, read A_{s2}
3. $A_s + A_n = A_{s1} + A_{s2}$
4. $A_s \geqslant A_{vf}$ from Table 5.20.2 or 5.20.3
5. $A_s \geqslant 120bd/f_y$
6. See section 5.9, 5.10 or 5.11 for additional reinforcement requirements
   $\phi = 0.85$

Values of A_{s1} or A_{s2}, sq in.

V_u or N_u	a/d or h/d											
	f_y = 40,000 psi						f_y = 60,000 psi					
	0.2	0.4	0.6	0.8	1.0	1.2	0.2	0.4	0.6	0.8	1.0	1.2
5	0.03	0.06	0.09	0.12	0.15	0.18	0.02	0.04	0.06	0.08	0.10	0.12
10	0.06	0.12	0.18	0.24	0.29	0.35	0.04	0.08	0.12	0.16	0.20	0.24
15	0.09	0.18	0.26	0.35	0.44	0.53	0.06	0.12	0.18	0.24	0.29	0.35
20	0.12	0.24	0.35	0.47	0.59	0.71	0.08	0.16	0.24	0.31	0.39	0.47
25	0.15	0.29	0.44	0.59	0.74	0.88	0.10	0.20	0.29	0.39	0.49	0.59
30	0.18	0.35	0.53	0.71	0.88	1.06	0.12	0.24	0.35	0.47	0.59	0.71
35	0.21	0.41	0.62	0.82	1.03	1.24	0.14	0.27	0.41	0.55	0.69	0.82
40	0.24	0.47	0.71	0.94	1.18	1.41	0.16	0.31	0.47	0.63	0.78	0.94
45	0.26	0.53	0.79	1.06	1.32	1.59	0.18	0.35	0.53	0.71	0.88	1.06
50	0.29	0.59	0.88	1.18	1.47	1.76	0.20	0.39	0.59	0.78	0.98	1.18
60	0.35	0.71	1.06	1.41	1.76	2.12	0.24	0.47	0.71	0.94	1.18	1.41
70	0.41	0.82	1.24	1.65	2.06	2.47	0.27	0.55	0.82	1.10	1.37	1.65
80	0.47	0.94	1.41	1.88	2.35	2.82	0.31	0.63	0.94	1.25	1.57	1.88
90	0.53	1.06	1.59	2.12	2.65	3.18	0.35	0.71	1.06	1.41	1.76	2.12
100	0.59	1.18	1.76	2.35	2.94	3.53	0.39	0.78	1.18	1.57	1.96	2.35
125	0.74	1.47	2.21	2.94	3.68	4.41	0.49	0.98	1.47	1.96	2.45	2.94
150	0.88	1.76	2.65	3.53	4.41	5.29	0.59	1.18	1.76	2.35	2.94	3.53
175	1.03	2.06	3.09	4.12	5.15	6.18	0.69	1.37	2.06	2.75	3.43	4.12
200	1.18	2.35	3.53	4.71	5.88	7.06	0.78	1.57	2.35	3.14	3.92	4.71
225	1.32	2.65	3.97	5.29	6.62	7.94	0.88	1.76	2.65	3.53	4.41	5.29
250	1.47	2.94	4.41	5.88	7.35	8.82	0.98	1.96	2.94	3.92	4.90	5.88
275	1.62	3.24	4.85	6.47	8.09	9.71	1.08	2.16	3.24	4.31	5.39	6.47
300	1.76	3.53	5.29	7.06	8.82		1.18	2.35	3.53	4.71	5.88	7.06

Table 5.20.6 Design strength of concrete brackets or corbels b = 8″, 10″, 12″, 14″

Design strength by Eqs. 5.11-1 or 5.11-2
for following criteria:

f'_c = 5000 psi, normal weight
f_y = 40,000 psi
N_u = 0.2 V_u

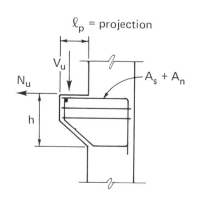

Values of ϕV_n, kips

b = 8″

$A_s + A_n$ / h =	4″ Projection				6″ Projection				8″ Projection				10″ Projection			
	6″	8″	10″	12″	10″	14″	18″	22″	10″	14″	18″	22″	10″	14″	18″	22″
2 #5	25	31	37	41	29	37			23	31			20	26		
2 #6		44	48	52	41	53	60	65	33	44	53	60	28	38	46	53
2 #7				62		66	73	79		60	72	79		51	62	72
2 #8						77	86	93		77	86	93		66	81	93
2 #9							98	107			98	107			98	107

b = 10″

$A_s + A_n$ / h =	4″ Projection				6″ Projection				8″ Projection				10″ Projection			
	6	8	10	12	10	14	18	22	10	14	18	22	10	14	18	22
2 #5	25	31	37	42	29				23				20			
2 #6		45	52	56	41	53	63		33	44	53		28	38	46	
2 #7			63	68	55	72	79	86	45	60	72	82	38	51	62	72
2 #8				79		85	94	101		77	93	101		66	81	93
2 #9						97	108	117		97	108	117		83	102	117

b = 12″

$A_s + A_n$ / h =	4″ Projection				6″ Projection				8″ Projection				10″ Projection			
	6	8	10	12	10	14	18	22	10	14	18	22	10	14	18	22
2 #5	25	31	37		29				23				20			
2 #6	35	45	53	60	41	53			33	44			28	38		
2 #7		61	68	73	55	72	85		45	60	72		38	51	62	
3 #6		66	72	77	62	79	90	97	50	66	79	90	42	56	69	79
2 #8			79	86	72	91	100	108	58	77	93	107	49	66	81	93
3 #7				93		99	109	118		89	107	118		76	93	107
2 #9				98		105	116	125		97	116	125		83	102	118
3 #8						116	128	139		116	128	139		99	121	139
3 #9							147	160			147	160			147	160

b = 14″

$A_s + A_n$ / h =	4″ Projection				6″ Projection				8″ Projection				10″ Projection			
	6	8	10	12	12	16	20	24	12	16	20	24	12	16	20	24
2 #5	25	31														
2 #6	35	45	53	61	47				39				33			
2 #7		61	72	77	64	79			53	66			45	57		
3 #6		67	76	82	71	87	99		58	73	85		50	63	74	
2 #8			85	91	83	102	111		68	86	100		58	74	87	
3 #7				92	96	111	121	129	79	99	115	129	67	85	100	114
2 #9				105	105	118	128	137	86	108	126	137	73	93	110	125
3 #8				116	116	130	142	153	103	128	142	153	87	110	131	148
3 #9						150	164	176		150	164	176		139	164	176

Table 5.20.6 (cont.) Design strength of concrete brackets or corbels   b = 16″, 18″, 20″, 22″, 24″

Values of ϕV_n, kips

$A_s + A_n$	6″ Projection				8″ Projection				10″ Projection				12″ Projection			
h =	14″	18″	22″	26″	14″	18″	22″	26″	14″	18″	22″	26″	14″	18″	22″	26″
b = 16″																
2 #7	72				60				51				44			
3 #6	79				66				56				49			
2 #8	93	110			77	93			66	81			58	71		
3 #7	108	122	131		89	107	123		76	93	107		67	82	95	
2 #9	117	130	140	149	97	117	134	149	83	102	118	131	73	90	104	118
3 #8	130	144	155	165	116	140	155	165	99	121	140	156	87	106	124	140
3 #9	150	166	180	192	146	166	180	192	125	152	176	192	109	134	157	176
4 #8	154	171	185	198	154	171	185	198	132	161	185	198	115	142	165	186
4 #9		197	213	228		197	213	228		197	213	228		179	209	228
h =	16	20	24	28	16	20	24	28	16	20	24	28	16	20	24	28
b = 18″																
2 #8	102				86				74				65			
3 #7	118	133			99	115			85	100			74	89		
2 #9	129	141			108	126			93	110			81	97		
5 #6	138	151	161		121	142	159		104	123	140		92	109	124	
3 #8	144	157	168		128	150	168		110	131	148		97	116	132	
3 #9	166	181	194	206	162	181	194	206	139	165	187	206	122	146	167	185
4 #8	171	187	201	213	171	187	201	213	147	174	198	213	129	154	176	196
4 #9	197	215	232	246	197	215	232	246	186	215	232	246	163	194	222	246
h =	18	22	26	30	18	22	26	30	18	22	26	30	18	22	26	30
b = 20″																
3 #7	127				107				93				82			
2 #9	139				117				102				90			
5 #6	151	162			132	151			114	132			101	117		
3 #8	157	169			140	160			121	140			106	124		
3 #9	181	196	208	220	176	196	208	220	152	176	197	215	134	157	176	194
4 #8	187	202	215	227	186	202	215	227	161	186	208	227	142	165	186	205
4 #9	215	233	249	263	215	233	249	263	203	233	249	263	179	209	235	259
5 #9	246	267	285	302	246	267	285	302	246	267	285	302	224	261	285	302
h =	20	24	28	32	20	24	28	32	20	24	28	32	20	24	28	32
b = 22″																
5 #6	162				142				123				109			
3 #8	169				150				131				116			
3 #9	196	210	222		189	210	222		165	187	207		146	167	185	
4 #8	202	217	229		200	217	229		174	198	218		154	176	196	
4 #9	233	251	266	280	233	251	266	280	220	249	266	280	194	222	247	269
5 #9	267	287	305	321	267	287	305	321	267	287	305	321	243	278	305	321
h =	22	26	30	34	22	26	30	34	22	26	30	34	22	26	30	34
b = 24″																
3 #9	210	223			202	223			176	197			157	176		
4 #8	217	230			213	230			186	208			165	186		
4 #9	251	267	282	295	251	267	282	295	235	263	282	295	209	235	259	280
5 #9	287	307	324	340	287	307	324	340	287	307	324	340	261	294	323	340

CONNECTIONS

Table 5.20.7 Design of structural steel haunches — concrete

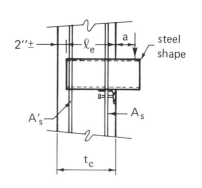

f'_c = 5000 psi; for other concrete strengths multiply values by $f'_c/5000$

Values are for design strength of concrete. Adequacy of structural steel section should be checked.

Additional design strength, ϕV_r, can be obtained with reinforcing bars — see Table 5.20.8

$V_u \leqslant \phi (V_c + V_r)$

Values of ϕV_c (kips)

Width of steel section, b	Shear span, a = 2″						Shear span a = 4″					
	t_c = 12″	t_c = 14″	t_c = 16″	t_c = 18″	t_c = 20″	t_c = 24″	t_c = 12″	t_c = 14″	t_c = 16″	t_c = 18″	t_c = 20″	t_c = 24″
3	24	30	36	42	47	59	21	26	32	37	43	54
3.5	28	35	42	49	55	69	24	30	37	43	50	63
4	32	40	48	55	63	79	27	35	42	50	57	72
4.5	36	45	54	62	71	89	31	39	47	56	64	81
5	40	50	60	69	79	99	34	43	53	62	71	90
5.5	44	55	66	76	87	108	38	48	58	68	78	99
6	48	60	72	83	95	118	41	52	63	74	86	108
6.5	53	65	78	90	103	128	45	56	68	80	93	117
7	57	70	83	97	111	138	48	61	74	87	100	127
7.5	61	75	89	104	119	148	51	65	79	93	107	136
8	65	80	95	111	126	158	55	69	84	99	114	145
8.5	69	85	101	118	134	167	58	74	89	105	121	154
9	73	90	107	125	142	177	62	78	95	111	128	163
9.5	77	95	113	132	150	187	65	82	100	118	136	172
10	81	100	119	139	158	197	69	87	105	124	143	181

Width of steel section, b	Shear span a = 6″						Shear span a = 8″					
	t_c = 12″	t_c = 14″	t_c = 16″	t_c = 18″	t_c = 20″	t_c = 24″	t_c = 12″	t_c = 14″	t_c = 16″	t_c = 18″	t_c = 20″	t_c = 24″
3	18	23	28	34	39	50	16	21	25	31	36	47
3.5	21	27	33	39	45	58	18	24	30	36	42	54
4	24	31	38	45	52	67	21	27	34	41	48	62
4.5	27	34	42	50	58	75	24	31	38	46	54	70
5	30	38	47	56	65	83	26	34	42	51	60	78
5.5	33	42	52	61	71	92	29	38	47	56	66	85
6	36	46	56	67	78	100	32	41	51	61	72	93
6.5	39	50	61	73	84	109	34	44	55	66	78	101
7	42	54	66	78	91	117	37	48	59	71	84	109
7.5	45	57	70	84	97	125	39	51	64	76	90	116
8	48	61	75	89	104	134	42	55	68	82	95	124
8.5	51	65	80	95	110	142	45	58	72	87	101	132
9	54	69	85	101	117	150	47	62	76	92	107	140
9.5	57	73	89	106	123	159	50	65	81	97	113	147
10	60	76	94	112	130	167	53	68	85	102	119	155

CONNECTIONS

Table 5.20.8 Design of structural steel haunches — reinforcement

	A_s		f_y = 40,000 psi						f_y = 60,000 psi					
			t_c = 12″	14″	16″	18″	20″	24″	t_c = 12″	14″	16″	18″	20″	24″
Shear span a = 2″	2 - #4	ϕV_r	9	9	10	10	10	10	14	14	14	15	15	15
		A'_s	.13	.12	.12	.11	.11	.10	.13	.12	.12	.11	.11	.10
	2 - #5	ϕV_r	14	15	15	15	15	16	21	22	22	23	23	24
		A'_s	.20	.19	.18	.17	.17	.16	.20	.19	.18	.17	.17	.16
	2 - #6	ϕV_r	20	21	21	22	22	22	30	31	32	32	33	33
		A'_s	.29	.27	.26	.25	.24	.22	.29	.27	.26	.25	.24	.22
	2 - #7	ϕV_r	27	28	29	29	30	30	41	42	43	44	45	46
		A'_s	.39	.37	.35	.34	.32	.31	.39	.37	.35	.34	.32	.31
	4 - #6	ϕV_r	40	41	42	43	44	45	60	62	63	65	65	67
		A'_s	.58	.54	.51	.49	.48	.45	.58	.54	.51	.49	.48	.45
	4 - #7	ϕV_r	55	56	58	59	59	61	82	85	87	88	89	91
		A'_s	.79	.74	.70	.67	.65	.61	.79	.74	.70	.67	.65	.61
Shear span a = 4″	2 - #4	ϕV_r	8	8	8	9	9	9	12	12	13	13	13	14
		A'_s	.17	.16	.15	.14	.14	.13	.17	.16	.15	.14	.14	.13
	2 - #5	ϕV_r	12	13	13	14	14	14	18	19	20	20	21	22
		A'_s	.27	.25	.23	.22	.21	.20	.27	.25	.23	.22	.21	.20
	2 - #6	ϕV_r	17	18	19	19	20	20	26	27	28	29	30	31
		A'_s	.38	.35	.33	.31	.30	.28	.38	.35	.33	.31	.30	.28
	2 - #7	ϕV_r	23	24	25	26	27	28	35	37	38	39	40	42
		A'_s	.52	.48	.45	.43	.41	.38	.52	.48	.45	.43	.41	.38
	4 - #6	ϕV_r	34	36	37	38	39	41	51	54	56	58	59	61
		A'_s	.76	.70	.66	.63	.60	.56	.76	.70	.66	.63	.60	.56
	4 - #7	ϕV_r	46	49	51	52	54	56	70	73	76	79	81	84
		A'_s	1.03	.96	.90	.86	.82	.76	1.03	.96	.90	.86	.82	.76
Shear span a = 6″	2 - #4	ϕV_r	7	7	8	8	8	9	10	11	11	12	12	13
		A'_s	.20	.19	.18	.17	.16	.15	.20	.19	.18	.17	.16	.15
	2 - #5	ϕV_r	10	11	12	12	13	13	16	17	18	18	19	20
		A'_s	.31	.29	.27	.26	.25	.23	.31	.29	.27	.26	.25	.23
	2 - #6	ϕV_r	15	16	17	17	18	19	22	24	25	26	27	28
		A'_s	.44	.41	.39	.37	.35	.33	.44	.41	.39	.37	.35	.33
	2 - #7	ϕV_r	20	22	23	24	24	26	30	32	34	36	37	39
		A'_s	.61	.56	.53	.50	.48	.44	.61	.56	.53	.50	.48	.44
	4 - #6	ϕV_r	30	32	33	35	36	38	44	47	50	52	54	57
		A'_s	.89	.83	.78	.74	.70	.65	.89	.83	.78	.74	.70	.65
	4 - #7	ϕV_r	40	43	45	47	49	51	60	65	68	71	73	77
		A'_s	1.21	1.13	1.06	1.01	.96	.89	1.21	1.13	1.06	1.01	.96	.89
Shear span a = 8″	2 - #4	ϕV_r	6	6	7	7	7	8	9	10	9	11	11	12
		A'_s	.23	.21	.20	.19	.18	.17	.23	.21	.20	.19	.18	.17
	2 - #5	ϕV_r	9	10	11	11	12	12	14	15	16	17	17	19
		A'_s	.35	.33	.31	.29	.28	.26	.35	.33	.31	.29	.28	.26
	2 - #6	ϕV_r	13	14	15	16	16	18	20	21	23	24	25	26
		A'_s	.50	.46	.44	.41	.40	.36	.50	.46	.44	.41	.40	.36
	2 - #7	ϕV_r	18	19	21	22	22	24	27	29	31	32	34	36
		A'_s	.68	.63	.60	.56	.54	.50	.68	.63	.60	.56	.54	.50
	4 - #6	ϕV_r	26	28	30	32	33	35	39	42	45	47	49	53
		A'_s	.99	.93	.87	.83	.79	.73	.99	.93	.87	.83	.79	.73
	4 - #7	ϕV_r	36	39	41	43	45	48	53	58	62	65	67	72
		A'_s	1.35	1.26	1.19	1.13	1.08	.99	1.35	1.26	1.19	1.13	1.08	.99

Table 5.20.9 Shear strength of welded headed studs

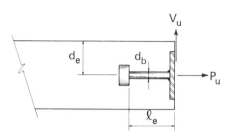

For combined shear and tension,
see Fig. 5.20.16

f'_c	Edge dist. d_e	Maximum Design Shear Strength, ϕV_c, Limited by Concrete Strength (kips)											
		Normal weight concrete ($\lambda = 1.0$)						Sand-lightweight concrete ($\lambda = 0.85$)					
		Stud diameter, d_b, in.						Stud diameter, d_b, in.					
		1/4	3/8	1/2	5/8	3/4	7/8	1/4	3/8	1/2	5/8	3/4	7/8
4000 psi	2	2.3	2.5	2.5	2.5	2.5	2.5	2.1	2.1	2.1	2.1	2.1	2.1
	3	2.3	4.9	4.9	4.9	4.9	4.9	2.3	4.2	4.2	4.2	4.2	4.2
	4	2.3	5.1	7.4	7.4	7.4	7.4	2.3	5.1	6.3	6.3	6.3	6.3
	5	2.3	5.1	9.0	9.9	9.9	9.9	2.3	5.1	8.4	8.4	8.4	8.4
	6	2.3	5.1	9.0	12.4	12.4	12.4	2.3	5.1	9.0	10.5	10.5	10.5
	7	2.3	5.1	9.0	14.1	14.8	14.8	2.3	5.1	9.0	12.6	12.6	12.6
	8	2.3	5.1	9.0	14.1	17.3	17.3	2.3	5.1	9.0	14.1	14.7	14.7
	9	2.3	5.1	9.0	14.1	19.8	19.8	2.3	5.1	9.0	14.1	16.8	16.8
	10	2.3	5.1	9.0	14.1	20.3	22.2	2.3	5.1	9.0	14.1	18.9	18.9
	11 or more	2.3	5.1	9.0	14.1	20.3	24.7	2.3	5.1	9.0	14.1	20.3	21.0
5000 psi	2	2.3	2.8	2.8	2.8	2.8	2.8	2.3	2.3	2.3	2.3	2.3	2.3
	3	2.3	5.1	5.5	5.5	5.5	5.5	2.3	4.7	4.7	4.7	4.7	4.7
	4	2.3	5.1	8.3	8.3	8.3	8.3	2.3	5.1	7.0	7.0	7.0	7.0
	5	2.3	5.1	9.0	11.1	11.1	11.1	2.3	5.1	9.0	9.4	9.4	9.4
	6	2.3	5.1	9.0	13.8	13.8	13.8	2.3	5.1	9.0	11.7	11.7	11.7
	7	2.3	5.1	9.0	14.1	16.6	16.6	2.3	5.1	9.0	14.1	14.1	14.1
	8	2.3	5.1	9.0	14.1	19.3	19.3	2.3	5.1	9.0	14.1	16.4	16.4
	9	2.3	5.1	9.0	14.1	20.3	22.1	2.3	5.1	9.0	14.1	18.8	18.8
	10	2.3	5.1	9.0	14.1	20.3	24.9	2.3	5.1	9.0	14.1	20.3	21.1
	11 or more	2.3	5.1	9.0	14.1	20.3	27.6	2.3	5.1	9.0	14.1	20.3	23.5
6000 psi	2	2.3	3.0	3.0	3.0	3.0	3.0	2.3	2.6	2.6	2.6	2.6	2.6
	3	2.3	5.1	6.1	6.1	6.1	6.1	2.3	5.1	5.1	5.1	5.1	5.1
	4	2.3	5.1	9.0	9.1	9.1	9.1	2.3	5.1	7.7	7.7	7.7	7.7
	5	2.3	5.1	9.0	12.1	12.1	12.1	2.3	5.1	9.0	10.3	10.3	10.3
	6	2.3	5.1	9.0	14.1	15.1	15.1	2.3	5.1	9.0	12.9	12.9	12.9
	7	2.3	5.1	9.0	14.1	18.2	18.2	2.3	5.1	9.0	14.1	15.4	15.4
	8	2.3	5.1	9.0	14.1	20.3	21.2	2.3	5.1	9.0	14.1	18.0	18.0
	9	2.3	5.1	9.0	14.1	20.3	24.2	2.3	5.1	9.0	14.1	20.3	20.6
	10	2.3	5.1	9.0	14.1	20.3	27.2	2.3	5.1	9.0	14.1	20.3	23.2
	11 or more	2.3	5.1	9.0	14.1	20.3	27.6	2.3	5.1	9.0	14.1	20.3	25.7
Max. Design Shear Strength, ϕV_s Limited by Steel Strength (kips)		1/4	3/8	1/2	5/8	3/4	7/8						
		2.2	5.0	8.8	13.8	19.9	27.1						

Table 5.20.10 Tensile strength of welded headed studs

		Maximum Design Tensile Strength, ϕP_c, Limited by Concrete Strength (kips)											
		Normal weight concrete[1]						Sand-lightweight concrete[1]					
Edge dist, d_e	Stud length ℓ_e	Stud diameter, d_b, in.						Stud diameter, d_b, in.					
		1/4	3/8	1/2	5/8	3/4	7/8	1/4	3/8	1/2	5/8	3/4	7/8
2 in.	2.5	6.4	6.9	7.5	8.0	8.0	8.3	5.4	5.9	6.4	6.8	6.8	7.0
	4.0	9.6	10.1	10.7	11.2	11.2	11.5	8.2	8.6	9.1	9.5	9.5	9.8
	5.0	11.7	12.3	12.8	13.3	13.3	13.6	10.0	10.4	10.9	11.3	11.3	11.6
	6.0	13.9	14.4	14.9	15.5	15.5	15.7	11.8	12.3	12.7	13.2	13.2	13.4
	7.0	16.0	16.5	17.1	17.6	17.6	17.9	13.6	14.1	14.5	15.0	15.0	15.2
	8.0	18.2	18.7	19.2	19.8	19.8	20.0	15.4	15.9	16.3	16.8	16.8	17.0
3 in.	2.5	8.0	8.7	9.3	10.0	10.0	10.3	6.8	7.4	7.9	8.5	8.5	8.8
	4.0	14.4	15.2	16.0	16.8	16.8	17.2	12.3	12.9	13.6	14.3	14.3	14.6
	5.0	17.6	18.4	19.2	20.0	20.0	20.4	15.0	15.7	16.3	17.0	17.0	17.4
	6.0	20.8	21.6	22.4	23.2	23.2	23.6	17.7	18.4	19.1	19.7	19.7	20.1
	7.0	24.0	24.8	25.6	26.4	26.4	26.8	20.4	21.1	21.8	22.5	22.5	22.8
	8.0	27.2	28.0	28.8	29.6	29.6	30.0	23.1	23.8	24.5	25.2	25.2	25.5
4 in.	2.5	8.0	8.7	9.3	10.0	10.0	10.3	6.8	7.4	7.9	8.5	8.5	8.8
	4.0	19.2	20.3	21.4	22.4	22.4	23.0	16.3	17.2	18.2	19.1	19.1	19.5
	5.0	23.5	24.6	25.6	26.7	26.7	27.2	20.0	20.9	21.8	22.7	22.7	23.1
	6.0	27.8	28.8	29.9	31.0	31.0	31.5	23.6	24.5	25.4	26.3	26.3	26.8
	7.0	32.0	33.1	34.2	35.2	35.2	35.8	27.2	28.1	29.0	29.9	29.9	30.4
	8.0	36.3	37.4	38.4	39.5	39.5	40.0	30.9	31.8	32.7	33.6	33.6	34.0
5 in.	2.5	8.0	8.7	9.3	10.0	10.0	10.3	6.8	7.4	7.9	8.5	8.5	8.8
	4.0	19.2	20.3	21.4	22.4	22.4	23.0	16.3	17.2	18.2	19.1	19.1	19.5
	5.0	29.4	30.7	32.0	33.4	33.4	34.0	25.0	26.1	27.2	28.4	28.4	28.9
	6.0	34.7	36.0	37.4	38.7	38.7	39.4	29.5	30.6	31.8	32.9	32.9	33.5
	7.0	40.0	41.4	42.7	44.0	44.0	44.7	34.0	35.2	36.3	37.4	37.4	38.0
	8.0	45.4	46.7	48.0	49.4	49.4	50.0	38.6	39.7	40.8	42.0	42.0	42.5
6 in.	2.5	8.0	8.7	9.3	10.0	10.0	10.3	6.8	7.4	7.9	8.5	8.5	8.8
	4.0	19.2	20.3	21.4	22.4	22.4	23.0	16.3	17.2	18.2	19.1	19.1	19.5
	5.0	29.4	30.7	32.0	33.4	33.4	34.0	25.0	26.1	27.2	28.4	28.4	28.9
	6.0	41.6	43.2	44.8	46.4	46.4	47.2	35.4	36.8	38.1	39.5	39.5	40.2
	7.0	48.0	49.6	51.3	52.9	52.9	53.7	40.8	42.2	43.6	44.9	44.9	45.6
	8.0	54.5	56.1	57.7	59.3	59.3	60.1	46.3	47.6	49.0	50.4	50.4	51.1
7 in.	2.5	8.0	8.7	9.3	10.0	10.0	10.3	6.8	7.4	7.9	8.5	8.5	8.8
	4.0	19.2	20.3	21.4	22.4	22.4	23.0	16.3	17.2	18.2	19.1	19.1	19.5
	5.0	29.4	30.7	32.0	33.4	33.4	34.0	25.0	26.1	27.2	28.4	28.4	28.9
	6.0	41.6	43.2	44.8	46.4	46.4	47.2	35.4	36.8	38.1	39.5	39.5	40.2
	7.0	56.1	57.9	59.8	61.7	61.7	62.6	47.6	49.2	50.8	52.4	52.4	53.2
	8.0	63.5	65.4	67.3	69.1	69.1	70.1	54.0	55.6	57.2	58.8	58.8	59.6
8 in.	2.5	8.0	8.7	9.3	10.0	10.0	10.3	6.8	7.4	7.9	8.5	8.5	8.8
	4.0	19.2	20.3	21.4	22.4	22.4	23.0	16.3	17.2	18.2	19.1	19.1	19.5
	5.0	29.4	30.7	32.0	33.4	33.4	34.0	25.0	26.1	27.2	28.4	28.4	28.9
	6.0	41.6	43.2	44.8	46.4	46.4	47.2	35.4	36.8	38.1	39.5	39.5	40.2
	7.0	56.1	57.9	59.8	61.7	61.7	62.6	47.6	49.2	50.8	52.4	52.4	53.2
	8.0	72.6	74.7	76.9	79.0	79.0	80.1	61.7	63.5	65.3	67.2	67.2	68.1

Maximum Design Tensile Strength, ϕP_{ns} Limited by Steel Strength (kips)	Diameter	1/4	3/8	1/2	5/8	3/4	7/8
	ϕP_s	2.7	6.0	10.6	16.6	23.9	32.5

(1) $f'_c = 5000$ psi, for other strengths multiply by $\sqrt{f'_c/5000}$

(2) For stud groups, also check Tables 5.20.11 - 5.20.15

(3) For combined shear and tension, see Fig. 5.20.16

CONNECTIONS

Table 5.20.11 Concrete design strength for stud or insert groups — Case 1

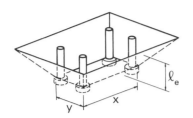

Case 1 — Not near a free edge

$$\phi P_c = \phi\, 4\lambda \sqrt{f'_c}\ A_o$$
$$A_o = xy + 2\ell_e(x+y)\sqrt{2} + 4\sqrt{2}\,\ell_e^{\,2}$$
$$\phi = 0.85$$
$$\lambda f'_c = 5000 \text{ psi}$$

for other values, multiply by $\lambda\sqrt{f'_c/5000}$

Maximum Tensile Strength, ϕP_c, of a Stud Group, kips

ℓ_e, in.	Dim. y, in.	Dimension x, in.											
		2	4	6	8	10	12	14	16	18	20	22	24
2.5	0	12	15	19	22	25	29	32	36	39	42	46	49
	2	16	21	25	29	34	38	42	47	51	56	60	64
	4	21	26	31	37	42	47	53	58	63	69	74	79
	6	25	31	38	44	50	56	63	69	75	82	88	94
	8	29	37	44	51	58	66	73	80	87	95	102	109
	10	34	42	50	58	67	75	83	91	99	108	116	124
	12	38	47	56	66	75	84	93	102	111	121	130	139
4	0	27	33	38	44	49	54	60	65	71	76	82	87
	2	34	40	46	53	59	66	72	78	85	91	98	104
	4	40	47	55	62	69	77	84	92	99	106	114	121
	6	46	55	63	71	80	88	96	105	113	121	130	138
	8	53	62	71	81	90	99	109	118	127	136	146	155
	10	59	69	80	90	100	110	121	131	141	151	162	172
	12	66	77	88	99	110	122	133	144	155	166	178	189
6	0	57	65	73	82	90	98	106	114	122	131	139	147
	2	66	75	84	94	103	112	121	130	139	148	157	167
	4	75	85	96	106	116	126	136	146	156	166	176	186
	6	84	96	107	118	129	140	151	162	173	184	195	206
	8	94	106	118	130	142	154	166	178	190	202	214	226
	10	103	116	129	142	155	168	181	194	206	219	232	245
	12	112	126	140	154	168	181	195	209	223	237	251	265
8	0	98	109	120	131	141	152	163	174	185	196	207	218
	2	110	122	133	145	157	169	181	193	204	216	228	240
	4	122	134	147	160	173	186	198	211	224	237	250	262
	6	133	147	161	175	189	202	216	230	244	257	271	285
	8	145	160	175	189	204	219	234	248	263	278	293	307
	10	157	173	189	204	220	236	251	267	283	298	314	330
	12	169	186	202	219	236	252	269	286	302	319	335	352
10	0	150	163	177	190	204	218	231	245	258	272	286	299
	2	164	179	193	208	222	237	252	266	281	295	310	324
	4	179	194	210	225	241	256	272	287	303	318	334	349
	6	193	210	226	243	259	276	292	309	325	342	358	375
	8	208	225	243	260	278	295	313	330	347	365	382	400
	10	222	241	259	278	296	314	333	351	370	388	406	425
	12	237	256	276	295	314	334	353	373	392	411	431	450
12	0	212	228	245	261	277	294	310	326	343	359	375	392
	2	229	247	264	281	299	316	333	350	368	385	402	420
	4	247	265	283	301	320	338	356	374	393	411	429	447
	6	264	283	302	322	341	360	379	398	418	437	456	475
	8	281	301	322	342	362	382	402	422	443	463	483	503
	10	299	320	341	362	383	404	425	446	468	489	510	531
	12	316	338	360	382	404	426	448	470	493	515	537	559

CONNECTIONS

Table 5.20.12 Concrete design strength for stud or insert groups — Case 2

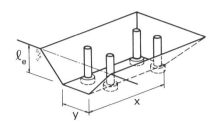

Case 2 — Near a free edge on one side

$$\phi P_c = \phi 4\lambda \sqrt{f'_c} \ A_o$$
$$A_o = xy + \ell_e(2x + y)\sqrt{2} + 2\sqrt{2}\ \ell_e^2$$
$$\phi = 0.85$$
$$\lambda f'_c = 5000 \text{ psi}$$

for other values, multiply by $\lambda \sqrt{f'_c/5000}$

Maximum Tensile Strength, ϕP_c, of a Stud Group, kips

ℓ_e, in.	Dim. y, in.	Dimension x, in.											
		2	4	6	8	10	12	14	16	18	20	22	24
2.5	0	8	11	14	18	21	25	28	31	35	38	42	45
	2	10	15	19	23	28	32	36	41	45	50	54	58
	4	13	18	24	29	34	40	45	50	56	61	66	72
	6	16	22	28	34	41	47	53	60	66	72	78	85
	8	18	26	33	40	47	55	62	69	76	84	91	98
	10	21	29	37	46	54	62	70	78	87	95	103	111
	12	24	33	42	51	60	69	79	88	97	106	115	124
4	0	16	22	27	33	38	44	49	54	60	65	71	76
	2	20	26	33	39	46	52	58	65	71	78	84	90
	4	24	31	38	46	53	60	68	75	83	90	97	105
	6	27	36	44	52	61	69	77	86	94	102	111	119
	8	31	40	50	59	68	77	87	96	105	115	124	133
	10	35	45	55	65	76	86	96	106	117	127	137	147
	12	38	50	61	72	83	94	106	117	128	139	151	162
6	0	33	41	49	57	65	73	82	90	98	106	114	122
	2	38	47	56	65	74	83	92	102	111	120	129	138
	4	43	53	63	73	83	93	103	113	123	133	144	154
	6	48	59	70	81	92	103	114	125	136	147	158	169
	8	53	65	77	89	101	113	125	137	149	161	173	185
	10	58	71	84	97	110	123	136	149	162	175	188	200
	12	63	77	91	105	119	133	146	160	174	188	202	216
8	0	54	65	76	87	98	109	120	131	141	152	163	174
	2	61	73	84	96	108	120	132	144	156	167	179	191
	4	67	80	93	106	118	131	144	157	170	182	195	208
	6	74	87	101	115	129	142	156	170	184	197	211	225
	8	80	95	109	124	139	154	168	183	198	213	227	242
	10	86	102	118	133	149	165	181	196	212	228	243	259
	12	93	109	126	143	159	176	193	209	226	243	259	276
10	0	82	95	109	122	136	150	163	177	190	204	218	231
	2	89	104	118	133	148	162	177	191	206	220	235	250
	4	97	113	128	144	159	175	190	206	221	237	252	268
	6	105	121	138	154	171	187	204	220	237	253	270	286
	8	113	130	148	165	182	200	217	235	252	270	287	305
	10	120	139	157	176	194	212	231	249	268	286	304	323
	12	128	148	167	186	206	225	244	264	283	302	322	341
12	0	114	131	147	163	180	196	212	228	245	261	277	294
	2	123	141	158	175	192	210	227	244	262	279	296	313
	4	132	151	169	187	205	224	242	260	278	297	315	333
	6	142	161	180	199	218	238	257	276	295	314	334	353
	8	151	171	191	211	231	252	272	292	312	332	352	373
	10	160	181	202	223	244	265	287	308	329	350	371	392
	12	169	191	213	235	257	279	302	324	346	368	390	412

CONNECTIONS

Table 5.20.13 Concrete design strength for stud or insert groups — Case 3

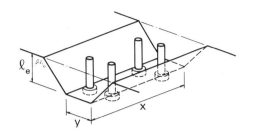

Case 3 — Near a free edge on 2 opposite sides

$\phi P_c = \phi 4\lambda \sqrt{f'_c} \, A_o$

$A_o = xy + 2\ell_e \, x \sqrt{2}$

$\phi = 0.85$

$\lambda f'_c = 5000$ psi

    for other values, multiply by $\lambda \sqrt{f'_c / 5000}$

Maximum Tensile Strength, ϕP_c, of a Stud Group, kips

ℓ_e, in.	Dim. y, in.	2	4	6	8	10	12	14	16	18	20	22	24
						Dimension x, in.							
2.5	0	3	7	10	14	17	20	24	27	31	34	37	41
	2	4	9	13	17	22	26	31	35	39	44	48	52
	4	5	11	16	21	27	32	37	43	48	53	59	64
	6	6	13	19	25	31	38	44	50	57	63	69	75
	8	7	14	22	29	36	43	51	58	65	72	80	87
	10	8	16	25	33	41	49	57	66	74	82	90	98
	12	9	18	28	37	46	55	64	73	83	92	101	110
4	0	5	11	16	22	27	33	38	44	49	54	60	65
	2	6	13	19	26	32	38	45	51	58	64	70	77
	4	7	15	22	29	37	44	52	59	66	74	81	88
	6	8	17	25	33	42	50	58	67	75	83	92	100
	8	9	19	28	37	46	56	65	74	84	93	102	111
	10	10	20	31	41	51	61	72	82	92	102	113	123
	12	11	22	34	45	56	67	78	90	101	112	123	135
6	0	8	16	24	33	41	49	57	65	73	82	90	98
	2	9	18	27	36	46	55	64	73	82	91	100	109
	4	10	20	30	40	50	60	71	81	91	101	111	121
	6	11	22	33	44	55	66	77	88	99	110	121	133
	8	12	24	36	48	60	72	84	96	108	120	132	144
	10	13	26	39	52	65	78	91	104	117	130	143	156
	12	14	28	42	56	70	84	98	111	125	139	153	167
8	0	11	22	33	44	54	65	76	87	98	109	120	131
	2	12	24	36	47	59	71	83	95	107	118	130	142
	4	13	26	38	51	64	77	90	102	115	128	141	154
	6	14	28	41	55	69	83	96	110	124	138	151	165
	8	15	29	44	59	74	88	103	118	133	147	162	177
	10	16	31	47	63	78	94	110	126	141	157	173	188
	12	17	33	50	67	83	100	117	133	150	166	183	200
10	0	14	27	41	54	68	82	95	109	122	136	150	163
	2	15	29	44	58	73	87	102	116	131	146	160	175
	4	16	31	47	62	78	93	109	124	140	155	171	186
	6	16	33	49	66	82	99	115	132	148	165	181	198
	8	17	35	52	70	87	105	122	140	157	174	192	209
	10	18	37	55	74	92	110	129	147	166	184	202	221
	12	19	39	58	77	97	116	136	155	174	194	213	232
12	0	16	33	49	65	82	98	114	131	147	163	180	196
	2	17	35	52	69	86	104	121	138	156	173	190	207
	4	18	36	55	73	91	109	128	146	164	182	201	219
	6	19	38	58	77	96	115	134	154	173	192	211	230
	8	20	40	60	81	101	121	141	161	181	202	222	242
	10	21	42	63	85	106	127	148	169	190	211	232	254
	12	22	44	66	88	110	133	155	177	199	221	243	265

PCI Design Handbook

CONNECTIONS

Table 5.20.14 Concrete design strength for stud or insert groups — Case 4

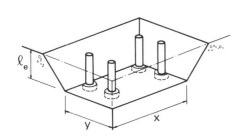

Case 4 — Near a free edge on 2 adjacent sides

$\phi P_c = \phi 4\lambda \sqrt{f'_c} \; A_o$

$A_o = xy + \ell_e (x + y)\sqrt{2} + \sqrt{2}\,\ell_e^2$

$\phi = 0.85$

$\lambda f'_c = 5000$ psi

for other values, multiply by $\lambda \sqrt{f'_c/5000}$

Maximum Tensile Strength, ϕP_c, of a Stud Group, kips

ℓ_e, in.	Dim. y, in.	Dimension x, in.											
		2	4	6	8	10	12	14	16	18	20	22	24
2.5	0	4	6	7	9	11	12	14	16	17	19	21	23
	2	6	9	12	14	17	20	22	25	28	30	33	36
	4	9	13	16	20	24	27	31	35	38	42	45	49
	6	12	16	21	26	30	35	39	44	48	53	58	62
	8	14	20	26	31	37	42	48	53	59	64	70	75
	10	17	24	30	37	43	50	56	63	69	76	82	89
	12	20	27	35	42	50	57	65	72	80	87	94	102
4	0	8	11	14	16	19	22	24	27	30	33	35	38
	2	12	16	19	23	27	30	34	38	41	45	49	52
	4	16	20	25	29	34	39	43	48	53	57	62	67
	6	19	25	30	36	42	47	53	58	64	70	75	81
	8	23	29	36	43	49	56	62	69	75	82	89	95
	10	27	34	42	49	57	64	72	79	87	94	102	109
	12	30	39	47	56	64	73	81	90	98	107	115	124
6	0	16	20	24	29	33	37	41	45	49	53	57	61
	2	21	26	31	36	42	47	52	57	62	67	72	77
	4	26	32	38	44	50	56	62	68	74	80	86	92
	6	31	38	45	52	59	66	73	80	87	94	101	108
	8	36	44	52	60	68	76	84	92	100	108	116	124
	10	42	50	59	68	77	86	95	104	113	122	130	139
	12	47	56	66	76	86	96	106	116	125	135	145	155
8	0	27	33	38	44	49	54	60	65	71	76	82	87
	2	34	40	46	53	59	66	72	78	85	91	98	104
	4	40	47	55	62	69	77	84	92	99	106	114	121
	6	46	55	63	71	80	88	96	105	113	121	130	138
	8	53	62	71	81	90	99	109	118	127	136	146	155
	10	59	69	80	90	100	110	121	131	141	151	162	172
	12	66	77	88	99	110	122	133	144	155	166	178	189
10	0	41	48	54	61	68	75	82	88	95	102	109	116
	2	49	56	64	72	80	87	95	103	111	118	126	134
	4	56	65	74	82	91	100	109	117	126	135	144	152
	6	64	74	83	93	103	113	122	132	142	151	161	171
	8	72	82	93	104	114	125	136	146	157	168	178	189
	10	80	91	103	114	126	138	149	161	172	184	196	207
	12	87	100	113	125	138	150	163	175	188	200	213	226
12	0	57	65	73	82	90	98	106	114	122	131	139	147
	2	66	75	84	94	103	112	121	130	139	148	157	167
	4	75	85	96	106	116	126	136	146	156	166	176	186
	6	84	96	107	118	129	140	151	162	173	184	195	206
	8	94	106	118	130	142	154	166	178	190	202	214	226
	10	103	116	129	142	155	168	181	194	206	219	232	245
	12	112	126	140	154	168	181	195	209	223	237	251	265

CONNECTIONS

Table 5.20.15 Concrete design strength for stud or insert groups — Case 5

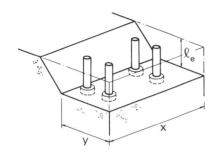

Case 5 — Near a free edge on 3 sides

$$\phi P_c = \phi 4\lambda \sqrt{f'_c}\ A_o$$
$$A_o = xy + \ell_e \times \sqrt{2}$$
$$\phi = 0.85$$
$$\lambda f'_c = 5000\ \text{psi}$$

for other values, multiply by $\lambda \sqrt{f'_c/5000}$

Maximum Tensile Strength, ϕP_c, of a Stud Group, kips

ℓ_e, in.	Dim. y, in.	\multicolumn{12}{c}{Dimension x, in.}										
		2	4	6	8	10	12	14	16	18	20	22	24
2.5	0	2	3	5	7	8	10	12	14	15	17	19	20
	2	3	5	8	11	13	16	19	21	24	27	29	32
	4	4	7	11	14	18	22	25	29	33	36	40	43
	6	5	9	14	18	23	28	32	37	41	46	50	55
	8	6	11	17	22	28	33	39	44	50	55	61	67
	10	7	13	20	26	33	39	46	52	59	65	72	78
	12	7	15	22	30	37	45	52	60	67	75	82	90
4	0	3	5	8	11	14	16	19	22	24	27	30	33
	2	4	7	11	15	18	22	26	29	33	37	40	44
	4	5	9	14	19	23	28	33	37	42	46	51	56
	6	6	11	17	22	28	34	39	45	50	56	62	67
	8	7	13	20	26	33	39	46	53	59	66	72	79
	10	8	15	23	30	38	45	53	60	68	75	83	90
	12	8	17	25	34	42	51	59	68	76	85	93	102
6	0	4	8	12	16	20	24	29	33	37	41	45	49
	2	5	10	15	20	25	30	35	40	45	50	55	60
	4	6	12	18	24	30	36	42	48	54	60	66	72
	6	7	14	21	28	35	42	49	56	63	70	77	84
	8	8	16	24	32	40	48	55	63	71	79	87	95
	10	9	18	27	36	44	53	62	71	80	89	98	107
	12	10	20	30	39	49	59	69	79	89	98	108	118
8	0	5	11	16	22	27	33	38	44	49	54	60	65
	2	6	13	19	26	32	38	45	51	58	64	70	77
	4	7	15	22	29	37	44	52	59	66	74	81	88
	6	8	17	25	33	42	50	58	67	75	83	92	100
	8	9	19	28	37	46	56	65	74	84	93	102	111
	10	10	20	31	41	51	61	72	82	92	102	113	123
	12	11	22	34	45	56	67	78	90	101	112	123	135
10	0	7	14	20	27	34	41	48	54	61	68	75	82
	2	8	16	23	31	39	47	54	62	70	78	85	93
	4	9	17	26	35	44	52	61	70	79	87	96	105
	6	10	19	29	39	48	58	68	77	87	97	107	116
	8	11	21	32	43	53	64	75	85	96	106	117	128
	10	12	23	35	46	58	70	81	93	104	116	128	139
	12	13	25	38	50	63	75	88	101	113	126	138	151
12	0	8	16	24	33	41	49	57	65	73	82	90	98
	2	9	18	27	36	46	55	64	73	82	91	100	109
	4	10	20	30	40	50	60	71	81	91	101	111	121
	6	11	22	33	44	55	66	77	88	99	110	121	133
	8	12	24	36	48	60	72	84	96	108	120	132	144
	10	13	26	39	52	65	78	91	104	117	130	143	156
	12	14	28	42	56	70	84	98	111	125	139	153	167

Fig. 5.20.16 Combined tension and shear — studs or inserts

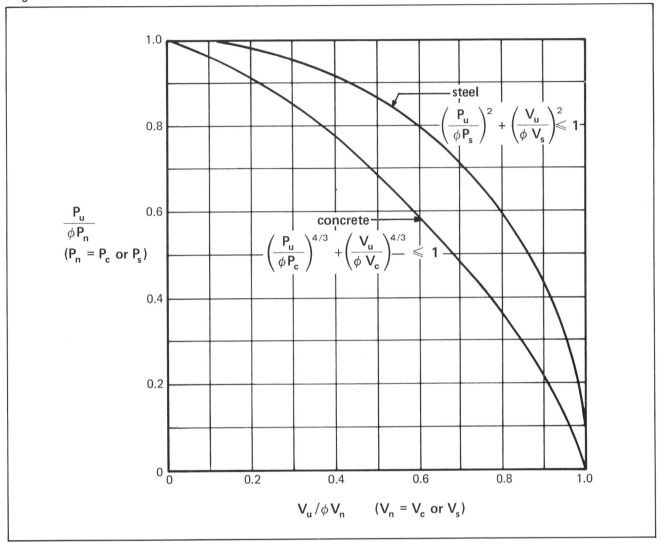

Table 5.20.17 Development length, ℓ_d, for deformed bar anchors, in.[1][2]

Concrete Strength f'_c	Normal Weight Concrete ($\lambda = 1.0$)					Sand-Lightweight Concrete ($\lambda = 0.85$)				
	Bar Diameter					Bar Diameter				
	1/4	3/8	1/2	5/8	3/4	1/4	3/8	1/2	5/8	3/4
3000	12	12	16	21	25	12	15	19	24	29
4000	12	12	14	18	21	12	13	17	21	25
5000	12	12	13	16	19	12	12	15	19	23
6000	12	12	12	15	17	12	12	14	17	21
7000	12	12	12	13	16	12	12	13	16	19
8000	12	12	12	13	15	12	12	12	15	18

(1) $f_y = 60,000$ psi; for values above 60,000 psi multiply by $\left(2 - \dfrac{60,000}{f_y}\right)$

(2) For top bars, multiply by 1.4

CONNECTIONS

Table 5.20.18 Design data for concrete loop inserts

Procedures:
1. Apply reduction factors, if required to ϕP_c values for concrete design strength
2. Check both concrete and insert strength

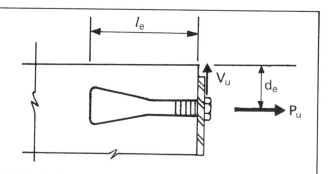

Concrete Design Strength

Shear, ϕV_c [1]				Pullout, ϕP_c [1] [2]	
d_e, in.	ϕV_c, lb	d_e, in.	ϕV_c, lb	l_e, in.	ϕP_c, lb $\left(\substack{\text{full}\\\text{cone}}\right)$
2	1280	10	18,280	2	4300
2.5	2340	11	20,400	3	9600
3	3400	12	22,520	4	17,100
3.5	4460	13	24,650	5	26,700
4	5520	14	26,780	6	38,400
4.5	6590	15	28,900	7	52,300
5	7650	16	31,060	8	68,300
5.5	8710	17	33,150	9	86,500
6	9780	18	35,280	10	106,800
6.5	10,840	19	37,400	11	129,200
7	11,900	20	39,520	12	153,700
7.5	12,960	21	41,650	13	180,400
8	14,020	22	43,780	14	209,200
8.5	15,090	23	45,900	15	240,200
9	16,150	24	48,020	18	345,900

(1) Multiply table values by 0.85 for sand-lightweight concrete

(2) Multiply pullout capacities by $\sqrt{f_c'/5000}$ for values other than 5000 psi.

Typical Insert Design Strength[3]

Capacity of Round Wire Used in Concrete Inserts			Capacity of Coil Bolts and Threaded Coil Rods			
Leg Wire Dia., in.	Wire Grade	Yield Strength, lb	Bolt Diameter (in.)	Min. Coil Penetration (in.)	Tensile Strength (P_s)	Shear Strength (V_s)
0.218	C1008	2000	1/2	1 1/2	13,500	8,100
0.223	C1038	3900	3/4	2	18,470	11,080
0.225	C1038	3700	1	2 1/2	37,870	22,720
0.240	C1008	2900	1 1/4	2 1/2	54,960	32,980
0.260	C1008	3550	1 1/2	3	83,340	50,000
0.281	C1035	6000				
0.306	C1035	6900	Design Strength of Machine Bolts used in "Ferrules" or "Weld Nuts"			
0.340	C1035	7500	Bolt Dia., (in.)	Bolt Grade	Tensile strength	Shear Strength	Ferrule Data
0.375	C1008	7450		(ASTM)	P_s (lb.)	V_s (lb.)	Threads/in.	Bolt Length
0.440	C1035	12,000	1/2	A307	4820	3330	13	1
			5/8	A307	7680	5220	11	1-1/8
			3/4	A307	11,360	7510	10	1-1/8
			1	A307	20,600	13,350	8	1-1/4

(3) Data supplied by manufacturers.

PCI Design Handbook

CONNECTIONS

Table 5.20.19 Shear strength of connection angles

$$t = \sqrt{\frac{4 V_u e_v}{\phi f_y b}} \text{ , in.}$$

ϕ = 0.90

b = width of angle, in.

f_y = yield strength of angle steel = 36,000 psi

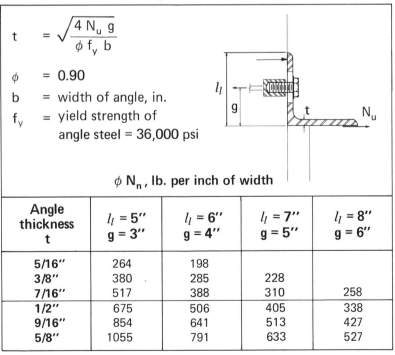

ϕV_n, lb. per inch of width

Angle thickness t	$e_v = 3/4''$	$e_v = 1''$	$e_v = 1\text{-}1/2''$	$e_v = 2''$	$e_v = 2\text{-}1/2''$
5/16''	1055	791	527	396	316
3/8''	1519	1139	759	570	456
7/16''	2067	1550	1034	775	620
1/2''	2700	2025	1350	1013	810
9/16''	3417	2563	1709	1281	1025
5/8''	4219	3164	2109	1582	1266

Table 5.20.20 Axial strength of connection angles

$$t = \sqrt{\frac{4 N_u g}{\phi f_y b}}$$

ϕ = 0.90

b = width of angle, in.

f_y = yield strength of angle steel = 36,000 psi

ϕN_n, lb. per inch of width

Angle thickness t	$l_l = 5''$ $g = 3''$	$l_l = 6''$ $g = 4''$	$l_l = 7''$ $g = 5''$	$l_l = 8''$ $g = 6''$
5/16''	264	198		
3/8''	380	285	228	
7/16''	517	388	310	258
1/2''	675	506	405	338
9/16''	854	641	513	427
5/8''	1055	791	633	527

CONNECTIONS

Table 5.20.21 Column base plate thickness requirements

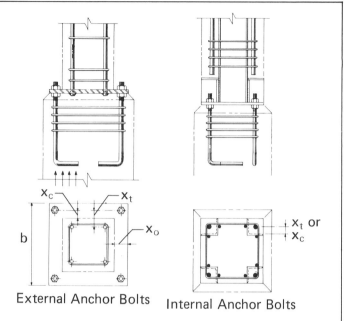

External Anchor Bolts

Internal Anchor Bolts

Thickness Required For Concrete Bearing

f_{bu} (psi)	$x_o = 3''$	$x_o = 4''$	$x_o = 5''$
500	5/8	3/4	1
1000	3/4	1	1 3/8
1500	1	1 3/8	1 5/8
2000	1 1/8	1 1/2	1 7/8
2500	1 1/4	1 5/8	2
3000	1 3/8	1 7/8	2 1/4
3500	1 1/2	2	2 1/2
4000	1 5/8	2 1/8	2 5/8

Thickness Required for Bolt Loading

Tension On External Anchor Bolts

b (in.)	No. & Diameter Of A 36 Or A 307 Anchor Bolts Per Side							
	2 – ¾'' $x_t = 3.75''$	2 – ¾'' $x_t = 4.25''$	2 – 1'' $x_t = 3.75''$	2 – 1'' $x_t = 4.25''$	2 – 1¼'' $x_t = 3.75''$	2 – 1¼'' $x_t = 4.25''$	2 – 1½'' $x_t = 3.75''$	2 – 1½'' $x_t = 4.25''$
12	1	1 1/8	1 3/8	1 1/2	1 3/4	1 7/8	2 1/8	2 1/4
14	1	1	1 3/8	1 3/8	1 5/8	1 3/4	2	2 1/8
16	7/8	1	1 1/4	1 3/8	1 1/2	1 5/8	1 7/8	2
18	7/8	1	1 1/8	1 1/4	1 1/2	1 1/2	1 3/4	1 7/8
20	7/8	7/8	1 1/8	1 1/8	1 3/8	1 1/2	1 5/8	1 3/4
22	3/4	7/8	1	1 1/8	1 3/8	1 3/8	1 5/8	1 5/8
24	3/4	3/4	1	1 1/8	1 1/4	1 3/8	1 1/2	1 5/8
26	3/4	3/4	1	1	1 1/4	1 1/4	1 1/2	1 1/2
28	3/4	3/4	1	1	1 1/8	1 1/4	1 3/8	1 1/2

Compression On Anchor Bolts Or Tension On Internal Anchor Bolts

b (in.)	No. & Diameter Of A 36 Or A 307 Anchor Bolts Per Side							
	2 – ¾'' $x_c = 1.5''$	2 – ¾'' $x_c = 2.0''$	2 – 1'' $x_c = 1.5''$	2 – 1'' $x_c = 2.0''$	2 – 1¼'' $x_c = 1.5''$	2 – 1¼'' $x_c = 2.0''$	2 – 1½'' $x_c = 1.5''$	2 – 1½'' $x_c = 2.0''$
12	3/4	3/4	7/8	1	1 1/4	1 3/8	1 3/8	1 5/8
14	3/4	3/4	7/8	1	1	1 1/4	1 1/4	1 1/2
16	3/4	3/4	3/4	7/8	1	1 1/8	1 1/4	1 3/8
18	3/4	3/4	3/4	7/8	1	1 1/8	1 1/8	1 1/4
20	3/4	3/4	3/4	7/8	7/8	1	1 1/8	1 1/4
22	3/4	3/4	3/4	3/4	7/8	1	1	1 1/8
24	3/4	3/4	3/4	3/4	7/8	1	1	1 1/8
26	3/4	3/4	3/4	3/4	3/4	7/8	1	1 1/8
28	3/4	3/4	3/4	3/4	3/4	7/8	7/8	1

CONNECTIONS

Fig. 5.21.1 Typical connection details

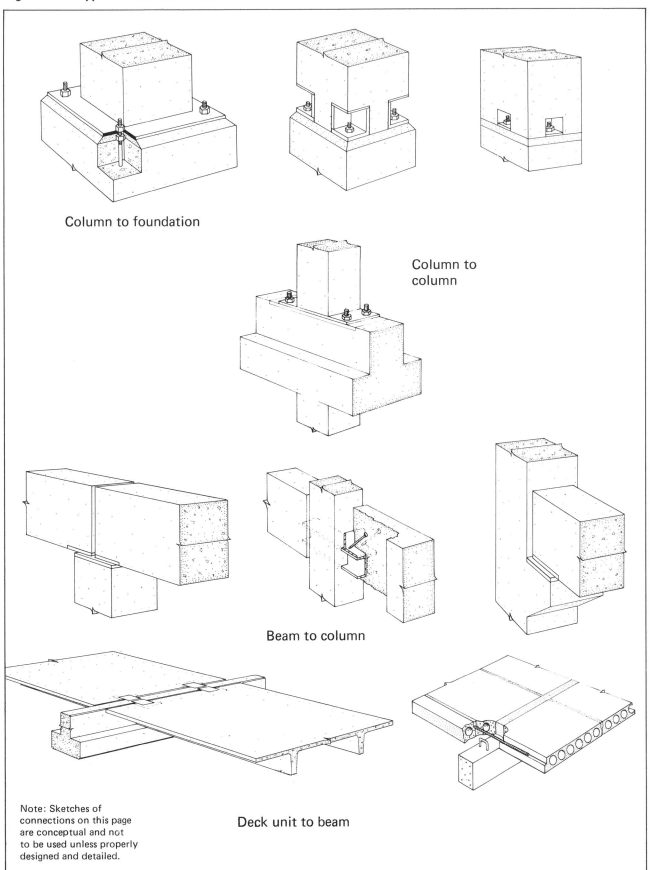

Column to foundation

Column to column

Beam to column

Deck unit to beam

Note: Sketches of connections on this page are conceptual and not to be used unless properly designed and detailed.

CONNECTIONS

Fig. 5.21.2 Typical connection details

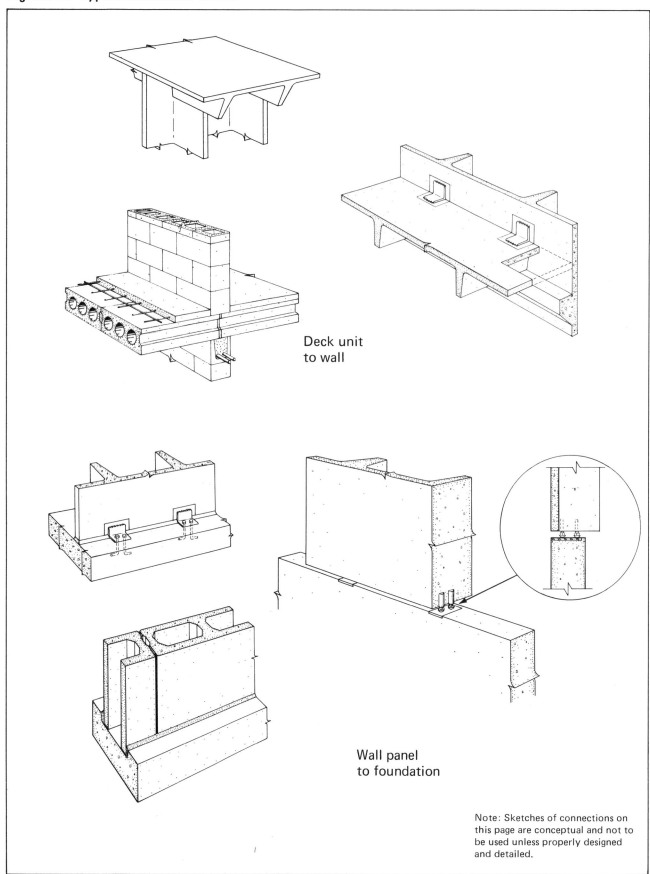

Deck unit
to wall

Wall panel
to foundation

Note: Sketches of connections on this page are conceptual and not to be used unless properly designed and detailed.

PCI Design Handbook

PART 6
RELATED CONSIDERATIONS

THERMAL PROPERTIES

6.1.1 Glossary

U = Overall coefficient of heat transmission or thermal *transmittance* (air-to-air); the time rate of heat flow usually expressed in Btu per (hr) (sq ft) (deg F temperature difference between air on the inside and air on the outside of a wall, floor, roof or ceiling). The term is applied to the usual combinations of materials, and also to single materials, such as window glass, and includes the surface conductance on both sides. This term is frequently called the U-value.

k = Thermal *conductivity*; the time rate of heat flow through a homogeneous material under steady-state conditions, through unit area, per unit temperature gradient in the direction perpendicular to an isothermal surface. Its unit is Btu (in.) per (hr) (sq ft) (deg F).

C = Thermal *conductance*; the time rate of heat flow expressed in Btu per (hr) (sq ft) (deg F average temperature difference between two surfaces). The term is applied to specific materials as used, either homogeneous or heterogeneous, for the thickness or construction stated, not per in. of thickness.

f = Film or *surface conductance*; the time rate of heat exchange by radiation, conduction and convection of a unit area of a surface with its surroundings. Its value is usually expressed in Btu per (hr) (sq ft of surface) (deg F temperature difference). Subscripts i and o are usually used to denote inside and outside surface conductances, respectively.

a = Thermal *conductance of an air space*; the time rate of heat flow through a unit area of an air space per unit temperature difference between the boundary surfaces. Its value is usually expressed in Btu per (hr) (sq ft of area) (deg F). The conductance of an air space is dependent on the temperature difference, the height, the depth and the position, character, and temperature of the boundary surfaces. Since the relationships are not linear, accurate values must be obtained by test and not by computation. The space must be fully described if the values are to be meaningful.

R = *Thermal resistance*; the reciprocal of a heat transmission coefficient, as expressed by U, C, f, or a. Its unit is (deg F) (hr) (sq ft) per Btu. For example, a wall with a U-value of 0.25 would have a resistance value of R = 1/U = 1/0.25 = 4.0.

Btu = *British thermal unit*; approximately the amount of heat to raise one pound of water from 59F to 60F.

Degree Day — A unit based on temperature difference and time, used in estimating fuel consumption and specifying nominal heating load of a building in winter. For any one day, when the mean temperature is less than 65F, there exist as many degree days as there are F degrees difference in temperature between the mean temperature for the day and 65F.

Dew-Point Temperature — The temperature at which condensation of water vapor begins for a given humidity and pressure as the temperature is reduced. The temperature corresponding to saturation (100% relative humidity) for a given absolute moisture content and given pressure.

Perm — A unit of permeance. A perm is 1 grain per (sq ft) (hr) (in. of mercury vapor pressure difference).

Permeability, water vapor — The property of a substance which permits the passage of water. It is equal to the permeance of 1 in. of a substance. Permeability is measured in perm inches. The permeability of a material varies with psychrometric conditions.

Permeance — The water vapor permeance of any sheet or assembly is the ratio of the water vapor flow to the vapor pressure difference between the two surfaces. Permeance is measured in perms.

Two commonly used test methods are the Wet Cup and Dry Cup Tests. Specimens are sealed over the tops of cups containing either water or desiccant; the surrounding atmosphere is maintained at 50% relative humidity, and weight changes are measured.

Relative Humidity — The ratio of moisture vapor present in air to the water vapor present in saturated air at the same temperature and pressure.

6.1.2 General

Thermal codes and standards prescribe in many different ways the heat transmission requirements for buildings. Therefore, it is important to have basic knowledge about the heat loss and heat gain of many materials. Some of the fundamentals and design aids* that are needed to analyze and compare the heat losses and heat gains through building envelopes are presented in this section.

Precast and prestressed concrete construction have a unique advantage with their thermal inertia and thermal storage properties. This advantage is recognized in some codes, however, procedures to account for the benefits of heavier materials are not usually given. Procedures are presented in Sect. 6.1.5.

The trend is toward more insulation with little regard given to its total impact on the energy saved. Before assuming that thick insulation is needed, mass effects, less glass area, reduced infiltration and controlled ventilation should be considered. Further considerations may include building orientation, exterior color, shading or reflections from adjacent structures, surrounding surfaces or vegetation, building aspect ratio, number of stories, and wind direction and speed.

Except where noted the information and design criteria that follow are taken from or derived from ASHRAE† *Handbook of Fundamentals*, hereafter referred to as the ASHRAE Handbook, and from the ASHRAE Standard 90-75, *Energy Conservation in New Building Design*, hereafter referred to as the ASHRAE Standard.

It is important to note that all design criteria are not given and the criteria used may change from time to time as the ASHRAE Standard and Handbook are revised. Therefore, as the design procedures are applied, it is essential to consult the applicable codes and revised references for specified values and procedures that govern in a particular area.

6.1.3 Thermal Properties of Materials, Surfaces and Air Spaces

The thermal properties of materials and air spaces are based on steady state tests. The tests establish the number of British thermal units (Btu) that pass from the warm side to the cool side per sq ft per hr per deg temperature difference either side of the item being tested. The results of the tests determine the conductivity, k, per inch thickness for homogeneous sections. For non-homogeneous compound sections and air spaces the tests determine the conductance, C, for the total thickness. The values k and C do not include surface conductances. The inside surface conductance, f_i, and the outside surface conductance, f_o, are considered separately.

The resistance method is used to determine the overall thermal effectiveness of wall, floor and roof sections. Therefore, the resistance, R, i.e., the reciprocal of k, C, f_i and f_o must be known. The R-values of construction materials are not influenced by the direction of heat flow. On the other hand the R of surfaces and air spaces differs depending on orientation, that is whether they are vertical, sloping or horizontal. Also the R-values of surfaces are affected by the velocity of air at the surfaces and by their reflective properties.

Tables 6.1.1 and 6.1.2 give the thermal resistances of surfaces and 3 1/2 in. air spaces. Values in Table 6.1.2 can be used for all air space thicknesses with very little change in the overall R of the section, with one exception. Where heat flow direction is down, follow the procedure given in footnote 1, Table 6.1.2.

Table 6.1.3 gives the thermal properties of most commonly used building materials. For glass, only U-values are given since the glass alone has almost no thermal resistance. The resistances of the surfaces of the glass contribute mostly to the U-value.

Table 6.1.4 gives the thermal properties of various weight concretes, including insulating concretes. It also gives the R-values of some of the commonly used prestressed concrete floor, roof and wall units.

Thermal conductances and resistances are usually determined and reported for building materials in their oven dry condition. Conductances and resistances for concrete are often reported in its normally dry condition as well as in its oven dry condition. Normally dry is the condition of concrete containing an equilibrium amount of free water after extended exposure to warm air at 35 to 50% relative humidity. Values given in the tables are based on concrete that is considered normally dry.

It should be noted that normally dry concrete in combination with insulation generally provides about the same R-value as equally insulated oven dry concrete. It should also be noted that normally dry concrete, because of its moisture content, has the ability to store a greater amount of heat than oven dry concretes, a property somewhat

---

*For a more complete presentation see "Thermal Design of Precast Concrete Building Envelopes," *PCI Journal*, January-February, 1978.

†American Society of Heating, Refrigerating and Air-Conditioning Engineers, Inc., New York, New York.

Table 6.1.1 Thermal resistances, R_f, of surfaces

Position of Surface	Direction of Heat Flow	Still air, R_{fi}			Moving air, R_{fo}	
		Nonreflective Surface	Reflective Surface		Nonreflective Surface	
			A [1]	B [2]	15 mph [3]	7½ mph [4]
Vertical	Horizontal	0.68	1.35	1.70	0.17	0.25
Horizontal	Up	0.61	1.10	1.32	0.17	0.25
	Down	0.92	2.70	4.55	0.17	0.25

(1) Aluminum painted paper   (3) Winter design
(2) Bright aluminum foil   (4) Summer design

Table 6.1.2 Thermal resistances, R_a, of air spaces [1]

Position of Air Space	Direction of Heat Flow	Air Space		Nonreflective Surfaces	Reflective Surfaces		
		Mean Temp. °F	Temp. Diff. °F		One Side [2]	One Side [3]	Both Sides [3]
Vertical	Horizontal (walls)	**Winter**					
		50	10	1.01	2.32	3.40	3.63
		50	30	0.91	1.89	2.55	2.67
	Horizontal (walls)	**Summer**					
		90	10	0.85	2.15	3.40	3.69
Horizontal	Up (roofs)	**Winter**					
		50	10	0.93	1.95	2.66	2.80
		50	30	0.84	1.58	2.01	2.09
	Down (floors)	50	—	1.23	3.97	8.72	10.37
	Down (roofs)	**Summer**					
		90	—	1.00	3.41	8.19	10.07

(1) For 3 1/2 in. air space thickness. The values with the exception of those for reflective surfaces, heat flow down, will differ about 10% for air space thickness of 3/4 in. to 16 in. Refer to Table 20, Chapter 20 of the ASHRAE Handbook for values of other thicknesses, reflective surfaces, heat flow down.
(2) Aluminum painted paper
(3) Bright aluminum foil

Table 6.1.3 Thermal properties of various building materials[1]

Material	Unit Weight, pcf	Resistance		Trans-mittance, U
		Per inch of thick-ness, 1/k	For thick-ness shown, R	
Insulation, rigid				
Cellular glass	8.5	2.63		
Fiberglass	4 to 9	4.00		
Mineral fiber, resin binder	15	3.45		
Mineral fiberboard, wet felted, roof insulation	16-17	2.94		
Wood, shredded, cemented in preformed slabs	22	1.67		
Polystyrene - cut cell surface	1.8	4.00		
Polystyrene - smooth skin surface	3.5	5.26		
Polystyrene - molded bead	1.0	3.57		
Polyurethane	1.5	6.25		
Miscellaneous				
Acoustical tile	18	2.50		
Carpet, fibrous pad			2.08	
Carpet, rubber pad			1.23	
Floor tile, asphalt, rubber, vinyl			0.05	
Gypsum board	50	0.90		
Particle board	50	1.06		
Plaster				
cement, sand agg.	116	0.20		
gyp, L.W. agg.	45	0.63		
gyp, sand agg.	105	0.17		
Roofing, 3/8 in. built-up	70		0.33	
Wood hard	45	0.91		
Wood soft	32	1.25		
Wood plywood	34	0.83		
Glass doors & windows				
Single, winter				1.13
Single, summer				1.06
Double, winter[2]				0.65
Double, summer[2]				0.61
Doors, metal				
Insulated, winter				0.19
Insulated, summer				0.18

(1) See Table 6.1.4 for all concretes, including insulating concrete for roof fill.
(2) 1/4'' air space.

Table 6.1.4 Thermal properties of concrete[1]

Description	Concrete Weight, pcf	Thick-ness, in.	Resistance, R	
			Per Inch of thick-ness, 1/k	For thick-ness shown, 1/C
Concretes including normal weight, lightweight and lightweight insulating concretes	145 140 130 120 110 100 90 80 70 60 50 40 30 20		0.075 0.083 0.11 0.14 0.19 0.24 0.30 0.37 0.45 0.52 0.67 0.83 1.00 1.43	
Normal weight tees[2] and solid slabs	145	2 3 4 5 6 8		0.15 0.23 0.30 0.38 0.45 0.60
Normal weight hollow core slabs	145	6 8 10 12		1.07 1.34 1.73 1.91
Structural lightweight tees[2] and solid slabs	110	2 3 4 5 6 8		0.38 0.57 0.76 0.95 1.14 1.52
Structural lightweight hollow core slabs	110	8 12		2.00 2.59

(1) Based on normally dry concrete (see Sect. 6.1.3).
(2) Thickness for tees is thickness of slab portion including topping, if used. The effect of the stems generally is not significant, therefore, their thickness and surface area may be dis-regarded.

PCI Design Handbook

beneficial when considering dynamic thermal response of concrete.

6.1.4 Computation of U-Values

The heat transmission values, U-values, of a building wall, floor or roof section are computed by adding together:

1. The R-values of the layers of materials in the section.
2. The R_{fi} and R_{fo} of the inside and outside surfaces.
3. The R_a-value of air spaces that may be within the section.

The reciprocal of the summation of all R's is the U-value. For example, a typical wall with an air space is calculated:

$$U = \frac{1}{R_{fi} + R_{materials} + R_a + R_{fo}}$$

where $R_{materials}$ is a summation, that is, $R_1 + R_2 + R_3$, etc., of all opaque materials in the wall.

For convenience to designers a number of wall and roof U-values has been computed. Tables 6.1.5, 6.1.6 and 6.1.7 contain winter and summer U-values for concrete wall and roof sections of various thicknesses and designs without added insulation or with added insulation having R-values 4, 6, 8, and 10 for walls, and 4, 8, and 12 for roofs. Wall tables can be applied to sandwich type panels as well as single wythe panels insulated on one side. Roof tables include the added effect of acoustical tile ceilings either applied directly to the bottom of the prestressed units or suspended below.

6.1.4.1 Design Examples — Using Tables 6.1.1 to 6.1.4

Example — Wall

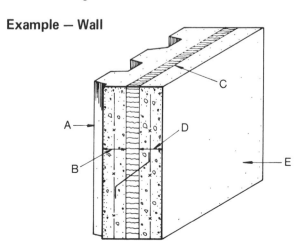

		R Winter	R Summer	Table
A.	Surface, outside	0.17	0.25	6.1.1
B.	Concrete, 2 in., (110 pcf)	0.38	0.38	6.1.4
C.	Polystyrene, 1½ in.	6.00	6.00	6.1.3
D.	Concrete, 2½ in., (110 pcf)	0.48	0.48	6.1.4
E.	Surface, inside	0.68	0.68	6.1.1
	Total R =	7.71	7.79	
	U = 1/R =	0.13	0.13	

Example — Roof

		R Winter	R Summer	Table
A.	Surface, outside	0.17	0.25	6.1.1
B.	Roofing, built-up	0.33	0.33	6.1.3
C.	Polystyrene, 2 in.	8.00	8.00	6.1.3
D.	Concrete, 2 in. (145 pcf)	0.15	0.15	6.1.4
E.	Surface, inside	0.61	0.92	6.1.1
	Total R =	9.26	9.65	
	U = 1/R =	0.11	0.10	

The U-values for heating may be modified to account for the effects of mass, given in Sect. 6.1.5.1. The U-values for cooling are used without modification. Mass effects for cooling are applied by using appropriate temperature differences, as explained in Sect. 6.1.5.2.

6.1.5 Thermal Storage Effects

For some time it has been known that walls and roofs of concrete or any massive material react to temperature variations slowly and therefore reduce heating and cooling loads, however, engineering application of this concept has been limited to only estimating mass effects. Computers now make it possible to better account for mass effects with hour by hour calcuations for 24 hours, a week, a month, or a year. Some 24-hour peak load computer studies* show that as the weight of walls increases the heat flow rates and peak loads decrease. Fig. 6.1.1 compares the heat gain through two roofs having different weights but with equal U-values and exposed to identical summer simulated temperatures.

---

*Catani, Mario J. and Goodwin, Stanley E., "Heavy Building Envelopes and Dynamic Thermal Response," *ACI Journal*, February 1976, pp. 83-86.

Table 6.1.5 U-values: walls of prestressed tees, hollow core slabs, solid and sandwich panels; winter and summer conditions[1]

Concrete Weight, pcf	Type of Wall Panel	Thickness t, and Resistance R, of the Concrete		Winter $R_{fo} = 0.17, R_{fi} = 0.68$					Summer $R_{fo} = 0.25, R_{fi} = 0.68$				
				Insulation Resistance, R									
		$t^{(2)}$	R	None	4	6	8	10	None	4	6	8	10
145	Solid walls, tees, and sandwich panels	2	0.15	1.00	.20	.14	.11	.09	.93	.20	.14	.11	.09
		3	0.23	.93	.20	.14	.11	.09	.87	.19	.14	.11	.09
		4	0.30	.87	.19	.14	.11	.09	.81	.19	.14	.11	.09
		5	0.38	.81	.19	.14	.11	.09	.76	.19	.14	.11	.09
		6	0.45	.77	.19	.14	.11	.09	.72	.19	.14	.11	.09
		8	0.60	.69	.18	.13	.11	.09	.65	.18	.13	.10	.09
	Hollow core slabs[3]	6(o)	1.07	.52	.17	.13	.10	.08	.50	.17	.13	.10	.08
		(f)	1.86	.37	.15	.11	.09	.08	.36	.15	.11	.09	.08
		8(o)	1.34	.46	.16	.12	.10	.08	.44	.16	.12	.10	.08
		(f)	3.14	.26	.13	.10	.08	.07	.25	.12	.10	.08	.07
		10(o)	1.73	.39	.15	.12	.09	.08	.38	.15	.12	.09	.08
		(f)	4.05	.20	.11	.09	.08	.07	.20	.11	.09	.08	.07
		12(o)	1.91	.36	.15	.11	.09	.08	.35	.15	.11	.09	.08
		(f)	5.01	.17	.10	.08	.07	.06	.17	.10	.08	.07	.06
110	Solid walls, tees, and sandwich panels	2	0.38	.81	.19	.14	.11	.09	.76	.19	.14	.11	.09
		3	0.57	.70	.18	.13	.11	.09	.67	.18	.13	.11	.09
		4	0.76	.62	.18	.13	.10	.09	.59	.18	.13	.10	.09
		5	0.95	.56	.17	.13	.10	.09	.53	.17	.13	.10	.08
		6	1.14	.50	.17	.13	.10	.08	.48	.16	.12	.10	.08
		8	1.52	.42	.16	.12	.10	.08	.41	.16	.12	.10	.08
	Hollow core slabs[3]	8(o)	2.00	.35	.15	.11	.09	.08	.34	.14	.11	.09	.08
		(f)	4.41	.19	.11	.09	.08	.07	.19	.11	.09	.08	.07
		12(o)	2.59	.29	.13	.11	.09	.07	.28	.13	.11	.09	.07
		(f)	6.85	.13	.09	.07	.06	.06	.13	.08	.07	.06	.06

(1) When insulations having other R values are used, U-values can be interpolated with adequate accuracy, or U can be calculated as shown in Sect. 6.1.4.
When a finish, air space or any material is added, the new U value is:

$$\frac{1}{\dfrac{1}{U \text{ from table}} + R \text{ of added finish, air space, or material}}$$

(2) Thickness for tees is thickness of slab portion. For sandwich panels, t is the sum of the thicknesses of the wythes.

(3) For hollow panels (o) and (f) after thickness designates cores open or cores filled with insulation.

Table 6.1.6 U-values: roofs with built-up roofing, winter conditions, heat flow upward[1]

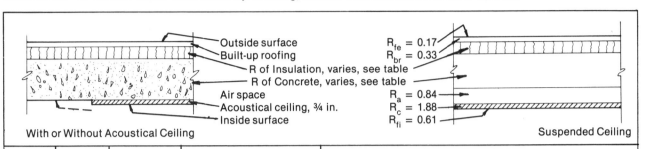

Concrete Weight pcf	Pre-stressed Concrete Member	Thickness t and R of Concrete		Without Ceiling				With Ceiling							
								Applied Direct				Suspended			
				Top Insulation Resistance, R											
		$t^{(2)}$	R	None	4	8	12	None	4	8	12	None	4	8	12
145	Solid slabs and tees	2	0.15	.79	.19	.11	.08	.32	.14	.09	.07	.25	.13	.08	.06
		3	0.23	.75	.19	.11	.07	.31	.14	.09	.07	.25	.12	.08	.06
		4	0.30	.71	.18	.11	.07	.30	.14	.09	.07	.24	.12	.08	.06
		5	0.38	.67	.18	.11	.07	.30	.14	.09	.06	.24	.12	.08	.06
		6	0.45	.64	.18	.10	.07	.29	.13	.09	.06	.23	.12	.08	.06
		8	0.60	.58	.18	.10	.07	.27	.13	.09	.06	.23	.12	.08	.06
	Hollow core slabs[3]	6(o)	1.07	.46	.16	.10	.07	.25	.12	.08	.06	.20	.11	.08	.06
		(f)	1.86	.34	.14	.09	.07	.21	.11	.08	.06	.18	.10	.07	.06
		8(o)	1.34	.41	.15	.10	.07	.23	.12	.08	.06	.19	.11	.08	.06
		(f)	3.14	.24	.12	.08	.06	.16	.10	.07	.06	.14	.09	.07	.05
		10(o)	1.73	.35	.15	.09	.07	.21	.11	.08	.06	.18	.10	.07	.06
		(f)	4.05	.19	.11	.08	.06	.14	.09	.07	.05	.13	.08	.06	.05
		12(o)	1.91	.33	.14	.09	.07	.20	.11	.08	.06	.17	.10	.07	.06
		(f)	5.01	.16	.10	.07	.06	.12	.08	.06	.05	.11	.08	.06	.05
110	Solid slabs and tees	2	0.38	.67	.18	.11	.07	.30	.14	.09	.06	.24	.12	.08	.06
		3	0.57	.60	.18	.10	.07	.28	.13	.09	.06	.23	.12	.08	.06
		4	0.76	.53	.17	.10	.07	.27	.13	.08	.06	.22	.12	.08	.06
		5	0.95	.49	.16	.10	.07	.25	.13	.08	.06	.21	.11	.08	.06
		6	1.14	.44	.16	.10	.07	.24	.12	.08	.06	.20	.11	.08	.06
		8	1.52	.38	.15	.09	.07	.22	.12	.08	.06	.19	.11	.08	.06
	Hollow core slabs[3]	8(o)	2.00	.32	.14	.09	.07	.20	.11	.08	.06	.17	.10	.07	.06
		(f)	4.41	.18	.11	.07	.06	.13	.09	.06	.05	.12	.08	.06	.05
		12(o)	2.59	.27	.13	.09	.06	.18	.11	.07	.06	.16	.10	.07	.05
		(f)	6.85	.13	.08	.06	.05	.10	.07	.06	.05	.09	.07	.05	.04

(1) When insulations having other R values are used, U-values can be interpolated with adequate accuracy, or U can be calculated as shown in Sect. 6.1.4.
When a finish, air space or any material is added, the new U value is:

$$\dfrac{1}{\dfrac{1}{U \text{ from table}} + R \text{ of added finish, air space, or material}}$$

(2) Thickness for tees is thickness of slab portion.
(3) For hollow panels (o) and (f) after thickness designates cores open or cores filled with insulation.

Table 6.1.7 U-values: roofs with built-up roofing, summer conditions, heat flow downward[1]

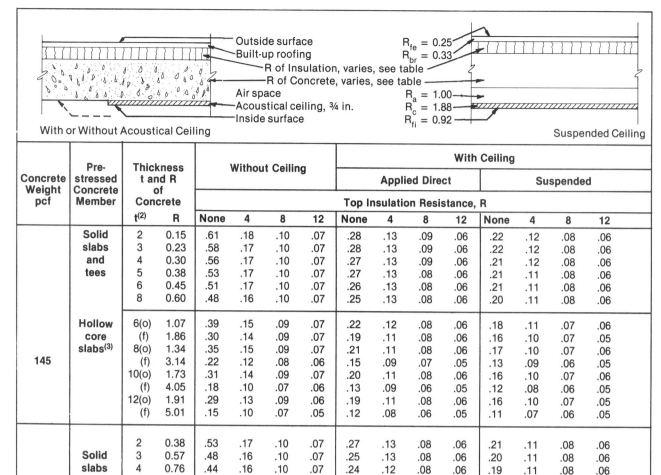

Concrete Weight pcf	Pre-stressed Concrete Member	Thickness t and R of Concrete t[2]	R	Without Ceiling None	4	8	12	With Ceiling Applied Direct None	4	8	12	Suspended None	4	8	12
								Top Insulation Resistance, R							
145	Solid slabs and tees	2	0.15	.61	.18	.10	.07	.28	.13	.09	.06	.22	.12	.08	.06
		3	0.23	.58	.17	.10	.07	.28	.13	.09	.06	.22	.12	.08	.06
		4	0.30	.56	.17	.10	.07	.27	.13	.09	.06	.21	.12	.08	.06
		5	0.38	.53	.17	.10	.07	.27	.13	.08	.06	.21	.11	.08	.06
		6	0.45	.51	.17	.10	.07	.26	.13	.08	.06	.21	.11	.08	.06
		8	0.60	.48	.16	.10	.07	.25	.13	.08	.06	.20	.11	.08	.06
	Hollow core slabs[3]	6(o)	1.07	.39	.15	.09	.07	.22	.12	.08	.06	.18	.11	.07	.06
		(f)	1.86	.30	.14	.09	.07	.19	.11	.08	.06	.16	.10	.07	.05
		8(o)	1.34	.35	.15	.09	.07	.21	.11	.08	.06	.17	.10	.07	.06
		(f)	3.14	.22	.12	.08	.06	.15	.09	.07	.05	.13	.09	.06	.05
		10(o)	1.73	.31	.14	.09	.07	.20	.11	.08	.06	.16	.10	.07	.06
		(f)	4.05	.18	.10	.07	.06	.13	.09	.06	.05	.12	.08	.06	.05
		12(o)	1.91	.29	.13	.09	.06	.19	.11	.08	.06	.16	.10	.07	.05
		(f)	5.01	.15	.10	.07	.05	.12	.08	.06	.05	.11	.07	.06	.05
110	Solid slabs and tees	2	0.38	.53	.17	.10	.07	.27	.13	.08	.06	.21	.11	.08	.06
		3	0.57	.48	.16	.10	.07	.25	.13	.08	.06	.20	.11	.08	.06
		4	0.76	.44	.16	.10	.07	.24	.12	.08	.06	.19	.11	.08	.06
		5	0.95	.41	.16	.10	.07	.23	.12	.08	.06	.19	.11	.07	.06
		6	1.14	.38	.15	.09	.07	.22	.12	.08	.06	.18	.10	.07	.06
		8	1.52	.33	.14	.09	.07	.20	.11	.08	.06	.17	.10	.07	.06
	Hollow core slabs[3]	8(o)	2.00	.29	.13	.09	.06	.19	.11	.07	.06	.16	.10	.07	.05
		(f)	4.41	.17	.10	.07	.06	.13	.08	.06	.05	.11	.08	.06	.05
		12(o)	2.59	.25	.12	.08	.06	.17	.10	.07	.06	.14	.09	.07	.05
		(f)	6.85	.12	.08	.06	.05	.10	.07	.05	.04	.09	.07	.05	.04

(1) When insulations having other R values are used, U-values can be interpolated with adequate accuracy, or U can be calculated as shown in Sect. 6.1.4.

When a finish, air space or any material is added, the new U value is:

$$\frac{1}{\dfrac{1}{U \text{ from table}} + R \text{ of added finish, air space, or material}}$$

(2) Thickness for tees is thickness of slab portion.
(3) For hollow panels (o) and (f) after thickness designates cores open or cores filled with insulation.

Fig. 6.1.1 Heat gain comparison for light and heavy roofs

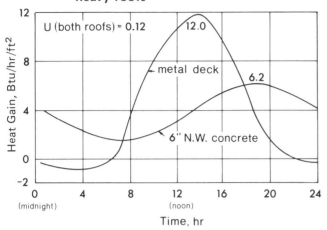

Fig. 6.1.2 Modification factor, M, for heating designs (for use in modifying steady state U-values)

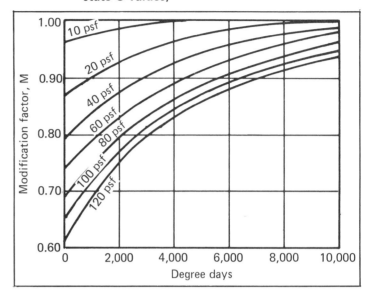

6.1.5.1 Heating Loads

The ASHRAE Handbook recognizes the effects of mass on a building's ability to retain heat, however it does not offer a way to account for the effect. Computer studies* of ten wall types each exposed to ten different weather conditions show that as the weight of walls increases, the heat flow outward decreases. As a result, the steady-state U-values can be modified to account for mass effects on walls and roofs. The modifications change as the number of heating degree days changes. The modification M-factors are given in Fig. 6.1.2.

When selecting the U-values for building walls or roofs, the effect of mass is reflected by the use of M-factors. Compare a building in Houston having 200,000 sq ft of lightweight walls with a similar building having precast walls. Assume U = 0.10 for the light, 5 psf wall (M-factor = 1.0), and a design temperature difference $\Delta t = 72 - 33 = 39$ F. The maximum heat loss is 200,000 x 0.10 x 39 = 780,000 Btu/hr. Assume the heating degree days are 1400 and the heavier concrete wall weighs 45 psf. From Fig. 6.1.2, M = 0.84. The revised maximum heat loss is then 200,000 x 0.10 x 0.84 x 39 = 655,000 Btu/hr. This is 16 percent less than the heat loss through the lightweight wall.

Assume that a code limits the heat loss through the wall by prescribing that steady-state U equals 0.10. Weight effects can be introduced to establish a modified U. In the above example, U modified equals:

$$\frac{U \text{ steady state}}{M} = \frac{0.10}{0.84} = 0.12$$

---

*"Mass, Masonry, Energy," The Masonry Industry Committee, Washington, D.C., 1976.

6.1.5.2 Cooling Loads

The effects of mass on cooling loads are reflected in the designs by the use of equivalent temperature differences, TD_{eq}, as given in Tables 6.1.8 and 6.1.9. The TD_{eqw} (walls) and TD_{eqr} (roofs) decrease as the weight of the sections increases.

Consider a building located anywhere within latitudes 0° to 50°, having 100,000 sq ft of metal curtain walls, weight 10 psf. From Table 6.1.8, the TD_{eqw} is 44 F. If the U-value of the walls is 0.10, the maximum hourly heat gain is 100,000 x 0.10 x 44 = 440,000 Btu. Change the wall to concrete, weight 50 psf, $TD_{eqw} = 30$, and also with a U-value of 0.10. The maximum hourly heat gain is 100,000 x 0.10 x 30 = 300,000 Btu. This is a reduction of 32 percent because of mass effects.

If a building code limits heat gain through walls, weight effects can be used to compare required U-values for lightweight and heavyweight walls. Use the above example for the lightweight walls where U = 0.10. The revised U for the 50 psf concrete wall equals:

$$\frac{TD_{eqw} \text{ (light)}}{TD_{eqw} \text{ (heavy)}} \times U = \frac{44}{30} \times 0.10 = 0.15$$

Now consider a building having a lightweight, 8 psf roof located anywhere within latitudes 0° to 50°. From Table 6.1.9, the TD_{eqr} is 70 F. Assume the roof contains 200,000 sq ft with a U of 0.13. The maximum hourly heat gain is then 200,000 x 0.13 x 70 = 1,820,000 Btu. If the roof is concrete weighing 45 psf, the TD_{eqr} is 50 F. The maximum hourly heat gain is 200,000 x 0.13 x 50 equalling 1,300,000 Btu. This is 29 percent less than the

Table 6.1.8 Wall temperature difference, TD_{eqw}

Weight of Wall lb / ft^2	TD_{eqw} °F
0 - 25	44
26 - 40	37
41 - 70	30
71 & above	23

Table 6.1.9 Roof temperature difference, TD_{eqr}

Weight of Roof lb / ft^2	TD_{eqr} °F
0 - 10	70
11 - 50	50
51 & above	40

heat gain through the lightweight roof.

Assume heat gain through a lightweight deck with a U = 0.13, such as in the above example, is limited to 1,820,000 Btu/hr. Weight effects can be introduced to establish a new U-value for concrete roofs. The revised U for the 45 psf concrete roof equals:

$$\frac{TD_{eqr} \text{ (light)}}{TD_{eqr} \text{ (heavy)}} \times U = \frac{70}{50} \times 0.13 = 0.18$$

6.1.6 Building Envelope Performance and Trade-Off Considerations

The ASHRAE Standard and some codes permit component trade-offs. That is, the stated overall U_o-value of any one assembly, such as a roof/ceiling, wall or floor, may be increased and the U_o-value for other components decreased, provided the overall heat transmission for the entire building envelope does not exceed the total allowed by the criteria. For buildings where the floors are not exposed to the outdoors the allowed overall envelope hourly heat loss is:

$$\Delta t \left[(U_{ow} A_{ow}) + (U_{or} A_{or}) \right]$$

where:

Δt = temperature difference, indoor to outdoor, °F

A_{ow} = overall area of walls, sq ft

A_{or} = overall area of roof, sq ft

U_{ow} = overall heat transmission value of walls, Btu/hr ft^2 °F

U_{or} = overall heat transmission value of roof, Btu/hr ft^2 °F

All values in the brackets of the equation can be altered from those given in the code or standard providing the summation of the values is not increased.

The trade-off concept is useful particularly for altering wall, floor or roof criteria without exceeding the total loss allowed for the envelope. The concept also permits different U_w-values on, for example, north and south exposures. Component trade-offs coupled with trade-off of elements of components, such as opaque vs. glass, permit a great deal of freedom in building envelope design. Also, with fewer prescriptive requirements, more efficient building designs are possible.

6.1.6.1 Design Example

Consider a building three stories high (25 feet) having a plan 80 ft x 200 ft located in Chicago, Illinois. From Table 6.1.10 the heating design temperature is 1F. With inside temperature t_i = 72F, Δt = 71 F.

Next assume that the code criteria limitations are U_{ow} = 0.265, U_{or} = 0.08. Since the ground floor is either over a heated basement or on grade, the heat loss is insignificant and therefore disregarded. The allowed envelope heat loss is then:

Envelope loss = Δt (wall loss + roof loss)

= 71 $(U_{ow} A_{ow} + U_{or} A_{or})$

= 71[(0.265 x 560 x 25) + (0.08 x 80 x 200)]

= 354,290 Btu/hr

(263,410 — walls
90,880 — roof)

Now assume there are windows only on one 200 ft side of the building and that the U_{ow} for that side is 0.30. The other three windowless sides, U_w = 0.12. The actual hourly loss through walls is 71[(0.30 x 200 x 25) + (0.12 x 360 x 25)] = 183,180 Btu. The difference between the $U_{ow} A_{ow}$ and this actual loss is 263,410 – 183,180 = 80,230 Btu/hr. This can be added to the allowance through the roof which would then be 90,880 + 80,230 = 171,110 Btu/hr.

From the revised loss allowed through the roof, a new U_{or} can be calculated as 171,110 ÷ (80 x 200 x 71) = 0.15 Btu/hr ft.^{2} This revised roof design which changes U_{or} from 0.08 to 0.15 means that the R is reduced by 12.5 – 6.67 = 5.83, representing 1 to 1 1/4 in. less insulation. Thus the trade-off concept is an important tool when considering and comparing total overall heat losses through envelopes.

Table 6.1.10 Outdoor temperatures, latitudes, and degree days

City	Latitude[1] deg.	min.	Winter temperatures[1] Med. of annual extremes	99%	97½%	Winter degree days[2]	Summer (design dry bulb) temperatures[1]
UNITED STATES							
Albuquerque, N.M.	35	00	6	14	17	4,400	94
Atlanta, Ga.	33	40	14	18	23	3,000	92
Baltimore, Md.	39	20	12	16	20	4,600	92
Birmingham, Ala.	33	30	14	19	22	2,600	94
Bismarck, N.D.	46	50	− 31	− 24	− 19	8,800	91
Boise, Ida.	43	30	0	4	10	5,800	93
Boston, Mass.	42	20	− 1	6	10	5,600	88
Burlington, Vt.	44	30	− 18	− 12	− 7	8,200	85
Charleston, W. Va.	38	20	1	9	14	4,400	90
Charlotte, N.C.	35	10	13	18	22	3,200	94
Casper, Wyo.	42	50	− 20	− 11	− 5	7,400	90
Chicago, Ill.	41	50	− 5	− 3	1	6,600	91
Cincinnati, Ohio	39	10	2	8	12	4,400	92
Cleveland, Ohio	41	20	− 2	2	7	6,400	89
Columbia, S.C.	34	00	16	20	23	2,400	96
Dallas, Texas	32	50	14	19	24	2,400	99
Denver, Colo.	39	50	− 9	− 2	3	6,200	90
Des Moines, Iowa	41	30	− 13	− 7	− 3	6,600	92
Detroit, Mich	42	20	0	4	8	6,200	88
Great Falls, Mont.	47	30	− 29	− 20	− 16	7,800	88
Hartford, Conn.	41	50	− 4	1	5	6,200	88
Houston, Texas	29	50	24	29	33	1,400	94
Indianapolis, Ind.	39	40	− 5	0	4	5,600	91
Jackson, Miss.	32	20	17	21	24	2,200	96
Kansas City, Mo.	39	10	− 2	4	8	4,800	97
Las Vegas, Nev.	36	10	18	23	26	2,800	106
Lexington, Ky.	38	00	0	6	10	4,600	92
Little Rock, Ark.	34	40	13	19	23	3,200	96
Los Angeles, Calif.	34	00	38	42	44	2,000	90
Memphis, Tenn	35	00	11	17	21	3,200	96
Miami, Fla.	25	50	39	44	47	200	90
Milwaukee, Wis.	43	00	− 11	− 6	− 2	7,600	87
Minneapolis, Minn.	44	50	− 19	− 14	− 10	8,400	89
New Orleans, La.	30	00	29	32	35	1,400	91
New York, N.Y.	40	50	6	11	15	5,000	91
Norfolk, Va.	36	50	18	20	23	3,400	91
Oklahoma City, Okla.	35	20	4	11	15	3,200	97
Omaha, Neb.	41	20	− 12	− 5	− 1	6,600	94
Philadelphia, Pa.	39	50	7	11	15	4,400	90
Phoenix, Ariz.	33	30	25	31	34	1,800	106
Pittsburgh, Pa.	40	30	1	7	11	6,000	88
Portland, Maine	43	40	− 14	− 5	0	7,600	85
Portland, Ore.	45	40	17	21	24	4,600	85
Portsmouth, N.H.	43	10	− 8	− 2	3	7,200	86
Providence, R.I.	41	40	0	6	10	6,000	86
Rochester, N.Y.	43	10	− 5	2	5	6,800	88
Salt Lake City, Utah	40	50	− 2	5	9	6,000	94
San Francisco, Calif.	37	50	38	42	44	3,000	77
Seattle, Wash.	47	40	22	28	32	5,200	79
Sioux Falls, S.D.	43	40	− 21	−− 14	− 10	7,800	92
St. Louis, Mo.	38	40	1	7	11	5,000	94
Tampa, Fla.	28	00	32	36	39	680	91
Trenton, N.J.	40	10	7	12	16	5,000	90
Washington, D.C.	38	50	12	16	19	4,200	92
Wichita, Kan.	37	40	− 1	5	9	4,600	99
Wilmington, Del.	39	40	6	12	15	5,000	90
ALASKA							
Anchorage	61	10	− 29	− 25	− 20	10,800	70
Fairbanks	64	50	− 59	− 53	− 50	14,280	78
CANADA							
Edmonton, Alta.	53	30	− 30	− 29	− 26	11,000	83
Halifax, N.S.	44	40	− 4	0	4	8,000	80
Montreal, Que.	45	30	− 20	− 16	− 10	9,000	86
Saskatoon, Sask.	52	10	− 37	− 34	− 30	11,000	86
St. John, Nwf.	47	40	1	2	6	8,600	77
Saint John, N.B.	45	20	− 15	− 12	− 7	8,200	79
Toronto, Ont.	43	40	− 10	− 3	1	7,000	87
Vancouver, B.C.	49	10	13	15	19	6,000	78
Winnipeg, Man.	49	50	− 31	− 28	− 25	10,800	87

(1) *Handbook of Fundamentals,* American Society of Heating, Refrigerating, and Air-Conditioning Engineers, Inc., New York, 1972.

(2) *Local Climatological Annual Summary,* U.S. Department of Commerce, Environmental Science Services Administration, Asheville, N.C.

6.1.7 Condensation Control

Moisture which condenses on the interior of a building is unsightly and can cause damage to the building or its contents. Even more undesirable is the condensation of moisture within a building wall or ceiling assembly where it is not readily noticed until damage has occurred. All air in buildings contains some water with warm air carrying more moisture than cold air. In many buildings moisture is added to the air by industrial processes, cooking, laundering, or humidifiers. If the inside surface temperature of a wall, floor or ceiling is too cold, the air contacting this surface will be cooled below its dew point temperature and leave its excess water on that surface. Condensation occurs on the surface with the lowest temperature.

Once condensation occurs, the relative humidity of the interior space of a building cannot be increased since any additional water vapor will simply condense on the cold surface. In effect, then, the inside temperature of an assembly limits the relative humidity which may be contained in an interior space.

6.1.7.1 Prevention of Condensation on Wall Surfaces

The U-value of a wall must be such that the surface temperature will not fall below the dew-point temperature of the room air in order to prevent condensation on the interior surface of a wall.

Fig. 6.1.3 gives U-values for any combination of outside temperatures and inside relative humidities above which condensation will occur on the interior surfaces. For example, if a building were located in an area with an outdoor design temperature of 0F and it was desired to maintain a relative humidity within the building of 25%, the wall must be designed so that all components have a U-value less than 0.80, otherwise there will be a problem with condensation. In many designs the desire to conserve energy will dictate the use of lower U-values than those required to avoid the condensation problem.

The degree of wall heat transmission resistance that must be provided to avoid condensation may be determined from the following relationship:

$$R_w = R_{fi} \frac{(t_i - t_o)}{(t_i - t_s)}$$

where:

R_w = thermal resistance of the wall, °F ft² hr/Btu

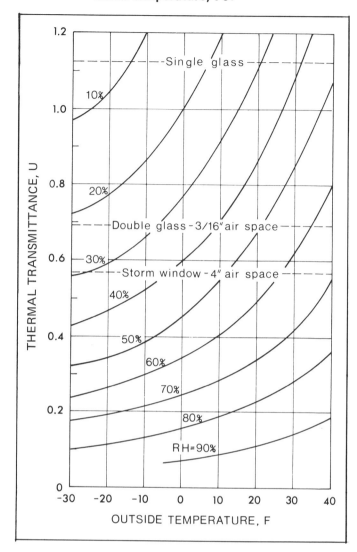

Fig. 6.1.3 Relative humidity at which visible condensation occurs on inside surfaces. Inside temperature, 70F°

R_{fi} = thermal resistance of the inside surface, °F ft² hr/Btu

t_i, t_o = respective inside, outside temperatures, °F

t_s = dew-point temperature, °F

Dew-point temperatures to the nearest deg F for various values of t_i and relative humidity are shown in Table 6.1.11.

Determine R_w when the room temperature and relative humidity to be maintained are 70F and 40%, and t_o during the heating season is –10F.

From Table 6.1.11, the dew-point temperature, t_s at 70F and 40% R.H. is 45F and from Table 6.1.1, R_{fi} = 0.68.

$$R_w = 0.68 \frac{[70 - (-10)]}{[70 - 45]} = 2.18$$

Table 6.1.11 Dew-point temperatures, F

Dry Bulb or Room Temperature	Relative Humidity — %									
	10	20	30	40	50	60	70	80	90	100
40	−9	5	13	19	24	28	31	34	37	40
45	−5	9	17	23	28	32	36	39	42	45
50	−1	13	21	27	32	37	41	44	47	50
55	3	17	25	31	37	41	45	49	52	55
60	6	20	29	36	41	46	50	54	57	60
65	10	24	33	40	46	51	55	58	62	65
70	13	28	37	45	51	56	60	63	67	70
75	17	31	42	49	55	60	65	68	72	75
80	20	36	46	54	60	65	69	73	77	80
85	23	40	50	58	65	70	74	78	82	85
90	27	44	55	62	69	74	79	82	86	90

6.1.7.2 Prevention of Condensation Within Wall Construction

Water vapor in air behaves as a gas and will diffuse through building materials at rates which depend on vapor permeabilities of the materials to water vapor and vapor pressure differentials. The colder the outside temperatures the greater the pressure of the water vapor in the warm inside air to reach the cooler, drier outside air. Also, leakage of moisture laden air into an assembly through small cracks may be a greater problem than vapor diffusion. The passage of water vapor through material is in itself generally not harmful. It becomes of consequence when, at some point along the vapor flow path, a temperature level is encountered that is below the dew-point temperature and condensation results.

Building materials have water vapor permeances from very low to very high, see Table 6.1.12. When properly used, low permeance materials keep moisture from entering a wall or roof assembly, and materials with higher permeance allow construction moisture and moisture which enters inadvertently or by design to escape.

When a material such as plaster or gypsum board has a permeance which is too high for the intended use, one or two coats of paint is frequently sufficient to lower the permeance to an acceptable level, or a vapor barrier can be used directly behind such products. Polyethylene sheet, aluminum foil and roofing materials are commonly used. Proprietary vapor barriers, usually combinations of foil and polyethylene or asphalt, are frequently used in freezer and cold storage construction.

Concrete is a relatively good vapor barrier. Permeance is a function of the water-cement ratio of the concrete. A low water-cement ratio, such as that used in most precast concrete members, results in concrete with low permeance.

Where climatic conditions demand insulation, a vapor barrier is generally necessary in order to prevent condensation. A closed cell insulation, if properly applied, will serve as its own vapor barrier. For other insulation materials a vapor barrier should be applied to the warm side of the insulation.

Table 6.1.12 Typical permeance and permeability values (dry cup and similar test methods)

Material	Perms, (M)	Perm-in., (μ)
Concrete	—	3.2
Woods	—	0.4 to 5.4
Foam plastics	—	0.4 to 6.0
Plaster on gypsum lath	20.00	
Gypsum wallboard	50.00	
Polyethylene, 2 mil	0.16	
Polyethylene, 10 mil	0.03	
Aluminum foil, 0.35 mil	0.05	
Aluminum foil, 1 mil	0.00	
Built-up roofing	0.00	
Coated roof sheet and aluminum foil	0.002	
Paint, 2 coats on wood, plaster or gypsum board		
Asphalt or oil base	0.4 to 3.0	
Water base	4 to 12	

ACOUSTICAL PROPERTIES

6.2.1 Glossary

Airborne Sound — sound that reaches the point of interest by propagation through air.

Background Level — the ambient sound pressure level existing in a space.

Decibel (dB) — a logarithmic unit of measure of sound pressure or sound power. Zero on the decibel scale corresponds to a standardized reference pressure (0.0002 microbar) or sound power (10^{-12} watt).

Flanking Transmission — transmission of sound by indirect paths other than through the primary barrier.

Frequency (Hz) — the number of complete vibration cycles per second.

Impact Insulation Class (IIC) — a single figure rating of the overall impact sound insulation merits of floor-ceiling assemblies in terms of a reference contour (ASTM E 492).

Impact Noise — the sound produced by one object striking another.

Loudness — the intensive attribute of an auditory sensation dependent on the sound pressure and frequency of the wave form.

Noise — unwanted sound.

Noise Reduction Coefficient (NRC) — the arithmetic average of the sound absorption coefficients at 250, 500, 1000 and 2000 Hz expressed to the nearest multiple of 0.05 (ASTM C 423).

Noise Reduction (NR) — the difference in decibels between the space-time average sound pressure levels produced in two enclosed spaces by one or more sound sources in one of them.

Preferred Noise Criteria (PNC) — a series of curves used as design goals to specify satisfactory background sound levels as they relate to particular use functions.

Reverberation — the persistence of sound in an enclosed or partially enclosed space after the source of sound has stopped.

Sabin — the unit of measure of sound absorption (ASTM C 423).

Sound Absorption Coefficient — the fraction of randomly incident sound energy absorbed or otherwise not reflected (ASTM C 423).

Sound Pressure Level (SPL) — the squared ratio, expressed in decibels, of the sound pressure under consideration to the standard reference pressure of 0.0002 microbar.

Sound Transmission Class (STC) — the single number rating system used to give a preliminary estimate of the sound insulation properties of a partition system (ASTM E 413).

Sound Transmission Loss (TL) — the difference, expressed on the decibel scale, of the airborne sound pressure incident on the sound barrier and that transmitted by the barrier and radiated on the other side (ASTM E 90).

Structure Borne Sound — sound that reaches the point of interest over at least part of its path by vibrations of a solid structure.

6.2.2 General

The basic purpose of architectural acoustics is to provide a satisfactory environment in which desired sounds are clearly heard by the intended listeners and unwanted sounds (noise) are isolated or absorbed.

Under most conditions, the architect/engineer can determine the acoustical needs of the space and then design the building to satisfy those needs. Good acoustical design utilizes both absorptive and reflective surfaces, sound barriers and vibration isolators. Some surfaces must reflect sound so that the loudness will be adequate in all areas where listeners are located. Other surfaces absorb sound to avoid echoes, sound distortion and long reverberation times. Sound is isolated from rooms where it is not wanted by selected wall and floor-ceiling constructions. Vibrations generated by mechanical equipment must be isolated from the structural frame of the building.

The problems of sound insulation are usually considerably more complicated than those of sound absorption. The former involves reductions of sound level, which are of greater orders of magnitude than can be achieved by absorption. These large reductions of sound level from space to space can be achieved only by continuous, impervious barriers. If the problem also involves structure borne sound, it may be necessary to introduce resilient layers or discontinuities into the barrier.

Sound absorbing materials and sound insulating materials are used for different purposes. There is not much sound absorption from an 8 in. concrete wall; similarly, high sound insulation is not available from a porous, lightweight material that may be applied to room surfaces. It is important to recognize that the basic mechanisms of sound absorption and sound insulation are quite different.

6.2.3 Sound Transmission Loss

Sound transmission loss measurements are made at 16 frequencies at one-third octave intervals covering the range from 125 to 4000 Hz. To simplify specification of desired performance characteristics the single number Sound Transmission Class (STC) was developed.

Airborne sound reaching a wall, floor or ceiling produces vibrations in the element and is radiated with reduced intensity on the other side. Airborne sound transmission loss of walls and floor-ceiling assemblies is a function of their weight, stiffness and vibration damping characteristics.

Weight is concrete's greatest asset when it is used as a sound insulator. For sections of similar design, but different weights, the STC increases approximately 6 units for each doubling of weight as shown in Fig. 6.2.1.

Fig. 6.2.1 Sound transmission class as a function of weight of floor or wall

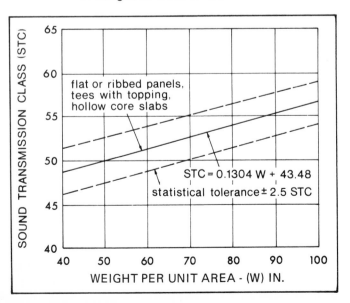

The acoustical test results of both airborne sound transmission loss and impact insulation of 4, 6 and 8 in. flat panels, a 14 in. double tee, and 6 and 8 in. hollow-core slabs are shown in Figs. 6.2.2, 6.2.3 and 6.2.4.

Table 6.2.1 presents the ratings for various precast concrete walls and floor-ceiling assemblies. The effects of various assembly treatments are shown in Table 6.2.2. The improvements are additive, but in some cases the total effect may be slightly less than the sum.

6.2.4 Impact Noise Reduction

As with the airborne standard, measurements of impact noise are made at 16 one-third octave intervals but in the range from 100 to 3150 Hz. For performance specification purposes the single number Impact Insulation Class (IIC) is used.

In general, thickness or unit weight of concrete does not greatly affect the transmission of impact sounds. Structural concrete floors in combination with resilient materials effectively control impact sounds.

Example: The performance of a 2 in. concrete topping, carpet and pad added to 8 in. hollow-core prestressed floor units is calculated as follows:

Materials	STC	IIC
Bare slab	50	28
2 in. concrete topping	3	0
Carpet and pad	0	50
Totals	53	78

6.2.5 Absorption of Sound

The sound absorption coefficient can be specified at individual frequencies or as an average of absorption coefficients (NRC).

A dense non-porous concrete surface typically absorbs 1 to 2 percent of incident of sound and has an NRC of 0.015. There are specially fabricated units with porous concrete surfaces which provide greater absorption. In the case where additional sound absorption of precast concrete is desired, a coating of acoustical material can be spray applied, acoustical tile can be applied with adhesive, or an acoustical ceiling can be suspended. Most of the spray applied fire retardant materials used to increase the fire resistance of precast concrete and other floor-ceiling systems can also be used to absorb sound. The NRC of the sprayed fiber types range from 0.25 to 0.75. Most cementitious types have an NRC from 0.25 to 0.50.

Fig. 6.2.2   Acoustical test data of solid flat concrete panels — normal weight concrete

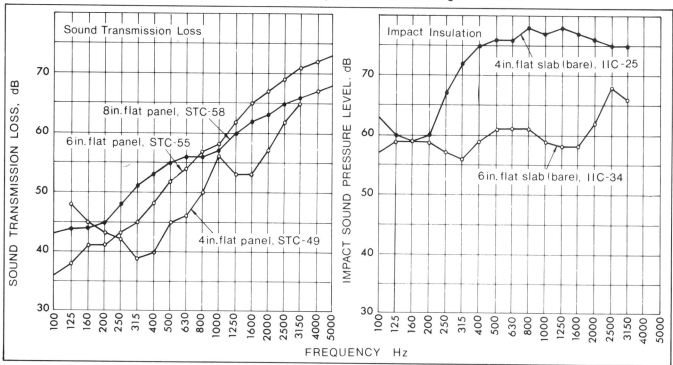

Fig. 6.2.3   Acoustical test data of 14 in. precast double tee system with 2 in. concrete topping — normal weight concrete

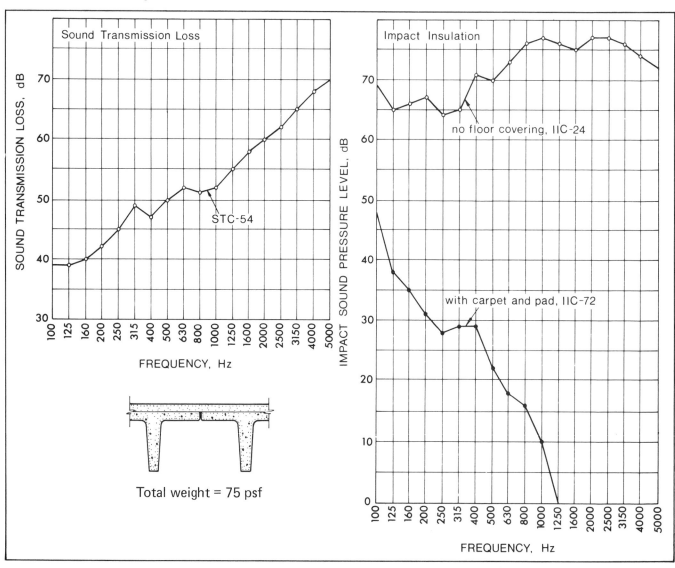

Total weight = 75 psf

Fig. 6.2.4 Acoustical test data of hollow-core panels — normal weight concrete

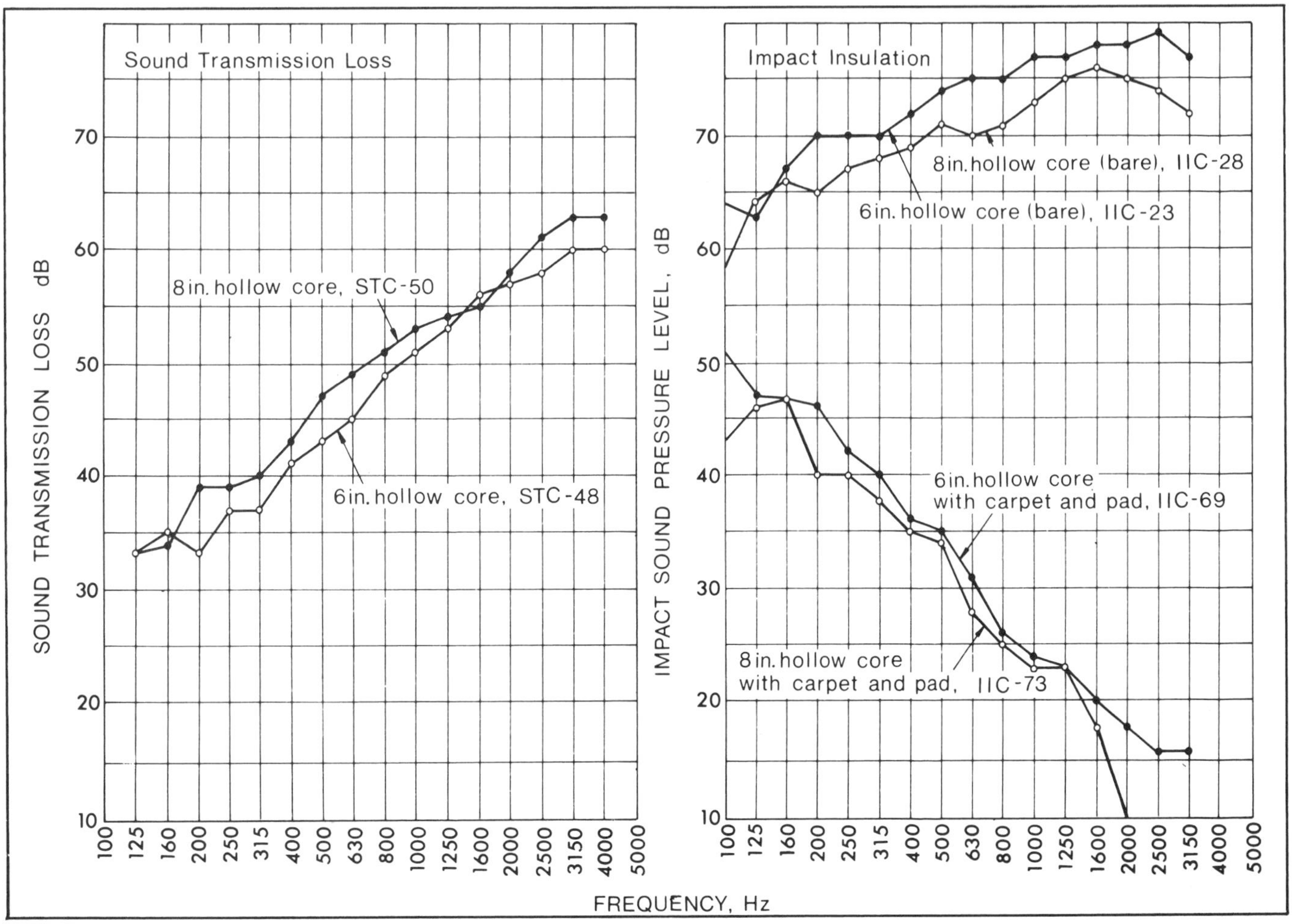

6.2.6 Acceptable Noise Criteria

As a rule, a certain amount of continuous sound can be tolerated before it becomes noise. An "acceptable" level neither disturbs room occupants nor interferes with the communication of wanted sound.

The most generally accepted and commonly used noise criteria today are expressed as the Preferred Noise Criteria (PNC) curves, Fig. 6.2.5.* These values are the result of extensive studies based on the human response to both sound pressure level and frequency and take into account the requirements for speech intelligibility. The figures in Table 6.2.3* represent general acoustical goals. They can also be compared with anticipated noise levels in specific rooms to assist in evaluating noise reduction problems.

Undesirable sounds may be from an exterior source such as automobiles or aircraft, or they may be generated as speech in an adjacent classroom or music in an adjacent apartment. They also may be direct impact-induced sound such as foot-falls on the floor above, rain impact on a lightweight roof construction or vibrating mechanical equipment.

Thus, the designer must always be ready to accept the task of analyzing the many potential sources of intruding sound as related to their frequency characteristics and the rates at which they occur. The level of toleration that is to be expected by those who will occupy the space must also be established. Fig. 6.2.6 gives the spectral characteristics of common exterior noise sources. Fig. 6.2.7 provides similar data on common interior noise sources.

With these criteria, the problem of sound isolation now must be solved, namely the reduction process between the high unwanted noise source and the desired ambient level. For this solution, two related yet mutually exclusive processes must be incorporated, i.e., sound transmission loss and sound absorption. For a more complete discussion see "Acoustical Properties of Precast Concrete,"

*L. L. Beranek, W. E. Blazier and J.J. Figwer; "Preferred Noise Criterion and Their Application to Rooms," *Journal of the Acoustical Society of America*, Vol. 50, Nov. 1971, pp. 1223-1228.

Table 6.2.1 Airborne sound transmission class and impact insulation class ratings from tests of precast concrete assemblies

Assembly No.		STC	IIC
	Wall Systems		
1	4 in. flat panel, 54 psf	49	—
2	6 in. flat panel, 75 psf	55	—
3	Assembly 2 with wood furring, 3/4 in. insulation and 1/2 in. gypsum board	58*	—
4	Assembly 2 with 1/2 in. space, 1 5/8 in. metal stud row, 1 1/2 in. insulation and 1/2 in. gypsum board	63*	—
5	8 in. flat panel, 95 psf	58	—
6	14 in. prestressed tees with 4 in. flange, 75 psf	54	—
	Floor-Ceiling Systems		
7	14 in. prestressed tees with 2 in. concrete topping, 75 psf	54	24
8	Assembly 7 with carpet and pad, 76 psf	54	72
9	Assembly 7 with resiliently suspended acoustical ceiling with 1 1/2 in. mineral fiber blanket above, 77 psf	59	51
10	Assembly 9 with carpet and pad, 78 psf	59	82
11	8 in. hollow-core prestressed units, 57 psf	50	28
12	Assembly 11 with carpet and pad, 58 psf	50	73
13	8 in. hollow-core prestressed units with 1/2 in. wood block flooring adhered directly, 58 psf	51	47
14	Assembly 13 except 1/2 in. wood block flooring adhered to 1/2 in. sound-deadening board underlayment adhered to concrete, 60 psf	52	55
15	Assembly 14 with acoustical ceiling, 62 psf	59	61
16	4 in. flat slabs, 54 psf	49	25
17	5 in. flat slabs, 60 psf	52*	24
18	6 in. flat slabs, 75 psf	55	34
19	8 in. flat slabs, 95 psf	58	34*
20	10 in. flat slabs, 120 psf	59*	31
21	5 in. flat slab concrete with carpet and pad, 61 psf	52*	68
22	10 in. flat slab concrete with carpet and pad, 121 psf	59*	74

*Estimated values

PCI Journal, Vol. 23, No. 2, March-April 1978, pp. 42-61.

6.2.7 Establishment of Noise Insulation Objectives

Often acoustical control is specified as to the minimum insulation values of the dividing partition system. Municipal building codes, lending institutions and the Department of Housing and Urban Development (HUD) list both airborne STC values and impact IIC values for different living environments. As an example, "HUD Minimum Property Standard — Multiple Housing" has the following requirements:

Location of Partition	STC
Between living units	45
Between living units and public space	50

Table 6.2.2 Typical improvements for wall, floor, and ceiling treatments used with precast concrete elements

Treatment	Increase in Ratings	
	Airborne (STC)	Impact (IIC)
Wall furring, 3/4 in. insulation & 1/2 in. gypsum board attached to concrete wall	3	
Separate metal stud system, 1 1/2 in. insulation in stud cavity & 1/2 in. gypsum board attached to concrete wall	5 to 10	
2 in. concrete topping (24 psf)	3	0
Carpets and pads	0	43 to 56
Vinyl tile	0	3
1/2 in. wood block adhered to concrete	0	20
1/2 in. wood block and resilient fiber underlayment adhered to concrete	4	26
Floating concrete floor on fiberboard	7	15
Wood floor, sleepers on concrete	5	15
Wood floor on fiberboard	10	20
Acoustical ceiling resiliently mounted	5	27
if added to floor with carpet	5	10
Plaster or gypsum board ceiling resiliently mounted	10	8
with insulation in space above ceiling	13	13
Plaster direct to concrete	0	0

Location of Floor-Ceiling	STC	IIC
Between living units	45	45
Between living units and public space	50	50

Once the objectives are established, the designer then should refer to available data, e.g., Fig. 6.2.1 or Table 6.2.1, and select the system which best meets these requirements. In this respect, concrete systems have superior properties and can with minimal effort comply with these criteria.

Design Example

Assume a precast prestressed concrete office building is to be erected adjacent to a major highway. Private and semiprivate offices will run along the perimeter of the structure. The first step is to

Table 6.2.3 Recommended category classification and suggested noise criteria range for steady background noise as heard in various indoor functional activity areas.

Type of space (and acoustical requirements)	PNC curve
Concert halls, opera houses, and recital halls (for listening to faint musical sounds)	10 to 20
Broadcast and recording studios (distant microphone pickup used)	10 to 20
Large auditoriums, large drama theaters, and churches (for excellent listening conditions)	Not to exceed 20
Broadcast, television, and recording studios (close microphone pickup only)	Not to exceed 25
Small auditoriums, small theaters, small churches, music rehearsal rooms, large meeting and conference rooms (for good listening), or executive offices and conference rooms for 50 people (no amplification)	Not to exceed 35
Bedrooms, sleeping quarters, hospitals, residences, apartments, hotels, motels, etc. (for sleeping, resting, relaxing)	25 to 40
Private or semiprivate offices, small conference rooms, classrooms, libraries, etc. (for good listening conditions)	30 to 40
Living rooms and similar spaces in dwellings (for conversing or listening to radio and TV)	30 to 40
Large offices, reception areas, retail shops and stores, cafeterias, restaurants, etc. (for moderately good listening conditions)	35 to 45
Lobbies, laboratory work spaces, drafting and engineering rooms, general secretarial areas (for fair listening conditions)	40 to 50
Light maintenance shops, office and computer equipment rooms, kitchens, and laundries (for moderately fair listening conditions)	45 to 55
Shops, garages, power-plant control rooms, etc. (for just acceptable speech and telephone communication). Levels above PNC 60 are not recommended for any office or communication situation.	50 to 60
For work spaces where speech or telephone communication is not required, but where there must be no risk of hearing damage	60 to 75

Fig. 6.2.5 Preferred noise criteria (PNC) curves

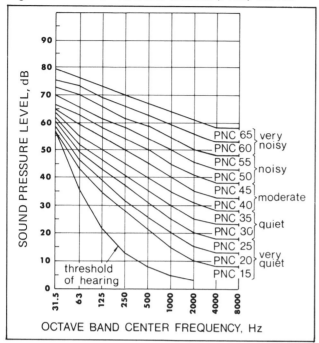

Fig. 6.2.6 Sound pressure levels — exterior noise sources

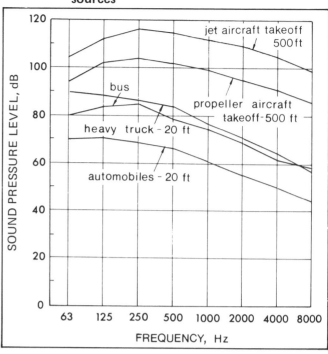

determine the degree of insulation required of the exterior wall system, as shown below.

The 500 Hz requirement, 38 dB, can be used as the first approximation of the wall STC category. However, if windows are planned for the wall, a system of about 50-55 STC should be selected (see following composite wall discussion). Individual transmission loss performance values of this system are then compared to the calculated needs and no deficiencies exist.

	Sound Pressure Level — (dB)							
Frequency (Hz)	63	125	250	500	1000	2000	4000	8000
Bus Traffic Source Noise (Fig. 6.2.6)	80	83	85	78	74	68	62	58
Private Office Noise Criteria — PNC 35 (Fig. 6.2.5)	55	50	45	40	35	32	28	28
Required Insulation	25	33	40	38	39	36	34	30
6 in. Precast Solid Concrete Wall (Fig. 6.2.2)		38	43	52	58	67	72	

The selected precast concrete wall should meet or exceed the insulation needs at all frequencies. However, to achieve the most efficient design conditions, certain limited deficiencies can be tolerated. Experience has shown that the maximum deficiencies are 3 dB at two frequencies or 5 dB on one frequency point.

Fig. 6.2.7 Sound pressure levels — interior noise sources

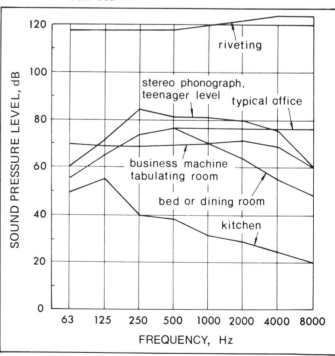

6.2.8 Composite Wall Considerations

Doors and windows are often the weak link in an otherwise effective sound barrier. Minimal effects on sound transmission loss will be achieved in most cases by a proper selection of glass, Table

Fig. 6.2.8 Chart for calculating the effective transmission loss of a composite barrier. (For purposes of approximation STC values can be used in place of TL values.)

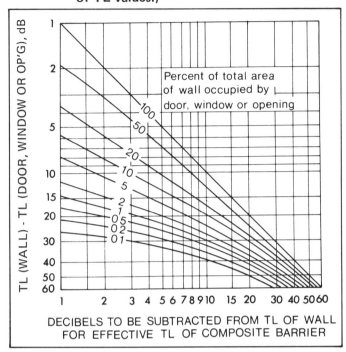

y-axis: TL (WALL) - TL (DOOR, WINDOW OR OP'G), dB

Percent of total area of wall occupied by door, window or opening

x-axis: DECIBELS TO BE SUBTRACTED FROM TL OF WALL FOR EFFECTIVE TL OF COMPOSITE BARRIER

6.2.4.* Mounting of the glass in its frame should be done with care to eliminate noise leaks and to reduce the glass plate vibrations.

Sound transmission loss of a door depends on its material and construction, and the sealing between the door and the frame.[†] There is a mass law dependence of STC on weight (psf) for both wood and steel doors. The approximate relationships are:

For steel doors: STC = 15 + 27 log W

For wood doors: STC = 12 + 32 log W

where W = weight of the door, psf

These relationships are purely empirical and a large deviation can be expected for any given door.

Fig. 6.2.8 can be used to calculate the effective acoustic isolation of a wall system which contains a composite of elements, each with known individual transmission loss data.

---

*H. J. Sabine, M. B. Lacher, D. R. Flynn, T. L. Quindry; "Acoustical and Thermal Performance of Exterior Residential Walls, Doors and Windows," National Bureau of Standards, U.S. Government Printing Office, Washington, D.C., 1975.

†IITRI; "Compendium of Materials for Noise Control," U.S. Department of Health, Education and Welfare, U.S. Government Printing Office, Washington, D.C., 1975.

Design Example

To complete the office building wall acoustical design from Sect. 6.2.7 assume the following:

1. The glazing area represents 10% of the exterior wall area.

2. The windows will be double glazed with a 38 STC acoustical insulation rating.

The problem now becomes the task of determining the combined effect of the concrete-glass combination and a redetermination of criteria compliance.

Frequency (Hz)	Sound Pressure Level — (dB)					
	125	250	500	1000	2000	4000
6 in. Precast Solid Concrete Wall (Fig. 6.2.2)	38	43	52	59	67	72
Double Glazed Windows (Table 6.2.4)	30	28	35	38	41	44
Correction (Fig. 6.2.8)	-2	-6	-7	-11	-15	-19
Combined Transmission Loss	36	37	45	48	52	53
Insulation Requirements	33	40	38	39	36	34
Deficiencies	—	3	—	—	—	—

The maximum deficiency is 3 dB and occurs at only one frequency point. The 6 in. precast concrete wall with double glazed windows will provide the required acoustical insulation.

Floor-ceiling assembly acoustical insulation requirements are determined in the same manner as for walls by using Figs. 6.2.2, 6.2.3, 6.2.4, 6.2.5 and 6.2.7.

6.2.9 Leaks and Flanking

The performance of a building section with an otherwise adequate STC can be seriously reduced by a relatively small hole or any other path which allows sound to bypass the acoustical barrier. All noise which reaches a space by paths other than through the primary barrier is called flanking. Common flanking paths are openings around doors or windows, at electrical outlets, telephone and television connections, and pipe and duct penetrations. Suspended ceilings in rooms where walls do not extend beyond the ceiling to the roof or floor above allow sound to travel to adjacent rooms.

Although not easily quantified, an inverse relationship exists between the performance of an element as a primary barrier and its propensity to transmit flanking sound. In other words, the probability of existing flanking paths in a concrete structure is much less than in one of steel or wood frame.

Table 6.2.4 Acoustical properties of glass

(a) Sound Transmission Class (STC)

Type and Overall Thickness	Inside Light	Construction Space	Outside Light	STC
1/8'' Plate or float	—	—	1/8''	23
1/4'' Plate or float	—	—	1/4''	28
1/2'' Plate or float	—	—	1/2''	31
1'' Insulated glass	1/4''	1/2'' Air Space	1/4''	31
1/4'' Laminated	1/8''	.030 Vinyl	1/8''	34
1 1/2'' Insulated glass	1/4''	1'' Air Space	1/4''	35
3/4'' Plate or float	—	—	3/4''	36
1'' Insulated glass	1/4''	1/2'' Air Space	1/4'' Laminated	38
1'' Plate or Float	—	—	1''	37
2 3/4'' Insulated glass	1/4''	2'' Air Space	1/2''	39
4 3/4'' Insulated glass	1/4''	4'' Air Space	1/2''	40
6 3/4'' Insulated glass	1/4''	6'' Air Space	1/4'' Laminated	42

(b) Transmission Loss (dB)

Frequency (Hz)															
125	160	200	250	315	400	500	630	800	1000	1250	1600	2000	2500	3150	4000
1/4 inch plate glass — 28 STC															
24	22	24	24	21	23	21	23	26	27	33	36	37	39	40	40
1 inch insulating glass with 1/2 inch air space — 31 STC															
25	25	22	20	24	27	27	30	32	33	35	34	29	31	33	36
1 inch insulating glass laminated with 1/2 inch air space — 38 STC															
30	29	26	28	31	34	35	37	37	38	38	40	41	40	41	44

6.2.10 Vibration Isolation

The isolation of vibrations produced by equipment with unbalanced operating or starting forces can frequently be accomplished by mounting the equipment on a heavy concrete slab placed on resilient supports. A slab of this type is called an inertia block.

Inertia blocks can provide a desirable low center of gravity and compensate for thrusts such as those generated by large fans. For equipment with less unbalanced weight, a "housekeeping" slab is sometimes used below the resilient mounts to provide a rigid support for the mounts and to keep them above the floor where they remain cleaner and easier to inspect. This slab may also be mounted on pads of precompressed glass fiber or neoprene.

If the static deflection of the floor is more than a small fraction of the static deflection of the resilient mounts, there is danger that the floor will act as part of the vibrating system. Prestressed concrete floors supporting such equipment can be built stiff to avoid this effect. A deflection much less than the otherwise satisfactory 1/360 of span is often desirable. Locating the equipment near the end of a span, away from its center, also reduces static deflection.

FIRE RESISTANCE

6.3.1 General

6.3.1.1 Notation

Note: Subscript θ indicates the property as affected by elevated temperatures.

a = depth of equivalent rectangular compression stress block

A_{ps} = area of prestressing steel

A_s = area of non-prestressed reinforcement

A_s^- = area of reinforcement in negative moment region

b = width of member

d = distance from centroid of prestressing steel to the extreme compression fiber

f_c' = compressive strength of concrete

f_{ps} = stress in the prestressing steel at nominal strength

f_{pu} = ultimate tensile strength of prestressing steel

h = total depth of a member

ℓ = span length

M_n = nominal moment strength

$M_{n\theta}^+,$ $M_{n\theta}^-$ = positive and negative nominal moment strength at elevated temperatures, respectively

R = fire endurance of a composite assembly

$R_1,$ R_2, R_n = fire endurance of individual courses

u = distance from prestressing steel to the fire exposed surface

w = uniform total load

w_d = uniform dead load

w_ℓ = uniform live load

$x, x_o,$ x_1, x_2 = horizontal distances as shown in Figs. 6.3.13, 6.3.14, and 6.3.15

θ_s = temperature of steel

ϕ = strength reduction coefficient

6.3.1.2 Introduction

Precast prestressed concrete members can be provided with any degree of fire resistance that may be required by building codes, insurance companies, and other authorities. The fire resistance of building assemblies is determined from standard fire tests defined by the American Society for Testing and Materials.

To insure that fire resistance requirements are satisfied, the engineer can use tabulated information provided by various authoritative bodies, such as Underwriters Laboratories, Inc., the American Insurance Association and model building codes. This information is based on the results of standard fire tests of assemblies that may include ceilings and other building components. The 1978 edition of the UL *Fire Resistance Directory* alone provides information on more than 120 assemblies incorporating precast prestressed concrete members.

In the absence of tabulated data, the fire resistance of precast prestressed concrete members and assemblies can be determined in most cases by calculation. These calculations are based on engineering principles and take into account the conditions of a standard fire test. This is known as the Rational Design Method of determining fire resistance. It is based on extensive research sponsored in part by the Prestressed Concrete Institute and conducted by the Portland Cement Association and other laboratories.

After a discussion on fire tests in Sect. 6.3.2 and 6.3.3, calculations using the Rational Design Method in many common situations are presented in the following sections. Brief explanations of the underlying principles are also given. For additional examples, design charts and a complete explanation of the method, refer to the PCI manual, MNL 124-77, *PCI Design for Fire Resistance of Precast Prestressed Concrete.*

6.3.2 Standard Fire Tests

The fire resistance of building components is measured in standard fire tests defined by ASTM E 119. During these tests the building assembly, such as a portion of floors, walls, roofs or columns, is subjected to increasing temperatures that vary with time as shown in Fig. 6.3.1. This time-temperature relation is used as a standard and it

Fig. 6.3.1 Standard time-temperature curve

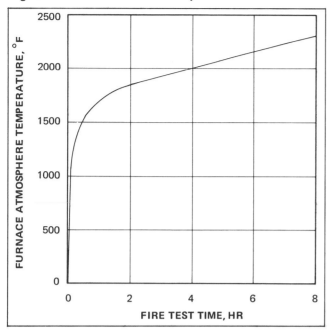

represents the combustion of about 10 lb of wood (with a heat potential of 8,000 Btu per lb) per sq ft of exposed area per hour of test. Actually, the fuel consumption to maintain the standard time-temperature relation during a fire test depends on the design of the furnace and on the test specimen. When fire-tested, assemblies with concrete members require considerably more fuel than other assemblies due to the favorable heat capacity. This fact is not recognized when evaluating fire resistance.

In addition to defining a standard time-temperature relationship, standard fire tests involve regulations concerning the size of the assemblies, the amount of applied load, the region of the assembly to be exposed to the simulated fire, and the end point criteria on which fire resistance (duration) is based.

The Standard, ASTM E 119-76, specifies the minimum sizes of specimens to be exposed in fire tests.* For floors and roofs, at least 180 sq ft must be exposed to fire from beneath, and neither dimension can be less than 12 ft. For tests of walls, either loadbearing or non-loadbearing, the minimum specified area is 100 sq ft with neither dimension less than 9 ft. The minimum length for columns is specified to be 9 ft, while for beams it is 12 ft.

During the fire tests of floors, roofs, beams, loadbearing walls, and columns, the maximum permissible superimposed load as required or permitted by nationally recognized standards is applied. A load other than the maximum load may

---

*Much valuable data have been developed for tests on specimens smaller than the ASTM minimum sizes (see Sect. 6.3.3.1).

be applied, but the test results then apply only to the restricted load condition.

Floor and roof specimens are exposed to fire from beneath, beams from the bottom and sides, walls from one side, and columns from all sides.

ASTM E 119-76 distinguishes between "restrained" and "unrestrained" assemblies and defines them as follows:

> "Floor and roof assemblies and individual beams in buildings shall be considered restrained when the surrounding or supporting structure is capable of resisting substantial thermal expansion throughout the range of anticipated elevated temperatures. Constructions not complying with this definition are assumed to be free to rotate and expand and shall therefore be considered as unrestrained."

ASTM E 119-76 includes a guide for classifying types of construction as restrained or unrestrained. The guide indicates that cast-in-place and most precast concrete constructions are considered to be restrained.

6.3.2.1 Fire Endurance, End Point Criteria, and Fire Rating

The *fire resistance* of an assembly is measured by its fire endurance defined as the period of time elapsed before a prescribed condition of failure or end point is reached during a standard fire test. A *fire rating* or *classification* is a legal term for a fire endurance required by a building code authority or determined in standard fire tests conducted by organizations such as Underwriters Laboratories, Inc.

The following end point criteria are defined by ASTM E 119:

1. Loadbearing specimens must sustain the applied loading. Collapse is an obvious end point (structural end point).

2. Holes, cracks, or fissures through which flames or gases hot enough to ignite cotton waste must not form (flame passage end point).

3. The temperature increase of the unexposed surface of floors, roofs, or walls must not exceed an average of 250°F or a maximum of 325°F at any one point (heat transmission end point).

4. In alternate tests of large steel beams (not loaded during test) the end point occurs when the steel temperature reaches an average of 1000°F or a maximum of 1200°F at any one point.

The additional end point criteria for unrestrained specimens are:

1. Structural steel members: temperature of the steel at any one section must not exceed an

average of 1100°F or a maximum of 1300°F.

2. Concrete structural members: average temperature of the tension steel at any section must not exceed 800°F for cold-drawn prestressing steel or 1100°F for reinforcing bars.

3. Multiple open-web steel joists: average temperature must not exceed 1100°F.

The additional end point criteria for restrained specimens are:

1. Beams more than 4 ft on centers: the above steel temperatures must not be exceeded for classifications of 1 hr or less; for classifications longer than 1 hr, the above temperatures must not be exceeded for the first half of the classification period or 1 hr, whichever is longer.

2. Beams 4 ft or less on centers and slabs are not subjected to steel temperature limitations.

Walls and partitions must meet the same structural, flame passage, and heat transmission end points described above. In addition, they must withstand a hose stream test (simulating, in a specified manner, a fire fighter's hose stream) and then support twice the superimposed load.

6.3.3 Fire Tests of Prestressed Concrete Assemblies

6.3.3.1 General

The first fire test of a prestressed concrete assembly in America was conducted in 1953 at the National Bureau of Standards. Since that time, more than 150 prestressed concrete assemblies have been subjected to standard fire tests in America. Although many of the tests were conducted for the purpose of deriving specific fire ratings, most of the tests were performed in conjunction with broad research studies whose objectives have been to understand the behavior of prestressed concrete subjected to fire. The knowledge gained from these tests has resulted in the development of (1) lists of fire resistive prestressed concrete building components, and (2) procedures for determining the fire endurance of prestressed concrete members by calculation.

Many different types of prestressed concrete elements have been fire tested. These elements include joists, double tees, mono-wing tees, single tees, solid slabs, hollow-core slabs, rectangular beams, ledger beams, and I-shaped beams. In addition, roofs with thermal insulation and loadbearing wall panels have also been tested. Nearly all of these elements have been exposed directly to fire, but a few tests have been conducted on specimens that received additional protection from the fire by spray-applied coatings, ceilings, etc.

6.3.3.2 Fire Tests of Flexural Elements

Tests have shown that the structural fire endurance of a flexural precast prestressed concrete element depends on several factors, the most important of which is the method of support, i.e., restrained or unrestrained. Other factors include size and shape of the element, thickness of cover (or more precisely, the distance between the centers of the prestressing tendons and the nearest fire-exposed surface), aggregate type, and load intensity. The fire endurance as determined by the criteria for temperature rise of the unexposed surface (heat transmission) depends primarily on the concrete thickness and aggregate type.

Reports of a number of tests sponsored by the Prestressed Concrete Institute have been issued by Underwriters Laboratories, Inc. Most of the reports have been reprinted by the Prestressed Concrete Institute, and the results of the tests are the basis for UL's listings and specifications for non-proprietary products such as double-tee and single-tee floors and roofs, wet-cast hollow-core and solid slabs, and prestressed concrete beams.

The Portland Cement Association (PCA) conducted many fire tests of prestressed concrete assemblies. PCA's unique furnaces have made it possible to study in depth the effects of support conditions. Four series of tests dealt with simply supported slabs and beams; two series dealt with continuous slabs and beams; and one major series dealt with the effects of restrained thermal expansion on the behavior during fire of prestressed concrete floors and roofs. PCA has also conducted a number of miscellaneous fire tests of prestressed and reinforced concrete assemblies. Reports of these tests have been published and are available from the Portland Cement Association.

In addition to the tests sponsored by PCI and PCA, a number of fire tests of proprietary products, such as hollow-core slabs, have been sponsored by their manufacturers. Most of these tests have been performed by Underwriters Laboratories, Inc., but some have been conducted by Ohio State University, the Fire Prevention Research Institute, and the National Bureau of Standards. Reports of proprietary tests are generally available from test sponsors. The UL *Fire Resistance Directory* lists many proprietary assemblies.

6.3.3.3 Fire Tests of Walls and Columns

Not all of the tests conducted by Underwriters Laboratories, Inc. result in listings in UL's publications; some tests are conducted for research purposes. One such test was conducted on a double-tee wall assembly. Fire was applied to the flat surface of the flange. The flange was only 1-1/2 in. thick. A load of about 10k per ft was applied at the top of the wall. The wall withstood a 2 hr fire and a subsequent hose stream test followed by a double load test without distress. Because the flange was only 1-1/2 in. thick, the heat transmission requirement was exceeded for most of the test. By providing adequate flange thickness or insulation, the heat transmission requirement would have been met in addition to the structural requirement.

Fire tests of loaded column assemblies (of any material) have not been conducted in America since the 1920's; therefore, prestressed concrete columns have not been fire tested. However, tests of reinforced and plain concrete columns indicate that the results are equally applicable to prestressed concrete columns.

6.3.4 Designing for Heat Transmission

ASTM E 119 imposes the heat transmission criterion, limiting the average temperature rise of the unexposed surface (the surface not exposed to fire) of floors, roofs, and walls to 250°F during the standard fire tests. Some jurisdictions waive or modify this requirement. For example, the Wisconsin Administrative Code modifies the criterion in Ind 51.042:

General Requirements:

"(5) The heat transmission requirements of ASTM E 119, with the exception of high hazard areas, penal and health care facilities and warehouses for combustile materials, may be reduced to one-half (1/2) of the hourly rating required by this code, but not less than one hour.

(a) The fire-resistive rating for structural integrity required by this code shall be maintained where the heat transmission criteria has been reduced."

6.3.4.1 Single Course Slabs

For concrete slabs, the temperature rise of the unexposed surface depends mainly on the thickness and aggregate type of the concrete. Other less important factors include unit weight, moisture condition, air content, and maximum aggregate size. Within the usual ranges, water-cement ratio, strength, and age have only insignificant effects.

Fig. 6.3.2 shows the fire endurance (heat transmission) of concrete slabs as influenced by aggregate type and thickness. For a hollow-core slab, this thickness may be obtained by dividing the net cross sectional area by its width. The curves represent air-entrained concrete made with air-dry aggregates having a nominal maximum size of 3/4 in. and fire tested when the concrete was at the standard moisture condition (75% R.H. at mid-depth). On the graph, concrete aggregates are designated as lightweight, sand-lightweight, carbonate, or siliceous. Lightweight aggregates include expanded clay, shale, slate, and fly ash which produce con-

Fig. 6.3.2 Fire endurance (heat transmission) of concrete slabs

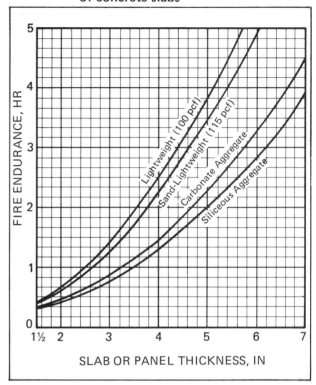

cretes having unit weights of about 95 to 105 pcf without sand replacement. Lightweight concretes, in which sand is used as part or all of the fine aggregate and weigh no more than about 120 pcf, are designated as sand-lightweight. Carbonate aggregates include limestone and dolomite, i.e., those consisting mainly of calcium and/or magnesium carbonate. Siliceous aggregates include quartzite, granite, basalt, and most hard rocks other than limestone and dolomite.

Fire endurance generally increases with a decrease in unit weight, but for structural concretes, the influence of aggregate type may overshadow the effect of unit weight.

Within the normal range of air contents, i.e., from non-air-entrained concrete to air contents up to about 6%, the influence of air content is insignificant. The fire endurance increases with an increase of air content above about 6% and the effect is more pronounced above about 10% particularly for lightweight concrete.

For normal weight concretes, fire endurance is improved by decreasing the maximum aggregate size. The reason for this is that the cement paste content increases with a decrease in aggregate size.

6.3.4.2 Multi-Course Assemblies

Floors and roofs often consist of concrete base slabs with overlays or undercoatings of other types of concrete or insulating materials. In addition, roofs generally have built-up roofing.

If the fire endurances of the individual courses are known, the fire endurance of the composite assembly can be estimated from the formula:

$$R = (R_1^{0.59} + R_2^{0.59} \ldots + R_n^{0.59})^{1.7}$$

(Eq. 6.3-1)

where R = the fire endurance of the composite assembly in minutes, and R_1, R_2, and R_n = the fire endurances of the individual courses in minutes. The following example illustrates the use of this equation:

Example 6.3.1:

Determine the fire endurance of a slab consisting of a 2 in. base slab of siliceous aggregate concrete

Fig. 6.3.3 Concrete slabs with gypsum wallboard ceilings

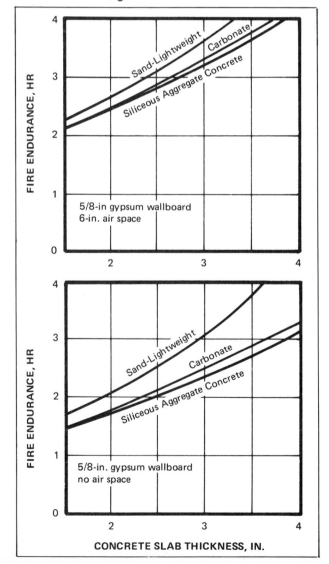

with a 2-1/2 in. topping of sand-lightweight concrete (115 pcf).

Solution:

From Fig. 6.3.2, the fire endurance of a 2 in. thick slab of siliceous aggregate concrete and 2-1/2 in. of sand-lightweight aggregate concrete are 26 min and 54 min, respectively.

$$R = [(26)^{0.59} + (54)^{0.59}]^{1.7}$$

$$R = (6.84 + 10.52)^{1.7} = 128 \text{ min} = 2 \text{ hr } 8 \text{ min}$$

Equation 6.3-1 has certain shortcomings in that it does not account for the location of the individual courses relative to the fired surface. Also, it is not possible to directly obtain the fire endurances of many insulating materials. Nevertheless, in a series of tests, the formula estimated the fire endurances within about 10% for most assemblies.

Fig. 6.3.3 shows the fire endurance of concrete slabs with 5/8 in. gypsum wallboard (Type X) for two cases: (1) a 6-in. air space between the wallboard and the slab, and (2) no space between the wallboard and slab. In Fig. 6.3.3 values above 3 hours are questionable because of the integrity of the wallboard after 3 hours of exposure. Materials and techniques of attaching the wallboard should be similar to those used in the UL test* on which the data are based.

A report on two-course floors and roofs† gives results of many fire tests. The report also shows graphically the fire endurances of assemblies consisting of various thicknesses of two materials. The graphs, several of which are reproduced in Figs. 6.3.4 through 6.3.7, can be used to estimate the required thicknesses of two-course materials for various fire endurances.

Example 6.3.2:

Determine the thickness of sprayed mineral fiber to be applied to the underside of the 2 in. thick flange of a sand-lightweight concrete double tee so that the fire endurance (heat transmission) will be 2 hours.

Solution:

From Fig. 6.3.5(b), the thickness must be about 7/8 in.

---

*"UL Report on Floor and Ceiling Assembly Consisting of Prestressed, Precast Concrete Double Tee Units with a Wallboard Ceiling," File R1319-131, February 21, 1973.

†Abrams, M.S. and Gustaferro, A. H., "Fire Endurance of Two-Course Floors and Roofs," *Journal of the American Concrete Institute,* February 1969.

Fig. 6.3.4 Combinations of base slabs and overlays of normal weight and lightweight concrete

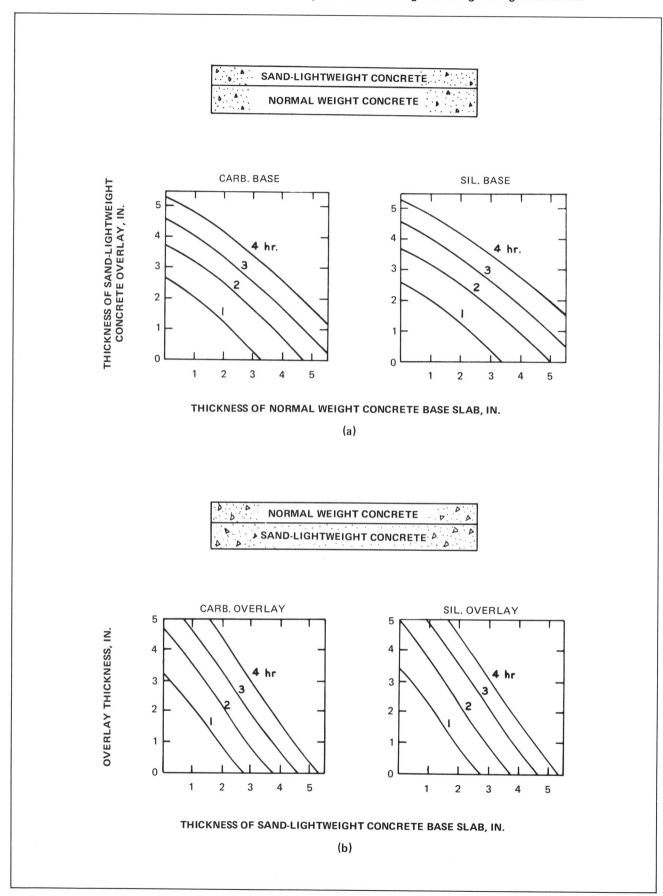

PCI Design Handbook

Fig. 6.3.5 Concrete slabs undercoated with various materials

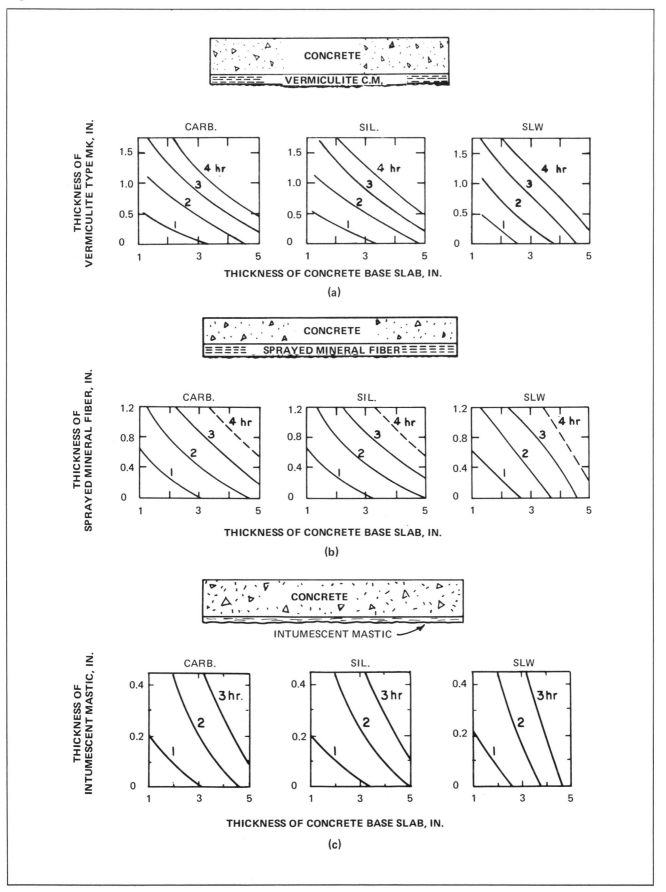

(a)

(b)

(c)

Fig. 6.3.6   Concrete base slabs with overlays of insulating concretes

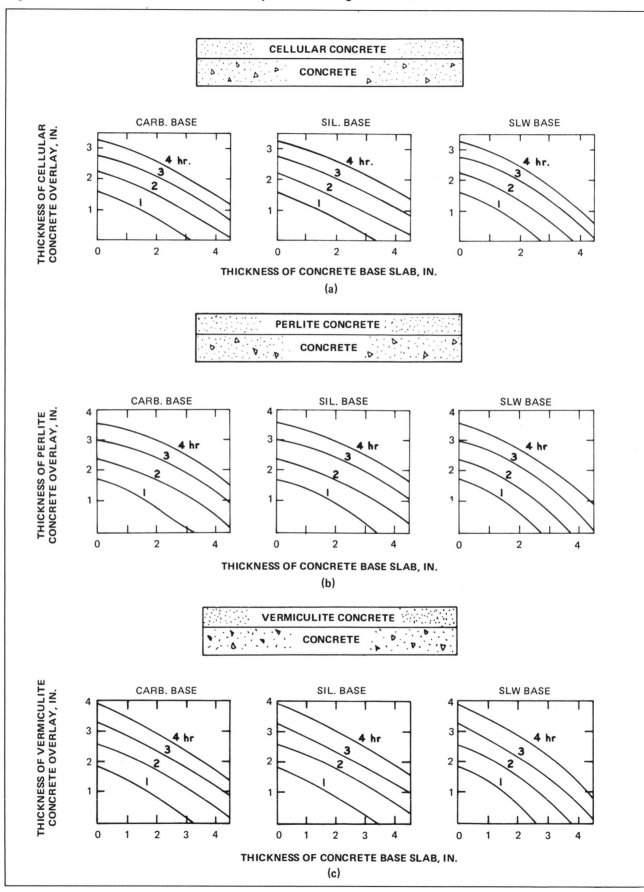

(a)

(b)

(c)

Fig. 6.3.7 Concrete roof slabs with insulations

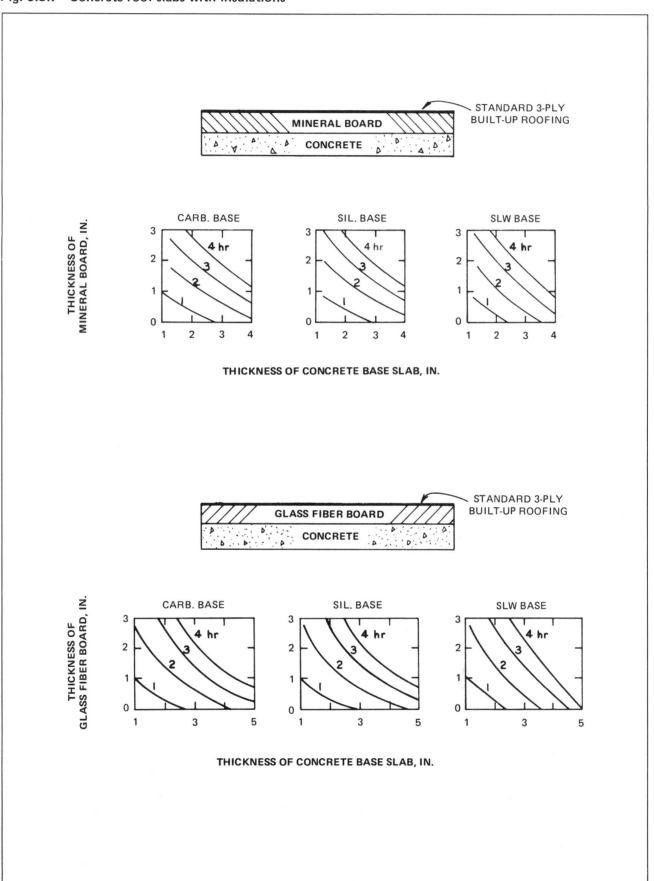

Tests of roof assemblies showed that the use of three-ply built-up roofing on one-course and two-course assemblies increased the fire endurance by 10 to 20 minutes.

Example 6.3.3:

Assume that 3-ply built-up roofing is to be applied to the assembly described in Example 6.3.2. What thickness of sprayed mineral fiber will be needed for a 2-hr fire endurance?

Solution:

Assume (conservatively) that the roofing will provide 10 minutes of fire endurance. Thus the concrete and sprayed mineral fiber must provide 1 hr 50 min. From Fig. 6.3.5(b), for a 2 in. sand-lightweight concrete slab for 1 hr 50 min, the thickness of sprayed mineral fiber must be about 3/4 in.

6.3.5 Designing for Structural Integrity

It was noted above that many fire tests and related research studies have been directed toward an understanding of the structural behavior of prestressed concrete subjected to fire. The information gained from that work has led to the development of calculation procedures which can be used in lieu of fire tests. The purpose of this section is to present an introduction to these calculation procedures. Because the method of support is the most important factor affecting structural behavior of flexural elements during fire, the discussion that follows deals with three conditions of support: simply supported members, continuous slabs and beams, and members in which restraint to thermal expansion occurs.

6.3.5.1 Simply Supported Members

Assume that a simply supported prestressed concrete slab is exposed to fire from below, that the ends of the slab are free to rotate, and that expansion can occur without restriction. Also assume that the reinforcement consists of straight strands located near the bottom of the slab. With the underside of the slab exposed to fire, the bottom will expand more than the top causing the slab to deflect downward; also, the strength of the steel and concrete near the bottom will decrease as the temperature rises. When the strength of the steel diminishes to that required to support the slab, flexural collapse will occur. In essence, the applied moment remains practically constant during the fire exposure, but the resisting moment capacity is reduced as the steel weakens.

Fig. 6.3.8 illustrates the behavior of a simply supported slab exposed to fire from beneath, as described above. Because strands are parallel to the axis of the slab, the design moment strength is constant throughout the length:

$$\phi M_n = \phi A_{ps} f_{ps} (d - a/2) \qquad \text{(Eq. 6.3-2)}$$

where:

ϕ = 0.90

A_{ps} = cross sectional area of the prestressing steel, sq in.

f_{ps} = stress in the prestressing steel at nominal strength, ksi

d = distance from the centroid of the prestressing steel to the extreme compression fiber, in.

a = depth of the equivalent rectangular compression stress block at nominal strength, in., and is equal to $A_{ps} f_{ps} / 0.85 f'_c b$, where f'_c is the compressive strength, ksi, of the concrete and b is the width of the slab, in.

M_n = nominal moment strength, in.-k

In lieu of an analysis based on strain compatibility the value of f_{ps} can be assumed to be:

$$f_{ps} = f_{pu} \left(1 - \frac{0.5 A_{ps} f_{pu}}{b d f'_c}\right) \qquad \text{(Eq. 6.3-3)}$$

Fig. 6.3.8 Moment diagrams for simply supported beam or slab

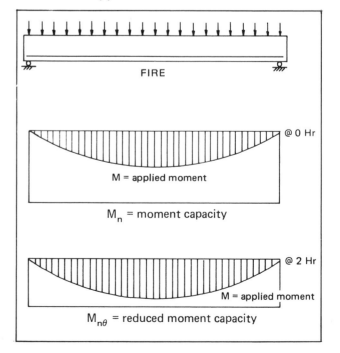

FIRE

@ 0 Hr

M = applied moment

M_n = moment capacity

@ 2 Hr

M = applied moment

$M_{n\theta}$ = reduced moment capacity

Fig. 6.3.9 Temperature-strength relationships for various steels

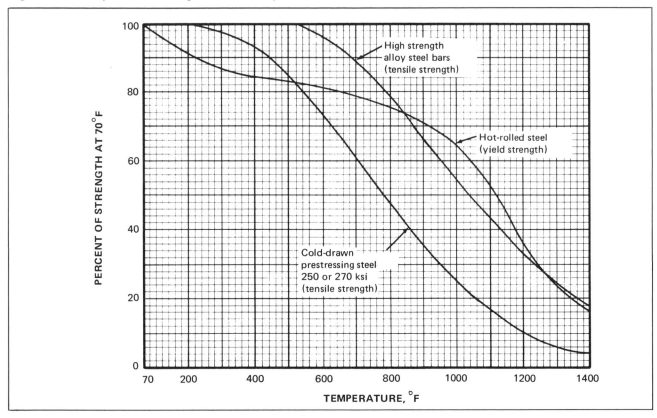

Fig. 6.3.10 Compressive strength of concrete at high temperatures

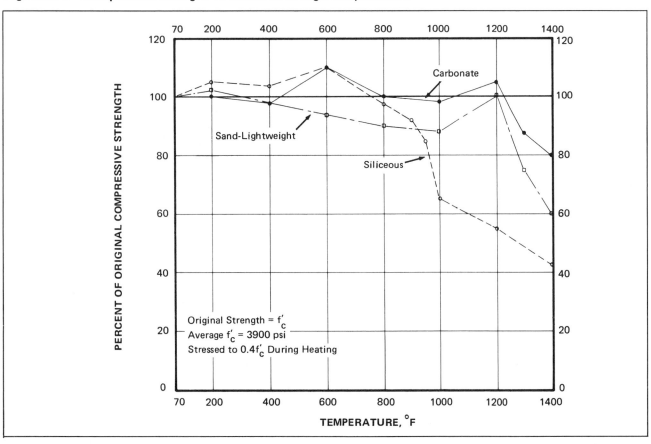

Fig. 6.3.11 Temperatures within concrete slabs during fire tests

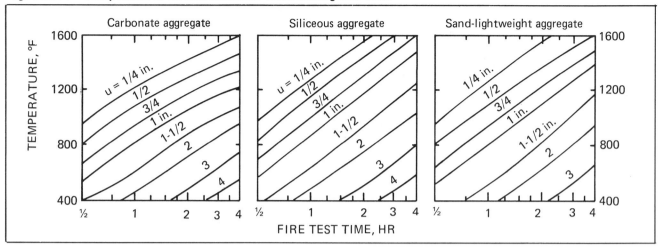

where f_{pu} is the ultimate tensile strength of the prestressing steel, ksi.

If the slab is uniformly loaded, the moment diagram will be parabolic with a maximum value at midspan of:

$$M = \frac{w\ell^2}{8} \qquad \text{(Eq. 6.3-4)}$$

where:

    w = dead plus live load per unit of length, k/in.

    ℓ = span length, in.

As the material strengths diminish with elevated temperatures, the retained nominal strength becomes:

$$M_{n\theta} = A_{ps}\, f_{ps\theta}\, (d - a_\theta/2) \qquad \text{(Eq. 6.3-5)}$$

in which θ signifies the effects of high temperatures. Note that A_{ps} and d are not affected, but f_{ps} is reduced. Similarly a_θ is reduced, but the concrete strength at the top of the slab, f'_c, is generally not reduced significantly because of its lower temperature. If, however, the compressive zone of the concrete is heated above 900°F, f'_c should also be reduced to calculate a_θ.

Flexural failure can be assumed to occur when $M_{n\theta}$ is reduced to M. Strength reduction factor, ϕ, is not applied because a safety factor is included in the required ratings. From this expression, it can be seen that the fire endurance depends on the applied loading and on the strength-temperature characteristics of the steel.

In turn, the duration of the fire before the "critical" steel temperature is reached depends on the protection afforded to the reinforcement.

To solve problems involving the above equa-

tions, it is necessary to utilize data on the strength-temperature relationships for steel and concrete, and information on temperature distributions within concrete members during fire exposures. Fig. 6.3.9 shows strengths of certain steels at elevated temperatures, and Fig. 6.3.10 shows similar data for various types of concrete.

Data on temperature distribution in concrete slabs during fire tests are shown in Fig. 6.3.11. Similar data for concrete beams and joists are more complex because beams are heated from the sides as well as from beneath. Fig. 6.3.12 shows temper-

Fig. 6.3.12 Temperatures on vertical centerlines of stemmed units — 2 hr exposure

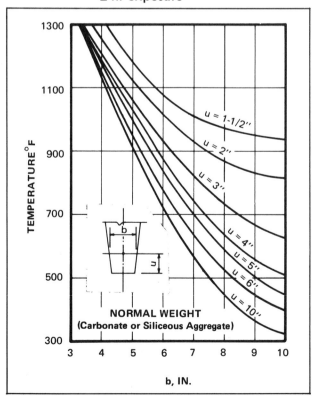

ature data for normal weight concrete joists and beams at 2 hr exposure. Note that the temperatures are given for points along the vertical centerline of rectangular beams or of members with tapered sides.

Example 6.3.4:

Determine the maximum safe superimposed load that can be supported by an 8 in. deep hollow-core slab with a simply supported unrestrained span of 25 ft and a fire endurance of 3 hr.

Given:

h = 8 in.; u = 1.75 in.; eight 1/2 in. 250 ksi strands; A_{ps} = 8(0.144) = 1.152 sq in.; b = 48 in.; d = 8 – 1.75 = 6.25 in.; w_d = 60 psf; carbonate aggregate concrete; ℓ = 25 ft; f'_c = 5000 psi.

Solution:

(a) Estimate strand temperature at 3 hr from Fig. 6.3.11: At 3 hr, carbonate aggregate, u = 1.75 in.
θ_s = 925°F.

(b) Determine $f_{pu\theta}$ from Fig. 6.3.9. For cold-drawn steel at 925°F, $f_{pu\theta}$ = 0.33 (f_{pu}) = 82.5 ksi

(c) Determine $M_{n\theta}$ and w

$$f_{ps} = f_{pu\theta}\left(1 - \frac{0.5\, A_{ps}\, f_{pu\theta}}{bdf'_c}\right)$$

$$= 82.5\left[1 - \frac{0.5(1.152)(82.5)}{48(6.25)(5)}\right] = 79.9\text{ ksi}$$

$$a_\theta = \frac{A_{ps}\, f_{ps\theta}}{0.85\, f'_c\, b} = \frac{1.152(79.9)}{0.85(5)(48)} = 0.45\text{ in.}$$

$$M_{n\theta} = A_{ps}\, f_{ps\theta}\, (d - a_\theta/2)$$
$$= 1.152(79.9)(6.25 - 0.45/2)/12$$
$$= 46.2\text{ ft-kips}$$

$$w = \frac{8M}{\ell^2} = \frac{8(46.2)}{(25)^2} = 0.591\text{ klf} = 148\text{ psf}$$

$$w_\ell = w - w_d = 148 - 60 = 88\text{ psf}$$

Example 6.3.5:

Provide 2 hr fire endurance (structurally) by adding strands and/or rebars to an 8DT16 + 2 for a 29 ft span with a live load of 40 psf. Simple support, no restraint; normal weight concrete, f'_c = 5 ksi; topping concrete, f'_c = 4 ksi; f_{pu} = 270 ksi;

b = 96 in.; member weight = 539 plf; strand locations as shown:

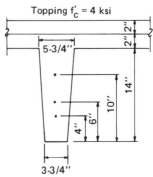

Solution:

A_{ps} = 6(0.153) = 0.918 in.^{2}

u = 6.67 in.

d = 18 – 6.67 = 11.33 in.

w_ℓ = 8(40) = 320 plf

w = 539 + 320 = 859 plf

(a) Estimate strand temperature at 2 hr from Fig. 6.3.12:

At c.g.s.; b = 3.75 + $\dfrac{6.67}{14}$(2) = 4.70 in.

Avg. θ_s = 1000°F

(b) Estimate $f_{pu\theta}$ from Fig. 6.3.9

$f_{pu\theta}$ = 0.25 f_{pu} = 0.25(270) = 67.5 ksi

(c) Calculate $M_{n\theta}$ and compare with M

$f_{ps\theta}$ = 67.0 ksi

$$a_\theta = \frac{A_{ps}\, f_{ps\theta}}{0.85\, f'_c\, b} = \frac{0.918(67.0)}{0.85(4)(96)} = 0.188\text{ in.}$$

$$M_{n\theta} = A_{ps}\, f_{ps\theta}\, (d - a_\theta/2)$$
$$= 0.918(67.0)(11.33 - 0.094)/12$$
$$= 57.6\text{ ft-kips}$$

$$M = w\ell^2/8 = 0.859(29)^2/8 = 90.3\text{ ft-kips}$$

(d) Try adding one No. 7 Grade 60 reinforcing bar at u = 4.69 in. in each stem.

(e) Estimate temperature and strength of the reinforcing bar from Fig. 6.3.12:

u = 4.69 in., b = 4.42 in.

θ_s = 1120°F

From Fig. 6.3.9 for hot-rolled steel:

$f_{y\theta}$ = 0.50 f_y = 30.0 ksi

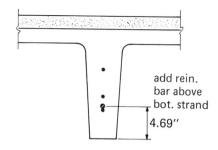

add rein.
bar above
bot. strand

4.69"

(f) Calculate $M_{n\theta}$

$$\text{adjusted } a_\theta = \frac{A_{ps} f_{ps\theta} + A_s f_{y\theta}}{0.85 \, b \, f'_c}$$

$$= \frac{0.918(67) + 2(0.60)(30.0)}{0.85(96)(4)}$$

$$= 0.30 \text{ in.}$$

Due to strand:

$$M_{n\theta} = A_{ps} f_{ps\theta} (d - a_\theta/2)$$

$$= (0.918)(67.0)(11.33 - 0.15)/12$$

$$= 57.3 \text{ ft-kips}$$

Due to rebars:

$$M_{n\theta} = A_s f_{y\theta} (d - a_\theta/2)$$

$$= 2(0.60)(30.0)(13.31 - 0.15)/12$$

$$= 39.5 \text{ ft-kips}$$

Total capacity = 57.3 + 39.5

$$= 96.8 \text{ ft-kips} > 90.3 \text{ ft-kips}$$

6.3.5.2 Continous Members

Continuous members undergo changes in stresses when subjected to fire. These stresses result from temperature gradients within the structural members, or changes in strength of the materials at high temperatures, or both.

Fig. 6.3.13 shows a two-span continous beam whose underside is exposed to fire. The bottom of the beam becomes hotter than the top and tends to expand more than the top. This differential temperature effect causes the ends of the beam to tend to lift from their supports thereby increasing the reaction at the interior support. This action results in a redistribution of moments, i.e., the negative moment at the interior support increases while the positive moments decrease.

During a fire, the negative moment reinforcement (Fig. 6.3.13) remains cooler than the positive moment reinforcement because it is better protected from the fire. In addition, the redistribution that occurs is sufficient to cause yielding of the negative moment reinforcement. Thus, a relatively large increase in negative moment can be accommodated throughout the test. The resulting decrease in positive moment means that the positive moment reinforcement can be heated to a higher temperature before failure will occur. Therefore, the fire endurance of a continous concrete beam is generally significantly longer than that of a simply supported beam having the same cover and the same applied loads.

It is possible to design the reinforcement in a continuous beam or slab for a particular fire endurance period. From Fig. 6.3.13 the beam can be expected to collapse when the positive moment capacity, $M_{n\theta}^+$, is reduced to the value of the maximum redistributed positive moment at a distance x_1 from the outer support.

Fig. 6.3.14 shows a uniformly loaded beam or slab continuous (or fixed) at one support and simply supported at the other. Also shown is the redistributed applied moment diagram at failure.

It can be shown that at the point of positive moment, x_1,

$$x_1 = \frac{\ell}{2} - \frac{M_{n\theta}^-}{w\ell} \qquad \text{(Eq. 6.3-6)}$$

Fig. 6.3.13 Moment diagram for two-span continuous beam

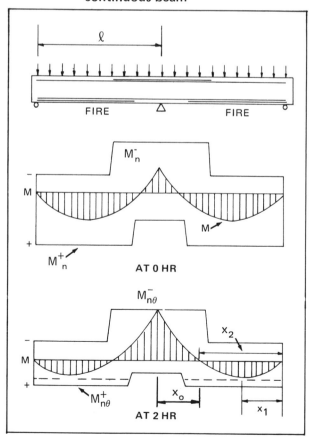

**Fig. 6.3.14 Uniformly loaded member
continuous at one support**

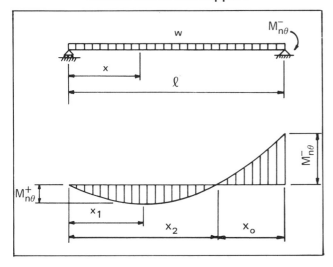

at $x = x_2$, $M_x = 0$ and $x_2 = 2 x_1$

$$x_0 = \frac{2M_{n\theta}^-}{w\ell} \qquad \text{(Eq. 6.3-7)}$$

$$M_{n\theta}^- = \frac{w\ell^2}{2} \pm w\ell^2 \sqrt{\frac{2M_{n\theta}^+}{w\ell^2}} \qquad \text{(Eq. 6.3-8)}$$

In most cases, redistribution of moments occur early during the course of a fire and the negative moment reinforcement can be expected to yield before the negative moment capacity has been reduced by the effects of fire. In such cases, the length of x_0 is increased, i.e., the inflection point moves toward the simple support. If the inflection point moves beyond the cut off points of the negative moment reinforcement, sudden failure may result.

Fig. 6.3.15 shows a symmetrical beam or slab in which the end moments are equal.

$$M_{n\theta}^- = w\ell^2/8 - M_{n\theta}^+ \qquad \text{(Eq. 6.3-9)}$$

$$\frac{wx_2^2}{8} = M_{n\theta}^+$$

$$x_2 = \frac{8M_{n\theta}^+}{w} \qquad \text{(Eq. 6.3-10)}$$

$$x_0 = \frac{1}{2}(\ell - x_2)$$

$$= \frac{\ell}{2} - \frac{1}{2}\sqrt{\frac{8M_{n\theta}^+}{w}} \qquad \text{(Eq. 6.3-11)}$$

To determine the maximum value of x_0, the value of w should be the minimum service load anticipated, and $(w\ell^2/8 - M_n^-)$ should be substituted for $M_{n\theta}^+$ in Eq. 6.3-11.

For any given fire endurance period, the value of $M_{n\theta}^+$ can be calculated by the procedures given in Sect. 6.3.5.1. Then the value of $M_{n\theta}^-$ can be calculated by the use of Eq. 6.3-8 or 6.3-9 and the necessary lengths of the negative moment reinforcement can be determind from Eq. 6.3-7 or 6.3-11. Use of these equations is illustrated in Example 6.3.6.

It should be noted that the amount of moment redistribution that can occur is dependent on the amount of negative moment reinforcement. Tests have clearly demonstrated that in most cases the negative moment reinforcement will yield, so the negative moment capacity is reached early during a fire test, regardless of the applied loading. The designer must exercise care to ensure that a secondary type of failure will not occur. To avoid a compression failure in the negative moment region, the amount of negative moment reinforcement should be small enough so that $\omega_\theta = A_s f_{y\theta} / b_\theta d_\theta f'_{c\theta}$, is less than 0.30, before and after reductions in f_y, b, d and f'_c are taken into account. Furthermore, the negative moment bars or mesh must be long enough to accommodate the complete redistributed moment and change in the inflection points. It should be noted that the worst condition occurs when the applied loading is smallest, such as the dead load plus partial or no live load. It is recommended that at least 20% of the maximum negative moment reinforcement extend throughout the span.

Example 6.3.6:

Design a floor using hollow-core slabs and topping for 22 ft span for 4 hr fire endurance. Service loads = 175 psf dead (including structure) and 150 psf live. Use 4 ft wide, 10 in. deep slabs with 2 in.

**Fig. 6.3.15 Symmetrical uniformly loaded
member continuous at both supports**

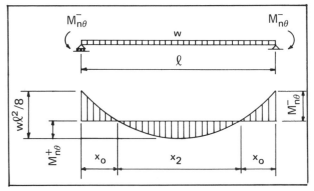

topping, carbonate aggregate concrete. Continuity can be achieved at both ends. Use f'_c (precast) = 5000 psi, f_{pu} = 250 ksi, and f'_c (topping) = 3000 psi, sixteen 3/8 in., 250 ksi strands at u = 1.75 in. Provide negative moment reinforcement needed for fire.

Solution:

A_{ps} = 16(0.080) = 1.28 sq in.

u = 1.75 in.

d = 12 − 1.75 = 10.25 in.

From Fig. 6.3.11, θ_s = 1010°F

From Fig. 6.3.9, $f_{pu\theta}$ = 0.24f_{pu} = 60 ksi

Using Table 3.9.1:*

$$\bar{\omega}_{p\theta} = \frac{16(0.080)(60)}{48(10.25)(3)} = 0.052$$

K_u = 132/0.9 = 147

$$M^+_{n\theta} = K_u bd^2 / 12,000$$
$$= 147(48)(10.25)^2 / 12,000$$
$$= 61.6 \text{ ft-kips/unit}$$
$$= 15.4 \text{ ft-kips/ft}$$

For simply supported members:

M = 0.325(22)²/8 = 19.7 ft-kips/ft

Req'd $M^-_{n\theta}$ = 19.7 − 15.4 = 4.3 ft-kips/ft

Assume $d - a_\theta/2$ = 10.25 in., and f_y = 60 ksi

$$A^-_s = \frac{4.3(12)}{60(10.25)} = 0.084 \text{ in.}^2/\text{ft}$$

Use 20% throughout span:

Try 6 × 6-W1.4 × W1.4 continuous plus 6 × 6-W2.9 × W2.9 over supports

A^-_s = 0.029 + 0.058 = 0.087 in.²/ft

Neglect concrete above 1400°F in negative moment region, i.e., from Fig. 6.3.11, neglect bottom 5/8 in. Also, concrete within compressive zone will be about 1350°F to 1400°F, so use $f'_{c\theta}$ = 0.81 f'_c (see Fig. 6.3.10)

= 4.05 ksi

Check $M^-_{n\theta}$, assuming that the temperature of the negative steel does not rise above 200°F. If greater than 200°F, steel strength should be reduced according to Fig. 6.3.9.

---

*The values for K_u in Table 3.9.1 include ϕ = 0.9. Since in the design for fire, ϕ = 1.0, the value of K_u must be divided by 0.9.

$$a_\theta = \frac{0.087(60)}{0.85(4.05)(12)} = 0.13 \text{ in.}$$

$$M^-_{n\theta} = 0.087(60)(10.37 - 0.065) / 12$$
$$= 4.48 \text{ ft-kips/ft}$$

With dead load + 1/2 live load, w = 0.25 klf, M = 15.12 ft-kips/ft, and M^-_n = 4.71 ft-kips/ft (calculated for room temperature)

M^+_{min} = 15.12 − 4.71 = 10.41 ft-kips/ft

From Eq. 6.3-11

$$\max x_o = \frac{22}{2} - \frac{1}{2}\sqrt{\frac{8(10.41)}{0.25}} = 1.87 \text{ ft}$$

Use 6 × 6-W1.4 × W1.4 continuous throughout plus 6 × 6-W2.9 × W2.9 for a distance of 3 ft from the support. Mesh must extend into walls which must be designed for the moment induced at the top.

6.3.5.3 Members Restrained Against Thermal Expansion

If a fire occurs beneath an interior portion of a large reinforced concrete slab, the heated portion will tend to expand and push against the surrounding part of the slab. In turn, the unheated part of the slab exerts compressive forces on the heated portion. The compressive force, or thrust, acts near the bottom of the slab when the fire test occurs but, as the fire progresses, the line of action of the thrust rises as the mechanical properties of the heated concretes change. This thrust is

Fig. 6.3.16 Axially restrained beam during fire exposure

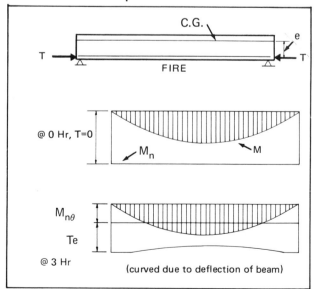

(curved due to deflection of beam)

generally great enough to increase the fire endurance significantly.

The effects of restraint to thermal expansion can be characterized as shown in Fig. 6.3.16. The thermal thrust acts in a manner similar to an external prestressing force, which, in effect, increases the positive moment capacity.

The increase in bending moment capacity is similar to the effect of added reinforcement located along the line of action of the thrust. It can be assumed that the added reinforcement has a yield strength (force) equal to the thrust. By this approach, it is possible to determine the magnitude and location of the required thrust to provide a given fire endurance.

The above explanation is greatly simplified because in reality restraint is quite complex, and can be likened to the behavior of a flexural member subjected to an axial force. Interaction diagrams similar to those for columns can be constructed for a given cross-section at a particular stage of a fire, e.g., 2 hr of a standard fire exposure.

The guidlelines in ASTM E 119-76 given for determining conditions of restraint are useful for preliminary design purposes. Basically, interior bays of multi-bay floors or roofs can be considered to be restrained and the magnitude and location of the thrust are generally of academic interest only.

6.3.6 Code and Economic Considerations

An important aspect of dealing with fire resistance is to understand what the benefits are to the owner of a building in the proper selection of materials incorporated in his structure. These benefits fall into two areas, code and economics.

Building codes are laws that must be satisfied regardless of any other considerations and the manner in which acceptance of code requirements is achieved is explained in the preceding pages. The designer, representing the owner, has no option in the code regulations, only in the materials and assemblies that meet these regulations.

Economic benefits are often overlooked by the designer/owner team at the time decisions are made on the structural system. Proper consideration of fire resistive construction through life-cycle cost analysis will provide the owner economic benefits over other types of construction in many areas, e.g., lower insurance costs, larger allowable gross areas under certain types of building construction, fewer stairwells and exits, increased value for loan purposes, longer mortgage terms, and better resale value. To ensure an owner of the best return on his investment, a life-cycle cost analysis using fire resistive construction must be prepared.

Beyond the theoretical considerations is the history of excellent performance of prestressed concrete in actual fires. Structural integrity has been provided, fires are contained in the area of origin, and, in many instances, repairs consist of "cosmetic" treatment only leading to early re-occupancy of structure.

COORDINATION WITH MECHANICAL, ELECTRICAL AND OTHER SUB-SYSTEMS

6.4.1 Introduction

Prestressed concrete is used in a wide variety of buildings, and its integration with lighting, mechanical, plumbing, and other services is of importance to the designer. Because of increased environmental demands, the ratio of costs for mechanical and electrical installations to total building cost has increased substantially in recent years. This section is intended to provide the designer with the necessary perspective to economically satisfy mechanical and electrical requirements, and to describe some standard methods of providing for the installation of other sub-systems.

6.4.2 Lighting and Power Distribution

For many applications, the designer can take advantage of the fire resistance, reflective qualities and appearance of prestressed concrete by leaving the columns, beams, and ceiling structure exposed. To achieve uniform lighting free from distracting shadows, the lighting system should parallel the stems of tee members.

By using a reflective paint and properly spaced high-output fluorescent lamps installed in a continuous strip, the designer can achieve a high level of illumination at a minimum cost. In special areas, lighting troffers can be enclosed with diffuser panels fastened to the bottom of the tee stems providing a flush ceiling. (See Fig. 6.4.1). By using reflective paints, these precast concrete lighting channels can be made as efficient as conventional fluorescent fixtures.

6.4.3 Electrified Floors

The increasing use of business machines, telephones, and other communication systems stresses the need for adequate and flexible means of supplying electricity and communication service. Since a cast-in-place topping is usually placed on prestressed floor members, conduit runs and floor outlets can be readily buried within this topping. With shallow height electrical systems, a comprehensive system can be provided in a reasonably thin topping. The total height of conduits for these comprehensive electrical systems is as little as 1-3/8 in. Most systems can easily be included in a 2 to 4 inch thick slab. Voids in hollow-core slabs can also be used as electrical raceways (see Fig. 6.4.2).

When the system is placed in a structural composite slab, the effect of ducts and conduits must be carefully examined and their location coordinated with reinforcing steel. Tests on slabs with buried ductwork have shown that structural strength is not normally impaired by the voids.

Because of the high load-carrying capacity of prestressed concrete members, it is possible to locate high-voltage substations, with heavy transformers, near the areas of consumption with little or no additional expense. For extra safety, distribution feeds can also be run within those channels created by stemmed members. Such measures also aid the economy of the structure by reducing the overall story height and minimizing maintenance expenses.

6.4.4 Ductwork

The designer may also utilize the space within stemmed members or the holes inside hollow-core slabs for distribution ducts for heating, air-conditioning, or exhaust systems. In stemmed members three sides of the duct are provided by the bottom of the flange and the sides of the stems. The bottom of the duct is completed by attaching a metal panel to the tee stems in the same fashion as the lighting diffusers (see Fig. 6.4.1).

Connections can be made by several means, among them powder-activated fasteners, cast-in inserts or reglets. Field installed devices generally offer the best economy and ensure placement in the exact location where the connecting devices will be required. Inserts should only be cast-in when they can be located in the design stage of the job, well in advance of casting the precast members.

With hollow-core slabs, additional duct work can be eliminated. These members have oval, round, or rectangular voids of varying size which can provide ducts or raceways for the various systems. Openings core drilled in the field can provide access and distribution. The voids in the slabs are aligned and connected to provide continuity of the system. Openings can also be provided in intermediate supporting beams such as inverted tees, to allow duct continuity.

If high velocity air movement is utilized, the enclosed space becomes a long plenum chamber with uniform pressure throughout its length. Diffusers are installed in the ceiling to distribute the

Fig. 6.4.1   Metal panels attached at the bottoms of the stems create ducts, and diffuser panels provide a flush ceiling

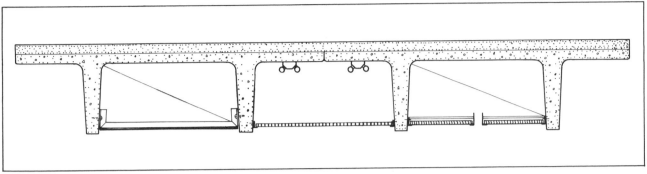

Fig. 6.4.2   Underfloor electrical ducts can be embedded within a concrete topping

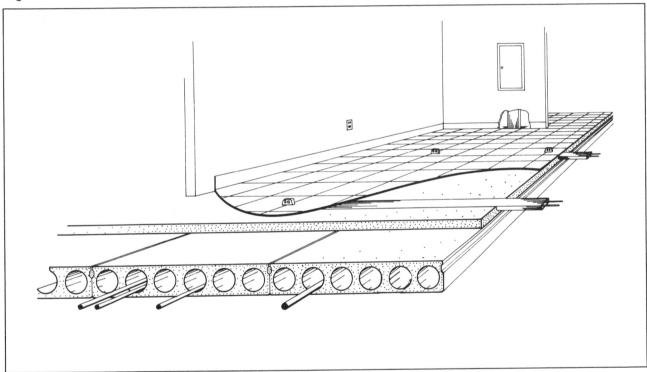

air. Branch runs, when required, can be standard ducts installed along the column lines.

When ceilings are required, proper selection of precast components can result in shallow ceiling spaces as shown in Fig. 6.4.3. This figure also illustrates the flexibility of space arrangements possible with long span prestressed concrete members.

Branch ducts of moderate size can also be accomodated by providing block-outs in the stems of tees or beams. To achieve best economy and performance in prestressed concrete members, particularly stemmed members, such block-outs should be repeated in size and location to handle all conditions demanded by mechanical, electrical, or plumbing runs. While this may lead to slightly larger openings in some cases the end

result will probably be more economical. It should also be noted that sufficient tolerance should also be allowed in sizing the openings to provide for necessary field assembly considerations.

Prestressed concrete box girders have been used to serve a triple function as air conditioning distribution ducts, conduit for utility lines and structural supporting members for the roof deck units. Conditioned air can be distributed within the void area of the girders and then introduced into the building work areas through holes cast into the sides and bottoms of the box girders. The system is balanced by plugging selected holes.

Vertical supply and return air trunks can be carried in the exterior walls, with only small ducts needed to branch out into the ceiling space. In

Fig. 6.4.3   Where ceilings are required, ducts, piping, and lighting fixtures can be accommodated within a shallow depth

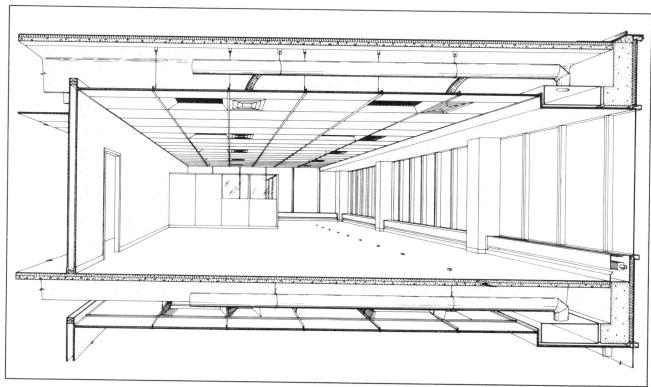

some cases the exterior wall cavities are replaced with three or four sided precast boxes stacked to provide vertical runs for the mechanical and electrical systems. These stacked boxes can also be used as columns or lateral bracing elements for the structure.

In some cases it may be required to provide openings through floor and roof units. Large openings are usually made by block-outs in the forms during the manufacture; smaller ones (up to about 8 in.) are usually field drilled. Openings in flanges of stemmed members should be limited to the "flat" portion of the flange, that is, beyond 1 in. of the edge of the stem on double tees and 3 in. of the edge of the stem on single tees. Angle headers are often used for framing large openings in hollow-core floor or roof systems. (See Fig. 6.4.4).

Fig. 6.4.4   Large openings in floors and roofs are made during manufacture of the units; small openings are field drilled. Some common types of openings are shown here

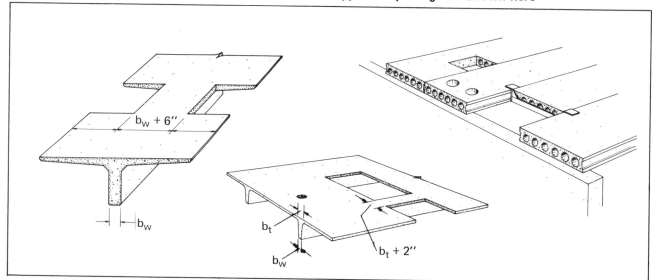

Fig. 6.4.5 Methods of attaching suspended ceilings, crane rails, and other sub-systems

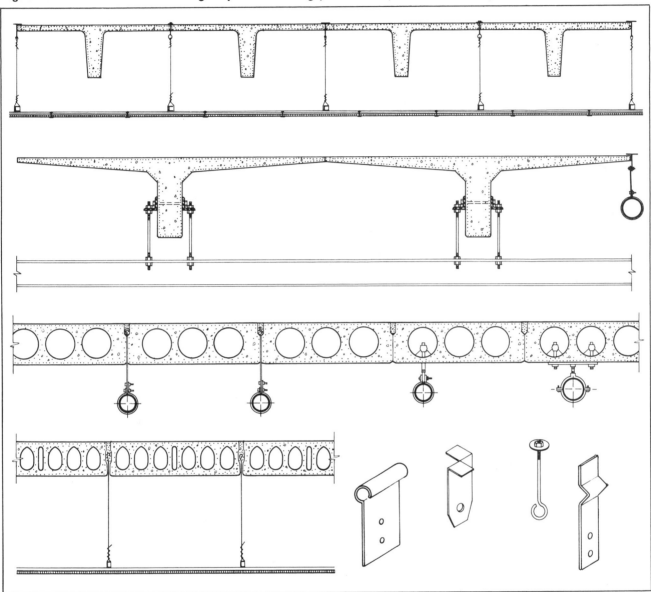

6.4.5 Other Sub-Systems

Suspended ceilings, crane rails, and other sub-systems can be easily accomodated with standard manufactured hardware items and embedded plates as shown in Fig. 6.4.5.

Architectural precast concrete wall panels can be adapted to combine with preassembled window or door units. Door or window frames properly braced to prevent bowing during concrete placement, can be cast in the panels and then the glazing or doors can be installed prior to or after delivery to the job site. If the glazing or doors are properly protected, they can also be cast into the panel at the plant. When casting in aluminum window frames, particular attention should be made to properly coat the aluminum so that it will not react with the concrete. It should be noted that repetition is one of the real keys to economy in a precast concrete wall assembly. Windows and doors should be located in identical places for all panels wherever possible.

Insulated wall panels can be produced by embedding an insulating material such as expanded polyurethane between layers of concrete. Normally two layers of concrete are separated by a one-inch thickness of insulating material and U-values of 0.15 or less can be achieved (see Section 6.1). These exterior panels are normally cast on flat beds or tilting tables. The inside surface of the concrete panel can be given a factory troweled finish followed by minor touchup work. The interior face is completed by painting, or by wall papering to achieve a finished wall. The formed surface of the panel can also be treated in a similar manner when used as the interior face.

6.4.6 Systems Building

As more and more complete systems buildings are built with precast and prestressed concrete, and as interest in this method of construction increases, we can expect that more of the building sub-systems will be prefabricated and pre-coordinated with the structure.

This leads to the conclusion that those parts of the structure that require the most labor skills should logically be prefabricated prior to installation in the field. The prefabricated components can be preassemblies of basic plumbing systems or electrical/mechanical systems plus lighting.

For housing systems, electrical conduits and boxes can be cast in the precast wall panels. This process requires coordination with the electrical

contractor; and savings on job site labor and time are possible. The metal or plastic conduit is usually pre-bent to the desired shape and delivered to the casting bed already connected to the electrical boxes. It is essential that all joints and connections be thoroughly sealed and the boxes enclosed prior to casting in order to prevent the system from becoming clogged. The wires are usually pulled through at the job site. Television antenna and telephone conduits have also been cast in using the same procedure.

To reduce on-site labor, prefabricated bathroom units or combination bathroom/kitchen modules have been developed (see Fig. 6.4.6). Such units include bathroom fixtures, kitchen cabinets and sinks, as well as wall, ceiling, and

Fig. 6.4.6 **Kitchen/bathroom modules can be pre-assembled on precast prestressed slabs ready for installation in systems buildings**

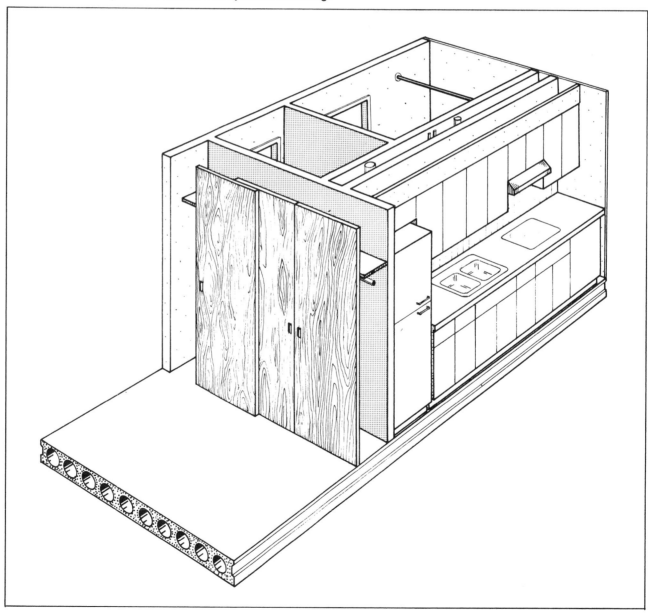

floor surfaces. Plumbing units are often connected and assembled prior to delivery to the job sites. These bathroom/kitchen modules can be molded plastic units or fabricated from drywall components. To eliminate a double floor, the module can be plant built on the structural member or the walls of the unit can be designed strong enough for all fixtures to be wall hung. In the latter case, the units are placed directly on a precast floor and in multi-story construction are located in a stack fashion with one bathroom directly over the one below. A block-out for a chase is provided in the precast floor and connections are made from each unit to the next to provide a vertical plumbing stack. Prefabricated wet-wall plumbing systems, as shown in Fig. 6.4.7,

incorporate preassembled piping systems using snap-on or no-hub connections made up of a variety of materials. These units only require a block-out in the prestressed flooring units and are also arranged in a stack fashion. Best economy results when bathrooms are backed up to each other, since a common vertical run can service two bathrooms.

Some core modules not only feature bath and kitchen components, but also HVAC components all packaged in one unit. These modules can also be easily accomodated in prestressed structural systems by placing them directly on the prestressed members with shimming and grouting as required.

Fig. 6.4.7   Prefabricated wet-wall plumbing systems incorporate pre-assembled piping

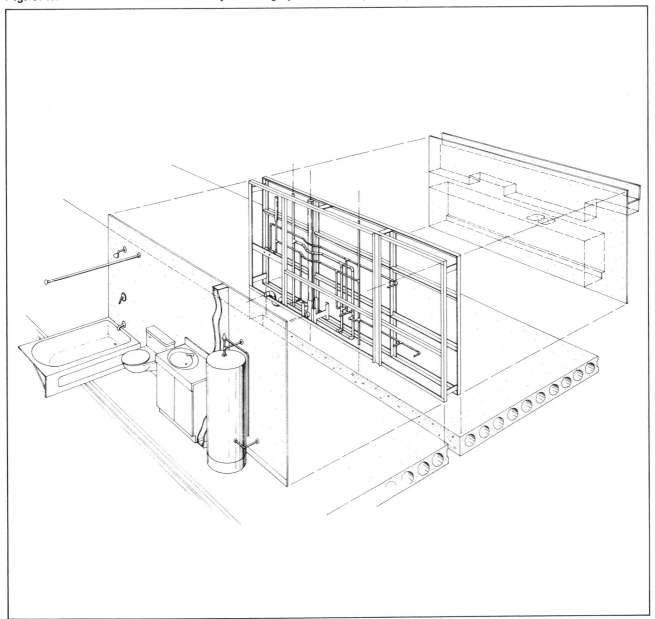

PART 7
SPECIFICATIONS AND REFERENCES

GUIDE SPECIFICATION FOR PRECAST, PRESTRESSED CONCRETE CONSTRUCTION FOR BUILDINGS

> This Guide Specification is intended to be used as a basis for the development of an office master specification or in the preparation of specifications for a particular project. In either case, this Guide Specification must be edited to fit the conditions of use.
>
> Particular attention should be given to the deletion of inapplicable provisions. Necessary items related to a particular project should be included. Also, appropriate requirements should be added where blank spaces have been provided.
>
> The Guide Specifications are on the left. *Notes to Specifiers are on the right.*

GUIDE SPECIFICATIONS

1. GENERAL

1.01 Description

A. Work included:

  1. These specifications cover precast and precast prestressed structural concrete construction, including product design not shown on contract drawings, manufacture, transportation, erection, and other related items such as anchorage, bearing pads, storage and protection of precast concrete.

B. Related work specified elsewhere:

  1. Cast-in-place concrete: Section _____

  2. Precast architectural concrete: Section _____

  3. Post-tensioning: Section _____

  4. Masonry bearing walls: Section _____

  5. Miscellaneous steel: Section _____

  6. Waterproofing: Section _____

  7. Flashing and sheet metal: Section _____

  8. Sealants and caulking: Section _____

  9. Painting: Section _____

  10. Holes for other trades: Sections _____

C. Work installed but furnished by others:

  1. Receivers or reglets for flashing: Section _____

  2. Elevator guides: Section _____

NOTES TO SPECIFIERS

1.01.A.   This Section is to be in Division 3 of Construction Specifications Institute format. Verify that plans clearly differentiate between this work and architectural precast concrete if both are on the same job. One may need to list items such as beams, purlins, girders, lintels, columns, slab or deck members, etc.

GUIDE SPECIFICATIONS	NOTES TO SPECIFIERS

1.02 Quality Assurance

A. Acceptable manufacturers: A company specializing in providing precast and/or precast prestressed concrete products and services normally associated with the industry for at least _____ years. When requested by the Architect/ Engineer, written evidence shall be submitted to show experience qualifications and adequacy of plant capability and facilities for performance of contract requirements.

B. Erector qualifications: Regularly engaged for at least _____ years in the erection of precast structural concrete similar to the requirements of this project.

C. Qualifications of welders: In accordance with AWS D1.1. Qualified within the past year.

D. Testing: In general compliance with applicable provisions of Prestressed Concrete Institute MNL-116, *Manual for Quality Control for Plants and Production of Precast Prestressed Concrete Products.*

E. Requirements of regulatory agencies: All local codes plus the following specifications, standards and codes are a part of these specifications:

1. ACI 318 — Building Code Requirements for Reinforced Concrete.

2. AWS D1.1 — Structural Welding Code.

3. AWS D12.1 — Reinforcing Steel Welding Code.

4. ASTM Specifications — As referred to in Part 2 — Products, of this Specification.

5. AASHTO Standard Specifications for Highway Bridges.

1.02.A. Usually 2 to 5 years. Plant certification, as provided in the PCI Plant Certification Program, is satisfactory evidence.

1.02.B. Usually 2 to 5 years.

1.02.E. Always include the specific year or edition of the specifications, codes and standards used in the design of the project and made part of the specifications.

For projects in Canada, the following standards from the National Building Code of Canada should be listed in addition to, or in place of the U.S. Standards, where appropriate:

CSA W47.1 Certification of Companies for Fusion Welding of Steel Structures

CSA W59.1 General Specification for Welding of Steel Structures

CSA W186 Welding of Reinforcing Bars in Reinforced Concrete Construction

CSA A23.1 Concrete Materials and Methods of Concrete Construction

CSA A23.2 Methods of Test for Concrete

CSA A23.3 Code for the Design of Concrete Structures for Buildings

CSA A23.3.1 Commentary on CSA Standard A23.3

CSA A23.4 Precast Concrete Materials and Construction

CSA A251 Qualification Code for Manufacturers of Architectural and Structural Precast Concrete

CSA S6 Design of Highway Bridges

Fire ratings are generally a code requirement. When required, fire-rated products shall be clearly identified on the design drawings.

1.03　Submittals

A. Shop drawings:

  1. Erection drawings

    a. Plans and/or elevations locating and defining all material furnished by manufacturer.

    b. Sections and details showing connections, cast-in items and their relation to the structure.

    c. Description of all loose, cast-in and field hardware.

    d. Field installed anchor location drawings.

    e. Erection sequences and handling requirements.

    f. All dead, live and other applicable loads used in the design.

  2. Production drawings

    a. Elevation view of each member.

    b. Sections and details to indicate quantities and position of reinforcing steel, anchors, inserts, etc.

    c. Lifting and erection inserts.

    d. Dimensions and finishes.

    e. Prestress for strand and concrete strengths.

    f. Estimated cambers.

    g. Method of transportation.

B. Product design criteria

  1. Loadings for design

    a. Initial handling and erection stresses.

    b. All dead and live loads as specified on the contract drawings.

    c. All other loads specified for member where they are applicable.

  2. Design calculations of products not completed on the contract drawings shall be performed by a registered engineer experienced in precast prestressed concrete design and submitted for approval upon request.

  3. Design shall be in accordance with applicable codes, ACI 318, or AASHTO Standard Specifications for Highway Bridges.

C. Permissible design deviations

  1. Design deviations will be permitted only after the Architect/Engineer's written approval of the manufacturer's proposed de-

1.03.A.2 Production drawings are normally submitted only upon request.

1.03.B and C.　The design and architectural drawings normally will be prepared using a local precast prestressed concrete manufacturer's design data and load tables. Dimensional changes which would not materially affect architectural and structural properties or details usually are permissible.

Most precast prestressed concrete is cast in continuous steel forms, therefore connection devices on the formed surfaces must be contained within the member since penetration of the form is impractical.

Camber will generally occur in prestressed concrete members having eccentricity of the stressing force. If camber considerations are important, check with your local prestressed concrete manufacturer to secure estimates of the amount of camber and of camber movement with time and temperature change.

Architectural details must recognize the existence of this camber and camber movement in connection with:

sign supported by complete design calculations and drawings.

2. Design deviations shall provide an installation equivalent to the basic intent without incurring additional cost to the owner.

(1) Closures to interior non-load bearing partitions.

(2) Closures parallel to prestressed concrete members (whether masonry, windows, curtain walls or others) must be properly detailed for appearance.

(3) Floor slabs receiving cast-in-place topping. The elevation of top of floor and amount of concrete topping must allow for camber of prestressed concrete members.

Design cambers less than obtained under normal design practices are possible but this usually requires the addition of tendons or non-prestressed steel reinforcement and price should be checked with the local manufacturer.

As the exact cross section of precast prestressed members might vary somewhat from producer to producer, permissible deviations in member shape from that shown on the contract drawings might enable more manufacturers to quote on the project. Manufacturing procedures also vary between plants and permissible modifications to connection details, inserts, etc., will allow the manufacturer to use devices he can best adapt to his manufacturing procedure.

Be sure that loads shown on the contract drawings are easily interpreted. For instance, on members which are to receive concrete topping, be sure to state whether all superimposed dead and live loads on precast prestressed members do or do not include the weight of the concrete topping.

It is best to list the live load, superimposed dead load, topping weight, and weight of the member, all as separate loads. Where there are two different live loads (e.g., roof level of a parking structure) indicate how they are to be combined.

D. Test reports: Reports of tests on concrete and other materials upon request.

2. PRODUCTS

2.01 Materials

A. Portland Cement:

   1. ASTM C150 — Type I or III.

B. Admixtures:

   1. Air-Entraining Admixtures: ASTM C260.

   2. Water Reducing, Retarding, Accelerating Admixtures: ASTM C494.

C. Aggregates:

   1. ASTM C33 or C330.

2.01.   Delete or add materials that may be required for the particular job.

D. Water:

Potable or free from foreign materials in amounts harmful to concrete and embedded steel.

E. Reinforcing Steel:

  1. Bars:

    Deformed Billet Steel: ASTM A615.

    Deformed Rail Steel: ASTM A616.

    Deformed Axle Steel: ASTM A617.

    Deformed Low Alloy Steel: ASTM A706.

  2. Wire:

    Cold Drawn Steel: ASTM A82.

  3. Wire Fabric:

    Welded Steel: ASTM A185.

    Welded Deformed Steel: ASTM A497.

F. Strand:

  1. Uncoated, 7-Wire, Stress-Relieved Strand: ASTM A416 — Grade 250K or 270K.

G. Anchors and Inserts:

  1. Materials:

    a. Structural Steel: ASTM A36.

    b. Malleable Iron

    c. Stainless Steel: ASTM A666.

  2. Finish:

    a. Shop primer: Manufacturer's standards.

    b. Hot Dipped Galvanized: ASTM A153.

    c. Zinc-Rich Coating: MIL-P-2135, self curing, one component, sacrificial.

    d. Cadmium Coating.

H. Grout:

  1. Cement grout: Portland cement, sand, and water sufficient for placement and hydration.

  2. Non-shrink grout: Premixed, packaged ferrous and non-ferrous aggregate shrink-resistant grout.

  3. Epoxy-resin grout: Two-component mineral-filled epoxy-polysulfide, FS MMM-G-560 _____, Type _____, Grade C.

I. Bearing Pads:

  1. Elastomeric: Conform to Division 2, Section 25 of AASHTO Standard Specifications for Highway Bridges.

2.01.E.1. When welding of bars is required, weldability must be established to conform to AWS D12.1.

2.01.G.1.b. Usually specified by type and manufacturer.

2.01.H. Indicate required strengths on contract drawings.

2.01.H.2. Non-ferrous grouts with a gypsum base should not be exposed to moisture. Ferrous grouts should not be used where possible staining would be undesirable.

2.01.H.3. Check with local suppliers to determine availability and types of epoxy-resin grouts.

2.01.I.1. Pads specified have a strength of 2500 psi. For many applications, commercial grade pads are adequate and are more economical, but

GUIDE SPECIFICATIONS

2. Tetraflouroethylene (TFE) reinforced with glass fibers and applied to stainless or structural steel plates.

J. Welded Studs: In accordance with AWS D1.1.

2.02 Concrete Mixes

A. 28-day compressive strength: Minimum of _____ psi.

B. Release strength: Minimum of _____ psi.

C. Use of calcium chloride, chloride ions or other salts is not permitted.

2.03 Manufacture

A. Manufacturing procedures shall be in general compliance with PCI MNL-116.

B. Manufacturing Tolerances:
   1. Standard tolerances:
      a. Length: ±3/4 in., or ±1/8 in. per 10 ft of length, whichever is greater.
      b. Cross sectional dimensions: Less than 24 in., ±3/8 in.; 24 to 36 in., ±1/2 in.; over 36 in., ±5/8 in.
      c. Flange thickness: (thin sections) ±1/4 in.
      d. Position of anchors and inserts: ±1 in. of center line location shown on drawings.
      e. Horizontal alignment (sweep): 1/2 in., or 1/8 in. per 10 ft length, whichever is greater. Maximum of 1 in. gap between two adjacent members due to sweep.
      f. End squareness: 1/2 in. maximum.
      g. Blockouts: ±1 in. of center line location shown on drawings.
      h. Camber deviation at midspan from design: ±3/16 in. per 10 ft length, ±5/8 in. maximum.
      i. Camber differential at midspan between adjacent members, after installation: 1/4 in. per 10 ft length, 3/4 in. maximum.
      j. Position of reinforcement designed primarily for connections: −1/4 in., +1/2

NOTES TO SPECIFIERS

strengths vary and should be determined in advance by the specifier.

Bearing strips of hard plastic or pressed, non-staining hardboard are acceptable for hollow-core or solid slabs and they are more economical than elastomeric bearing pads.

2.01.I.2 ASTM D2116 applies only to basic TFE resin molding and extrusion material in powder or pellet form. Physical and mechanical properties must be specified by naming manufacturer or other methods.

2.02.A. and B. Verify with local manufacturer. Minimum of 5000 psi for prestressed products is normal practice, with minimum release strength of 3500 psi.

2.03.B.1. These tolerances are less restrictive than tolerances in PCI MNL-116 for some members. If specifier has only double tees, single tees, tee or keystone joists, hollow-core slabs, building beams, or piles, he may state that manufacturer's tolerances shall comply with PCI MNL-116.

in. (minus represents a reduction in cover).

2. Special Tolerances (if indicated on structural drawings):

a. Length: ±1/8 in. per 10 ft, ±1/4 in. maximum.

b. Cross sectional dimensions: Less than 24 in., ±1/4 in.; 24 to 36 in. ±3/8 in.; over 36 in., ±1/2 in.

c. Thickness: ±1/4 in.

d. Position of anchors and inserts: ±1/2 in. of center line location shown on drawings.

e. Horizontal alignment (sweep): 1/4 in., or 1/8 in. per 10 ft length, whichever is greater. Maximum of 1/2 in. gap between two adjacent members due to sweep.

f. End squareness: 3/8 in. maximum.

g. Blockouts: ±1/2 in. off center line locations shown on drawings.

h. Out of square: 1/8 in. per 6 ft measured on the diagonal.

i. Warpage, after installation: 1/8 in. per 6 ft length, or 3/8 in., whichever is greater.

C. Finishes

1. Standard Underside: Resulting from casting against approved forms using good industry practice in cleaning of forms, design of concrete mix, placing and curing. Small surface holes caused by air bubbles, normal color variations, normal form joint marks, and minor chips and spalls will be tolerated, but no major or unsightly imperfections, honeycomb, or other defects will be permitted.

2. Standard Top: Result of vibrating screed and additional hand finishing at projections. Normal color variations, minor indentations, minor chips and spalls will be permitted. No major imperfections, honeycomb, or defects will be permitted.

3. Exposed Vertical Ends: Strands shall be recessed and the ends of the member will receive sacked finish.

2.03.B.2 Special tolerances involve additional expense to the manufacturer and will result in higher costs to the project. If special tolerances are required for some or all of the products, special notes should be made on the structural drawings noting which pieces or dimensions require special tolerances. Tolerances tighter than the special tolerances indicated can only be obtained through the use of non-standard special forms and will involve considerable additional cost. Please consult the manufacturer before specifying such tolerances.

2.03.C. Other formed finishes which may be specified are:

Commercial Finish. Concrete may be produced in forms that impart a texture to the concrete (e.g., plywood or lumber). Fins and large protrusions shall be removed and large holes shall be filled. All faces shall have true, well-defined surfaces. Any exposed ragged edges shall be corrected by rubbing or grinding.

Architectural Grade B Finish. All air pockets and holes over 1/4 in. in diameter shall be filled with a sand-cement paste. All form offsets or fins over 1/8 in. shall be ground smooth.

Architectural Grade A Finish. In addition to the requirements for Architectural Grade B Finish, all exposed surfaces shall be coated with a neat cement paste using an acceptable float. After thin pastecoat has dried, the surface shall be rubbed vigorously with burlap to remove loose particles. These requirements are not applicable to extruded products using zero-slump concrete in their process.

4. Special Finish: If required, listed as follows:

D. Openings: Primarily on thin sections, the manufacturer shall provide for those openings 10 in. round or square or larger as shown on the structural drawings. Other openings shall be located and field drilled or cut by the trade requiring them after the precast prestressed products have been erected. Openings shall be approved by Architect/Engineer before drilling or cutting.

E. Patching: Patching will be acceptable providing the structural adequacy of the product and the appearance are not impaired.

F. Fasteners: The manufacturer shall cast in structural inserts, bolts and plates as detailed or required by the contract drawings.

3. EXECUTION

3.01 Product Delivery, Storage, and Handling

A. Delivery and Handling

1. Precast concrete members shall be lifted and supported during manufacturing, stockpiling, transporting and erection operations only at the lifting or supporting points, or both, as shown on the contract and shop drawings, and with approved lifting devices. All lifting devices shall have a minimum safety factor of 4.

2. Transportation, site handling, and erection shall be performed with acceptable equipment and methods, and by qualified personnel.

B. Storage:

1. Store all units off ground.

2. Place stored units so that identification marks are discernible.

3. Separate stacked members by battens across full width of each bearing point.

4. Stack so that lifting devices are accessible and undamaged.

5. Do not use upper member of stacked tier as storage area for shorter member or heavy equipment.

2.03.C.4. Special finishes, if required, should be described in this section of the specifications and noted on the contract drawings, pointing out which members require special finish. Such finishes will involve additional cost and consultation with the manufacturer is recommended. A sample of such finishes should be made available for review prior to bidding.

2.03.D This paragraph requires other trades to field drill holes needed for their work, and such trades should be alerted to this requirement through proper notation in their sections of the specifications. Some manufacturers prefer to install openings smaller than 10 in. which is acceptable if their locations are properly identified on the structural drawings.

2.03.F. Exclude this requirement from extruded sections.

3.02 Erection

A. Site Access: The General Contractor shall be responsible for providing suitable access to the building and firm level bearing for the hauling and erection equipment to operate under their own power.

B. Preparation: The General Contractor shall be responsible for:

1. Providing true, level bearing surfaces on all field placed bearing walls and other field placed supporting members.

2. Placement and accurate alignment of anchor bolts, plates or dowels in column footings, grade beams and other field placed supporting members.

3. All shoring required for composite beams and slabs. Shoring shall have a minimum load factor of 1.5 x (dead load plus construction loads).

C. Installation: Installation of precast prestressed concrete shall be performed by the manufacturer or a competent erector. Members shall be lifted by means of suitable lifting devices at points provided by the manufacturer. Temporary shoring and bracing, if necessary, shall comply with manufacturer's recommendations.

D. Alignment: Members shall be properly aligned and leveled as required by the approved shop drawings. Variations between adjacent members shall be reasonably leveled out by jacking, loading, or any other feasible method as recommended by the manufacturer and acceptable to the Architect/Engineer.

3.03 Field Welding

A. Field welding is to be done by qualified welders using equipment and materials compatible to the base material.

3.04 Attachments

A. Subject to approval of the Architect/Engineer, precast prestressed products may be drilled or "shot" provided no contact is made with the prestressing steel. Should spalling occur, it shall be repaired by the trade doing the drilling or the shooting.

3.05 Inspection and Acceptance

A. Final inspection and acceptance of erected precast prestressed concrete shall be made by Architect/Engineer to verify conformance with plans and specifications.

3.02.B. Construction tolerances for cast-in-place concrete, masonry, etc., should be specified in those sections of the specifications.

3.02.D. The following erection tolerances may be specified if other requirements do not control: Individual pieces are considered plumb, level and aligned if the error does not exceed 1:500 excluding structural deformations caused by loads.

CODE OF STANDARD PRACTICE
FOR PRECAST CONCRETE

The precast concrete industry has grown very rapidly and certain practices relating to the design, manufacture and erection of precast concrete have become standard in many areas of North America. This "Code of Standard Practice" is a compilation of these practices, and others deemed worthy of consideration, in the form of recommendations for the guidance of those involved with the use of structural and architectural precast concrete.

The primary goal of this Code is to build better understanding by suggesting standards which more clearly define procedures and responsibilities, thus resulting in fewer problems for everyone involved in the planning, preparation and completion of any project.

As the precast concrete industry continues to evolve, and it becomes apparent that additional practices have become standard in the industry or that current standards require modification, it is the intent of the Prestressed Concrete Institute to enlarge and revise this Code.

1. DEFINITIONS OF PRECAST CONCRETE

1.1 Structural Precast Concrete

Structural precast concrete usually includes beams, tees, joists, purlins, girders, lintels, columns, posts, piers, piles, slab or deck members, and wall panels. In order to avoid misunderstandings, it is important that the contract documents for each project list all the elements that are considered to be structural precast concrete.

1.2 Architectural Precast Concrete

Architectural precast concrete usually includes all precast elements that require architectural finishes and/or exhibit decorative exposed surfaces. Typical architectural precast concrete elements include wall panels, window wall panels, mullions and column covers.

In order to avoid misunderstandings, it is important that the contract documents for each project list all the elements that are considered to be architectural precast concrete.

1.3 Prestressed Concrete

Both types of precast concrete may be prestressed or non-prestressed. All structural precast concrete products referred to herein which are prestressed, are specifically referred to as prestressed concrete.

2. SAMPLES, MOCKUPS, AND QUALIFICATION OF MANUFACTURERS

2.1 Samples and Mockups

Samples, mockups, etc., are rarely required for structural prestressed concrete. If samples are required, they should be described in the contract documents and the samples should be manufactured in accordance with Section 5.1.4, *PCI Architectural Precast Concrete.**

2.2 Qualification of Manufacturer

Manufacture, transportation, erection and testing should be accomplished by a company, firm, corporation, or similar organization specializing in providing precast products and services normally associated with structural and architectural precast concrete construction.

The manufacturer may be requested to list similar and comparable work successfully completed by him, and adequacy of plant capability and facilities for performance of contract requirements.

Standards of performance are given in the PCI quality control manuals MNL 116-77 and MNL 117-77. Current certification under the PCI Plant Certification Program is normally accepted as fulfilling experience and plant capability requirements.

3. CONTRACT DOCUMENTS AND DESIGN RESPONSIBILITY

3.1 Contract Documents

Prior to initiation of the engineering-drafting function, the manufacturer should have the following contract documents at his disposal:

1. Architectural drawings.

2. Structural drawings.

3. Electrical, mechanical and plumbing drawings (if pertinent).

4. Specifications (complete with addenda).

Other pertinent drawings may also be desirable, such as shop drawings from other trades, roofing requirements, alternates, etc.

---

*Available from Prestressed Concrete Institute

3.2 Design Responsibilities

It is the responsibility of the owner* to keep the manufacturer supplied with up-to-date documents and written information. The manufacturer should not be held responsible for problems arising from the use of outdated or obsolete contract documents. If updated documents are furnished, it may also be necessary to modify the contract.

The contract documents should clearly define the following:

1. Items furnished by manufacturer.

2. Size, location and function of all openings, blockouts, and cast-in items.

3. Production and erection schedule requirements and restrictions.

4. Design intent including connections and reinforcement.†

5. Allowable tolerances and deviations. Normal field tolerances should be recommended by the manufacturer.

6. Dimension, material and quantity requirements.

7. General and supplemental general conditions.

8. Any other special requirements and conditions.

9. Site plan showing storage areas to be used, parking areas for trucks and equipment, etc.

Other design responsibility relationships are described in Table 3.10.1, *PCI Architectural Precast Concrete.*

4. ERECTION AND PRODUCTION DRAWINGS

4.1 Erection Drawings

The information provided in the contract documents is used by the manufacturer to prepare erection drawings for approval and field use. They contain:

1. Plans and/or elevations locating and dimensioning all members furnished by manufacturer.

2. Sections and details showing connections, fin-

ishes, openings, blockouts and cast-in items and their relationship to the structure.

3. Description of all loose and cast-in hardware including designation of who furnishes it.

4. Drawings showing location of anchors installed in the field.

5. Erection sequences and handling requirements.

4.2 Production Drawings

The contract documents are also used to prepare production drawings for manufacturing showing all dimensions together with locations and quantities for all cast-in materials (reinforcement, inserts, etc.) and completely defining all finish requirements.

Normal practices for the preparation of drawings for architectural precast concrete are described in the *PCI Architectural Precast Concrete Drafting Handbook.*

4.3 Discrepancies

When discrepancies or omissions occur on the contract documents, the manufacturer has the responsibility to check with the engineer to resolve the problem. If this is not possible, the following procedures are normally followed:

1. Contract terms govern over specifications and drawings.

2. Specifications govern over drawings.

3. Structural drawings govern over architectural drawings.

4. Written dimensions govern over scale dimensions.

5. Sections govern over plans or elevations.

6. Details govern over sections.

Graphic verification should be requested for any unclear condition.

4.4 Approvals

Completed erection drawings, usually in reproducible form, should be submitted for approval. The exact sequence is dictated by construction schedules and erection sequences, and is determined when the contract is awarded.

Production drawings should not be started prior to receipt of approved or approved-as-noted erection drawings. Production drawings should be submitted for approval only when so requested.

Corrections should be noted on the reproducible erection drawings and copies made for distri-

---

*The owner of the proposed structure or his designated representatives, who may be the architect, engineer, general contractor, public authority or others contracting with the precast manufacturer.

†When the manufacturer accepts design responsibility, the area or amount of responsibility must be clearly defined in the contract documents. The engineer or architect of record must be identified and it is understood that all designs are submitted through him for his approval and acceptance. The manufacturer's responsibility can be limited to member design only or it may include the entire structure.

bution.

The following approval interpretation is normal practice:

1. **Approved** — The approvers* have completely checked and verified the drawings for conformance with contract documents and all expected loading conditions. Such approval should not relieve the manufacturer from responsibility for his design when that responsibility is placed upon him by the contract. The manufacturer may then proceed with production drawings without resubmittal. Erection drawings may then be released for field use and plant use.

2. **Approved as Noted** — Same as above except that noted changes should be made and corrected erection drawings issued. Production drawings and production may be started after noted changes have been made.

3. **Not Approved** — Drawings must be corrected and resubmitted. Production drawings should not be started until "approved" or "approved as noted" erection drawings are returned.

5. MATERIALS

The relevant ASTM Standards that apply to materials for a project should be listed in the contract documents together with any special requirements that are not included in the ASTM Standards.

Note: Additional information regarding material specifications can be found in "Guide Specification for Precast Prestressed Concrete Construction for Buildings" (see pp. 7–2 to 7–10) and "Guide Specification for Architectural Precast Concrete."†

6. TESTS AND INSPECTIONS

6.1 Tests of Materials

Manufacturers generally keep the test records required by the PCI manuals for quality control (MNL-116 and MNL-117). The contract documents may require the precast concrete manufacturer to make these records available for inspection by the owner's representative upon his request.

When the manufacturer is required to submit copies of test records to the owner and/or required to perform or have performed tests **not** required by MNL-116 and MNL-117, these special testing requirements should be clearly described in the contract documents along with the responsibility for payment.

6.2 Inspections

On certain projects the owner may require inspection of precast concrete products in the manufacturer's yard by persons other than the manufacturer's own quality control personnel. Such inspections are normally made at the owner's expense. The contract documents should describe how, when and by whom the inspections are to be made, and who is to pay for them. Alternatively, the owner may accept plant certification in lieu of outside inspection, as provided in the PCI Plant Certification Program.

6.3 Fire Rated Products

If the manufacturer is expected to provide a fire rated product and/or labels, these requirements should be clearly stated in the contract documents.

7. FINISHES

Finishes on precast concrete products, both structural and architectural, are probably the cause of more misunderstandings between the various members of the building team than any other question concerning product quality.

It is therefore extremely important that the contract documents describe clearly and completely the required finishes for all surfaces of all members, and that the erection drawings also include this information. When finish is not specified, the standard finish described in "Guide Specification for Precast Prestressed Concrete Construction for Buildings" should normally be furnished.

For descriptions of the usual finishes for structural precast concrete, see "Guide Specification for Precast Prestressed Concrete Construction for Buildings," and for architectural precast concrete, see Sect. 4.3, *PCI Architectural Precast Concrete.* Where special or critical requirements exist or where large expanses of exposed precast will occur on a project, samples are essential and, if required, should be so stated and described in the contract documents.

8. DELIVERY OF MATERIALS

8.1 Manner of Delivery

The manufacturer should deliver the precast concrete to the erector* in a manner to facilitate the speed of erection of the building or as mutually agreed upon between the owner, manufacturer and erector. Special requirements of the owner for the delivery of materials or the mode of transport, should be stated in the contract documents.

---

*The contract should state who has approval authority.

†*PCI Journal,* November-December, 1977.

---

*The erector may be either the manufacturer or a subcontractor engaged by the manufacturer or general contractor.

8.2 Marking and Shipping of Materials

The precast concrete members should be separately marked in accordance with erection drawings in such a manner as to distinguish varying pieces and to facilitate erection of the structure. Any members which require a sequential erection should be properly marked.

The owner should give the manufacturer sufficient time to fabricate and ship any special plates, bolts, anchorage devices, etc., contractually agreed to be furnished by the manufacturer.

8.3 Precautions During Delivery

Special protection or precautions beyond that required in MNL-116 and MNL-117 should not be expected unless stated in the bid invitation or specifications. The manufacturer is not responsible for the product after delivery to the site unless required by the contract documents.

8.4 Access to Jobsite

Free and easy access to the delivery site should be provided to the manufacturer, including backfilling and compacting, adequate drainage and snow removal, so that delivery trucks can operate under their own power.

8.5 Unloading Time Allowance

Delivery of product includes a reasonable unloading time allowance. Any delay beyond a reasonable time is normally paid for by the party which is responsible for the delay.

9. ERECTION

9.1 Special Erection Requirements

When the owner requires a particular method or sequence of erection, this information should be stated in the contract documents.

9.2 Tolerances

Some variation is to be expected in the overall dimensions of any building or other structure. It is common practice for the manufacturer and erector to work within the tolerances recommended by the American Concrete Institute and Prestressed Concrete Institute.

The owner, by whatever agencies he may elect, immediately upon completion of the erection, should determine if the work is plumb, level, aligned and properly fastened. Discrepancies should immediately be brought to the attention of the erector so that proper corrective action can be taken.

The work of the manufacturer and erector is complete once the building has been properly plumbed, leveled and aligned within the established tolerances. Acceptance for this work should be secured from the authorized representative of the general contractor (see Section 11.2).

9.3 Foundations, Piers, Abutments and Other Bearing Surfaces

The invitation to bid should state the anticipated time when all foundations, piers and abutments will be ready and accessible to the erector. Final scheduling should be coordinated with the general contractor.

9.4 Building Lines and Bench Marks

The precast manufacturer should be furnished a drawing on which all building lines and bench marks at the site of the structure are accurately located.

9.5 Anchor Bolts and Bearing Plates

The precast manufacturer does not normally furnish or install anchor bolts, plates, etc. that are to be installed in cast-in-place concrete or masonry for connection with precast members, however, if this is to be the responsibility of the precast manufacturer, it should be so defined in the specifications. It is important that such items be installed true to line and grade, and that installation be completed in time to avoid delays or interference with the precast erection.

Erectors should check both line and grade in sufficient time before erection is scheduled to permit any necessary corrections. Corrections, if any, should be made by the general contractor before erection begins

9.6 Utilities

Water and electricity should be furnished for erection and grouting operations by the owner.

9.7 Working Space

The owner should furnish adequate, well drained, convenient working space for the erector and access for his equipment necessary to assemble the structure. The owner should provide adequate storage space for the precast products to enable the erector to operate at the speed required to meet the established schedule. Unusual hazards such as high voltage lines, buried utilities, or areas of restricted access should be declared in the invitation to bid.

9.8 Materials of Other Trades

Other building materials or work of other trades preferably should not be built up above the bearing of the precast concrete until after erection of the precast.

9.9 Correction of Errors

Corrections of minor misfits are considered a part of erection even if the precast concrete is not erected by the manufacturer. Any error in manufacturing which prevents proper connection or fitting should be immediately reported to the manufacturer and the engineer and/or owner so that corrective action can be taken.

9.10 Field Assembly

The size of assembled pieces of precast concrete may be limited by transportation requirements for weight and clearance dimensions. Unless agreed upon between the manufacturer and owner, the manufacturer should provide for such field connections that will meet required loads and forces without altering the function or appearance of the structure.

All loose materials for temporary and permanent connection of structural precast members are normally furnished by the erector. The manufacturer furnishes only those items embedded in the precast. Temporary guys, braces, falsework, and cribbing are the property of the erector and are removed only by the erector or with the erector's approval upon completion of the erection of the structure, unless otherwise agreed.

9.11 Blockouts, Cuts and Alterations

Neither the manufacturer nor the erector is responsible for the blockouts, cuts or alterations by or for other trades unless so specified in the contract documents. Whenever such additional work is required, all information regarding size, location and number of alterations is furnished by the owner prior to preparation of the precast production and erection drawings.

The general contractor is responsible for warning other trades against indiscriminate cutting of prestressed concrete members.

9.12 Temporary Floors and Access

The precast concrete manufacturer or erector is not required to furnish temporary flooring for access unless so specified in the contract documents.

9.13 Painting, Grouting, Caulking and Closure Panels

Painting, grouting, caulking and placing of closure panels between stems of flanged concrete members are services not ordinarily supplied by the manufacturer or erector. If any of these services are required of the manufacturer, it should be stated in the contract documents.

9.14 Patching

A certain amount of patching of product is to be expected to repair minor spalls and chips. Required patching should meet the finish requirements of the project and color should be reasonably matched. Responsibility for accomplishing this work should be resolved between the manufacturer and erector.

9.15 Safety

Safety procedures for the erection of the precast concrete members is the responsibility of the erector and must be in accordance with all local, state or Federal rules and regulations which have jurisdiction in the area where the work is to be performed, but not less than required in ANSI Standard A 10.9, *American National Standard Safety Requirements for Concrete Construction and Masonry Work.**

9.16 Security Measures

Security protection at the job site should be the responsibility of the general contractor.

10. INTERFACE WITH OTHER TRADES

Coordination of the requirements for other trades to be included in the precast should be the responsibility of the owner unless clearly defined otherwise in the contract documents.

The PCI manuals for quality control (MNL-116 and MNL-117) specify manufacturing tolerances for precast concrete members. Interfaces with other materials and trades must take these tolerances into account. Unusual requirements or allowances for interfacing should be stated in the contract documents.

11. WARRANTY AND ACCEPTANCE

11.1 Warranties

Warranties of product and workmanship have become a widely accepted practice in this industry as in most others. Warranties given by the precast concrete manufacturer and erector should indicate that their product and work meet the design criteria and specifications for the project.

In no case should the warranty of the manufacturer and erector be in excess of of the warran-

---

*American National Standards Institute, New York, New York.

ty required by the project specifications. Warranties should in all instances include a time limit and it is recommended that this should not exceed one year.

In order to protect the interests of all parties concerned, warranties should also state that any deviations in the designed use of the product, modifications of the product by the owner and/or contractor or changes in other products used in conjunction with the manufacturer's product will cause said warranty to become null and void.

Warranty may be included as a part of the conditions of the contract agreement, or it may be presented in letter form, as requested by the owner. A sample warranty follows:

Sample warranty

Manufacturer warrants that all materials furnished have been manufactured in accordance with the design criteria and specifications for this project. Manufacturer further warrants that if erection of said material is to be performed by those subject to his control and direction, work will be completed in accordance with the same design criteria and specifications.

In no event shall manufacturer be held responsible for any damages, liability or costs of any kind or nature occasioned by or arising out of the actions or omissions of others, or for work, including design, done by others; or for material manufactured, supplied or installed by others; or for inadequate construction of foundations, bearing walls, or other units to which materials furnished by the precast manufacturer are attached or affixed.

This warranty ceases to be in effect beyond the date of _____. Should any defect develop during the contract warranty period, which can be directly attributed to defect in quality of product or workmanship, precast manufacturer shall, upon written notice, correct defects or replace products without expense to owner and/or contractor.

COMPANY NAME

_____
Signature   Title

11.2 Acceptance

Manufacturer should request approval and acceptance for all materials furnished and all work completed by him periodically as deemed necessary in order to adequately protect the interests of everyone involved in the project. In most cases, the size and nature of the project will dictate the proper intervals for securing approval and acceptance. Periodic approval in writing should be considered when it appears that such action will minimize possible problems which would seriously af-

fect the progress of the project. A sample acceptance form follows:

Sample acceptance form

FIELD INSPECTION REPORT

Project # _____

On this _____ day of _____, 19 _____,

_____ of _____
Company Field Superintendent   Precast Manufacturer

and _____ of _____
General Contractor Superintendent   General Contractor

_____ have inspected _____

_____
portion of building being inspected

All of the work performed by the above indicated company in the above described portion of the project has been performed to the satisfaction of the above named General Contractor's Superintendent with the exception of the following: _____

_____

The above named General Contractor's Superintendent hereby releases the above indicated Precast Manufacturer of its responsibility to perform any other work in the above described portion of the project except as detailed herein.

The above mentioned Precast Manufacturer in turn hereby releases the above described portion of the project to the General Contractor and his subcontractors for their work.

_____   _____
Precaster's Superintendent   Gen'l. Contr. Superintendent

12. CONTRACT ADMINISTRATION

12.1 General Statement

Information relative to invoicing, payment, bonding and other data pertinent to a project or material sale should be specifically provided for in the major provisions of the contract documents or in the special terms and conditions applicable to all contractual agreements between manufacturer and owner.

Contract agreements may vary widely from area to area, but the objective should be the same in all instances. The contract agreement should be written to protect the interests of all parties concerned and at the same time, be specific enough in content to avoid misunderstandings once the project begins.

The intent of this section is to recommend those things which ought to be considered, but not necessarily the form in which they should be expressed. The final statement of policies should be the result of careful consideration of all pertinent factors as well as of the normal practices in the area.

12.2 Retentions

Although retentions have been used for many years as a means of assuring a satisfactory job performance, it is apparent that they directly contribute to the cost of construction, frequently lead to disputes, and often result in job delays. In view of the unfavorable consequences of retentions and possible abuse, it is recommended that the following procedure be followed:

1. Wherever possible, retentions should be eliminated and bonding should be used as the single, best source of protection. This should apply to prime contractors and subcontractors equally.

2. Where there are no bonding requirements, the retention percentage should be as low as possible. It is recommended that this be not more than 5 percent.

3. The percentage level of any retention should be the same for subcontractors as for prime contractors on a job.

4. Release of retained funds and final payment, as well as computing the point of reduction of the retention, should be done on a line item basis, that is, each contractor or subcontractor's work considered as a separate item and the retention reduced by 50 percent upon substantial completion and the balance released within 30 days after final completion of his own work.

5. Retained funds should be held in an escrow account with interest accruing to the benefit of the party to whom the funds are due.

6. When materials are furnished FOB plant or jobsite, it is recommended that there be no retentions.

12.3 Contract Agreement

1. Contract agreement should fully describe the project involved, including job location, project name, name of owner/developer, architect or other principals and all reference numbers identifying job relation information such as plans, specifications, addenda, bid number, etc.

2. Contract agreement should fully describe the materials to be furnished and/or all work to be completed by the seller.

3. All exclusions should be stated to avoid the possibility of any misunderstanding.

4. Price quoted should be stated to eliminate any possibility of misunderstanding.

5. Reference should be made to the terms and conditions governing the proposed contract agreement. The terms and conditions may best be stated on the reverse side of the contract form. Special terms or conditions should be stated in sufficient detail to avoid the possibility of misunderstanding.

6. The terms of payment should be specifically detailed so there is no doubt in anyone's mind as to the intent. Special care should be exercised where the terms of payment will differ from those normally in effect or where they deviate from the general terms and conditions appearing on the reverse side of the contract form.

7. A statement of policy should be made with reference to the inclusion or exclusion of taxes in the stated price.

8. The proposal form stating the full intent and conditions under which the project will be performed may contain an acceptance clause to be signed by the purchaser. At such time as said acceptance clause is signed, the proposal form then becomes the contract agreement.

9. Seller should clearly state the limits of time within which an accepted proposal will be recognized as a binding contract. To protect all concerned, this time limit should not be extended beyond a reasonable period.

10. A statement indicating the classification of labor to perform the work in the field is advisable to eliminate later dispute over jurisdiction of work performed.

12.4 Terms and Conditions

The terms and conditions stated on the proposal contract agreement should include, but are not necessarily limited to, the following:

1. **Lien Laws** — Where the lien laws of a state specifically require advance notice of intent, it is advisable to include the required statement in the general terms and conditions.

2. **Specifications** — Seller should make a specific declaration of material and/or work specifications, but normally this should not be in excess of the specifications required by the contract agreement.

3. **Contract Control** — A statement should be made indicating that the agreement when duly signed by both parties supersedes and invalidates any verbal agreement and can only be modified in writing with the approval of those signing the original agreement.

4. **Terms of Payment** — Terms of payment should be specifically stated either on the face of the contract or in the general terms and conditions. Mode and frequency of invoicing should be so stated, indicating time within which payment is expected.

5. **Late Payment Charges** — The contract may provide for legal interest charges for late payments not made in accordance with contract terms, and if this is desired, it should be stated in the general terms and conditions. A statement indicating seller is entitled to reasonable attorney's fees and related costs should collection proceedings be necessary may also be included.

6. **Overtime Work** — Prices quoted in the proposal should be based on an 8-hour day and a 5-day week under prevailing labor regulations. Provisions should be included in the contract agreement to provide for recovery of overtime costs plus a reasonable markup when the seller is requested to provide such service.

7. **Financial Responsibility** — General terms and conditions may indicate the right of the seller to suspend or terminate material delivery and/ or work on a project if there is a reasonable doubt of the ability of the purchaser to fulfill his financial responsibility.

8. **Payment for Inventory**

   (a) It has become common practice to include in the contract terms and conditions provisions for the invoicing and payment of all materials stored at the plant or jobsite when deliveries or placement of said materials are delayed for more than a stipulated time beyond the originally scheduled date because of purchaser's inability either to accept delivery of materials or to provide proper job access.

   (b) Under certain conditions, it may be necessary to purchase special materials and to produce components well in advance of job requirements to insure timely deliveries. When job requirements are of such a nature, it is advisable to include provisions for payment of such raw and finished inventories stored in seller's plant or on jobsites on a current basis.

9. **Payment for Suspended or Discontinued Projects** — The terms and conditions should provide that in the event of a discontinued or suspended project, seller shall be entitled to payment for all material manufactured including costs, overhead and profit, and not previously billed as well as reasonable engineering and other costs incurred.

10. **Job Extras** — Requests for job extras should be confirmed in writing. Invoicing should be presented immediately following completion of the extra work with payment subject to the terms and conditions of the contract agreement, or as otherwise stated in the change order.

11. **Claims for Shortages, Damages or Delays** — Seller should, upon immediate notification in writing on the face of the delivery ticket of rejected material or shortage, acknowledge and furnish replacement material at no cost to purchaser. It is normal practice that the seller should not be responsible for any loss, damage, detention or delay caused by fire, accident, labor dispute, civil or military authority, insurrection, riot, flood or by occurrences beyond his control.

12. **Back Charges** — Back charges should not be binding on the seller, unless the condition is promptly reported in writing, and opportunity is given seller to inspect and correct the problem.

13. **Permits, Fees and Licenses** — Costs of permits, fees, licenses and other similar expenses are normally assumed by the purchaser.

14. **Bonds** — Cost of bonds is normally assumed by the purchaser.

15. **Taxes** — Federal, State, County or Municipal, Occupation or similar taxes which may be imposed are normally paid by the purchaser.

16. **Insurance** — Seller shall carry Workmen's Compensation, Public Liability, Property Damage and Auto Insurance and certificates of insurance will be furnished to purchaser upon request. Additional coverage required over and above that provided by the seller is normally paid by the purchaser.

17. **Services** — Heat, water, light, electricity, toilet, telephone, watchmen and general services of a similar nature are normally the responsiblity of the purchaser unless specifically stated otherwise in the contract agreement.

18. **Safety Equipment** — The purchaser is normally responsible for necessary barricades, guard

rails and warning lights for the protection of vehicular and pedestrian traffic and seller's equipment. Purchaser is also normally responsible for furnishing, installing and maintaining all safety appliances and devices required on the project under U.S. Department of Labor, *Safety and Health Regulations for Construction,* as well as all other safety regulations imposed by other agencies having jurisdiction over the project.

19. **Warranty** — Seller should provide specific information relative to warranties given, including limitations, exclusions and methods of settlement. Warranties should not be in excess of warranty required by a specific project.

20. **Title** — Contract should provide for proper identification of title to material furnished. It is normal practice for title and risk of loss or damage to the product furnished to pass to the purchaser at the point of delivery, except in cases of FOB factory, in which event title to and risk of loss of damage to the product normally should pass to purchaser at factory pickup.

21. **Shop Drawings Approval** — Seller should prepare and submit to purchaser for approval all shop drawings* necessary to describe the work to be completed. Shop drawings approval should constitute final agreement to quantity and general description of material to be supplied. No work should be done upon material to be furnished by seller until approved shop drawings and erection drawings are in his possession.

22. **Delivery** — Delivery times or schedules set forth in contract agreements should be computed from the date of delivery to the seller of approved shop drawings. Where materials are specified to be delivered FOB to jobsite, the purchaser should provide labor, cranes or other equipment to remove the materials from the trucks and should pay seller for truck expense for time at the jobsite in excess of a specified time for each truck. On shipments to be delivered by trucks, delivery should be made as near to the construction site as the truck can travel under its own power. In the event delivery is required beyond the curb line, the purchaser should assume full liability for damages to sidewalks, driveways or other properties and should secure in advance all necessary permits or licenses to effect such deliveries.

23. **Builder's Risk Insurance** — Purchaser should provide Builder's Risk Insurance without cost to seller, protecting seller's work, materials and equipment at the site from loss or damage caused by fire or the standard perils of extended coverage, including vandalism and malicious acts.

24. **Erection** — Purchaser should assure that the proposed project will be accessible to all necessary equipment including cranes and trucks, and that the operation of this equipment will not be impeded by construction materials, water, presence of wires, pipes, poles, fences or framings. Purchaser should further indemnify and save harmless the seller and his respective representatives, including subcontractors, vendors, assigns and successors from any and all liability, fine, penalty or other charge, cost or expense and defend any action or claim brought against seller for any failures by purchaser to provide suitable access for work to be performed. Seller also reserves the right to discontinue the work for failure of purchaser to provide suitable access and the purchaser should be responsible for all expenses and costs incurred.

25. **Exclusions of Work to be Performed** — Unless otherwise stated in the contract, all shoring, forming, framing, cutting holes, openings for mechanical trades and other modifications of seller's products should not be performed by the seller nor are they included in the contract price. Seller should not be held responsible for modifications made by others to his product unless said modifications are previously approved by him.

26. **Sequence of Erection** — Sequence of erection should be as agreed upon between seller and purchaser and expressly stated in the contract agreement. Purchaser should have ready all foundations, bearing walls or other units to which seller's material is to be affixed, connected or placed, prior to start of erection. Purchaser should be responsible for the accuracy of all job dimensions, bench marks, and true and level bearing surfaces. Claims or expenses arising from the purchaser's neglect to fulfill this responsibility should be assumed by the purchaser.

27. **Arbitration** — In view of the many difficulties and misunderstandings which may occur due to misinterpretation of contractual documents, it is recommended that the seller stipulate that all claims, disputes and other matters in question, arising out of or related to the

---

*See Section 4 for definition of shop drawings.

contract, be decided by arbitration in accordance with the Construction Industry Arbitration Rules of the American Arbitration Association then obtaining, or some other rules acceptable to both parties. The location for such arbitration should be stipulated.

28. **Contract Form** — Contract documents should stipulate policy governing acceptance of proposal on other than the seller's form. In the event purchaser does not accept the seller's proposal and/or contract agreement, but requires the execution of a contract on his own form, it is advisable that seller stipulate in writing on the contract agreement that the contract will be fulfilled according to his proposal originally submitted. All identifying information such as proposal number, dates, etc. should be included so there can be no question of the document referred to.

PARTIAL BIBLIOGRAPHY
PRECAST AND PRESTRESSED CONCRETE

1. Kurt Billig, *Prestressed Reinforced Concrete;* (London: Knapp, Drewett & Sons, Ltd., 1944).

2. P. W. Abeles, *Principles and Practices of Prestressed Concrete;* (London: Crosby Lockwood & Sons, Ltd., 1949).

3. *Proceedings of the First United States Conference on Prestressed Concrete;* (Massachusetts Institute of Technology, Cambridge, Massachusetts, 1951).

4. A. E. Komendant, *Prestressed Concrete Structures;* (New York: McGraw-Hill Book Co., 1952).

5. F. Walley, *Prestressed Concrete Design and Construction;* (London: Her Majesty's Stationery Office, 1953).

6. Gustave Magnel, *Prestressed Concrete;* (London: Concrete Publications, Ltd., Third Edition, 1954, 345 pp).

7. H. J. Cowan, *The Theory of Prestressed Concrete Design — Statically Determinate Structures;* (New York: St. Martin's Press, 1956, 264 pp).

8. University of California, *Proceedings, World Conference on Prestressed Concrete;* (San Francisco: Lithotype Process Co., July, 1957).

9. R. H. Evans and E. W. Bennett, *Prestressed Concrete;* (New York: Chapman and Hall, Inc., 1958, 293 pp).

10. Y. Guyon, *Prestressed Concrete, Vol. I;* (New York: John Wiley & Sons, Inc., 1953, Fourth Impression, 1960, 559 pp). *Prestressed Concrete, Vol. II — Statically Indeterminate Structures;* (New York: John Wiley & Sons, Inc., 1960, 741 pp).

11. William H. Connolly, *Design of Prestressed Concrete Beams;* (New York: F. W. Dodge Corp., 1960, 252 pp).

12. H. Kent Preston, *Practical Prestressed Concrete;* (New York: McGraw-Hill Book Co., Inc., 1960, 340 pp).

13. James R. Libby, *Prestressed Concrete — Design and Construction;* (New York: The Ronald Press Co., 1961, 468 pp).

14. Michael Chi and Frank A. Biberstein, *Theory of Prestressed Concrete;* (Englewood Cliffs, N.J.: Prentice-Hall, Inc., 1963, 252 pp).

15. T. Y. Lin, *Design of Prestressed Concrete Structures;* (New York: John Wiley & Sons, Inc., Second Edition, 1963, 614 pp).

16. J. D. Harris and I. C. Smith, *Basic Design and Construction in Prestressed Concrete;* (London: Chatto and Windus, 1963, 265 pp).

17. Laurence Cazaly and M. W. Huggins, *Canadian Prestressed Concrete Institute Handbook;* (Don Mills, Ontario, Canada: T. H. Best Printing Co., Ltd., 1964, 501 pp).

18. B. J. Bell, *Practical Prestressed Concrete Design;* (New York: Pitman Publishing Corp., 1964, 353 pp).

19. Fritz Leonhardt, *Prestressed Concrete — Design and Construction;* (Berlin-Munich: Wilhelm Ernst & Sohn, Second Edition, 1964, 677 pp).

20. H. Kent Preston, *Prestressed Concrete for Architects and Engineers;* (New York: McGraw-Hill Book Co., 1964, 196 pp).

21. H. J. Cowan, *Reinforced and Prestressed Concrete in Torsion;* (New York: St. Martin's Press, 1965, 138 pp).

22. P. W. Abeles, *Introduction to Prestressed Concrete, Vol. I and Vol. II;* (London: Concrete Publications, Ltd., Vol. I, 1964, 384 pp; Vol. II, 1966, 355 pp. — available in the U.S. and Canada from Frederick Ungar Publishing Co., Inc., New York).

23. Henry J. Cowan and Peter R. Smith, *The Design of Prestressed Concrete in Accordance with the S.A.A. Code;* (Sydney: Angus and Robertson, Ltd., 1966, 212 pp).

24. H. Kent Preston and Norman J. Sollenberger, *Modern Prestressed Concrete;* (New York: McGraw-Hill Book Co., 1967, 337 pp).

25. C. B. Wilby, *Prestressed Concrete Beams, Design and Logical Analysis;* (New York: American Elsevier Publishing Co., Inc., 1969, 97 pp).

26. Tihamer Koncz, *Manual of Precast Concrete Construction,* Vol 1, Principles, Roof and Floor Units, Wall Panels (248 pp); Vol. 2, Industrial Shed-Type and Low-Rise Buildings, Special Structures (427 pp); Vol. 3, Systems Building with Large Panels (368 pp); (Wiesbaden: Bauverlag GmbH, 1968, 1971 and 1970).

27. Colin O'Connor, *Design of Bridge Superstructures;* (New York: Wiley-Interscience, 1971, 552 pp).

28. Ben C. Gerwick, Jr., *Construction of Prestressed Concrete Structures;* (New York: Wiley-Interscience, 1971, 411 pp).

29. Y. Guyon, *Limit-State Design of Prestressed Concrete,* Vol. 1 — The Design of the Section; (New York: John Wiley & Sons, 1972, 485 pp). Vol. 2 — The Design of the Member; (New York: John Wiley & Sons, 1974, 469 pp).

30. P. W. Abeles, B. K. Bardhan-Roy and F. H. Turner, *Prestressed Concrete Designer's Handbook;* (England: Cement and Concrete Association, Second Edition, 1976, 548 pp).

31. S. C. C. Bate and E. W. Bennett, *Design of Prestressed Concrete;* (New York: John Wiley & Sons, Inc., 1976, 138 pp).

32. James R. Libby and Norman D. Perkins, *Modern Prestressed Concrete Highway Bridge Superstructures;* (New York: Van Nostrand Reinhold Company, 1977, 262 pp).

33. James R. Libby, *Modern Prestressed Concrete;* (New York: Van Nostrand Reinhold Company, Second Edition, 1977, 516 pp).

34. Arthur Nilson, *Design of Prestressed Concrete;* (New York: John Wiley & Sons, Inc., 1978, 560 pp).

PART 8
GENERAL DESIGN INFORMATION

DESIGN INFORMATION

8.1.1 Dead weights of floors, ceilings, roofs, and walls

Floorings	Weight (psf)
Normal weight concrete topping, per inch of thickness	12
Sand-lightweight (120 pcf) concrete topping, per inch	10
Lightweight (90-100 pcf) concrete topping, per inch	8
7/8" hardwood floor on sleepers clipped to concrete without fill	5
1 1/2" terrazzo floor finish directly on slab	19
1 1/2" terrazzo floor finish on 1" mortar bed	30
1" terrazzo finish on 2" concrete bed	38
3/4" ceramic or quarry tile on 1/2" mortar bed	16
3/4" ceramic or quarry tile on 1" mortar bed	22
1/4" linoleum or asphalt tile directly on concrete	1
1/4" linoleum or asphalt tile on 1" mortar bed	12
3/4" mastic floor	9
Hardwood flooring, 7/8" thick	4
Subflooring (soft wood), 3/4" thick	2 1/2
Asphaltic concrete, 1 1/2" thick	18

Ceilings	
1/2" gypsum board	2
5/8" gypsum board	2 1/2
3/4" plaster directly on concrete	5
3/4" plaster on metal lath furring	8
Suspended ceilings	2
Acoustical tile	1
Acoustical tile on wood furring strips	3

Roofs	
Five-ply felt and gravel (or slag)	6 1/2
Three-ply felt and gravel (or slag)	5 1/2
Five-ply felt composition roof, no gravel	4
Three-ply felt composition roof, no gravel	3
Asphalt strip shingles	3
Rigid insulation, per inch	1/2
Gypsum, per inch of thickness	4
Insulating concrete, per inch	3

Walls	Un-Plastered	One side Plastered	Both sides Plastered
4" brick wall	40	45	50
8" brick wall	80	85	90
12" brick wall	120	125	130
4" hollow normal weight concrete block	28	33	38
6" hollow normal weight concrete block	36	41	46
8" hollow normal weight concrete block	51	56	61
12" hollow normal weight concrete block	59	64	69
4" hollow lightweight block or tile	19	24	29
6" hollow lightweight block or tile	22	27	32
8" hollow lightweight block or tile	33	38	43
12" hollow lightweight block or tile	44	49	54
4" brick 4" hollow normal weight block backing	68	73	78
4" brick 8" hollow normal weight block backing	91	96	101
4" brick 12" hollow normal weight block backing	119	124	129
4" brick 4" hollow lt-wt. block or tile backing	59	64	69
4" brick 8" hollow lt-wt. block or tile backing	73	78	83
4" brick 12" hollow lt-wt. block or tile backing	84	89	94
Windows, glass, frame and sash	8		
4" stone	55		
Steel or wood studs, lath, 3/4" plaster	18		
Steel or wood studs, 5/8" gypsum board each side	6		
Steel or wood studs, 2 layers 1/2" gypsum board each side	9		

DESIGN INFORMATION

8.1.2 Recommended minimum floor live loads*

Uniformly Distributed Loads

Occupancy or Use	Live Load (psf)
Apartments (*see* Residential)	
Armories and drill rooms	150
Assembly halls and other places of assembly:	
Fixed seats	60
Movable seats	100
Platforms (assembly)	100
Balcony (exterior)	100
On one and two family residences only and not exceeding 100 sq ft	60
Bowling alleys, poolrooms, and similar recreational areas	75
Corridors:	
First floor	100
Other floors, same as occupancy served except as indicated	
Dance halls and ballrooms	100
Dining rooms and restaurants	100
Dwellings (*see* Residential)	
Fire escapes	100
On multi- or single-family residential buildings only	40
Garages (passenger cars only)**	50
For trucks and buses use AASHTO lane loads	
Grandstands (*see* Reviewing stands)	
Gymnasiums, main floors and balconies	100
Hospitals:	
Operating rooms, laboratories	60
Private rooms	40
Wards	40
Corridors, above first floor	80
Hotels (*see* Residential)	
Libraries:	
Reading rooms	60
Stack rooms (books & shelving at 65 pcf) but not less than	150
Corridors, above first floor	80
Manufacturing:	
Light	125
Heavy	250
Marquees	75
Office buildings:	
Offices	50
Lobbies	100
Corridors, above first floor	80
File and computer rooms require heavier loads based upon anticipated occupancy	
Penal institutions:	
Cell blocks	40
Corridors	100
Residential:	
Multifamily houses:	
Private apartments	40
Public rooms	100
Corridors	80
Dwellings:	
First floor	40
Second floor and habitable attics	30
Uninhabitable attics	20

Occupancy or Use	Live Load (psf)
Residential (cont.)	
Hotels:	
Guest rooms	40
Public rooms	100
Corridors serving public rooms	100
Corridors	80
Reviewing stands and bleachers	100
Schools:	
Classrooms	40
Corridors	80
Sidewalks, vehicular driveways, and yards, subject to trucking	250
Skating rinks	100
Stairs and exitways	100
Storage warehouse:	
Light	125
Heavy	250
Stores:	
Retail:	
First floor, rooms	100
Upper floors	75
Wholesale	125
Theaters:	
Aisles, corridors, and lobbies	100
Orchestra floors	60
Balconies	60
Stage floors	150
Yards and terraces, pedestrians	100

Concentrated Loads

Location	Load (lb)
Elevator machine room grating (on area of 4 sq in)	300
Finish light floor plate construction (on area of 1 sq in)	200
Garages	**
Office floors	2000
Scuttles, skylight ribs, and accessible ceilings	200
Sidewalks	8000
Stair treads (on area of 4 sq in at center of tread)	300

**Floors in garages or portions of buildings used for storage of motor vehicles shall be designed for the uniformly distributed live loads shown or the following concentrated loads: (1) for passenger cars accommodating not more than nine passengers, 2000 pounds acting on an area of 20 sq in; (2) mechanical parking structures without slab or deck, passenger cars only, 1500 pounds per wheel; (3) for trucks or buses, maximum axle load on an area of 20 sq in.

*Source: American National Standard ANSI A58.1-1972
Local building codes take precedence.

DESIGN INFORMATION

8.1.3 Beam design equations and diagrams

(1) Simple Beam — uniformly distributed load

$$R = V \dots = \frac{wl}{2}$$

$$V_x \dots = w\left(\frac{l}{2} - x\right)$$

$$M \text{ max. (at center)} \dots = \frac{wl^2}{8}$$

$$M_x \dots = \frac{wx}{2}(l - x)$$

$$\Delta \text{ max. (at center)} \dots = \frac{5 wl^4}{384 EI}$$

$$\Delta_x \dots = \frac{wx}{24 EI}(l^3 - 2lx^2 + x^3)$$

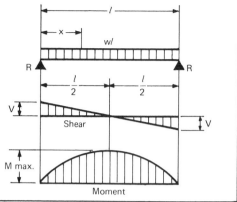

(2) Simple Beam — concentrated load at center

$$R = V \dots = \frac{P}{2}$$

$$M \text{ max. (at point of load)} \dots = \frac{Pl}{4}$$

$$M_x \left(\text{when } x < \frac{l}{2}\right) \dots = \frac{Px}{2}$$

$$\Delta \text{ max. (at point of load)} \dots = \frac{Pl^3}{48 EI}$$

$$\Delta_x \left(\text{when } x < \frac{l}{2}\right) \dots = \frac{Px}{48 EI}(3l^2 - 4x^2)$$

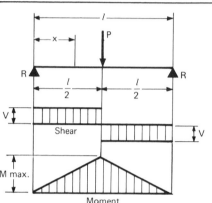

(3) Simple Beam — concentrated load at any point

$$R_1 = V_1 \text{ (max. when } a < b) \dots = \frac{Pb}{l}$$

$$R_2 = V_2 \text{ (max. when } a > b) \dots = \frac{Pa}{l}$$

$$M \text{ max. (at point of load)} \dots = \frac{Pab}{l}$$

$$M_x \text{ (when } x < a) \dots = \frac{Pbx}{l}$$

$$\Delta \text{ max. } \left(\text{at } x = \sqrt{\frac{a(a+2b)}{3}} \text{ when } a > b\right) \dots = \frac{Pab(a+2b)\sqrt{3a(a+2b)}}{27 EI\, l}$$

$$\Delta a \text{ (at point of load)} \dots = \frac{Pa^2 b^2}{3 EI\, l}$$

$$\Delta_x \text{ (when } x < a) \dots = \frac{Pbx}{6 EI\, l}(l^2 - b^2 - x^2)$$

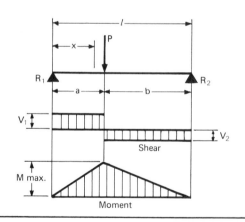

(4) Simple Beam — two equal concentrated loads symmetrically placed

$$R = V \dots = P$$

$$M \text{ max. (between loads)} \dots = Pa$$

$$M_x \text{ (when } x < a) \dots = Px$$

$$\Delta \text{ max. (at center)} \dots = \frac{Pa}{24 EI}(3l^2 - 4a^2)$$

$$\Delta_x \text{ (when } x < a) \dots = \frac{Px}{6 EI}(3la - 3a^2 - x^2)$$

$$\Delta_x \text{ (when } x > a \text{ and } < (l - a)) \dots = \frac{Pa}{6 EI}(3lx - 3x^2 - a^2)$$

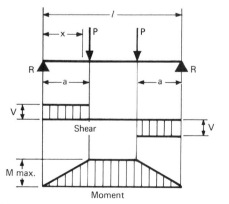

8.1.3 (Cont.) Beam design equations and diagrams

(5)

Simple Beam — two unequal concentrated loads unsymmetrically placed

$R_1 = V_1$ .... $= \dfrac{P_1 (l - a) + P_2 b}{l}$

$R_2 = V_2$ .... $= \dfrac{P_1 a + P_2 (l - b)}{l}$

V_x (when $x > a$ and $< (l - b)$) .... $= R_1 - P_1$

M_1 (max. when $R_1 < P_1$) .... $= R_1 a$

M_2 (max. when $R_2 < P_2$) .... $= R_2 b$

M_x (when $x < a$) .... $= R_1 x$

M_x $\left(\text{when } x > a \text{ and} < (l - b)\right)$ .... $= R_1 x - P_1 (x - a)$

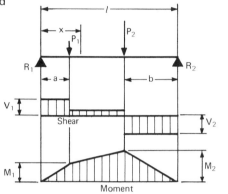

(6)

Simple Beam — uniform load partially distributed

$R_1 = V_1$ (max. when $a < c$) .... $= \dfrac{wb}{2l} (2c + b)$

$R_2 = V_2$ (max. when $a > c$) .... $= \dfrac{wb}{2l} (2a + b)$

V_x (when $x > a$ and $< (a + b)$) .... $= R_1 - w (x - a)$

M max. $\left(\text{at } x = a + \dfrac{R_1}{w}\right)$ .... $= R_1 \left(a + \dfrac{R_1}{2w}\right)$

M_x (when $x < a$) .... $= R_1 x$

M_x (when $x > a$ and $< (a + b)$) .... $= R_1 x - \dfrac{w}{2} (x - a)^2$

M_x (when $x > (a + b)$) .... $= R_2 (l - x)$

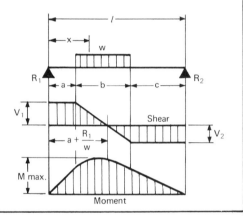

(7)

Simple Beam — load increasing uniformly to one end

$R_1 = V_1$ .... $= \dfrac{W}{3}$

$R_2 = V_2$ max. .... $= \dfrac{2W}{3}$

V_x .... $= \dfrac{W}{3} - \dfrac{Wx^2}{l^2}$

M max. $\left(\text{at } x = \dfrac{l}{\sqrt{3}} = .5774l\right)$ .... $= \dfrac{2Wl}{9\sqrt{3}} = .1283\,Wl$

M_x .... $= \dfrac{Wx}{3l^2} (l^2 - x^2)$

$\triangle$ max. $\left(\text{at } x = l \sqrt{1 - \sqrt{\dfrac{8}{15}}} = .5193l\right)$ .... $= .01304 \dfrac{Wl^3}{EI}$

$\triangle_x$ .... $= \dfrac{Wx}{180\,EI\,l^2}(3x^4 - 10l^2 x^2 + 7l^4)$

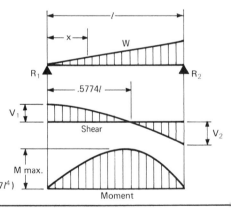

(8)

Simple Beam — load increasing uniformly to center

$R = V$ .... $= \dfrac{W}{2}$

V_x $\left(\text{when } x < \dfrac{l}{2}\right)$ .... $= \dfrac{W}{2l^2} (l^2 - 4x^2)$

M max.   (at center) .... $= \dfrac{Wl}{6}$

M_x $\left(\text{when } x < \dfrac{l}{2}\right)$ .... $= Wx \left(\dfrac{1}{2} - \dfrac{2x^2}{3l^2}\right)$

$\triangle$ max.   (at center) .... $= \dfrac{Wl^3}{60\,EI}$

$\triangle_x$ .... $= \dfrac{Wx}{480\,EI\,l^2} (5l^2 - 4x^2)^2$

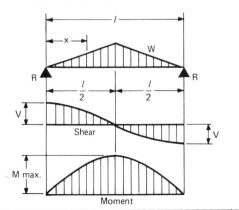

8.1.3 (Cont.) Beam design equations and diagrams

(9) Beam overhanging one support — uniformly distributed load

$R_1 = V_1$ .... $= \dfrac{w}{2l}(l^2 - a^2)$

$R_2 = V_2 + V_3$ .... $= \dfrac{w}{2l}(l + a)^2$

V_2 .... $= wa$

V_3 .... $= \dfrac{w}{2l}(l^2 + a^2)$

V_x (between supports) .... $= R_1 - wx$

V_{x_1} (for overhang) .... $= w(a - x_1)$

$M_1 \left(\text{at } x = \dfrac{l}{2}\left[1 - \dfrac{a^2}{l^2}\right] \right)$ .... $= \dfrac{w}{8l^2}(l + a)^2 (l - a)^2$

M_2 (at R_2) .... $= \dfrac{wa^2}{2}$

M_x (between supports) .... $= \dfrac{wx}{2l}(l^2 - a^2 - xl)$

M_{x_1} (for overhang) .... $= \dfrac{w}{2}(a - x_1)^2$

Δ_x (between supports) .... $= \dfrac{wx}{24\,EI\,l}(l^4 - 2l^2 x^2 + lx^3 - 2a^2 l^2 + 2a^2 x^2)$

Δ_{x_1} (for overhang) .... $= \dfrac{wx_1}{24\,EI}(4a^2 l - l^3 + 6a^2 x_1 - 4ax_1{}^2 + x_1{}^3)$

---

(10) Beam overhanging one support — uniformly distributed load on overhang

$R_1 = V_1$ .... $= \dfrac{wa^2}{2l}$

$R_2 = V_1 + V_2$ .... $= \dfrac{wa}{2l}(2l + a)$

V_2 .... $= wa$

V_{x_1} (for overhang) .... $= w(a - x_1)$

M max. (at R_2) .... $= \dfrac{wa^2}{2}$

M_x (between supports) .... $= \dfrac{wa^2 x}{2l}$

M_{x_1} (for overhang) .... $= \dfrac{w}{2}(a - x_1)^2$

Δ max. $\left(\text{between supports at } x = \dfrac{l}{\sqrt{3}}\right)$ .... $= \dfrac{wa^2 l^2}{18\sqrt{3}\,EI} = .03208\,\dfrac{wa^2 l^2}{EI}$

Δ max. (for overhang at $x_1 = a$) .... $= \dfrac{wa^3}{24\,EI}(4l + 3a)$

Δ_x (between supports) .... $= \dfrac{wa^2 x}{12\,EI\,l}(l^2 - x^2)$

Δ_{x_1} (for overhang) .... $= \dfrac{wx_1}{24\,EI}(4a^2 l + 6a^2 x_1 - 4ax_1{}^2 + x_1{}^3)$

---

(11) Beam overhanging one support — uniformly distributed load between supports

$R = V$ .... $= \dfrac{wl}{2}$

V_x .... $= w\left(\dfrac{l}{2} - x\right)$

M max. (at center) .... $= \dfrac{wl^2}{8}$

M_x .... $= \dfrac{wx}{2}(l - x)$

Δ max. (at center) .... $= \dfrac{5wl^4}{384\,EI}$

Δ_x .... $= \dfrac{wx}{24\,EI}(l^3 - 2lx^2 + x^3)$

Δ_{x_1} .... $= \dfrac{wl^3 x_1}{24\,EI}$

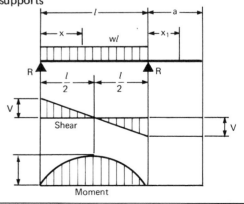

DESIGN INFORMATION

8.1.3 (Cont.) Beam design equations and diagrams

(12)

Beam overhanging one support — concentrated load at any point between supports

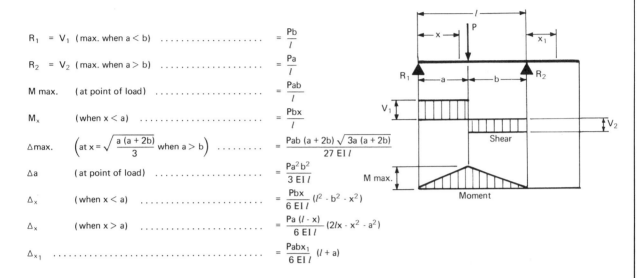

$R_1 = V_1$ (max. when a < b) .... $= \dfrac{Pb}{l}$

$R_2 = V_2$ (max. when a > b) .... $= \dfrac{Pa}{l}$

M max. (at point of load) .... $= \dfrac{Pab}{l}$

M_x (when x < a) .... $= \dfrac{Pbx}{l}$

Δmax. $\left(\text{at } x = \sqrt{\dfrac{a\,(a+2b)}{3}} \text{ when } a > b\right)$ .... $= \dfrac{Pab\,(a+2b)\sqrt{3a\,(a+2b)}}{27\,EI\,l}$

Δa (at point of load) .... $= \dfrac{Pa^2 b^2}{3\,EI\,l}$

Δ_x (when x < a) .... $= \dfrac{Pbx}{6\,EI\,l}\,(l^2 - b^2 - x^2)$

Δ_x (when x > a) .... $= \dfrac{Pa\,(l-x)}{6\,EI\,l}\,(2lx - x^2 - a^2)$

Δ_{x_1} .... $= \dfrac{Pabx_1}{6\,EI\,l}\,(l+a)$

---

(13)

Beam overhanging one support — concentrated load at end of overhang

$R_1 = V_1$ .... $= \dfrac{Pa}{l}$

$R_2 = V_1 + V_2$ .... $= \dfrac{P}{l}\,(l+a)$

V_2 .... $= P$

M max. (at R_2) .... $= Pa$

M_x (between supports) .... $= \dfrac{Pax}{l}$

M_{x_1} (for overhang) .... $= P\,(a - x_1)$

Δmax. $\left(\text{between supports at } x = \dfrac{l}{\sqrt{3}}\right)$ .... $= \dfrac{Pal^2}{9\sqrt{3}\,EI} = .06415\,\dfrac{Pal^2}{EI}$

Δmax. (for overhang at $x_1 = a$) .... $= \dfrac{Pa^2}{3\,EI}\,(l+a)$

Δ_x (between supports) .... $= \dfrac{Pax}{6\,EI\,l}\,(l^2 - x^2)$

Δ_{x_1} (for overhang) .... $= \dfrac{Px_1}{6\,EI}\,(2al + 3ax_1 - x_1^2)$

DESIGN INFORMATION

8.1.3 (Cont.) Beam design equations and diagrams

(14)

Cantilever Beam — uniformly distributed load

$R = V$ .... $= wl$

V_x .... $= wx$

M max. (at fixed end) .... $= \dfrac{wl^2}{2}$

M_x .... $= \dfrac{wx^2}{2}$

Δ max. (at free end) .... $= \dfrac{wl^4}{8\,EI}$

Δ_x .... $= \dfrac{w}{24\,EI}\,(x^4 - 4l^3 x + 3l^4)$

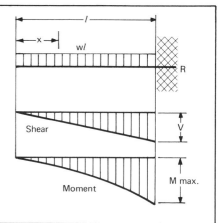

(15)

Cantilever Beam — concentrated load at free end

$R = V$ .... $= P$

M max. (at fixed end) .... $= Pl$

M_x .... $= Px$

Δ max. (at free end) .... $= \dfrac{Pl^3}{3\,EI}$

Δ_x .... $= \dfrac{P}{6\,EI}\,(2l^3 - 3l^2 x + x^3)$

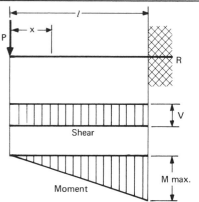

(16)

Cantilever Beam — concentrated load at any point

$R = V$ .... $= P$

M max. (at fixed end) .... $= Pb$

M_x (when $x > a$) .... $= P\,(x - a)$

Δ max. (at free end) .... $= \dfrac{Pb^2}{6\,EI}\,(3l - b)$

Δa (at point of load) .... $= \dfrac{Pb^3}{3\,EI}$

Δ_x (when $x < a$) .... $= \dfrac{Pb^2}{6\,EI}\,(3l - 3x - b)$

Δ_x (when $x > a$) .... $= \dfrac{P\,(l - x)^2}{6\,EI}\,(3b - l + x)$

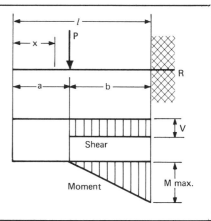

(17)

Cantilever Beam — load increasing uniformly to fixed end

$R = V$ .... $= W$

V_x .... $= W\,\dfrac{x^2}{l^2}$

M max. (at fixed end) .... $= \dfrac{Wl}{3}$

M_x .... $= \dfrac{Wx^3}{3l^2}$

Δ max. (at free end) .... $= \dfrac{Wl^3}{15\,EI}$

Δ_x .... $= \dfrac{W}{60\,EI\,l^2}\,(x^5 - 5l^4 x + 4l^5)$

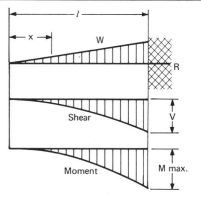

8.1.3 (Cont.) Beam design equations and diagrams

(18) Beam fixed at one end, supported at other — uniformly distributed load

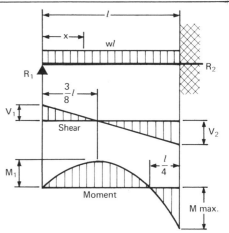

$R_1 = V_1$ .... $= \dfrac{3wl}{8}$

$R_2 = V_2 \text{ max.}$ .... $= \dfrac{5wl}{8}$

V_x .... $= R_1 - wx$

$M \text{ max.}$ .... $= \dfrac{wl^2}{8}$

M_1 $\left(\text{at } x = \dfrac{3}{8}l\right)$ .... $= \dfrac{9}{128}wl^2$

M_x .... $= R_1 x - \dfrac{wx^2}{2}$

$\Delta \text{max.}$ $\left(\text{at } x = \dfrac{l}{16}(1 + \sqrt{33}) = .4215l\right)$ ...... $= \dfrac{wl^4}{185\,EI}$

Δ_x .... $= \dfrac{wx}{48\,EI}(l^3 - 3lx^2 + 2x^3)$

---

(19) Beam fixed at one end, supported at other — concentrated load at center

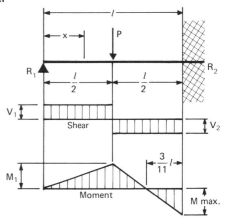

$R_1 = V_1$ .... $= \dfrac{5P}{16}$

$R_2 = V_2 \text{ max.}$ .... $= \dfrac{11P}{16}$

$M \text{ max.}$ (at fixed end) .... $= \dfrac{3Pl}{16}$

M_1 (at point of load) .... $= \dfrac{5Pl}{32}$

M_x $\left(\text{when } x < \dfrac{l}{2}\right)$ .... $= \dfrac{5Px}{16}$

M_x $\left(\text{when } x > \dfrac{l}{2}\right)$ $= P\left(\dfrac{l}{2} - \dfrac{11x}{16}\right)$

$\Delta \text{max.}$ $\left(\text{at } x = l\sqrt{\dfrac{1}{5}} = .4472l\right)$ .... $= \dfrac{Pl^3}{48\,EI\sqrt{5}} = .009317\,\dfrac{Pl^3}{EI}$

Δ_x (at point of load) .... $= \dfrac{7Pl^3}{768\,EI}$

Δ_x $\left(\text{when } x < \dfrac{l}{2}\right)$ .... $= \dfrac{Px}{96\,EI}(3l^2 - 5x^2)$

Δ_x $\left(\text{when } x > \dfrac{l}{2}\right)$ .... $= \dfrac{P}{96\,EI}(x - l)^2(11x - 2l)$

---

(20) Beam fixed at one end, supported at other — concentrated load at any point

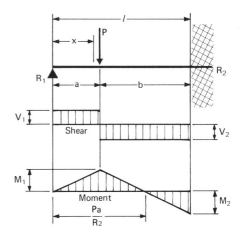

$R_1 = V_1$ .... $= \dfrac{Pb^2}{2l^3}(a + 2l)$

$R_2 = V_2$ .... $= \dfrac{Pa}{2l^3}(3l^2 - a^2)$

M_1 (at point of load) .... $= R_1 a$

M_2 (at fixed end) .... $= \dfrac{Pab}{2l^2}(a + l)$

M_x (when $x < a$) .... $= R_1 x$

M_x (when $x > a$) .... $= R_1 x - P(x - a)$

$\Delta \text{max.}$ $\left(\text{when } a < .414l \text{ at } x = l\dfrac{l^2 + a^2}{3l^2 - a^2}\right)$ ...... $= \dfrac{Pa}{3\,EI}\dfrac{(l^2 - a^2)^3}{(3l^2 - a^2)^2}$

$\Delta \text{max.}$ $\left(\text{when } a > .414l \text{ at } x = l\sqrt{\dfrac{a}{2l + a}}\right)$ ..... $= \dfrac{Pab^2}{6\,EI}\sqrt{\dfrac{a}{2l + a}}$

Δa (at point of load) .... $= \dfrac{Pa^2b^3}{12\,EIl^3}(3l + a)$

Δ_x (when $x < a$) .... $= \dfrac{Pb^2x}{12\,EIl^3}(3al^2 - 2lx^2 - ax^2)$

Δ_x (when $x > a$) .... $= \dfrac{Pa}{12\,EIl^3}(l - x)^2(3l^2x - a^2x - 2a^2l)$

DESIGN INFORMATION

8.1.3 (Cont.) Beam design equations and diagrams

(21)

Beam fixed at both ends — uniformly distributed loads

$R = V$ .... $= \dfrac{wl}{2}$

V_x .... $= w\left(\dfrac{l}{2} - x\right)$

M max. (at ends) .... $= \dfrac{wl^2}{12}$

M_1 (at center) .... $= \dfrac{wl^2}{24}$

M_x .... $= \dfrac{w}{12}(6lx - l^2 - 6x^2)$

$\triangle$ max. (at center) .... $= \dfrac{wl^4}{384\ EI}$

$\triangle_x$ .... $= \dfrac{wx^2}{24\ EI}(l - x)^2$

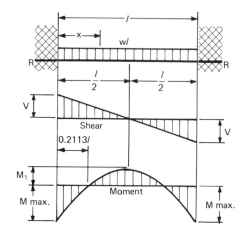

(22)

Beam fixed at both ends — concentrated load at any point

$R_1 = V_1$ (max. when $a < b$) .... $= \dfrac{Pb^2}{l^3}(3a + b)$

$R_2 = V_2$ (max. when $a > b$) .... $= \dfrac{Pa^2}{l^3}(a + 3b)$

M_1 (max. when $a < b$) .... $= \dfrac{Pab^2}{l^2}$

M_2 (max. when $a > b$) .... $= \dfrac{Pa^2 b}{l^2}$

Ma (at point of load) .... $= \dfrac{2Pa^2 b^2}{l^3}$

M_x (when $x < a$) .... $= R_1 x - \dfrac{Pab^2}{l^2}$

$\triangle$ max. $\left(\text{when } a > b \text{ at } x = \dfrac{2al}{3a + b}\right)$ .... $= \dfrac{2Pa^3 b^2}{3\ EI\ (3a + b)^2}$

$\triangle a$ (at point of load) .... $= \dfrac{Pa^3 b^3}{3\ EI\ l^3}$

$\triangle_x$ (when $x < a$) .... $= \dfrac{Pb^2 x^2}{6\ EI\ l^3}(3al - 3ax - bx)$

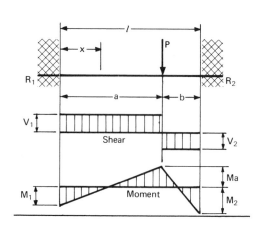

DESIGN INFORMATION

8.1.4 Camber (deflection) and rotation coefficients for prestress force and loads*

Prestress Pattern	Equivalent Moment or Load	Equivalent Loading	Camber (+ ↑)	End Rotation	
(1)	$M = Pe$		$+\dfrac{Ml^2}{16\,EI}$	$+\dfrac{Ml}{3\,EI}$	$-\dfrac{Ml}{6\,EI}$
(2)	$M = Pe$		$+\dfrac{Ml^2}{16\,EI}$	$+\dfrac{Ml}{6\,EI}$	$-\dfrac{Ml}{3\,EI}$
(3)	$M = Pe$		$+\dfrac{Ml^2}{8\,EI}$	$+\dfrac{Ml}{2\,EI}$	$-\dfrac{Ml}{2\,EI}$
(4)	$N = \dfrac{4\,Pe'}{l}$		$+\dfrac{Nl^3}{48\,EI}$	$+\dfrac{Nl^2}{16\,EI}$	$-\dfrac{Nl^2}{16\,EI}$
(5)	$N = \dfrac{Pe'}{bl}$		$+\dfrac{b\,(3-4b^2)\,Nl^3}{24\,EI}$	$+\dfrac{b\,(1-b)\,Nl^2}{2\,EI}$	$-\dfrac{b\,(1-b)\,Nl^2}{2\,EI}$
(6)	$w = \dfrac{8\,Pe'}{l^2}$		$+\dfrac{5wl^4}{384\,EI}$	$+\dfrac{wl^3}{24\,EI}$	$-\dfrac{wl^3}{24\,EI}$
(7)	$w = \dfrac{8\,Pe'}{l^2}$		$+\dfrac{5\,wl^4}{768\,EI}$	$+\dfrac{9\,wl^3}{384\,EI}$	$-\dfrac{7\,wl^3}{384\,EI}$
(8)	$w = \dfrac{8\,Pe'}{l^2}$		$+\dfrac{5\,wl^4}{768\,EI}$	$+\dfrac{7\,wl^3}{384\,EI}$	$-\dfrac{9\,wl^3}{384\,EI}$
(9)	$w = \dfrac{4\,Pe'}{(0.5-b)\,l^2}$, $\;w_1 = \dfrac{w}{b}(0.5-b)$		$\left[\dfrac{5}{8} - \dfrac{b}{2}(3-2b^2)\right]\dfrac{wl^4}{48\,EI}$	$+\dfrac{(1-b)(1-2b)\,wl^3}{24\,EI}$	$-\dfrac{(1-b)(1-2b)\,wl^3}{24\,EI}$
(10)	$w = \dfrac{4\,Pe'}{(0.5-b)\,l^2}$, $\;w_1 = \dfrac{w}{b}(0.5-b)$		$\left[\dfrac{5}{16} - \dfrac{b}{4}(3-2b^2)\right]\dfrac{wl^4}{48\,EI}$	$\left[\dfrac{9}{8} - b(2-b)^2\right]\dfrac{wl^3}{48\,EI}$	$-\left[\dfrac{7}{8} + b(2-b^2)\right]\dfrac{wl^3}{48\,EI}$
(11)	$w = \dfrac{4\,Pe'}{(0.5-b)\,l^2}$, $\;w_1 = \dfrac{w}{b}(0.5-b)$		$\left[\dfrac{5}{16} - \dfrac{b}{4}(3-2b^2)\right]\dfrac{wl^4}{48\,EI}$	$\left[\dfrac{7}{8} - b(2-b^2)\right]\dfrac{wl^3}{48\,EI}$	$-\left[\dfrac{9}{8} + b(2-b)^2\right]\dfrac{wl^3}{48\,EI}$

*The tabulated values apply to the effects of prestressing. By adjusting the directional notation, they may also be used for the effects of loads.

DESIGN INFORMATION

8.1.5 Moments in beams with fixed ends

Loading	Moment at A	Moment at center	Moment at B
(1)	$-\dfrac{Pl}{8}$	$+\dfrac{Pl}{8}$	$-\dfrac{Pl}{8}$
(2)	$-Pla(1-a)^2$		$-Pla^2(1-a)$
(3)	$-\dfrac{2Pl}{9}$	$+\dfrac{Pl}{9}$	$-\dfrac{2Pl}{9}$
(4)	$-\dfrac{5Pl}{16}$	$+\dfrac{3Pl}{16}$	$-\dfrac{5Pl}{16}$
(5)	$-\dfrac{Wl}{12}$	$+\dfrac{Wl}{24}$	$-\dfrac{Wl}{12}$
(6)	$-\dfrac{Wl(1+2a-2a^2)}{12}$	$+\dfrac{Wl(1+2a+4a^2)}{24}$	$-\dfrac{Wl(1+2a-2a^2)}{12}$
(7)	$-\dfrac{Wl(3a-2a^2)}{12}$	$+\dfrac{Wla^2}{6}$	$-\dfrac{Wl(3a-2a^2)}{12}$
(8)	$-\dfrac{Wla(6-8a+3a^2)}{12}$		$-\dfrac{Wla^2(4-3a)}{12}$
(9)	$-\dfrac{5Wl}{48}$	$+\dfrac{3Wl}{48}$	$-\dfrac{5Wl}{48}$
(10)	$-\dfrac{Wl}{10}$		$-\dfrac{Wl}{15}$

W = Total load on beam

PCI Design Handbook

DESIGN INFORMATION

8.1.6 Moving load placement for maximum moment and shear

(1) Simple Beam — one concentrated moving load

$$R_1 \text{ max.} = V_1 \text{ max. (at } x = 0) \ldots\ldots\ldots = P$$

$$M \text{ max. (at point of load, when } x = \frac{\ell}{2}) \ldots = \frac{P\ell}{4}$$

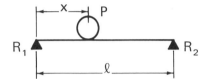

(2) Simple Beam — two equal concentrated moving loads

$$R_1 \text{ max.} = V_1 \text{ max. (at } x = 0) \ldots\ldots\ldots\ldots = P \left(2 - \frac{a}{\ell}\right)$$

$$M \text{ max.} \begin{cases} \left[\begin{array}{l} \text{when } a < (2 - \sqrt{2})\ \ell = .586\ell \\ \text{under load 1 at } x = \frac{1}{2}\left(\ell - \frac{a}{2}\right) \end{array}\right] = \frac{P}{2\ell}\left(\ell - \frac{a}{2}\right)^2 \\[20pt] \left[\begin{array}{l} \text{when } a > (2 - \sqrt{2})\ \ell = .586\ell \\ \text{with one load at center of span} \end{array}\right] = \frac{P\ell}{4} \end{cases}$$

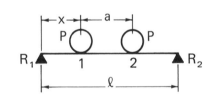

(3) Simple Beam — two unequal concentrated moving loads

$$R_1 \text{ max.} = V_1 \text{ max. (at } x = 0) \ldots\ldots\ldots\ldots = P_1 + P_2 \frac{\ell - a}{\ell}$$

$$M \text{ max.} \begin{cases} \left[\text{under } P_1, \text{ at } x = \frac{1}{2}\left(\ell - \frac{P_2 a}{P_1 + P_2}\right)\right] = (P_1 + P_2)\frac{x^2}{\ell} \\[15pt] \left[\begin{array}{l} M \text{ max. may occur with larger} \\ \text{load at center of span and other} \\ \text{load off span} \end{array}\right] = \frac{P_1\ell}{4} \end{cases}$$

$$P_1 > P_2$$

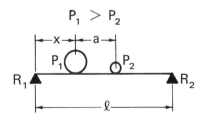

CONCRETE MATERIAL PROPERTIES

8.2.1 Table of concrete stresses

f'_c	$0.45 f'_c$	$0.6 f'_c$	$\sqrt{f'_c}$	$0.6\sqrt{f'_c}$	$2\sqrt{f'_c}$	$3.5\sqrt{f'_c}$	$4\sqrt{f'_c}$	$5\sqrt{f'_c}$	$6\sqrt{f'_c}$	$7.5\sqrt{f'_c}$	$12\sqrt{f'_c}$
3000	1350	1800	55	33	110	192	219	274	329	411	657
3500	1575	2100	59	35	118	207	237	296	355	444	710
4000	1800	2400	63	38	126	221	253	316	379	474	759
4500	2025	2700	67	40	134	235	268	335	402	503	805
5000	2250	3000	71	42	141	247	283	354	424	530	849
5500	2475	3300	74	44	148	260	297	371	445	556	890
6000	2700	3600	77	46	155	271	310	387	465	581	930
6500	2925	3900	81	48	161	281	322	403	484	605	967
7000	3150	4200	84	50	167	293	335	418	502	627	1004
7500	3375	4500	87	52	173	303	346	433	519	650	1039
8000	3600	4800	89	54	179	313	358	447	537	671	1073

8.2.2 Concrete modulus of elasticity as affected by unit weight and strength

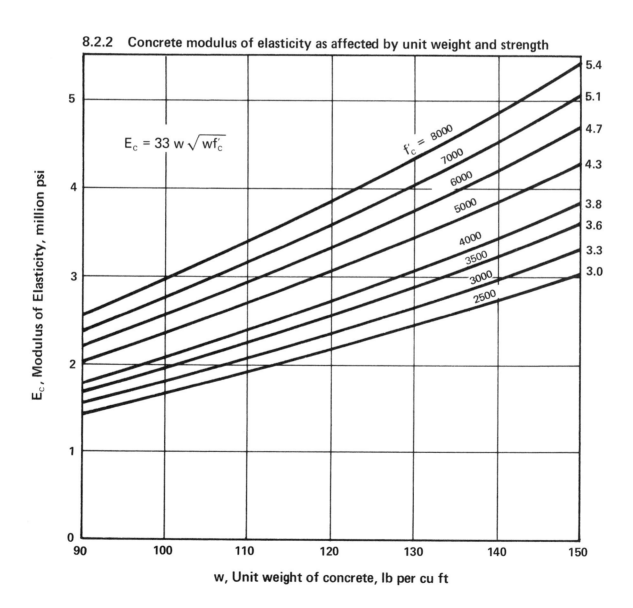

$$E_c = 33\,w\sqrt{wf'_c}$$

$f'_c = 8000$

7000

6000

5000

4000

3500

3000

2500

E_c, Modulus of Elasticity, million psi

w, Unit weight of concrete, lb per cu ft

MATERIAL PROPERTIES
PRESTRESSING STEEL

8.2.3   Properties and design strengths of prestressing strand and wire

Seven-Wire Strand, f_{pu} = 270 ksi

Nominal Diameter, in.	3/8	7/16	1/2	9/16	0.600
Area, sq in.	0.085	0.115	0.153	0.192	0.215
Weight, plf	0.29	0.40	0.53	0.65	0.74
0.7 f_{pu} A_{ps}, kips	16.1	21.7	28.9	36.3	40.7
0.8 f_{pu} A_{ps}, kips	18.4	24.8	33.0	41.4	46.5
f_{pu} A_{ps}, kips	23.0	31.0	41.3	51.8	58.1

Seven-Wire Strand, f_{pu} = 250 ksi

Nominal Diameter, in.	1/4	5/16	3/8	7/16	1/2	0.600
Area, sq in.	0.036	0.058	0.080	0.108	0.144	0.215
Weight, plf	0.12	0.20	0.27	0.37	0.49	0.74
0.7 f_{pu} A_{ps}, kips	6.3	10.2	14.0	18.9	25.2	37.6
0.8 f_{pu} A_{ps}, kips	7.2	11.6	16.0	21.6	28.8	43.0
f_{pu} A_{ps}, kips	9.0	14.5	20.0	27.0	36.0	53.8

Three- and Four-Wire Strand, f_{pu} = 250 ksi

Nominal Diameter, in.	1/4	5/16	3/8	7/16
No. of wires	3	3	3	4
Area, sq in.	0.036	0.058	0.075	0.106
Weight, plf	0.13	0.20	0.26	0.36
0.7 f_{pu} A_{ps}, kips	6.3	10.2	13.2	18.6
0.8 f_{pu} A_{ps}, kips	7.2	11.6	15.0	21.2
f_{pu} A_{ps}, kips	9.0	14.5	18.8	26.5

Prestressing Wire

Diameter	0.105	0.120	0.135	0.148	0.162	0.177	0.192	0.196	0.250	0.276
Area, sq in.	0.0087	0.0114	0.0143	0.0173	0.0206	0.0246	0.0289	0.0302	0.0491	0.0598
Weight, plf	0.030	0.039	0.049	0.059	0.070	0.083	0.098	0.10	0.17	0.20
Ult. strength, f_{pu}, ksi	279	273	268	263	259	255	250	250	240	235
0.7 f_{pu} A_{ps}, kips	1.70	2.18	2.68	3.18	3.73	4.39	5.05	5.28	8.25	9.84
0.8 f_{pu} A_{ps}, kips	1.94	2.49	3.06	3.64	4.26	5.02	5.78	6.04	9.42	11.24
f_{pu} A_{ps}, kips	2.43	3.11	3.83	4.55	5.33	6.27	7.22	7.55	11.78	14.05

MATERIAL PROPERTIES
PRESTRESSING STEEL

8.2.4   Properties and design strengths of prestressing bars

Smooth Prestressing Bars, f_{pu} = 145 ksi

Nominal Diameter, in.	3/4	7/8	1	1 1/8	1 1/4	1 3/8
Area, sq in.	0.442	0.601	0.785	0.994	1.227	1.485
Weight, plf	1.50	2.04	2.67	3.38	4.17	5.05
0.7 $f_{pu} A_{ps}$, kips	44.9	61.0	79.7	100.9	124.5	150.7
0.8 $f_{pu} A_{ps}$, kips	51.3	69.7	91.0	115.3	142.3	172.2
$f_{pu} A_{ps}$, kips	64.1	87.1	113.8	144.1	177.9	215.3

Smooth Prestressing Bars, f_{pu} = 160 ksi

Nominal Diameter, in.	3/4	7/8	1	1 1/8	1 1/4	1 3/8
Area, sq in.	0.442	0.601	0.785	0.994	1.227	1.485
Weight, plf	1.50	2.04	2.67	3.38	4.17	5.05
0.7 $f_{pu} A_{ps}$, kips	49.5	67.3	87.9	111.3	137.4	166.3
0.8 $f_{pu} A_{ps}$, kips	56.6	77.0	100.5	127.2	157.0	190.1
$f_{pu} A_{ps}$, kips	70.7	96.2	125.6	159.0	196.3	237.6

Deformed Prestressing Bars

Nominal Diameter, in.	5/8	1	1	1 1/4	1 1/4	1 1/2
Area, sq. in.	0.28	0.852	0.852	1.295	1.295	1.630
Weight, plf	0.98	2.96	2.96	4.55	4.55	5.74
Ult. strength, f_{pu}, ksi	157	150	160	150	160	150
0.7 $f_{pu} A_{ps}$, kips	30.5	89.5	95.4	136.0	145.0	171.2
0.8 $f_{pu} A_{ps}$, kips	34.8	102.2	109.1	155.4	165.8	195.6
$f_{pu} A_{ps}$, kips	43.5	127.8	136.3	194.3	207.2	244.5

Stress-strain characteristics (all prestressing bars):

For design purposes, following assumptions are satisfactory:

E_s = 29,000 ksi

f_y = 0.95 f_{pu}

MATERIAL PROPERTIES
PRESTRESSING STEEL

8.2.5 Typical stress-strain curve, 7-wire prestressing strand

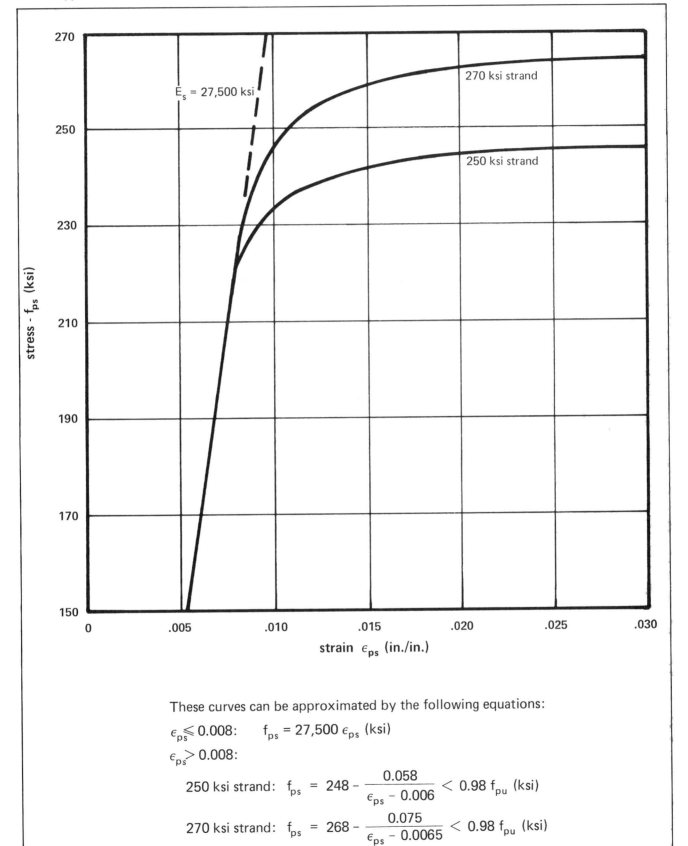

These curves can be approximated by the following equations:

$\epsilon_{ps} \leqslant 0.008$: $f_{ps} = 27{,}500\,\epsilon_{ps}$ (ksi)

$\epsilon_{ps} > 0.008$:

250 ksi strand: $f_{ps} = 248 - \dfrac{0.058}{\epsilon_{ps} - 0.006} < 0.98\,f_{pu}$ (ksi)

270 ksi strand: $f_{ps} = 268 - \dfrac{0.075}{\epsilon_{ps} - 0.0065} < 0.98\,f_{pu}$ (ksi)

8.2.6 Reinforcing bar data

		ASTM STANDARD REINFORCING BARS		
Bar Size Designation	Weight (lb per foot)	NOMINAL DIMENSIONS		
		Diameter (in.)	Area (sq. in.)	Perimeter (in.)
#3	0.376	0.375	0.11	1.178
#4	0.668	0.500	0.20	1.571
#5	1.043	0.625	0.31	1.963
#6	1.502	0.750	0.44	2.356
#7	2.044	0.875	0.60	2.749
#8	2.670	1.000	0.79	3.142
#9	3.400	1.128	1.00	3.544
#10	4.303	1.270	1.27	3.990
#11	5.313	1.410	1.56	4.430
#14	7.65	1.693	2.25	5.32
#18	13.60	2.257	4.00	7.09

STANDARD HOOKS
STIRRUP AND TIE-HOOKS

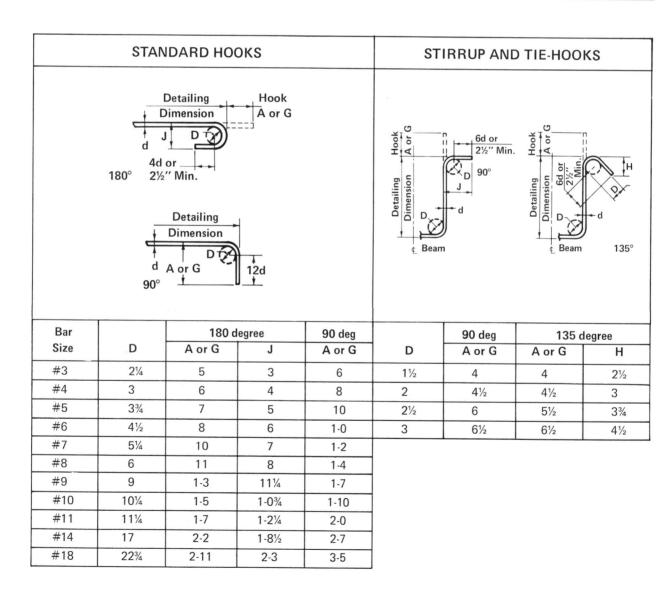

Bar Size	D	180 degree		90 deg	D	90 deg	135 degree	
		A or G	J	A or G		A or G	A or G	H
#3	2¼	5	3	6	1½	4	4	2½
#4	3	6	4	8	2	4½	4½	3
#5	3¾	7	5	10	2½	6	5½	3¾
#6	4½	8	6	1-0	3	6½	6½	4½
#7	5¼	10	7	1-2				
#8	6	11	8	1-4				
#9	9	1-3	11¼	1-7				
#10	10¼	1-5	1-0¾	1-10				
#11	11¼	1-7	1-2¼	2-0				
#14	17	2-2	1-8½	2-7				
#18	22¾	2-11	2-3	3-5				

8.2.7 Required development and lap lengths

f_y = 40,000 psi

Tension:

$$\ell_d = 0.04\, A_b \frac{f_y}{\sqrt{f'_c}} \; ; \text{ min } 0.0004\, f_y\, d_b \text{ or 12 in.}$$

Compression development length:

$$\ell_d = 0.02\, f_y \frac{d_b}{\sqrt{f'_c}} \; ; \text{ min } 0.0003\, f_y\, d_b \text{ or 8 in.}$$

Compression Splice:

Compression ℓ_d; min $0.0005\, f_y\, d_b$ or 12 in.

A_b = area of individual bar, in.^{2}
d_b = diameter of bar, in.

For limitations, see ACI 318-77, Chapter 12

Multiply table values by:

For tension only —
1.4   for top reinforcement
1.33 for "all lightweight" concrete
1.18 for "sand-lightweight" concrete
0.8   for bar spacing 6" or more
      (3" from member face)

For tension or compression —

$$\frac{A_{s\ req'd}}{A_{s\ provided}} \quad \text{when } A_s \text{ is greater than required}$$

0.75 for reinforcement in spirals

Development and lap lengths in inches

Bar size	f'_c = 3000 psi					f'_c = 4000 psi					f'_c = 5000 psi					Min. Comp. Splice
	Tension				Compression	Tension				Compression	Tension				Compression	
	ℓ_d	$1.3\ell_d$	$1.7\ell_d$	$2.0\ell_d$	ℓ_d	ℓ_d	$1.3\ell_d$	$1.7\ell_d$	$2.0\ell_d$	ℓ_d	ℓ_d	$1.3\ell_d$	$1.7\ell_d$	$2.0\ell_d$	ℓ_d	
3	12.0	12.0	12.0	12.0	8.0	12.0	12.0	12.0	12.0	8.0	12.0	12.0	12.0	12.0	8.0	12.0
4	12.0	12.0	13.6	16.0	8.0	12.0	12.0	13.6	16.0	8.0	12.0	12.0	13.6	16.0	8.0	12.0
5	12.0	13.0	17.0	20.0	9.1	12.0	13.0	17.0	20.0	8.0	12.0	13.0	17.0	20.0	8.0	12.5
6	12.9	16.8	21.9	25.8	11.0	12.0	15.6	20.4	24.0	9.4	12.0	15.6	20.4	24.0	9.0	15.0
7	17.5	22.8	29.9	35.0	12.8	15.2	19.8	25.9	30.4	11.0	14.0	18.2	23.8	28.0	10.5	17.5
8	23.0	29.9	39.0	46.0	14.6	20.0	26.0	33.8	40.0	12.7	17.9	23.3	30.2	35.8	12.0	20.0
9	29.2	38.0	49.6	58.4	16.5	25.3	32.9	43.0	50.6	14.3	22.6	29.4	38.4	45.2	13.5	22.6
10	37.1	48.2	62.9	74.2	18.6	32.1	41.7	54.5	64.2	16.1	28.7	37.3	48.7	57.4	15.2	25.4
11	45.6	59.3	77.5	91.2	20.6	39.5	51.3	67.2	78.9	17.8	35.3	45.9	60.1	70.6	16.9	28.2

Bar size	f'_c = 6000 psi					f'_c = 7000 psi					f'_c = 8000 psi					Min. Comp. Splice
	Tension				Compression	Tension				Compression	Tension				Compression	
	ℓ_d	$1.3\ell_d$	$1.7\ell_d$	$2.0\ell_d$	ℓ_d	ℓ_d	$1.3\ell_d$	$1.7\ell_d$	$2.0\ell_d$	ℓ_d	ℓ_d	$1.3\ell_d$	$1.7\ell_d$	$2.0\ell_d$	ℓ_d	
3	12.0	12.0	12.0	12.0	8.0	12.0	12.0	12.0	12.0	8.0	12.0	12.0	12.0	12.0	8.0	12.0
4	12.0	12.0	13.6	16.0	8.0	12.0	12.0	13.6	16.0	8.0	12.0	12.0	13.6	16.0	8.0	12.0
5	12.0	13.0	17.0	20.0	8.0	12.0	13.0	17.0	20.0	8.0	12.0	13.0	17.0	20.0	8.0	12.5
6	12.0	15.6	20.4	24.0	9.0	12.0	15.6	20.4	24.0	9.0	12.0	15.6	20.4	24.0	9.0	15.0
7	14.0	18.2	23.8	28.0	10.5	14.0	18.2	23.8	28.0	10.5	14.0	18.2	23.8	28.0	10.5	17.5
8	16.3	21.2	27.6	32.6	12.0	16.0	20.8	27.2	32.0	12.0	16.0	20.8	27.2	32.0	12.0	20.0
9	20.7	26.9	35.1	41.4	13.5	19.1	24.8	32.5	38.2	13.5	18.0	23.4	30.7	36.0	13.5	22.6
10	26.2	34.1	44.5	52.4	15.2	24.3	31.6	41.2	48.6	15.2	22.7	29.5	38.5	45.4	15.2	25.4
11	32.2	41.9	54.8	64.4	16.9	29.8	38.7	50.8	59.6	16.9	27.9	36.3	47.5	55.8	16.9	28.2

8.2.7 (Cont.) Required development and lap lengths

$f_y = 60,000$ psi

Tension:

$$\ell_d = 0.04\, A_b \frac{f_y}{\sqrt{f'_c}}\,; \text{ min } 0.0004\, f_y\, d_b \text{ or } 12 \text{ in.}$$

Compression development length:

$$\ell_d = 0.02\, f_y \frac{d_b}{\sqrt{f'_c}}\,; \text{ min } 0.0003\, f_y\, d_b \text{ or } 8 \text{ in.}$$

Compression Splice:

Compression ℓ_d; min $0.0005\, f_y\, d_b$ or 12 in.

A_b = area of individual bar, in.^{2}
d_b = diameter of bar, in.

For limitations, see ACI 318-77, Chapter 12

Multiply table values by:
For tension only —
1.4 for top reinforcement
1.33 for "all lightweight" concrete
1.18 for "sand-lightweight" concrete
0.8 for bar spacing 6" or more
    (3" from member face)

For tension or compression —

$\dfrac{A_{s\ req'd}}{A_{s\ provided}}$ when A_s is greater than required

0.75 for reinforcement in spirals

Development and lap lengths in inches

Bar size	f'_c = 3000 psi					f'_c = 4000 psi					f'_c = 5000 psi					Min. Comp. Splice
	Tension				Compression	Tension				Compression	Tension				Compression	
	ℓ_d	$1.3\ell_d$	$1.7\ell_d$	$2.0\ell_d$	ℓ_d	ℓ_d	$1.3\ell_d$	$1.7\ell_d$	$2.0\ell_d$	ℓ_d	ℓ_d	$1.3\ell_d$	$1.7\ell_d$	$2.0\ell_d$	ℓ_d	
3	12.0	12.0	15.3	18.0	8.2	12.0	12.0	15.3	18.0	8.0	12.0	12.0	15.3	18.0	8.0	12.0
4	12.0	15.6	20.4	24.0	11.0	12.0	15.6	20.4	24.0	9.4	12.0	15.6	20.4	24.0	9.0	15.0
5	15.0	19.5	25.5	30.0	13.7	15.0	19.5	25.5	30.0	11.9	15.0	19.5	25.5	30.0	11.2	18.8
6	19.2	25.0	32.9	38.4	16.4	18.0	23.4	30.6	36.0	14.2	18.0	23.4	30.6	36.0	13.5	22.5
7	26.3	34.2	44.8	52.6	19.2	22.8	29.6	38.8	45.6	16.6	21.0	27.3	35.7	42.0	15.8	26.2
8	34.6	45.0	58.5	69.2	21.9	30.0	39.0	50.7	60.0	19.0	26.8	34.8	45.3	53.6	18.0	30.0
9	43.8	56.9	74.4	87.6	24.7	37.9	49.3	64.5	75.8	21.4	33.9	44.1	57.7	67.8	20.3	33.8
10	55.6	72.3	94.4	111.2	27.8	48.2	62.7	81.7	96.4	24.1	43.1	56.0	73.1	86.2	22.9	38.1
11	68.4	88.9	116.3	136.8	30.9	59.2	77.0	100.7	118.4	26.8	52.9	68.8	90.1	105.8	25.4	42.3

Bar size	f'_c = 6000 psi					f'_c = 7000 psi					f'_c = 8000 psi					Min. Comp. Splice
	Tension				Compression	Tension				Compression	Tension				Compression	
	ℓ_d	$1.3\ell_d$	$1.7\ell_d$	$2.0\ell_d$	ℓ_d	ℓ_d	$1.3\ell_d$	$1.7\ell_d$	$2.0\ell_d$	ℓ_d	ℓ_d	$1.3\ell_d$	$1.7\ell_d$	$2.0\ell_d$	ℓ_d	
3	12.0	12.0	15.3	18.0	8.0	12.0	12.0	15.3	18.0	8.0	12.0	12.0	15.3	18.0	8.0	12.0
4	12.0	15.6	20.4	24.0	9.0	12.0	15.6	20.4	24.0	9.0	12.0	15.6	20.4	24.0	9.0	15.0
5	15.0	19.5	25.5	30.0	11.2	15.0	19.5	25.5	30.0	11.2	15.0	19.5	25.5	30.0	11.2	18.8
6	18.0	23.4	30.6	36.0	13.5	18.0	23.4	30.6	36.0	13.5	18.0	23.4	30.6	36.0	13.5	22.5
7	21.0	27.3	35.7	42.0	15.8	21.0	27.3	35.7	42.0	15.8	21.0	27.3	35.7	42.0	15.8	26.2
8	24.4	31.7	41.4	48.8	18.0	24.0	31.2	40.8	48.0	18.0	24.0	31.2	40.8	48.0	18.0	30.0
9	31.0	40.3	52.6	62.0	20.3	28.7	37.3	48.7	57.4	20.3	27.0	35.1	46.0	54.0	20.3	33.8
10	39.3	51.1	66.7	78.6	22.9	36.4	47.3	61.8	72.8	22.9	34.1	44.3	57.8	68.2	22.9	38.1
11	48.3	62.8	82.2	96.6	25.4	44.7	58.1	76.1	89.4	25.4	41.9	54.5	71.2	83.8	25.4	42.3

MATERIAL PROPERTIES
WELDED WIRE FABRIC

Table 8.2.8 Common stock styles of welded wire fabric

Style Designation		Steel Area		Approx. Weight
		sq in. per ft		
Old Designation (By Steel Wire Gage)	New Designation (By W-Number)	Longit.	Trans.	lb per 100 sq ft
6x6-10x10	6x6-W1.4xW1.4	.029	.029	21
4x12-8x12**	4x12-W2.1xW0.9	.062	.009	25
6x6-8x8	6x6-W2.1xW2.1	.041	.041	30
4x4-10x10	4x4-W1.4xW1.4	.043	.043	31
4x12-7x11**	4x12-W2.5xW1.1	.074	.011	31
6x6-6x6*	6x6-W2.9xW2.9	.058	.058	42
4x4-8x8	4x4-W2.1xW2.1	.062	.062	44
6x6-4x4*	6x6-W4.0xW4.0	.080	.080	58
4x4-6x6	4x4-W2.9xW2.9	.087	.087	62
6x6-2x2*	6x6-W5.5xW5.5***	.110	.110	80
4x4-4x4*	4x4-W4.0xW4.0	.120	.120	85
4x4-3x3*	4x4-W4.7xW4.7	.141	.141	102
4x4-2x2*	4x4-W5.5xW5.5***	.165	.165	119

* Commonly available in 8 ft x 12 ft or 8 ft x 15 ft sheets.

** These items may be carried in sheets by various manufacturers in certain parts of the U.S. and Canada.

*** Exact W-Number size for 2 gage is 5.4.

order length

overall width
order width

Side overhangs may be varied as required and do not need to be equal. Overhang lengths limited only by overall sheet width.

tranverse wire
longitudinal wire

End overhangs may differ. The sum of the two end overhangs, however, should equal the transverse wire spacing.

Industry Method of Designating Style:

Example: 6 x 12   W16 x W8

longitudinal wire spacing / longitudinal wire size

transverse wire spacing / transverse wire size

MATERIAL PROPERTIES
WELDED WIRE FABRIC

Table 8.2.9 Special welded wire fabric for double tee flanges*

Application	Style Designation	Steel Area sq. in. per ft.		Approx. Weight lb per 100 sq. ft.
		Longit.	Trans.	
8-ft wide DT, 2-in. flange	12 X 6-W1.4 X W2.5	.014	.050	23
10-ft wide DT, 2-in. flange	12 X 6-W2.0 X W4.0	.020	.080	35
10-ft wide DT, 2½-in. flange	12 X 6-W1.4 X W2.9	.014	.058	27

*See "Standardization of Welded Wire Fabric," *PCI Journal*, July/Aug., 1976

Table 8.2.10 Wires used in welded wire fabric

Wire Size Number		Nominal Diameter In.	Nominal Weight Plf	Area — sq in. per ft of width						
				Center to Center Spacing, in.						
Smooth	Deformed			2	3	4	6	8	10	12
W31	D31	0.628	1.054	1.86	1.24	.93	.62	.465	.372	.31
W30	D30	0.618	1.020	1.80	1.20	.90	.60	.45	.36	.30
W28	D28	0.597	.952	1.68	1.12	.84	.56	.42	.336	.28
W26	D26	0.575	.934	1.56	1.04	.78	.52	.39	.312	.26
W24	D24	0.553	.816	1.44	.96	.72	.48	.36	.288	.24
W22	D22	0.529	.748	1.32	.88	.66	.44	.33	.264	.22
W20	D20	0.504	.680	1.20	.80	.60	.40	.30	.24	.20
W18	D18	0.478	.612	1.08	.72	.54	.36	.27	.216	.18
W16	D16	0.451	.544	.96	.64	.48	.32	.24	.192	.16
W14	D14	0.422	.476	.84	.56	.42	.28	.21	.168	.14
W12	D12	0.390	.408	.72	.48	.36	.24	.18	.144	.12
W11	D11	0.374	.374	.66	.44	.33	.22	.165	.132	.11
W10.5		0.366	.357	.63	.42	.315	.21	.157	.126	.105
W10	D10	0.356	.340	.60	.40	.30	.20	.15	.12	.10
W9.5		0.348	.323	.57	.38	.285	.19	.142	.114	.095
W9	D9	0.338	.306	.54	.36	.27	.18	.135	.108	.09
W8.5		0.329	.289	.51	.34	.255	.17	.127	.102	.085
W8	D8	0.319	.272	.48	.32	.24	.16	.12	.096	.08
W7.5		0.309	.255	.45	.30	.225	.15	.112	.09	.075
W7	D7	0.298	.238	.42	.28	.21	.14	.105	.084	.07
W6.5		0.288	.221	.39	.26	.195	.13	.097	.078	.065
W6	D6	0.276	.204	.36	.24	.18	.12	.09	.072	.06
W5.5		0.264	.187	.33	.22	.165	.11	.082	.066	.055
W5	D5	0.252	.170	.30	.20	.15	.10	.075	.06	.05
W4.5		0.240	.153	.27	.18	.135	.09	.067	.054	.045
W4	D4	0.225	.136	.24	.16	.12	.08	.06	.048	.04
W3.5		0.211	.119	.21	.14	.105	.07	.052	.042	.035
W3		0.195	.102	.18	.12	.09	.06	.045	.036	.03
W2.9		0.192	.098	.174	.116	.087	.058	.043	.035	.029
W2.5		0.178	.085	.15	.10	.075	.05	.037	.03	.025
W2.1		0.162	.070	.126	.084	.063	.042	.031	.025	.021
W2		0.159	.068	.12	.08	.06	.04	.03	.024	.02
W1.5		0.138	.051	.09	.06	.045	.03	.022	.018	.015
W1.4		0.135	.049	.084	.056	.042	.028	.021	.017	.014

SECTION PROPERTIES

8.3 Properties of Geometric Sections

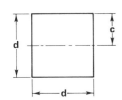

SQUARE

Axis of moments through center

$$A = d^2$$
$$c = \frac{d}{2}$$
$$I = \frac{d^4}{12}$$
$$S = \frac{d^3}{6}$$
$$r = \frac{d}{\sqrt{12}} = .288675\, d$$

RECTANGLE

Axis of moments on diagonal

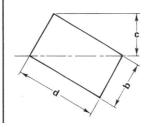

$$A = bd$$
$$c = \frac{bd}{\sqrt{b^2 + d^2}}$$
$$I = \frac{b^3 d^3}{6\,(b^2 + d^2)}$$
$$S = \frac{b^2 d^2}{6\sqrt{b^2 + d^2}}$$
$$r = \frac{bd}{\sqrt{6\,(b^2 + d^2)}}$$

SQUARE

Axis of moments on base

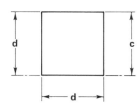

$$A = d^2$$
$$c = d$$
$$I = \frac{d^4}{3}$$
$$S = \frac{d^3}{3}$$
$$r = \frac{d}{\sqrt{3}} = .577350\, d$$

RECTANGLE

Axis of moments any line through center of gravity

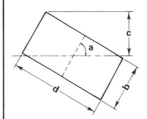

$$A = bd$$
$$c = \frac{b \sin a + d \cos a}{2}$$
$$I = \frac{bd\,(b^2 \sin^2 a + d^2 \cos^2 a)}{12}$$
$$S = \frac{bd\,(b^2 \sin^2 a + d^2 \cos^2 a)}{6\,(b \sin a + d \cos a)}$$
$$r = \sqrt{\frac{b^2 \sin^2 a + d^2 \cos^2 a}{12}}$$

SQUARE

Axis of moments on diagonal

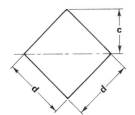

$$A = d^2$$
$$c = \frac{d}{\sqrt{2}} = .707107\, d$$
$$I = \frac{d^4}{12}$$
$$S = \frac{d^3}{6\sqrt{2}} = .117851\, d^3$$
$$r = \frac{d}{\sqrt{12}} = .288675\, d$$

HOLLOW RECTANGLE

Axis of moments through center

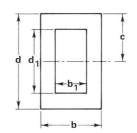

$$A = bd - b_1 d_1$$
$$c = \frac{d}{2}$$
$$I = \frac{bd^3 - b_1 d_1^3}{12}$$
$$S = \frac{bd^3 - b_1 d_1^3}{6d}$$
$$r = \sqrt{\frac{bd^3 - b_1 d_1^3}{12\,A}}$$

RECTANGLE

Axis of moments through center

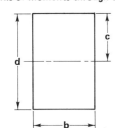

$$A = bd$$
$$c = \frac{d}{2}$$
$$I = \frac{bd^3}{12}$$
$$S = \frac{bd^2}{6}$$
$$r = \frac{d}{\sqrt{12}} = .288675\, d$$

EQUAL RECTANGLES

Axis of moments through center of gravity

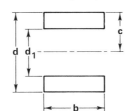

$$A = b\,(d - d_1)$$
$$c = \frac{d}{2}$$
$$I = \frac{b\,(d^3 - d_1^3)}{12}$$
$$S = \frac{b\,(d^3 - d_1^3)}{6d}$$
$$r = \sqrt{\frac{d^3 - d_1^3}{12(d - d_1)}}$$

RECTANGLE

Axis of moments on base

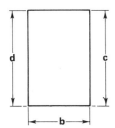

$$A = bd$$
$$c = d$$
$$I = \frac{bd^3}{3}$$
$$S = \frac{bd^2}{3}$$
$$r = \frac{d}{\sqrt{3}} = .577350\, d$$

UNEQUAL RECTANGLES

Axis of moments through center of gravity

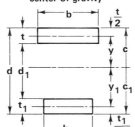

$$A = bt + b_1 t_1$$
$$c = \frac{\frac{1}{2}\, bt^2 + b_1 t_1\,(d - \frac{1}{2}\, t_1)}{A}$$
$$I = \frac{bt^3}{12} + bty^2 + \frac{b_1 t_1^3}{12} + b_1 t_1 y_1^2$$
$$S = \frac{I}{c} \qquad S_1 = \frac{I}{c_1}$$
$$r = \sqrt{\frac{I}{A}}$$

SECTION PROPERTIES

8.3 (cont.) Properties of Geometric Sections

TRIANGLE
Axis of moments through center of gravity

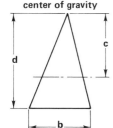

$$A = \frac{bd}{2}$$

$$c = \frac{2d}{3}$$

$$I = \frac{bd^3}{36}$$

$$S = \frac{bd^2}{24}$$

$$r = \frac{d}{\sqrt{18}}$$

HALF CIRCLE
Axis of moments through center of gravity

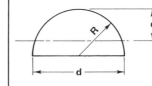

$$A = \frac{\pi R^2}{2}$$

$$c = R\left(1 - \frac{4}{3\pi}\right)$$

$$I = R^4\left(\frac{\pi}{8} - \frac{8}{9\pi}\right)$$

$$S = \frac{R^3}{24}\frac{(9\pi^2 - 64)}{(3\pi - 4)}$$

$$r = R\frac{\sqrt{9\pi^2 - 64}}{6\pi}$$

TRIANGLE
Axis of moments on base

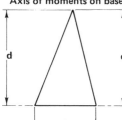

$$A = \frac{bd}{2}$$

$$c = d$$

$$I = \frac{bd^3}{12}$$

$$S = \frac{bd^2}{12}$$

$$r = \frac{d}{\sqrt{6}}$$

SEGMENT OF A CIRCLE
Axis of moments through circle center

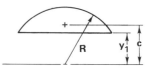

$$I = \frac{\pi R^4}{8} + \frac{y_1}{2}\sqrt{(R^2 - y_1^2)^3}$$

$$- \frac{R^2}{4}\left(y_1\sqrt{R^2 - y_1^2} + R^2 \sin^{-1}\frac{y_1}{R}\right)$$

$$A = \frac{\pi R^2}{2} - y_1\sqrt{R^2 - y_1^2}$$

$$- R^2 \sin^{-1}\left(\frac{y_1}{R}\right)$$

$$c = \frac{2(R^2 - y_1^2)^{3/2}}{3} / A$$

TRAPEZOID
Axis of moments through center of gravity

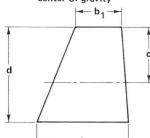

$$A = \frac{d(b + b_1)}{2}$$

$$c = \frac{d(2b + b_1)}{3(b + b_1)}$$

$$I = \frac{d^3(b^2 + 4bb_1 + b_1^2)}{36(b + b_1)}$$

$$S = \frac{d^2(b^2 + 4bb_1 + b_1^2)}{12(2b + b_1)}$$

$$r = \frac{d}{6(b + b_1)}$$

$$\times \sqrt{2(b^2 + 4bb_1 + b_1^2)}$$

PARABOLA

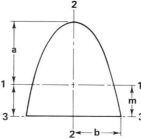

$$A = \frac{4}{3}ab$$

$$m = \frac{2}{5}a$$

$$I_1 = \frac{16}{175}a^3b$$

$$I_2 = \frac{4}{15}ab^3$$

$$I_3 = \frac{32}{105}a^3b$$

CIRCLE
Axis of moments through center

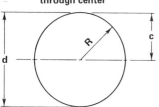

$$A = \frac{\pi d^2}{4} = \pi R^2$$

$$c = \frac{d}{2} = R$$

$$I = \frac{\pi d^4}{64} = \frac{\pi R^4}{4}$$

$$S = \frac{\pi d^3}{32} = \frac{\pi R^3}{4}$$

$$r = \frac{d}{4} = \frac{R}{2}$$

HALF PARABOLA

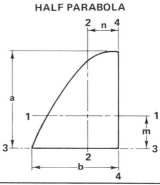

$$A = \frac{2}{3}ab$$

$$m = \frac{2}{5}a$$

$$n = \frac{3}{8}b$$

$$I_1 = \frac{8}{175}a^3b$$

$$I_2 = \frac{19}{480}ab^3$$

$$I_3 = \frac{16}{105}a^3b$$

$$I_4 = \frac{2}{15}ab^3$$

HOLLOW CIRCLE
Axis of moments through center

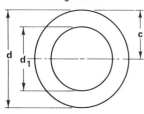

$$A = \frac{\pi(d^2 - d_1^2)}{4}$$

$$c = \frac{d}{2}$$

$$I = \frac{\pi(d^4 - d_1^4)}{64}$$

$$S = \frac{\pi(d^4 - d_1^4)}{32d}$$

$$r = \frac{\sqrt{d^2 + d_1^2}}{4}$$

COMPLEMENT OF HALF PARABOLA

$$A = \frac{1}{3}ab$$

$$m = \frac{7}{10}a$$

$$n = \frac{3}{4}b$$

$$I_1 = \frac{37}{2100}a^3b$$

$$I_2 = \frac{1}{80}ab^3$$

SECTION PROPERTIES

8.3 (cont.) Properties of geometric shapes

PARABOLIC FILLET IN RIGHT ANGLE

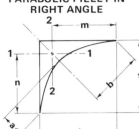

$$a = \frac{t}{2\sqrt{2}}$$

$$b = \frac{t}{\sqrt{2}}$$

$$A = \frac{1}{6}t^2$$

$$m = n = \frac{4}{5}t$$

$$I_1 = I_2 = \frac{11}{2100}t^4$$

*HALF ELLIPSE

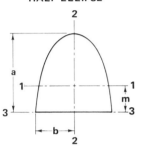

$$A = \frac{1}{2}\pi ab$$

$$m = \frac{4a}{3\pi}$$

$$I_1 = a^3 b\left(\frac{\pi}{8} - \frac{8}{9\pi}\right)$$

$$I_2 = \frac{1}{8}\pi ab^3$$

$$I_3 = \frac{1}{8}\pi a^3 b$$

*QUARTER ELLIPSE

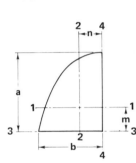

$$A = \frac{1}{4}\pi ab$$

$$m = \frac{4a}{3\pi}$$

$$n = \frac{4b}{3\pi}$$

$$I_1 = a^3 b\left(\frac{\pi}{16} - \frac{4}{9\pi}\right)$$

$$I_2 = ab^3\left(\frac{\pi}{16} - \frac{4}{9\pi}\right)$$

$$I_3 = \frac{1}{16}\pi a^3 b$$

$$I_4 = \frac{1}{16}\pi ab^3$$

*To obtain properties of half circles, quarter circle and circular complement, substitute a = b = R.

*ELLIPTIC COMPLEMENT

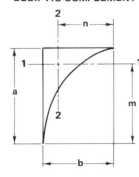

$$A = ab\left(1 - \frac{\pi}{4}\right)$$

$$m = \frac{a}{6\left(1 - \frac{\pi}{4}\right)}$$

$$n = \frac{b}{6\left(1 - \frac{\pi}{4}\right)}$$

$$I_1 = a^3 b\left(\frac{1}{3} - \frac{\pi}{16} - \frac{1}{36\left(1 - \frac{\pi}{4}\right)}\right)$$

$$I_2 = ab^3\left(\frac{1}{3} - \frac{\pi}{16} - \frac{1}{36\left(1 - \frac{\pi}{4}\right)}\right)$$

REGULAR POLYGON

Axis of moments through center

$$n = \text{Number of sides}$$

$$\phi = \frac{180^\circ}{n}$$

$$a = 2\sqrt{R^2 - R_1^{\,2}}$$

$$R = \frac{a}{2\sin\phi}$$

$$R_1 = \frac{a}{2\tan\phi}$$

$$A = \frac{1}{4}na^2\cot\phi = \frac{1}{2}nR^2\sin 2\phi = nR_1^{\,2}\tan\phi$$

$$I_1 = I_2 = \frac{A(6R^2 - a^2)}{24} = \frac{A(12R_1^{\,2} + a^2)}{48}$$

$$r_1 = r_2 = \sqrt{\frac{6R^2 - a^2}{24}} = \sqrt{\frac{12R_1 + a^2}{48}}$$

BEAMS AND CHANNELS

Transverse force oblique through center of gravity

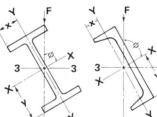

$$I_3 = I_x\sin^2\phi + I_y\cos^2\phi$$

$$I_4 = I_x\cos^2\phi + I_y\sin^2\phi$$

$$f_b = M\left(\frac{y}{I_x}\sin\phi + \frac{x}{I_y}\cos\phi\right)$$

where M is bending moment due to force F.

ANGLE

Axis of moments through center of gravity

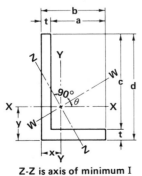

Z-Z is axis of minimum I

$$\tan 2\theta = \frac{2K}{I_y - I_x}$$

$$A = t(b + c) \qquad x = \frac{b^2 + ct}{2(b + c)} \qquad y = \frac{d^2 + at}{2(b + c)}$$

$$K = \text{Product of Inertia about X-X \& Y-Y}$$

$$= \mp \frac{abcdt}{4(b + c)}$$

$$I_x = \frac{1}{3}\left(t(d - y)^3 + by^3 - a(y - t)^3\right)$$

$$I_y = \frac{1}{3}\left(t(b - x)^3 + dx^3 - c(x - t)^3\right)$$

$$I_z = I_x\sin^2\theta + I_y\cos^2\theta + K\sin 2\theta$$

$$I_w = I_x\cos^2\theta + I_y\sin^2\theta - K\sin 2\theta$$

K is negative when heel of angle, with respect to c. g., is in 1st or 3rd quadrant, positive when in 2nd or 4th quadrant.

METRIC CONVERSION

Table 8.4 Conversion to International System of Units (SI)

SI Base Units

Quantity	Name	Symbol
length	meter	m
mass	kilogram	kg
time	second	s
electric current	ampere	A
thermodynamic temperature	kelvin	K
amount of substance	mole	mol
luminous intensity	candela	cd

SI Supplementary Units

Quantity	Name	Symbol
plane angle	radian	rad
solid angle	steradian	sr

SI Derived Units

Quantity	Name	Symbol	In Terms of Other Units	In Terms of Base Units
frequency	hertz	Hz	—	s^{-1}
force	newton	N	—	$m \cdot kg \cdot s^{-2}$
pressure, stress	pascal	Pa	N/m^2	$m^{-1} \cdot kg \cdot s^{-2}$
energy, work, quantity of heat	joule	J	$N \cdot m$	$m^2 \cdot kg \cdot s^{-2}$
power	watt	W	J/s	$m^2 \cdot kg \cdot s^{-3}$

SI Derived Units

Quantity	Description	In Terms of Other Units	In Terms of Base Units
area	square meter	—	m^2
volume	cubic meter	—	m^3
density, mass density	kilogram per cubic meter	—	kg/m^3
specific volume	cubic meter per kilogram	—	m^3/kg
moment of force	newton meter	$N \cdot m$	$m^2 \cdot kg \cdot s^{-2}$
heat capacity	joule per kelvin	J/K	$m^2 \cdot kg \cdot s^{-2} \cdot K^{-1}$
specific heat capacity	joule per kilogram kelvin	$J/kg \cdot K$	$m^2 \cdot s^{-2} \cdot K^{-1}$
thermal conductivity	watt per meter kelvin	$W/m \cdot K$	$m \cdot kg \cdot s^{-3} K^{-1}$

METRIC CONVERSION

Table 8.4 (Cont.) Conversion to International System of Units (SI)

Other Units to Use with SI

Quantity	Name	Symbol	Value in SI Units
time	minute	min	1 min = 60s
	hour	h	1 h = 3600s
plane angle	degree	°	1° = (π/180) rad
	minute	'	1' = (π/10 800) rad
temperature	degree Celsius	°C	1°C (interval) = 1 K 0°C = 273.15 K
mass	tonne	t	1 t = 1000 kg

Conversion Factors

U.S. Customary to SI

To convert from	to	multiply by
Length		
foot	meter (m)	0.3048
inch	millimeter (mm)	25.4
Area		
square foot	square meter (m²)	0.0929
square inch	square millimeter (mm²)	645.2
Volume (capacity)		
cubic foot	cubic meter (m³)	0.02832
gallon (U.S. liquid)	cubic meter (m³)	0.003785
Force		
kip	kilonewton (kN)	4.448
pound	newton (N)	4.448
Pressure or Stress (Force per Area)		
kip/square inch (ksi)	megapascal (MPa)	6.895
pound/square foot	kilopascal (kPa)	0.04788
pound/square inch (psi)	kilopascal (kPa)	6.895
pound/square inch (psi)	megapascal (MPa)	0.006895

METRIC CONVERSION

Table 8.4 (Cont.) Conversion to International System of Units (SI)

To convert from	to	multiply by
Bending Moment or Torque		
inch-pound	newton-meter (N•m)	0.1130
foot-pound	newton-meter (N•m)	1.356
foot-kip	newton-meter (N•m)	1356
Mass		
pound (avdp)	kilogram (kg)	0.4536
ton (short, 2000 lb)	kilogram (kg)	907.2
ton (short, 2000 lb)	tonne (t)	0.9072
Mass per Volume		
pound/cubic foot	kilogram/cubic meter (kg/m³)	16.02
pound/cubic yard	kilogram/cubic meter (kg/m³)	0.5933
Temperature		
deg Fahrenheit (F)	deg Celsius (C)	$C = (F - 32)\,5/9$
deg Fahrenheit (F), interval	deg Celsius (C), interval	5/9
Other		
section modulus in.³	mm³	16,387
moment of inertia in.⁴	mm⁴	416,231
Coefficient of heat transfer, Btu/ft² hr °F	W/m²•°C	5.678
Modulus of elasticity, psi	MPa	0.006895
Thermal conductivity, Btu/in./ft² hr °F	W/m•°C	0.1442
Thermal expansion in./in. °F	mm/mm•°C	1.800
$f_c = \sqrt{f'_c}$, psi	MPa	$0.083036 \sqrt{f'_c}$

INDEX